Photovoltaic Systems Engineering
Fourth Edition

Photovoltaic Systems Engineering
Fourth Edition

Roger Messenger and
Homayoon "Amir" Abtahi

CRC Press is an imprint of the
Taylor & Francis Group, an **informa** business

MATLAB® is a trademark of The MathWorks, Inc. and is used with permission. The MathWorks does not warrant the accuracy of the text or exercises in this book. This book's use or discussion of MATLAB® software or related products does not constitute endorsement or sponsorship by The MathWorks of a particular pedagogical approach or particular use of the MATLAB® software.

CRC Press
Taylor & Francis Group
6000 Broken Sound Parkway NW, Suite 300
Boca Raton, FL 33487-2742

© 2017 by Taylor & Francis Group, LLC
CRC Press is an imprint of Taylor & Francis Group, an Informa business

No claim to original U.S. Government works

Printed by CPI on sustainably sourced paper

International Standard Book Number-13: 978-1-4987-7277-8 (Hardback)

This book contains information obtained from authentic and highly regarded sources. Reasonable efforts have been made to publish reliable data and information, but the author and publisher cannot assume responsibility for the validity of all materials or the consequences of their use. The authors and publishers have attempted to trace the copyright holders of all material reproduced in this publication and apologize to copyright holders if permission to publish in this form has not been obtained. If any copyright material has not been acknowledged please write and let us know so we may rectify in any future reprint.

Except as permitted under U.S. Copyright Law, no part of this book may be reprinted, reproduced, transmitted, or utilized in any form by any electronic, mechanical, or other means, now known or hereafter invented, including photocopying, microfilming, and recording, or in any information storage or retrieval system, without written permission from the publishers.

For permission to photocopy or use material electronically from this work, please access www.copyright.com (http://www.copyright.com/) or contact the Copyright Clearance Center, Inc. (CCC), 222 Rosewood Drive, Danvers, MA 01923, 978-750-8400. CCC is a not-for-profit organization that provides licenses and registration for a variety of users. For organizations that have been granted a photocopy license by the CCC, a separate system of payment has been arranged.

Trademark Notice: Product or corporate names may be trademarks or registered trademarks, and are used only for identification and explanation without intent to infringe.

Visit the Taylor & Francis Web site at
http://www.taylorandfrancis.com

and the CRC Press Web site at
http://www.crcpress.com

We did not inherit this world from our parents....
We are borrowing it from our children.

(Author unknown)

It is our fervent hope that the engineers
who read this book
will dedicate themselves to the creation of a world
where children and grandchildren will be left
with air they can breathe and water they can drink,
where humans and the rest of nature
will nurture one another.

Contents

Preface xxi
Disclaimer xxv
Acknowledgments xxvii
Authors xxix
Abbreviations xxxi

Chapter 1
Background 1

1.1 Introduction 1
1.2 Population and Energy Demand 2
1.3 Current World Energy Use Patterns 2
1.4 Exponential Growth 6
    1.4.1 Introduction 6
    1.4.2 Compound Interest 6
    1.4.3 Doubling Time 7
    1.4.4 Accumulation 8
    1.4.5 Resource Lifetime in an Exponential Environment 9
1.5 Hubbert's Gaussian Model 10
1.6 Net Energy, BTU Economics, and the Test for Sustainability 12
1.7 Direct Conversion of Sunlight to Electricity with PV 13
1.8 Energy Units 15
References 19
Suggested Reading 20

Chapter 2
The Sun 23

2.1 Introduction 23
2.2 The Solar Spectrum 23
2.3 Effect of Atmosphere on Sunlight 25
2.4 Sunlight Specifics 27
    2.4.1 Introduction 27
    2.4.2 Definitions 28
    2.4.3 The Orbit and Rotation of the Earth 29
    2.4.4 Tracking the Sun 31
    2.4.5 Measuring Sunlight 34
        2.4.5.1 Precision Measurements 34
        2.4.5.2 Less Precise Measurements 35
2.5 Capturing Sunlight 35
    2.5.1 Maximizing Irradiation on the Collector 35
    2.5.2 Shading 38
        2.5.2.1 Field Measurement of Shading Objects 38
        2.5.2.2 Computational Methods of Determining Shading ... 40

vii

	2.5.3	Special Orientation Considerations 41
		2.5.3.1 Horizontal Mounting 41
		2.5.3.2 Non-South-Facing Mounting 41

References 46
Suggested Reading 47

Chapter 3
Introduction to PV Systems 49

3.1 Introduction 49
3.2 The PV Cell 50
3.3 The PV Module 54
3.4 The PV Array 56
3.5 Energy Storage 57
    3.5.1 Introduction 57
    3.5.2 The Lead-Acid Storage Battery 58
        3.5.2.1 Chemistry of the Lead-Acid Cell 58
        3.5.2.2 Properties of the Lead-Acid Storage Battery 60
    3.5.3 Lithium-Ion Battery Technologies 64
    3.5.4 Nickel-Based Battery Systems 66
        3.5.4.1 Nickel–Cadmium Batteries 66
        3.5.4.2 Nickel–Zinc Batteries 67
        3.5.4.3 Nickel–Metal Hydride Batteries 67
    3.5.5 Emerging Battery Technologies 68
    3.5.6 Hydrogen Storage 70
    3.5.7 The Fuel Cell 71
    3.5.8 Other Storage Options 71
3.6 PV System Loads 72
3.7 PV System Availability: Traditional Concerns and New Concerns 73
    3.7.1 Traditional Concerns as Applied to Stand-Alone Systems 73
    3.7.2 New Concerns about PV System Availability 77
3.8 Associated System Electronic Components 78
    3.8.1 Introduction 78
    3.8.2 Charge Controllers 78
        3.8.2.1 Charging Considerations 79
        3.8.2.2 Discharging Considerations 82
    3.8.3 Maximum Power Point Trackers and Linear Current Boosters 82
    3.8.4 Inverters 86
        3.8.4.1 Square Wave Inverters 87
        3.8.4.2 Modified Sine Wave Inverters 89
        3.8.4.3 PWM Inverters 91
        3.8.4.4 Transformerless Inverters 95
        3.8.4.5 Other Desirable Inverter Features 95
3.9 Generators 97
    3.9.1 Introduction 97

CONTENTS

	3.9.2	Types and Sizes of Generators	97
	3.9.3	Generator Operating Characteristics	98
		3.9.3.1 Rotation Speed	98
		3.9.3.2 Efficiency versus Electrical Load	99
		3.9.3.3 Fuel Types	100
		3.9.3.4 Altitude Effects	101
		3.9.3.5 Waveform Harmonic Content	101
		3.9.3.6 Frequency Stability	101
		3.9.3.7 Amplitude Stability	101
		3.9.3.8 Noise Level	102
		3.9.3.9 Type of Starting	102
		3.9.3.10 Overload Characteristics	102
		3.9.3.11 Power Factor Considerations	102
	3.9.4	Generator Maintenance	102
	3.9.5	Generator Selection	103
3.10	Balance of System Components		103
	3.10.1	Introduction	103
	3.10.2	Switches, Circuit Breakers, Fuses, and Receptacles	104
	3.10.3	Ground Fault, Arc Fault, Surge, and Lightning Protection	105
		3.10.3.1 Grounding versus Grounded and Ground Fault Protection	105
		3.10.3.2 Arc Fault Protection	105
		3.10.3.3 Surge Protection	106
	3.10.4	Inverter Bypass Switches and Source Circuit Combiner Boxes	106
		3.10.4.1 Inverter Bypass Switches	106
		3.10.4.2 Source Circuit Combiner Boxes	106
	3.10.5	Grounding Devices	107
	3.10.6	Rapid Shutdown	107
References			111
Suggested Reading			112

Chapter 4
Grid-Connected Utility-Interactive Photovoltaic Systems ... 113

4.1	Introduction		113
4.2	Applicable Codes and Standards		114
	4.2.1	The *National Electrical Code*	114
		4.2.1.1 Introduction	114
		4.2.1.2 Voltage Drop and Wire Sizing	116
	4.2.2	IEEE Standard 1547-2003	119
		4.2.2.1 Introduction	119
		4.2.2.2 Specific Requirements	119
		4.2.2.3 Comparison of PV Inverters to Mechanically Rotating Generators	123
		4.2.2.4 Islanding Analysis	123

	4.2.3	Other Issues	125
		4.2.3.1 Aesthetics	125
		4.2.3.2 Electromagnetic Interference	126
		4.2.3.3 Surge Protection	126
		4.2.3.4 Structural Considerations	127
4.3	Design Considerations for Straight Grid-Connected PV Systems		128
	4.3.1	Determining System Energy Output	128
	4.3.2	Array Installation	130
	4.3.3	Inverter Selection and Mounting	130
	4.3.4	Other Installation Considerations	132
4.4	Design of a System Based on Desired Annual System Performance		132
	4.4.1	Array Sizing	132
	4.4.2	Inverter Selection	133
	4.4.3	Module Selection	134
	4.4.4	Balance of System	136
		4.4.4.1 Wiring from Array to Rooftop Junction Box	137
		4.4.4.2 Wire and Conduit from Rooftop Junction Box to Inverter	138
		4.4.4.3 Rapid Shutdown	141
		4.4.4.4 Ground Fault and Arc Fault Protection	142
		4.4.4.5 DC and AC Disconnects and Overcurrent Protection	142
		4.4.4.6 Point of Utility Connection	143
		4.4.4.7 Final System Electrical Schematic Diagram	144
4.5	Design of a System Based upon Available Roof Space		145
	4.5.1	Array Selection	145
	4.5.2	Inverter Selection	149
	4.5.3	Balance of System	149
		4.5.3.1 Wiring from Array to Rooftop Junction Box	149
		4.5.3.2 Rapid Shutdown, Ground Fault, and Arc Fault Protection	151
		4.5.3.3 DC and AC Disconnects and Overcurrent Protection	151
		4.5.3.4 Point of Utility Connection	152
		4.5.3.5 Estimating System Annual Performance	152
		4.5.3.6 Final System Electrical Schematic Diagram	153
	4.5.4	Extension of Design to Lower Wind Speed Region	153
4.6	Design of a Microinverter-Based System		154
	4.6.1	Introduction	154
	4.6.2	System Design	155
	4.6.3	Bells and Whistles (i.e., Monitoring Possibilities)	156
4.7	Design of a Nominal 20 kW System That Feeds a Three-Phase Distribution Panel		157
	4.7.1	Introduction	157
	4.7.2	Inverter	157
	4.7.3	Modules	158
	4.7.4	System DC Wiring	159

	4.7.5	System AC Wiring	161
		4.7.5.1 Wire and Overcurrent Protection Sizing	161
		4.7.5.2 Voltage Drop Calculations	162
	4.7.6	Annual System Performance Estimate	163
4.8	Design of a Nominal 500-kW System		163
	4.8.1	Introduction	163
	4.8.2	Inverter	163
	4.8.3	Modules and Array	164
	4.8.4	Configuring the Array	164
		4.8.4.1 Sizing the Array	164
		4.8.4.2 Combiner Boxes	165
		4.8.4.3 Array Layout	166
	4.8.5	Wire Sizing and Voltage Drop Calculations	168
		4.8.5.1 Source-Circuit Calculations	168
		4.8.5.2 Combiner to Recombiner Wiring (PV Output Circuits)	170
		4.8.5.3 Disconnects, GFDI, and Overcurrent Protection on the DC Side of the Inverter	171
	4.8.6	AC Wire Sizing, Disconnects, and Overcurrent Protection	171
	4.8.7	Arc Flash Calculations	172
		4.8.7.1 Introduction	172
		4.8.7.2 AC Voltage Sources	173
		4.8.7.3 DC Current Sources	174
4.9	System Commissioning		176
4.10	System Performance Monitoring		178
References			183
Suggested Reading			184

Chapter 5
Mechanical Considerations 185

5.1	Introduction		185
5.2	Important Properties of Materials		185
	5.2.1	Introduction	185
	5.2.2	Stress and Strain	186
	5.2.3	Strength of Materials	190
	5.2.4	Column Buckling	190
	5.2.5	Thermal Expansion and Contraction	191
	5.2.6	Chemical Corrosion and Ultraviolet Degradation	193
	5.2.7	Properties of Steel	195
	5.2.8	Properties of Aluminum	196
5.3	Establishing Mechanical System Requirements		198
	5.3.1	Mechanical System Design Process	198
	5.3.2	Functional Requirements	198
	5.3.3	Operational Requirements	199

	5.3.4	Constraints 200
	5.3.5	Trade-Offs 200
5.4	Design and Installation Guidelines 201	
	5.4.1	Standards and Codes 201
	5.4.2	Building Code Requirements 202
5.5	Forces Acting on PV Arrays 203	
	5.5.1	Structural Loading Considerations 203
	5.5.2	Dead Loads 204
	5.5.3	Live Loads 204
	5.5.4	Wind Loads 204
	5.5.5	Snow Loads 209
	5.5.6	Other Loads 210
5.6	Array Mounting System Design 210	
	5.6.1	Introduction 210
	5.6.2	Objectives in Designing the Array Mounting System 210
		5.6.2.1 Minimizing Installation Costs 210
		5.6.2.2 Building Integration Considerations 212
		5.6.2.3 Costs and Durability of Array-Roof Configurations 213
	5.6.3	Enhancing Array Performance 213
		5.6.3.1 Irradiance Enhancement 213
		5.6.3.2 Shading 214
		5.6.3.3 Array Cooling 214
		5.6.3.4 Protection from Vandalism 214
	5.6.4	Roof-Mounted Arrays 215
		5.6.4.1 Standoff Mounting 215
		5.6.4.2 Rack Mounting 216
		5.6.4.3 Integrated Mounting 217
		5.6.4.4 Direct Mounting 218
	5.6.5	Ground-Mounted Arrays 218
		5.6.5.1 Rack Mounting 218
		5.6.5.2 Pole Mounting 218
		5.6.5.3 Tracking-Stand Mounting 219
	5.6.6	Aesthetics 220
5.7	Computing Mechanical Loads and Stresses 221	
	5.7.1	Introduction 221
	5.7.2	Withdrawal Loads 221
	5.7.3	Tensile Stresses 222
	5.7.4	Buckling 222
5.8	Standoff, Roof Mount Examples 223	
	5.8.1	Introduction to ASCE 7 Wind Load Analysis Tabular Method 223
	5.8.2	Array Mount Design, High Wind Speed Case 227
	5.8.3	Array Mount Design, Lower Wind Speed Case 229
	5.8.4	Exposure C, Exposure D, and Other Correction Factors 231
References 233		
Suggested Reading 234		

Chapter 6
Battery-Backup Grid-Connected Photovoltaic Systems 235

- 6.1 Introduction 235
- 6.2 Battery-Backup Design Basics 237
  - 6.2.1 Introduction 237
  - 6.2.2 Load Determination 237
  - 6.2.3 Inverter Sizing 238
  - 6.2.4 Battery Sizing 238
  - 6.2.5 Sizing the Array 239
- 6.3 A Single Inverter 120-V Battery-Backup System Based on Standby Loads 241
  - 6.3.1 Determination of Standby Loads 241
  - 6.3.2 Inverter Selection 241
  - 6.3.3 Battery Selection 242
  - 6.3.4 Array Sizing 244
  - 6.3.5 Charge Controller and Module Selection 245
  - 6.3.6 BOS Selection and Completion of the Design 246
    - 6.3.6.1 Array Mounting Equipment 246
    - 6.3.6.2 Rooftop Junction Box 247
    - 6.3.6.3 Source-Circuit Combiner Box and Surge Arrestor 248
    - 6.3.6.4 Wire and Circuit Breaker Sizing—DC Side 248
    - 6.3.6.5 Wire and Circuit Breaker Sizing—AC Side 251
    - 6.3.6.6 Wiring of Standby Loads 251
    - 6.3.6.7 Equipment Grounding Conductor and Grounding Electrode Conductor Sizing 253
  - 6.3.7 Programming the Inverter and the Charge Controller 253
  - 6.3.8 Fossil-Fuel Generator Connection Options 256
- 6.4 A 120/240-V Battery-Backup System Based on Available Roof Space 257
  - 6.4.1 Introduction 257
  - 6.4.2 Module Selection and Source Circuit Design 257
  - 6.4.3 Source-Circuit Combiner Box and Charge Controller Selection 258
  - 6.4.4 Inverter Selection 260
  - 6.4.5 Determination of Standby Loads and Battery Selection 260
  - 6.4.6 BOS Selection and Completion of Design 262
    - 6.4.6.1 Rooftop Junction Box 262
    - 6.4.6.2 Source-Circuit Combiner Box and Surge Arrestors 263
    - 6.4.6.3 Wire and Circuit Breaker Sizing—DC Side 263
    - 6.4.6.4 Wire and Circuit Breaker Sizing—AC Side 265
    - 6.4.6.5 Wiring of Standby Loads 265
    - 6.4.6.6 Equipment Grounding Conductor and Grounding Electrode Conductor Sizing 266
    - 6.4.6.7 Rapid Shutdown 266
- 6.5 An 18-kW Battery-Backup System Using Inverters in Parallel 267
  - 6.5.1 Introduction 267

		6.5.2	Inverter and Charge Controller Selection 268
			6.5.2.1 Inverter Selection 268
			6.5.2.2 Charge Controllers 270
		6.5.3	Module Selection and Array Layout 270
			6.5.3.1 Module Selection 271
			6.5.3.2 Array Layout 272
			6.5.3.3 Array Performance 273
		6.5.4	Battery and BOS Selection 275
		6.5.5	Wire Sizing 275
		6.5.6	Final Design 277
	6.6	AC-Coupled Battery-Backup Systems 280
		6.6.1	Introduction 280
		6.6.2	A 120/240-V Battery-Backup Inverter with 240-V Straight Grid-Connected Inverter 281
		6.6.3	A 120-V Battery-Backup Inverter with a 240-V Straight Grid-Connected Inverter 283
		6.6.4	A 120/208-V Three-Phase AC-Coupled System 285
	6.7	Battery Connections 286
		6.7.1	Lead-Acid Connections 286
		6.7.2	Other Battery Systems 291
References 294

Chapter 7
Stand-Alone Photovoltaic Systems 295

7.1	Introduction 295
7.2	The Simplest Configuration: Module and Fan 297
7.3	A PV-Powered Water Pumping System 298
	7.3.1 Introduction 298
	7.3.2 Selection of System Components 299
	7.3.3 Design Approach for Simple Pumping System 301
	7.3.3.1 Pump Selection 301
	7.3.3.2 Battery Selection 302
	7.3.3.3 Module Selection 302
	7.3.3.4 Charge Controller Selection 303
	7.3.3.5 BOS and Completion of the System 303
7.4	A PV-Powered Parking Lot Lighting System 304
	7.4.1 Determination of the Lighting Load 304
	7.4.2 Parking Lot Lighting Design 306
	7.4.2.1 Introduction 306
	7.4.2.2 Determination of Lamp Wattage and Daily Load Presented by the Fixture 307
	7.4.2.3 Determination of Battery Storage Requirements 308
	7.4.2.4 Determination of Array Size 308

CONTENTS

		7.4.2.5	Charge Controller and Inverter Selection	309
		7.4.2.6	Final System Schematic	309
		7.4.2.7	Structural Comment	310
7.5	A Cathodic Protection System			311
	7.5.1	Introduction		311
	7.5.2	System Design		312
7.6	A Portable Highway Advisory Sign		315
	7.6.1	Introduction		315
	7.6.2	Determination of Available Average Power		316
	7.6.3	Determination of Battery Requirements		317
	7.6.4	Additional Observations and Considerations		317
7.7	A Critical Need Refrigeration System		317
	7.7.1	Introduction		317
	7.7.2	Load Determination		318
	7.7.3	Battery Sizing		318
	7.7.4	Array Sizing		320
	7.7.5	Charge Controller and Inverter Selection		321
	7.7.6	BOS Component Selection		321
	7.7.7	Overall System Design		322
7.8	A PV-Powered Mountain Cabin		323
	7.8.1	Introduction		323
	7.8.2	Load Determination		324
	7.8.3	Battery Selection		327
	7.8.4	Array Sizing and Tilt		328
	7.8.5	Charge Controller Selection		330
	7.8.6	Inverter Selection		330
	7.8.7	Excess Electrical Production		331
	7.8.8	BOS Component Selection		332
		7.8.8.1	Wire, Circuit Breaker, and Switch Selection	332
		7.8.8.2	Other Items	333
7.9	A Hybrid-Powered, Off-Grid Residence		334
	7.9.1	Introduction		334
	7.9.2	Summary of Loads		336
	7.9.3	Battery Selection		337
	7.9.4	Array Design		339
	7.9.5	Generator Selection		342
	7.9.6	Generator Operating Hours and Operating Cost		342
	7.9.7	Charge Controller and Inverter Selection		344
	7.9.8	Wire, Circuit Breaker, and Disconnect Selection		346
	7.9.9	BOS Component Selection		348
	7.9.10	Total System Design		348
7.10	Summary of Design Procedures		349
References			354
Suggested Reading			354

Chapter 8
Economic Considerations 355

8.1 Introduction 355
8.2 Life-Cycle Costing 356
    8.2.1 The Time Value of Money 356
    8.2.2 Present Worth Factors and Present Worth 357
    8.2.3 Life-Cycle Cost 360
    8.2.4 Annualized LCC 364
    8.2.5 Unit Electrical Cost 365
    8.2.6 LCOE Analysis 365
8.3 Borrowing Money 367
    8.3.1 Introduction 367
    8.3.2 Determination of Annual Payments on Borrowed Money 367
    8.3.3 The Effect of Borrowing on LCC 370
8.4 Payback Analysis 371
8.5 Externalities 372
    8.5.1 Introduction 372
    8.5.2 Subsidies 374
    8.5.3 Externalities and PV 375
References 377
Suggested Reading 378

Chapter 9
Externalities and Photovoltaics 379

9.1 Introduction 379
9.2 Externalities 380
9.3 Environmental Effects of Energy Sources 381
    9.3.1 Introduction 381
    9.3.2 Air Pollution 382
        9.3.2.1 The Clean Air Act and the U.S. Environmental Protection Agency 382
        9.3.2.2 Greenhouse Gases and the Greenhouse Effect 383
    9.3.3 Water and Soil Pollution 384
    9.3.4 Infrastructure Degradation 385
    9.3.5 Quantifying the Cost of Externalities 385
        9.3.5.1 The Cost of CO_2 385
        9.3.5.2 Sequestering CO_2 with Trees 386
        9.3.5.3 Attainment Levels as Commodities 387
        9.3.5.4 Subsidies 388
    9.3.6 Health and Safety as Externalities 389
9.4 Externalities Associated with PV Systems 389
    9.4.1 Environmental Effects of PV System Implementation 389
    9.4.2 Environmental Effects of PV System Deployment and Operation 391

  9.4.3 Environmental Impact of Large-Scale Solar PV Installations 392
  9.4.4 Environmental Effects of PV System Decommissioning............. 393
References.. 394

Chapter 10
The Physics of Photovoltaic Cells 397

10.1 Introduction.. 397
10.2 Optical Absorption.. 397
  10.2.1 Introduction.. 397
  10.2.2 Semiconductor Materials 398
  10.2.3 Generation of EHP by Photon Absorption.................................... 399
  10.2.4 Photoconductors... 402
10.3 Extrinsic Semiconductors and the PN Junction 404
  10.3.1 Extrinsic Semiconductors 404
  10.3.2 The PN Junction 406
    10.3.2.1 Drift and Diffusion.. 406
    10.3.2.2 Junction Formation and Built-In Potential 407
    10.3.2.3 The Illuminated PN Junction 410
    10.3.2.4 The Externally Biased PN Junction 412
10.4 Maximizing PV Cell Performance 414
  10.4.1 Introduction.. 414
  10.4.2 Minimizing the Reverse Saturation Current................................... 415
  10.4.3 Optimizing Photocurrent 416
    10.4.3.1 Minimizing Reflection of Incident Photons 416
    10.4.3.2 Maximizing Minority Carrier Diffusion Lengths 417
    10.4.3.3 Maximizing Junction Width... 419
    10.4.3.4 Minimizing Surface Recombination Velocity 421
    10.4.3.5 A Final Expression for the Photocurrent....................... 422
  10.4.4 Minimizing Cell Resistance Losses 424
10.5 Exotic Junctions 426
  10.5.1 Introduction.. 426
  10.5.2 Graded Junctions... 427
  10.5.3 Heterojunctions 428
  10.5.4 Schottky Junctions 428
  10.5.5 Multijunctions 431
  10.5.6 Tunnel Junctions 432
References.. 434

Chapter 11
Evolution of Photovoltaic Cells and Systems... 435

11.1 Introduction.. 435
11.2 Silicon PV Cells 437
  11.2.1 Production of Pure Silicon... 437
  11.2.2 Single-Crystal Silicon Cells... 438

		11.2.2.1	Fabrication of the Wafer	438
		11.2.2.2	Fabrication of the Junction	439
		11.2.2.3	Contacts	441
		11.2.2.4	Antireflective Coating (ARC)	444
		11.2.2.5	Modules	445
	11.2.3	A High-Efficiency Si Cell with All Contacts on the Back		445
	11.2.4	Multicrystalline Silicon Cells		447
	11.2.5	Other Thin Silicon Cells		447
	11.2.6	Amorphous Silicon Cells		448
		11.2.6.1	Introduction	448
		11.2.6.2	Fabrication	449
		11.2.6.3	Cell Performance	451
11.3	Gallium Arsenide Cells			451
	11.3.1	Introduction		451
	11.3.2	Production of Pure Cell Components		452
		11.3.2.1	Gallium	452
		11.3.2.2	Arsenic	452
		11.3.2.3	Germanium	453
	11.3.3	Fabrication of the Gallium Arsenide Cell		453
	11.3.4	Cell Performance		455
11.4	CIGS Cells			456
	11.4.1	Introduction		456
	11.4.2	Production of Pure Cell Components		457
		11.4.2.1	Copper	457
		11.4.2.2	Indium	458
		11.4.2.3	Selenium	458
		11.4.2.4	Cadmium	458
		11.4.2.5	Sulfur	459
		11.4.2.6	Molybdenum	459
	11.4.3	Fabrication of the CIS Cell		459
	11.4.4	Cell Performance		460
11.5	CdTe Cells			462
	11.5.1	Introduction		462
	11.5.2	Production of Pure Tellurium		462
	11.5.3	Production of the CdTe Cell		463
	11.5.4	Cell Performance		464
11.6	Emerging Technologies			465
	11.6.1	New Developments in Silicon Technology		465
	11.6.2	CIS-Family-Based Absorbers		467
	11.6.3	Other III–V and II–VI Emerging Technologies		468
	11.6.4	Other Technologies		469
		11.6.4.1	Thermophotovoltaic Cells	469
		11.6.4.2	Intermediate Band Solar Cells	469
		11.6.4.3	Supertandem Cells	470
		11.6.4.4	Hot Carrier Cells	470

		11.6.4.5	Optical Up- and Down-Conversion	470
		11.6.4.6	Organic PV Cells	471
		11.6.4.7	Concentrating PV Cells	472
		11.6.4.8	Perovskites	473
		11.6.4.9	Quantum Dot and Dye-Sensitive Solar Cells	475
11.7	New Developments in System Design			475
	11.7.1	Micro Grids		475
	11.7.2	Smart Grids		476
	11.7.3	Inverter Performance Enhancement		478
	11.7.4	Module Performance Enhancement		478
11.8	Summary			479
References				480

Appendix: Design Review Checklist 485
Index 487

Preface

The goal of this textbook is to present a comprehensive engineering basis for photovoltaic (PV) system design so that the engineer can understand the *what*, the *why*, and the *how* associated with the electrical, mechanical, economic, and aesthetic aspects of PV system design. This book is intended to *educate* the engineer in the design of PV systems so that when engineering judgment is needed, the engineer will be able to make intelligent decisions based upon a clear understanding of the parameters involved. This goal differentiates this textbook from the many design and installation manuals that are currently available that *train* the reader *how* to do it but not *why*.

Widespread acceptance of the first three editions, coupled with significant growth and new ideas in the PV industry, along with additional years of experience with PV system design and installation for the authors, and, for that matter, a bit of nudging from the publisher, have led to the publication of this Fourth Edition. In recent years, annual installed renewable capacity has exceeded annual installed nonrenewable capacity in the United States. So who knows, there might even be jobs waiting out there for many who learn the material in this book.

The *what* question is addressed in the first three chapters, which present an updated background of energy production and consumption, some mathematical background for understanding energy supply and demand, a summary of the solar spectrum, how to locate the sun, and how to optimize the capture of its energy, as well as the various components that are used in PV systems. Chapter 3 has been shortened a bit so actual design work can begin by about the fifth week of a three-credit course, but it also introduces lithium and a few other emerging battery technologies.

The *why* and *how* questions are dealt with in the remaining chapters in which every effort is made to explain why certain PV designs are done in certain ways as well as how the design process is implemented. Included in the *why* part of the PV design criteria are economic and environmental issues that are discussed in Chapters 8 and 9. Chapter 5 has been embellished with additional practical considerations added to the theoretical background associated with mechanical and structural design.

Chapters 4, 6, and 7 have once again have been updated, including updated homework problems, to incorporate the most recently available technology and design and installation practice. In particular, Chapter 4 incorporates more emphasis on higher voltage systems, new developments in IEEE 1547, new fire code requirements, and *National Electrical Code* changes in the 2014 Edition, such as rapid shutdown and arc fault protection. New sections on Arc Flash Calculations and System Commissioning have been added to Chapter 4.

By the end of Chapter 4, instructors can assign relevant design problems earlier in the course to avoid having designs due at final exam time. We have found when significant design projects are due just before final exams, students tend to spend time on design projects that might better be spent studying for the final exam, especially if the final exam counts more toward the overall course grade. Paradoxically, the better designers have sometimes ended up with lower course grades.

Since the publication of the previous edition in 2009, a nearly overwhelming wealth of new research on old and new technologies has been underway. We try to cover the highlights of some of this activity in Chapter 11 but have eliminated some of the historical content to make room for the new stuff.

The Appendix presents a recommended format for submittal of a PV design package for permitting or for design review that has been widely used in Florida by the authors and their associates.

A modified top-down approach is used in the presentation of the material. The material is organized to present a relatively quick exposure to all of the building blocks of PV systems followed by design, design, and design. Even the physics of PV cells of Chapter 10 and the material on emerging technologies of Chapter 11 are presented with a design flavor. The focus is on adjusting the parameters of PV cells to optimize their performance as well as on presenting the physical basis of PV cell operation.

Homework problems are incorporated that require both analysis and design, since the ability to perform analysis is the precursor to being able to understand how to implement good design. Many of the problems have multiple answers, such as "Calculate the number of daylight hours on the day you were born in the city of your birth." We have eliminated a few homework problems based on old technology and added a number of new problems based on contemporary technology. Hopefully there are a sufficient number to enable students to test their understanding of the material.

We recommend the course be presented so that by the end of Chapter 4 students will be able to engage in a comprehensive, relevant design project and by the end of Chapter 7 they will be able to design relatively complex systems. We like to assign two design projects: a straight grid-connected system based on Chapter 4 material and a battery-backup grid-connected system based on Chapter 6 material. At the discretion of the instructor, an additional design of a stand-alone system might be considered at the end of Chapter 7, or the stand-alone design might be assigned rather than the battery-backup grid-connected system.

While it is possible to cover all the material in this textbook in a three-credit semester course, it may be necessary to skim over some of the topics. This is where the discretion of the instructor enters the picture. For example, each of the design examples of Chapter 4 introduces something new, but a few examples might be left as exercises for the reader with a preface by the instructor as to what is new in the example. Alternatively, by summarizing the old material in each example and then focusing on the new material, the *why* of the new concepts can be emphasized.

The order of presentation of the material actually seems to foster a genuine reader interest in the relevance and importance of the material. Subject matter covers a wide range of topics, from chemistry to circuit analysis to electronics, solid-state device theory and economics. *The material is presented at a level that can best be understood by those who have reached the upper division at the engineering undergraduate level and have also completed the coursework in circuits and in electronics.*

PREFACE

We recognize that the movement to reduce credit toward the bachelor's degree has left many programs with less flexibility in the selection of undergraduate elective courses, but note the material in this textbook can also be used for a beginning graduate level course.

While the primary purpose of this material is for classroom use, with an emphasis on the electrical components of PV systems, we have endeavored to present the material in a manner sufficiently comprehensive that it will also serve practicing engineers as a useful reference book.

The course can be successfully taught as an Internet course with a preference for live participation open to all. Those remote students who were sufficiently motivated to keep up with the course generally reported that they found the text to be very readable and a reasonable replacement for lectures. We highly recommend that if the Internet is tried that quizzes be given frequently to coerce students into feeling that this course is just as important as their linear systems analysis course. Informal discussion sessions can also be useful in this regard.

The PV field is evolving rapidly. While every effort has been made to present contemporary material in this work, the fact that it has evolved over a period of a year almost guarantees that by the time it is adopted, some of the material will be outdated. For engineers who wish to remain current in the field, many of the references and websites listed will keep them up to date. Proceedings of the many PV conferences, symposia, and workshops, along with manufacturers' data, are especially helpful.

This textbook should provide engineers with the intellectual tools needed for understanding new technologies and new ideas in this rapidly emerging field. The authors hope that at least one in every 4.6837 students will make their own contribution to the PV knowledge pool.

We apologize at the outset for the occasional presentation of information that may be considered to be practical or, perhaps, even interesting or useful. We fully recognize that engineering students expect the material in engineering courses to be of a highly theoretical nature with little apparent practical application. We have made every effort to incorporate heavy theory to satisfy this appetite whenever possible.

MATLAB® is a registered trademark of The MathWorks, Inc. For product information, please contact:

The MathWorks, Inc.
3 Apple Hill Drive
Natick, MA 01760-2098 USA
Tel: 508 647 7000
Fax: 508-647-7001
E-mail: info@mathworks.com
Web: www.mathworks.com

Disclaimer

It's been said that no two snowflakes are the same. And, for that matter, it is unlikely that the weather on any 2 days at any location will be exactly the same. The minute-by-minute temperature and cloud cover will differ between any 2 days, whether they are one after the other or the same date on two successive or nonsuccessive years. As a result, estimating the performance of a PV system is, at best, an inexact science, since performance is critically weather dependent.

However, the longer the period included in a prediction or simulation of PV system performance, the more reliable the prediction becomes, since weather conditions tend to average out over the long term. Still, predicting the future is not nearly as reliable as recording the past. As a result, *when system performance is simulated in this text* using the System Advisory Model (SAM) developed by the U.S. National Renewable Energy Laboratory, because of the careful modeling tools used by NREL, SAM is one of the most reasonable tools to use to estimate annual energy production for a PV system. But no claim has been made that SAM is exact or perfect, and thus when SAM is used to estimate system performance, it must be done with the understanding of the following disclaimer:

"USER AGREES TO INDEMNIFY DOE/NREL/ALLIANCE AND ITS SUBSIDIARIES, AFFILIATES, OFFICERS, AGENTS, AND EMPLOYEES AGAINST ANY CLAIM OR DEMAND, INCLUDING REASONABLE ATTORNEYS' FEES, RELATED TO USER'S USE OF THE DATA. THE DATA ARE PROVIDED BY DOE/NREL/ALLIANCE "AS IS," AND ANY EXPRESS OR IMPLIED WARRANTIES, INCLUDING BUT NOT LIMITED TO THE IMPLIED WARRANTIES OF MERCHANTABILITY AND FITNESS FOR A PARTICULAR PURPOSE ARE DISCLAIMED. IN NO EVENT SHALL DOE/NREL/ALLIANCE BE LIABLE FOR ANY SPECIAL, INDIRECT OR CONSEQUENTIAL DAMAGES OR ANY DAMAGES WHATSOEVER, INCLUDING BUT NOT LIMITED TO CLAIMS ASSOCIATED WITH THE LOSS OF DATA OR PROFITS, THAT MAY RESULT FROM AN ACTION IN CONTRACT, NEGLIGENCE OR OTHER TORTIOUS CLAIM THAT ARISES OUT OF OR IN CONNECTION WITH THE ACCESS, USE OR PERFORMANCE OF THE DATA."

In other words, using SAM for educational purposes is just fine, but blaming SAM if a system is built and does not work as predicted by SAM is not okay.

Acknowledgments

Thanks to a long list of PV pioneers for making this publication possible. Only a few years ago, PV was considered to be an emerging technology. Some thought it would never make it on a large scale. But others with more persistence have persevered over the years such that PV is no longer a hobby but a mature, widely accepted, large-scale, worldwide technology that is growing faster than most advocates ever imagined.

One of these PV pioneers is Jerry Ventre, who coauthored the first three editions of this text. Jerry has been retired for a while now and has spent much of his time doing volunteer work with organizations that focus on PV job growth. It has been an honor and a privilege to work with Jerry over the years, and we wish him the best in his current and future activities.

We are convinced that it is virtually impossible to undertake and complete a project such as this without the encouragement, guidance, and assistance from a host of friends, family, and colleagues.

For this edition, we again asked many questions of many people as we rounded up information for the wide range of topics contained herein. A wealth of information flowed our way from the National Renewable Energy Laboratory (NREL) as well as from many manufacturers and distributors of a diverse range of PV system components. We feel confident that had we asked, we would have been able to obtain additional cooperation and materials from many more.

Special thanks to Kimandy Lawrence of VB Engineering for the many discussions during and after normal work hours on various nuances associated with certain systems, including, but not limited to, ac-coupled battery-backup systems, three-phase battery-backup PV systems, microinverter systems, and other experimental designs as well as the latest administrative and permitting issues. We would also like to thank Ms. Hadis Moradi, PhD Candidate, for her review and comments on the latest developments in microgrid systems.

Last, but not least, we thank Mimi Abtahi and Jane Caputi for encouraging us to work on this manuscript rather than play golf, tennis, watch TV, or pay attention to them.

Roger Messenger
Homayoon "Amir" Abtahi
August 2016

Authors

Roger Messenger is a professor emeritus of electrical engineering at Florida Atlantic University in Boca Raton, Florida. He earned a PhD in electrical engineering at the University of Minnesota and is a Registered Professional Engineer, a former Certified Electrical Contractor, and a former NABCEP Certified PV Installer who has enjoyed working on field installations as much as he enjoys teaching classes or working on the design of a system or contemplating the theory of operation of a system or commissioning a system. His research work has ranged from electrical noise in gas discharge tubes to deep impurities in silicon to energy conservation to PV system design and performance. He worked on the development and promulgation of the original Code for Energy Efficiency in Building Construction in Florida and has conducted extensive field studies of energy consumption and conservation in buildings and swimming pools.

Since his retirement from Florida Atlantic University in 2005, he has worked as vice president for engineering at VB Engineering, Inc., in Boca Raton and as senior associate at FAE Consulting in Boca Raton, Florida. While at VB Engineering, he directed the design of several hundred PV designs, including the 5808-module, 4-acre, 1-MW system on the roof of the Orange County Convention Center in Orlando, Florida. While at FAE Consulting, he led the design of an additional 6 MW of systems that were installed. He has also been active in the Florida Solar Energy Industries Association and the Florida Alliance for Renewable Energy, has served as a peer reviewer for the U.S. Department of Energy, and has served on the Florida Solar Energy Center Advisory Board. He has conducted numerous seminars and webinars on designing, installing, and inspecting PV systems.

Homayoon "Amir" Abtahi is an associate professor of mechanical engineering at Florida Atlantic University in Boca Raton, Florida. He earned a PhD in mechanical engineering from the Massachusetts Institute of Technology in 1981 and joined Florida Atlantic University in 1983. In addition to his academic activity, he has a wealth of practical experience, much of which has been obtained as a volunteer. He is a Registered Professional Engineer in Florida and is a member of ASME, IEEE, ASHRAE, and SAE. He has held LEED Certification since 2007, is ESTIDAMA Certified in the United Arab Emirates, and is a Certified General Contractor and a Certified Solar Contractor in the state of Florida. His interests range widely from PV to PEM fuel cells, integrated capacitor/battery power modules, and atmospheric water generation.

In 1985, he installed the first solar-power system in Venezuela and was responsible for the first application of solar power for posthurricane emergency power and lighting and Ham radio communication operations in the aftermath of Hurricane Hugo in St. Croix in 1989 and Hurricane Marilyn in St. Thomas in 1995. In 1989, he published the first comprehensive catalog of 12-V appliances for use with PV systems.

Recently, he has been involved with PV installations in the Caribbean, South America, Bangladesh, and India. From 2008 to 2010, he was responsible for design and installation of over 100 residential and 20 commercial/industrial PV systems. Over the past 15 years, he has had responsibility for the design and installation of 1 million BTUD of solar hot water and solar process heat. Along with PV and thermal applications, he has had experience with heat exchangers, MEP plan review, LEED projects, tracking PV, micro-turbines, parabolic trough solar, and other hybrid applications.

Abbreviations

AFCI	arc fault circuit interrupter
AGM	absorbent glass mat
AGS	automatic generator start
AM	air mass
ARC	anti-reflective coating
AWG	American Wire Gauge
BTU	British Thermal Unit
CEC	California Energy Commission
DH	daylight hours
DPP	differential power processor
EHP	electron-hole pair
EVA	ethylene vinyl acetate
GFDI	ground fault detection and interruption
LCB	linear current booster
MOSFET	metal oxide semiconductor field effects transistor
MPPT	maximum power point tracker
NEC	National Electrical Code
NFPA	National Fire Protection Association
NOCT	nominal operating cell temperature
Pa	In finance: Cumulative present worth factor
Pa	In engineering: Unit of pressure-Pascal
Pr	present worth factor
PW	present worth
PUC	point of utility connection
SFS	Sandia Frequency Shift
SOC	state of charge
STC	standard test conditions
SVS	Sandia Voltage Shift
THD	total harmonic distortion
UL	Underwriters Laboratory

CHAPTER 1

Background

1.1 INTRODUCTION

On June 26, 1997, the U.S. Million Solar Roofs Initiative (MSRI) was announced at the United Nations Special Session on Environment and Development in New York [1]. It was proclaimed that "Now we will work with businesses and communities to use the sun's energy to reduce our reliance on fossil fuels by installing solar panels on 1 million more roofs around our nation by 2010. Capturing the sun's warmth can help us to turn down the Earth's temperature." In 1997, little concern was being expressed in the public sector about energy problems and perhaps even less discussion related to global warming, so this statement by President Clinton carries special significance from a 2016 perspective. Now, less than 20 years later as this chapter goes to press, both topics are receiving significant attention. By the end of 2007, it is estimated that more than 600,000 solar systems had been installed around the United States [2] and the State of California had introduced its own MSRI [3]. The 600,000 installations included photovoltaic (PV), solar water heating, and solar pool heating systems. In the 12 months of 2013, over 145,000 residential PV systems were installed in the United States, in addition to over 1 GW of PV installed by U.S. utilities [4].

But installations have not been limited to the United States. Germany, China, Japan, Spain, and many other countries have also had even more aggressive renewable energy programs [2,3]. In 2013, China and Japan were the largest markets for PV [4] and the United States was falling behind.

In the United States, more and more states are now adopting renewable portfolio standards (RPSs), which set goals for the percentage of the electrical energy mix to be provided by renewable sources by a certain date. Implementation of these goals will require engineers who understand the how and why of PV system design. RPSs, coupled with carefully devised incentive programs, significant reductions in installed system cost, and significant improvements in reliability have now provided the momentum for a sustained effort in the deployment of solar technologies well beyond the year 2030. In fact, the MSRI may need to be extended to a 100 Million Solar Roofs Initiative to meet the sustainable energy needs of future generations. According to *CNN Money*, the solar industry added 35,000 jobs in 2015 [5].

This book is dedicated to the engineers and technicians who have been and may become involved in turning this dream into reality.

1.2 POPULATION AND ENERGY DEMAND

The human population of the Earth has now passed 7.3 billion [6], and all of these inhabitants need the energy necessary to sustain their lives. Exactly how much energy is required to meet these needs and exactly what sources of energy will meet these needs will be questions to be addressed by present and future generations. One certainty, however, is that developing nations will be increasing their per capita energy use. For example, in 1997, the Peoples Republic of China was building electrical generating plants at the rate of 300 MW per week. These plants used relatively inexpensive, old, inefficient, coal-fired technology and provided electricity for predominantly inefficient end uses [7]. By 2008, pollution from factories and power plants in China had become so intense that some Olympic athletes had expressed reluctance to compete in the 2008 Olympic Games in Beijing for health reasons. The potential consequences to the planet of continued increase in the use of fossil fuels are profound [8,9]. Before proceeding with the details of PV power systems, a promising source of energy for the future, it is instructive to look at the current technical and economic energy picture. This will hopefully enable the reader to better assess the contributions that engineers will need to make toward a sustainable energy future for the planet.

1.3 CURRENT WORLD ENERGY USE PATTERNS

Figure 1.1 shows the increase in worldwide energy production by source between 1989 and 2014 [10]. In 2000, worldwide annual primary energy consumption was 397.40 quads [11]. The developed countries of the world consumed approximately 75% of this energy, while nearly 2 billion people in developing countries, mostly within the tropics, remained without electricity. In 2005, worldwide primary energy consumption increased to 462 quads [12] and 2 billion people still lived without electricity.

There can be a time delay between market forces and market responses. Note that the production of crude oil continued upward after the 1973 oil embargo and the subsequent significant crude oil price increases during the remaining 1970s and early 1980s. During this period, high crude oil prices spurred the development of energy efficiency legislation, such as the National Energy Conservation and Policy Act, codes for energy efficiency in building construction and increased vehicle fleet mileage requirements. Consumers also responded by reducing energy use by lowering thermostats and installing insulation and other energy conservation measures. The result was lower crude oil production for a period in the mid-1980s, since the demand was lower. During this same period, more efficient use of electricity resulted in the cancellation of nuclear plant construction, resulting in a significant decrease

BACKGROUND

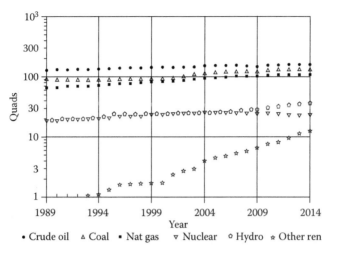

Figure 1.1 Growth of worldwide energy production by source. (Data from BP Statistical Review of World Energy, 64th Ed., June 2015, http://www.bp.com/content/dam/bp/pdf/energy-economics/statistical-review-2015/bp-statistical-review-of-world-energy-2015-full-report.pdf.)

in the growth rate of nuclear-produced electricity. Finally, concern over oil price control and embargoes prompted a switch from crude oil to coal and natural gas for use in fossil-fired electrical generation. Figure 1.2 shows the global mix of energy sources in the year 2013 [13]. The "renewables" category includes sources such as hydroelectric, wind, biomass, geothermal, and PV.

Figures 1.3 [13] and 1.4 [14] illustrate that the world faces a challenge of mammoth proportions as developing countries strive to achieve energy equity with

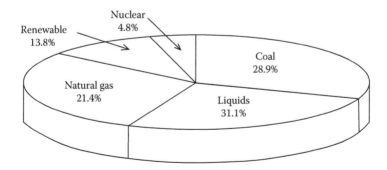

Figure 1.2 2013 global energy production mix by fuel type. (Data from International Energy Agency, 2015 key world energy statistics, 2013 total primary energy supply by source, http://www.iea.org/publications/freepublications/publication/KeyWorld_Statistics_2015.pdf.)

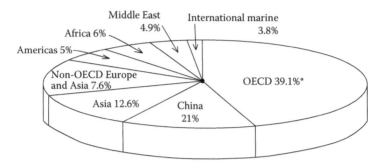

*Organization for economic cooperation and development

Figure 1.3 2013 distribution of energy users. (Data from International Energy Agency, 2015 key world energy statistics, 2013 total primary energy supply by source, http://www.iea.org/publications/freepublications/publication/KeyWorld_Statistics_2015.pdf.)

the developed countries. Note that energy equity is simply another term for the attempt to achieve comparable standards of living. But achieving a higher standard of living can carry with it a price. The price includes not only monetary obligations, but also the potential for significant environmental degradation if energy equity is pursued via the least expensive, first-cost options. Regrettably, this is the most probable scenario, since it is already underway in regions such as Eastern Europe and Asia. In fact, it is probably more likely that use of least-cost energy options may lead to comparable per capita energy use, but may simultaneously degrade the standard of living by producing air not suitable for breathing and

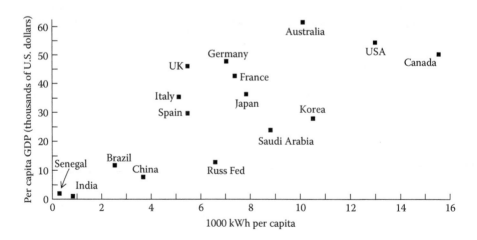

Figure 1.4 2013 worldwide per capita GDP versus per capita kW. (Data from The World Bank Data: Electric Power Consumption (kWh per Capita) and per Capita GDP, 2013, http://data.worldbank.org/indicator/EG.USE.ELEC.KH.PC.)

water not suitable for drinking. These issues will be dealt with in more detail in Chapter 9.

What is missing in Figure 1.3 is the efficiency with which the energy is consumed in these regions. Figure 1.4 is particularly interesting because it clearly shows examples of countries where energy is used least and most efficiently [14]. Those countries with the highest ratios of per capita GDP to per capita kWh have developed more energy-efficient economies than those with low ratios. Some have shifted from predominantly manufacturing economies to predominantly information economies, with the result of a smaller fraction of energy use for manufacturing. Countries at the low end of the scale tend to be farming-oriented with the use of mostly manual labor or manufacturing oriented with very low wages of workers and low first-cost energy sources.

As developing countries increase their manufacturing capabilities, using cheap but polluting local energy sources, there may be pressure to relax pollution control standards within the developed world in order to maintain the ability to compete with production from developing countries. Opposition to free-trade agreements has been partially based on such environmental concerns, since such agreements forbid any country to impose import tariffs on goods produced in countries having weak environmental standards. So the developed countries are challenged to export efficiency to the developing countries. In fact, one of the arguments the United States made against the signing of the Kyoto Treaty on the reduction of greenhouse gases is that it does not require equal restrictions on greenhouse gas emissions in developing countries. The reasoning here is that it puts the United States at an economic disadvantage, but in Chapter 9, the argument will be made that efficient use of energy, with minimal environmental degradation, leads to economic advantage. As of 2016, however, the situation has changed, with the United States and China both strongly supporting the Paris climate agreement of November–December 2015. Both countries, along with nearly 200 others [15], appear to have concluded that energy efficiency is good for business and the economy as well as for the environment.

But what does this discussion have to do with PV power production? Simple. As will be shown in Chapter 9, PV energy sources are very clean, but current PV deployment costs have only recently become competitive with the initial installed direct costs of fossil sources of electrical generation in most cases. It means that the consumer must be familiar with life-cycle costing and that the engineer must be able to create the most cost-effective PV solution. It also means that a significant amount of research and development must be done to ensure the continuation of the decrease in the price of PV generation. It also means that work must continue in the effort to put a price tag on environmental degradation caused by energy sources, so this price can be factored into the total cost to society of any energy source.

The bottom line is that there remains a significant amount of work in research, development, and public education to be done in the energy field, and particularly in the field of PV. And much of this work can best be done by knowledgeable engineers who, for example, among other things, understand the concept of exponential growth.

1.4 EXPONENTIAL GROWTH

1.4.1 Introduction

Exponential growth is probably most familiar to the electrical engineer in the diode equation, in which I and V are the diode current and voltage, respectively, I_o is the reverse saturation current and kT/q is 26 mV at T = 300 K, or, specifically,

$$I = I_o(e^{(qV/kT)} - 1). \tag{1.1}$$

While this equation is fundamental to the performance of PV cells, many other physical processes are also characterized by exponential growth.

Exponential growth is commonly referred to as compound interest. Almost everyone has heard about it, but few understand the ramifications of constant annual percentage increase in a quantity, whether it be money, population, or energy supply or demand. One of the first to warn of the dangers of exponential growth was Malthus in 1798 [16]. He warned that population growth would exceed the ability to feed the population. The Malthusian theory is often the subject of ridicule of growth enthusiasts [17]. The intent of this discussion is neither to support nor to discount the predictions of Malthus, but merely to illustrate an important mathematical principle that engineers often overlook. The application of the principles of exponential growth is widespread in society, so the principles of exponential growth should be just as important to a well-informed engineer as is the second law of thermodynamics. For those who may have missed the second law of thermodynamics in either chemistry or thermodynamics class, it is a statement that in every process less energy comes out than what is put in. In other words, there is no free lunch.

1.4.2 Compound Interest

Compound interest is the process of compounding simple interest. If a quantity N_o is subject to an interest rate i, with i expressed as a fraction (i.e., i = %/100), the quantity will increase (or decrease, if i < 0) to a value of

$$N(1) = N_o(1+i) \tag{1.2}$$

after one prescribed time period has elapsed. If the quantity present after the prescribed time period is allowed to remain and to continue to accumulate at the same rate, then the quantity is subject to compound interest and the amount present after n time periods will be

$$N(n) = N_o(1+i)^n. \tag{1.3}$$

BACKGROUND 7

To show that this formula is a form of the exponential function, one need only to recall that

$$y^x = e^{x \ln y}.$$

Hence,

$$N(n) = N_o e^{n \ln(1+i)}. \tag{1.4}$$

Some special properties of the exponential function can now be considered.

1.4.3 Doubling Time

To determine the time, D, it will take for the original quantity to double, one need only to set $N(n) = 2N_o$ and solve for n. The result (and the answer to problem 1.1) is

$$n = \frac{\ln 2}{\ln(1+i)} = D. \tag{1.5}$$

For small values of i, $\ln(1+i)$ can be approximated as $\ln(1+i) \cong i$. Noting that $\ln 2 = 0.693$ leads to the formula so popular in the financial world, that is,

$$D = n \cong \frac{0.7}{i}. \tag{1.6}$$

Hence, for an interest rate of 7% per year (i = 0.07), the doubling time will be 10 years. For an interest rate of 10%, the doubling time will be approximately 7 years. However, as the interest rate exceeds 10%, the approximation becomes less valid, and the exact expression should be used for accurate results. In the case where the interest rate is negative, it should be obvious that no doubling can occur. The authors have proven this to be the case with various investments in the stock market.

An important property of the exponential function is that the doubling process continues for all time. Hence, if the doubling time is 10 years, then the quantity will double again in another 10 years, so it will now be 4 times its original value. In another 10 years it will double again to 8 times its original value. After 40 years, the quantity will be 16 times its original value. Figure 1.5 shows this exponential increase.

Note that if the function $y = Ae^{bx} = A10^{bx \log e}$ is plotted with linear coordinates, the familiar exponential curve appears, as in Figure 1.5a. If the logarithm of each side is taken, the result is

$$\log y = \log A + (b \log e)x. \tag{1.7}$$

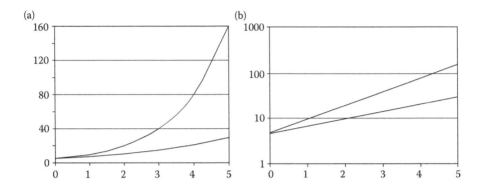

Figure 1.5 Examples of exponential functions plotted on linear and on semilogarithmic coordinates. (a) Linear vertical axis. (b) Logarithmic vertical axis.

Hence, if log y is plotted as a function of x, the graph will be linear with slope b log e, as shown in Figure 1.5b. Figure 1.5b shows that plotting log y versus x is a convenient way to check for an exponential relationship between two variables. Notice also that the vertical axis can be labeled either in terms of log y or, simply, in terms of y. When the axis is labeled in terms of y, it is understood that the axis is linear in the logarithm of y. For the person who prefers to let computers do the work, either Excel or MATLAB® can conveniently plot data and produce a least mean square curve fit to the data along with an estimate of the goodness of the fit. When a set of data is available, all one needs to do is to list the independent and dependent variables side-by-side on an Excel spreadsheet and then use the Chart Wizard to plot an x–y scatter plot of the data. By selecting of the chart option, and then selecting the plot that shows data but no connecting lines, one can then click on one of the data points on the chart, which will select the data set on the graph. Then go to "Chart" on the pull-down menu, and select "Add Trendline." This option opens a dialog box that offers a choice of six trend/regression types. To obtain the best fit, select a regression type, then select "Options" and choose to add the equation and the R^2 value to the graph. An R^2 value of 1 indicates a perfect fit of the data to the curve by the equation that is displayed. Low R^2 values suggest choosing a different regression type. Several good user guides for Excel are available for those who would like to further explore the use of Excel as a convenient analysis tool for use in PV system design or other technical endeavors [18,19]. MATLAB and other math programs are also powerful tools.

For the reader who has not had the pleasure of using Excel to find the least mean square fit to a data set, several problems are available at the end of this chapter. It is anticipated that the reader will use Excel frequently as a tool for the design of PV systems in later chapters.

1.4.4 Accumulation

Another important property of the exponential function is the amount of accumulation (or depletion) of a quantity in a doubling time. Noting that integration of

BACKGROUND

a function yields the area of the curve under the function, which is the accumulation of the quantity being integrated, it is a straightforward math exercise to integrate Equation 1.4 to show that *the amount accumulated (or depleted) in a doubling time is equal to the amount accumulated (or depleted) in all previous history* (see Problem 1.1).

The implications of this result are far reaching. For example, a prominent political figure once noted that there was still as much oil underground in the United States as what had been pumped out since pumping first started more than 140 years ago. This was at a time when oil extraction was increasing at a rate of approximately 7% per year. If the extraction had continued to increase at 7% per year, which has a doubling time of approximately 10 years, in the next doubling time all of the remaining petroleum would have been extracted. Many other important public figures have made similar statements that tend to assign a linear nature to the exponential function [20]. Could this be an argument for engineers to run for public office?

In fact, extraction did not continue to increase at this rate. With regard to the use of resources, Hubbert [21] developed a model that incorporates a Gaussian function for depletion, which seems to have more validity than the exponential model. The rising edge of the Gaussian function, however, is conveniently approximated by an exponential.

The accumulation formula, of course, may also apply to the deployment of new technology. Since the early 1990s, PV power production has been increasing at a very impressive rate. Problems 1.15 and 1.16 offer an opportunity to explore the relevance of this observation if this rate of increase continues.

1.4.5 Resource Lifetime in an Exponential Environment

The previous discussion of exponential growth has been based on total amounts of a quantity at any given time. If the time derivative of the exponential expression for quantity is taken, the rate of change of the quantity is obtained. Since the derivative of an exponential is also an exponential, the same rules apply to the derivative as to the function. Distinguishing between amount present and rate of use (or increase) is important when determining the lifetime of a resource. Hence, when considering an exponential expression, one needs to establish whether it refers to barrels or barrels per day, or, perhaps, megawatts or megawatts per year of PV deployment.

The final concept to explore relating to the exponential function is the lifetime of a resource under conditions of exponential increase. It is common to predict the lifetime of a resource under the current rate of consumption. This involves simple arithmetic, since if there are Z (quantity) widgets left to use and if we use X (rate of use) of them per year, then the widgets will last for Y years, where $Y = Z/X$. But what happens to the expected lifetime of the widget if people decide that they really like widgets and they decide to use them at an increasing rate of 100i% per year, noting that if i represents a fraction, then 100i converts the fraction to a percentage? This problem can be solved by assuming that C_o represents the present rate of consumption of a resource and Y_o represents the estimated lifetime of the resource at the present rate of consumption. Then, if consumption increases by 100i% per year, the rate of consumption at any point in time, x, is given by

$$C(x) = C_o(1+i)^x. \qquad (1.8)$$

The total accumulated consumption over a period of m years, TOT, can be found from previous formulas, or, more formally, by evaluating

$$\text{TOT} = \int_0^m C(x)dx = \int_0^m C_o e^{x \ln(1+i)} dx$$

$$= \frac{C_o}{\ln(1+i)} (e^{m \ln(1+i)} - 1).$$

Next, set the total to equal the estimated amount remaining ($C_o Y_o$) and solve for m, since this will yield the number of years for the total consumption to equal the amount remaining. The result is

$$m = \frac{\ln[Y_o \ln(1+i) + 1]}{\ln(1+i)}. \qquad (1.9)$$

As an example of the use of this result, assume that a resource is estimated to last for another 100 years at present consumption rates ($Y_o = 100$), but that consumption will increase at the rate of 5% per year (i = 0.05). Using these numbers in the above formula gives m = 36.31 years. If the estimate is off by a factor of 10, and there is really a 1000-year supply left at current consumption rates, then m = 80.09 years.

As perhaps a more reassuring example of the use of this result, it is also possible that the consumption of a resource might decline at a constant percentage per year. This could happen if the resource were to be replaced by another resource, for example. For the above example, with $Y_o = 100$ years and an annual decrease of 0.5% (i.e., i = −0.005), the new lifetime becomes 139 years, and if i = −0.01, the resource will last forever.

Hence, two important observations emerge from the lifetime formula:

1. If annual consumption of a resource increases exponentially, it is not important how much is thought to remain; it will be consumed much faster than one can imagine.
2. If annual consumption decreases exponentially, it is possible to extend the lifetime of a resource to forever.

1.5 HUBBERT'S GAUSSIAN MODEL

In 1956, M. King Hubbert, who was employed by the Shell Oil Company, published his now acclaimed theory of resource depletion [21]. Simply put, Hubbert

BACKGROUND

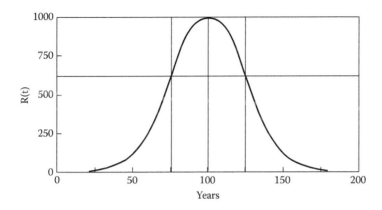

Figure 1.6 Gaussian function with $R_m = 1000$, $t_0 = 100$, and $s = 25$.

reasoned that the life of a finite resource follows a Gaussian curve described by the equation, often referred to as the error function or normal curve,

$$R(t) = R_m e^{-(t-t_0)^2/2s^2}, \qquad (1.10)$$

where $R(t)$ represents the consumption rate at a given time, t, R_m represents the maximum consumption rate, and s represents a shape factor for the curve, commonly known as the standard deviation.

He also observed that production of petroleum lagged discovery by approximately 40 years. Figure 1.6 shows a plot of Equation 1.10, and Figure 1.7 [21] shows

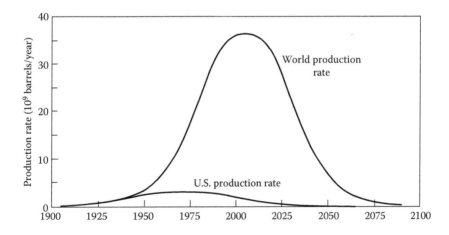

Figure 1.7 Hubbert's predictions for U.S. and world petroleum production. (Adapted from Hubbert, M. K., *American Petroleum Institute, Drilling and Production Practice*, 23, 1956, 7–25, with permission.)

Hubbert's 1956 curves relating to petroleum production. The rising edge represents a nearly exponential function, until it nears the peak of the curve, where leveling occurs, followed by nearly exponential decay. Since the curve of Figure 1.6 plots consumption rate, note that according to Hubbert's theory, when half a resource is consumed, the consumption will have leveled at its maximum value and then will begin its nearly exponential decline. According to Hubbert's model, domestic petroleum production in the United States would peak near 1970 and world petroleum production would peak near the turn of the century. Indeed, as 2008 approached, it appeared that Hubbert may have been correct. According to projections of the U.S. Department of Energy, Energy Information Agency, both domestic and Alaskan petroleum production would continue to decline through the year 2015 [22]. Current data, however, show healthy U.S. domestic production as a result of enhanced hydraulic fracturing and horizontal drilling (fracking) of petroleum and natural gas, a glitch that was not anticipated in 1956.

Interestingly enough, it also turns out with the Hubbert theory that if an error is made in estimating the amount of a resource present, the peak of the curve will not be shifted by a significant amount. Perhaps, the most significant conclusion of the Hubbert theory, however, is that the consumption of a resource follows a smooth curve rather than an abrupt one that involves unabated consumption until suddenly no more is left. Hopefully, approaching the peak sends a message to find a replacement. For that matter, exceeding 400 ppm of CO_2 in the atmosphere presents an even more compelling reason for a carbon replacement. In fact, the rapid increase in deployment of renewables as shown in Figure 1.1 suggests that perhaps the replacement has already been found.

1.6 NET ENERGY, BTU ECONOMICS, AND THE TEST FOR SUSTAINABILITY

The net energy associated with an energy source is simply the difference between the energy required to obtain and convert the source to useful energy and the actual useful energy obtained from the source. For example, in order to be able to burn a barrel of oil, it is necessary to find the oil, extract the oil, transport the oil, refine the oil, and construct the facility for burning the oil. The refined oil must then be transported to the burning site and, presumably, after the oil is burned, any environmental damage resulting from the extraction, transportation, refining, and burning should be repaired. Energy is involved in all of these steps. The bottom line is that if it takes more than a barrel of oil worth of energy to obtain and convert the energy available in a barrel of oil, one should question seriously whether it makes sense to burn the oil in the first place.

In some cases, it may make sense to expend the energy to get the resource. Suppose, for example, that another use were discovered for oil, such as providing an essential chemical for the cure of cancer. Then it would make sense to expend energy from sources other than oil, even if the energy exceeded the energy value of the oil, in order to make the oil available for the more important use. Another situation would be to use a lower grade or quality form of energy to produce a higher grade or quality form of energy. Sometimes such an action might make energy sense.

For example, burning coal to produce electricity takes about three units of coal energy to produce one unit of electrical energy. Until television sets that run directly from coal are invented, this inefficient process of converting coal energy to electrical energy will probably continue, in spite of the known environmental problems with coal.

The concept of net energy was introduced by Odum and Odum in 1976 [22,23]. They incorporated the net energy concept into a new standard for economics that they felt made better sense than the gold standard. They called it the BTU standard. The BTU standard simply recognizes that everything has an energy content. Henderson [24] has written extensively on the concept of BTU economics. The reader is encouraged to read Odum and Odum and Henderson during a term break for enlightening discussions of how the economic system might be changed to an energy-based standard.

For the purposes of this book, the test for sustainability for an energy source will include two factors. The first will be whether the source is finite. A finite source is generally termed nonrenewable, while an infinite source is termed renewable. The second will be whether the source has positive net energy. That is, once energy is expended to produce the source, will the source then generate more energy than was required for its production?

The idea that a source can generate more energy than was expended in its production may seem inconsistent with the second law of thermodynamics. However, if we allow the use of energy from a very large reservoir as a supply of energy to be converted by the source, then the source becomes nearly infinite. In the case of the sun, which is expected to survive for another 4 billion years or so [25], we have such a reservoir. Thus, for example, if a PV cell can generate more electrical energy over its lifetime than was expended in its production and deployment and ultimately in its disposal, including environmental energy costs, then the cell would be considered to have positive net energy.

The concept of net energy will be considered in the context of PV cell production and in the discussion of environmental effects of energy sources.

1.7 DIRECT CONVERSION OF SUNLIGHT TO ELECTRICITY WITH PV

Becquerel first discovered that sunlight can be converted directly into electricity in 1839, when he observed the photogalvanic effect [26]. Then, in 1876, Adams and Day found that selenium has PV properties. When Planck postulated the quantum nature of light in 1900, the door was opened for other scientists to build on this theory. It was in 1930 that Wilson proposed the quantum theory of solids, providing a theoretical linkage between the photon and the properties of solids. Ten years later, Mott and Schottky developed the theory of the solid-state diode, and in 1949, Bardeen, Brattain, and Shockley invented the bipolar transistor. This invention, of course, revolutionized the world of solid-state devices. The first solar cell, developed by Chapin, Fuller, and Pearson, followed in 1954. It had an efficiency of 6%. Only 4 years later, the first solar cells were used on the Vanguard I orbiting satellite.

One might wonder why it took so long to develop the PV cell. The answer lies in the difficulty in producing sufficiently pure materials to obtain a reasonable level of cell efficiency. Prior to the development of the bipolar transistor and the advent of the space program, there was little incentive for preparing highly pure semiconductor materials. Coal and oil were meeting the world's need for electricity, and vacuum tubes were meeting the needs of the electronics industry. But since vacuum tubes and conventional power sources were impractical for space use, solid state gained its foothold.

PV cells are made of semiconductor materials and are assembled into modules of 36 or more cells. This observation is significant, since this means the same industry that has, in the past 60 years, progressed from the development of the bipolar transistor to integrated circuits containing millions of transistors is also involved in the development of PV cells. Figure 1.8 [27] shows the decline in the cost of PV modules since 1977. Much of the initial cost reduction has been due to process improvement in the production of the cells. At this point, the limiting factor is becoming the energy cost of the cells. For a few years following 2005, refined silicon prices increased significantly due to a worldwide shortage, but then fell again in 2008 as supply caught up with demand. Hence, the challenge of the future will be to reduce the energy content of the cell production process while maintaining or increasing cell performance, efficiency, and reliability.

Figure 1.9 [28–32] shows world PV shipments in megawatts from 1975 to 2014. Note that the data are plotted on semilogarithmic coordinates. The actual data since 1995 are given in Problem 1.17, so the reader can generate a plot and determine the goodness of fit and the rate of growth over this period. A further important observation is that in 1995, 45% of the world's PV modules were manufactured in the United

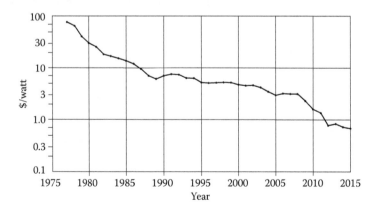

Figure 1.8 Decline in cost per watt for PV modules. (Data from Shahan, Z., 13 charts on solar panel cost and growth trends, Clean Technica, September 4, 2014, http://cleantechnica.com/2014/09/04/solar-panel-cost-trends-10-charts/.)

BACKGROUND

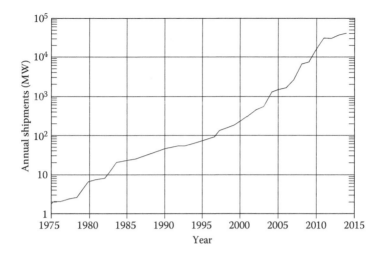

Figure 1.9 Worldwide PV shipments, 1975–2014. (Data from Growth of Photovoltaics, World wide, Wikipedia, https://en.wikipedia.org/wiki/Growth_of_photovoltaics#Deployment_by_country; *International Marketing Data and Statistics*, 22nd Ed., Euromonitor PIC, London, 1998; Maycock, P. D., The world PV market, production increases 36%, *Renewable Energy World*, July–August 2002, 147–161; Earth Policy Institute, Eco-economy indicators: SOLAR POWER—Data, www.earth-policy.org/Indicators/Solar/2007_data.htm; Maycock, P. D., *PV News*, 22(3), March 2003.)

States, while Europe, Japan, and the rest of the world manufactured 80% of the world's PV modules in 2002 [30] and 93% of the world's PV cells in 2007 [33]. Just as the United States has allowed the manufacture of consumer electronics to transfer to other countries, it appears that the United States is also allowing the same to happen with the PV industry. It will be interesting to observe this trend in the future.

As an initial test as to whether to continue with this book, it is useful to determine whether the net energy associated with PV cells is positive. Indeed, current PV cells can produce more than a 10:1 return on the energy invested in their production, and future improvements in production technology and practice may result in values up to 45:1 [34]. Hence, it appears to be worth investigating this technology in more detail. In order to compare various energy sources, it is helpful to compare the units with which the available energy in these sources is measured.

1.8 ENERGY UNITS

Energy is measured in a number of ways, including the calorie, the BTU, the quad, the foot-pound (ft-lb), and the kilowatt hour (kWh). When pounds (lb) and kilograms (kg) are used in this text, they will be treated as weight, rather than mass. Thus, a 220 lb person can also be described as a 100 kg person. For the benefit of those who may not have memorized the appendices of their freshman physics books, we repeat the definitions of a number of these energy-related quantities for an Earth-based system at or about a temperature of 27°C [10,35].

> 1 calorie is the heat needed to raise the temperature of 1 g of water 1°C.
> 1 BTU is the heat needed to raise the temperature of 1 lb of water 1°F.
> 1 quad is 1 quadrillion (10^{15}) BTU.
> 1 ft-lb is the energy expended in raising 1 lb through a height of 1 ft.
> 1 kWh is the energy expended by 1 kW operating for 1 hour.
> 1 MTOE (million tons of oil equivalent) is the energy available from 1 million tons of crude oil.
> 1 Mbd (million barrels per day) is the energy available from 1 million barrels of crude oil.
> 1 Bcm (billion cubic meters) is the energy available from 1 billion cubic meters of natural gas.

With these definitions, and with the numerical values of MTOE, million barrels of oil equivalent (MBOE), and Bcm, the following equivalencies can be determined:

> 1 BTU = 252 calories
> 1 kWh = 3413 BTU = 2,655,000 ft-lb
> 1 ft-lb = 0.001285 BTU
> 1 quad = 10^{15} BTU = 2.930×10^{11} kWh
> 1 MTOE = 0.0397 quad
> 1 MBOE = 0.00541 quad
> 1 Bcm = 0.0357 quad

It should be noted that MTOE, MBOE, and Bcm represent the *total* amount of energy stored in the specified quantities of coal, oil, and natural gas. When these fuels are burned to produce electricity, the laws of thermodynamics limit the percentage of the available energy that can be converted to electricity to approximately 40%. In addition, some of the heat lost in the process is sometimes recovered for other use. So, in converting MTOE, for example, to kWh of useful electricity, an overall efficiency factor is needed for the process.

Because the emphasis of this text will be on electrical generation, and since the kWh is the common unit for electrical energy, the equivalence between kWh and ft-lb is especially noteworthy. For example, suppose a 150-lb person wished to generate 1 kWh, assuming a system with 100% efficiency. One way would be to climb to the top of a 17,700-ft mountain to create 1 kWh of potential energy. Then, by returning to sea level by way of a chair, connected via a pulley system to a generator, the person's potential energy could be converted to electrical energy. This kWh could then be sold at wholesale for about 6 cents. Another somewhat simpler method is to burn approximately 11 fluid ounces of petroleum to produce steam to turn a steam turbine as shown in Figure 1.10.

Still another method is to deploy about 2 m² of PV cells. This system will produce about 1 kWh per day for 25 years or more with no stops for refueling, no noise, minimal maintenance, and no release of CO_2, SO_2, or NO_2 while the electricity is being produced.

BACKGROUND

Figure 1.10 Several ways to produce a kWh of electricity.

PROBLEMS

1.1 Prove Equation 1.5.
1.2 Calculate the approximate and exact doubling times for annual percentage increases of 5%, 10%, 15%, and 20%.
1.3 The human population of the Earth is approximately 7 billion and is increasing at approximately 2% per year. The diameter of the Earth is approximately 8000 miles and the surface of the Earth is approximately two-thirds water. Calculate the population doubling time, then set up a spreadsheet that will show (a) the population, (b) the number of square feet of land area per person, and (c) the length of the side of a square that will produce the required area per person. Carry out the spreadsheet for 15 doubling times, assuming that the rate of population increase remains constant. What conclusions can you draw from this exercise?
1.4 Show that in an exponential growth scenario, the amount accumulated in a doubling time equals the amount accumulated in all previous history.
1.5 Assume there is enough coal left to last for another 300 years at current consumption rates.
   a. Determine how long the coal will last if its use is increased at a rate of 5% per year.
   b. If there is enough coal to last for 10,000 years at current consumption rates, then how long will it last if its use increases by 5% per year?
   c. Can you predict any other possible consequences if coal burning increases at 5% per year for the short or long term?
   d. Determine the annual percentage reduction in coal consumption to ensure that coal will last forever, assuming the 300-year lifetime at present consumption rates.

1.6 The half-life is the time it takes to decay exponentially to half the original amount. If the half-life of a radioactive isotope is 500 years, how many years will it take for an amount of the isotope to decay to 1% of its original value? Assume the isotope decays exponentially.

1.7 If a colony of bacteria lives in a jar and doubles in number every day, and it takes 30 days to fill the jar with bacteria,
   a. How long does it take for the jar to be half full?
   b. How long before the bacteria notice they have a problem? (You may want to pretend you are a bacterium.)
   c. If on the 30th day, three more jars are found, how much longer will the colony be able to continue to multiply at its present rate?

1.8 An enterprising young engineer enters an interesting salary agreement with an employer. She agrees to work for a penny the first day, 2 cents the second, 4 cents the third, and so on, each day doubling the amount of the previous day. Set up a spreadsheet that will show her daily and cumulative earnings for her first 30 days of employment.

1.9 Burning a gallon of petroleum produces approximately 25 lb of carbon dioxide and burning a ton of coal produces approximately 7000 lb of carbon dioxide.
   a. If a barrel of petroleum contains 42 gallons, if the world consumes 80 million barrels of petroleum per day and if the atmosphere weighs 14.7 lb per square inch of Earth surface area, calculate the weight of carbon dioxide generated each year from burning petroleum and compare this amount with the weight of the atmosphere.
   b. If a total of 16 million tons of coal are burned every day on the Earth, calculate the weight of carbon dioxide generated each year from coal burning and compare it with the weight of the atmosphere.

1.10 Assume a world population of 7.3 billion and a U.S. population of 323 million.
   a. Look up the present total annual U.S. primary energy consumption. Then determine the total world energy consumption in quads if the rest of the world were to use the same per capita energy as in the United States
   b. If the energy source mix were to remain the same as the present mix in achieving the scenario of Part a, what would be the percentage increase in CO_2 emissions?

1.11 Obtain data on worldwide energy consumption by sector from the U.S. Department of Energy, Energy Information Administration website. Plot the data and estimate annual percentage growth rates for the regions reported and then for the world.

1.12 The following measurements of x(t) are made:

t	0	1	2.3	3.0	4.5	5.2	6.5	8.0
x	2.1	8.4	31	65	360	850	3700	20,000

Construct a semilog plot of x(t) either manually or with a computer, and determine whether the function appears to have an exponential dependence. If so, determine x(t).

1.13 a. What does the area under the Gaussian curve represent?
    b. Show that 68% of the area under the Gaussian curve lies within one standard deviation, s, of maximum value of the function.
    c. What percentage of the area lies within 2 s?

1.14 Determine R_m, t_o, and s for the worldwide graph of Figure 1.7.

1.15 Look up actual U.S. and world petroleum production figures and plot them on Hubbert's curves to compare the actual production with the theoretical production.
1.16 Based on the data of Figure 1.9,
   a. Estimate the year when PV shipments will reach 100 GW.
   b. Estimate the year when PV shipments will reach 1000 GW.
   c. Estimate the year when PV shipments will reach 2700 GW.
1.17 The Wikipedia [28] reports the following worldwide PV production figures. Plot the data on an Excel graph, establish an equation to represent the data, and then answer the three questions posed in Problem 1.16. Compare the results of the two problems.

Year	2003	2004	2005	2006	2007	2008	2009	2010	2011	2012	2013	2014
GW	0.57	1.08	1.39	1.55	2.52	6.66	7.34	17.2	30.1	30.0	38.4	40.1

REFERENCES

1. Remarks by President William Clinton in Address to the United Nations Special Session on Environment and Development, June 26, 1997.
2. Sherwood, L., U.S. solar market trends 2007, Interstate Renewable Energy Council, August 2008, www.irecusa.org/fileadmin/user_upload/NationalOutreachPubs/IREC SolarMarketTrendsAugust2008_2.pdf.
3. Dorn, J. G., Solar cell production jumps 50 percent in 2007, Earth Policy Institute, December 2007, www.earth-policy.org/Indicators/Solar/2007.htm.
4. Sherwood, L., U.S. solar market trends 2013, Interstate Renewable Energy Council, July 2014.
5. Gillespie, P., Powering your world: Solar energy jobs double in 5 years, CNN Money, January 12, 2016, http://money.cnn.com/2016/01/12/news/economy/solar-energy.
6. United States Census, U.S. and world population clock, http://www.census.gov/popclock/ (accessed April 13, 2016).
7. Lindley, D., An overview of renewable energy in China, *Renewable Energy World*, November 1998, 65–69.
8. Gore, A., *An Inconvenient Truth: The Planetary Emergency of Global Warming and What We Can Do about It*, Rodale, Emmaus, PA, 2006.
9. Intergovernmental Panel on Climate Change, Fourth Assessment Report, *Climate Change 2007*, www.ipcc.ch/.
10. *BP Statistical Review of World Energy*, 64th Ed., June 2015, http://www.bp.com/content/dam/bp/pdf/energy-economics/statistical-review-2015/bp-statistical-review-of-world-energy-2015-full-report.pdf.
11. *Annual Energy Review*, 2001, U.S. Department of Energy, Energy Information Administration, Washington, DC, www.eia.doe.gov/emeu/aer.
12. U.S. Department of Energy, Energy Information Administration, Washington, DC, International energy outlook 2008, September 2008, www.eia.doe.gov/oaif/ieo/index/html.
13. International Energy Agency, 2015 key world energy statistics, 2013 total primary energy supply by source, http://www.iea.org/publications/freepublications/publication/KeyWorld_Statistics_2015.pdf.
14. The World Bank Data: Electric Power Consumption (kWh per Capita) and per Capita GDP, 2013, http://data.worldbank.org/indicator/EG.USE.ELEC.KH.PC.

15. COP 21, 2015 Paris Climate Conference, Sustainable Innovation Forum 2015, UNEP Climate Action, http://www.cop21paris.org/about/cop21.
16. Malthus, T. R., *An Essay on the Principle of Population, as It Affects the Future Improvement of Society, with Remarks on the Speculations of Mr. Godwin, M. Condorcet, and Other Writers*, Printed for J. Johnson in St. Paul's Church-Yard, London, 1798.
17. Bahr, H. M., Chadwick, B. A., and Thomas, D. L., Eds., *Population, Resources, and the Future: Non-Malthusian Perspectives*, Brigham Young University Press, Provo, UT, 1974.
18. Bloch, S. C., *Excel for Engineers and Scientists*, 2nd Ed., John Wiley & Sons, Hoboken, NJ, 2003.
19. Gottfried, B., *Spreadsheet Tools for Engineers Using Excel, Including Excel 2002*, McGraw-Hill, New York, 2003.
20. Bartlett, A. A., Forgotten fundamentals of the energy crisis, *Am. J. Phys.* 46, 1978, 876–888.
21. Hubbert, M. K., Nuclear energy and fossil fuels, *American Petroleum Institute, Drilling and Production Practice*, 23, 1956, 7–25.
22. Odum, H. T. and Odum, E. C., *Energy Basis for Man and Nature*, McGraw-Hill, New York, 1976.
23. Odum, H. T. and Odum, E. C., *Energy Basis for Man and Nature*, 2nd Ed., McGraw-Hill, New York, 1981.
24. Henderson, H., *The Politics of the Solar Age: Alternatives to Economics*, Anchor Press/Doubleday, Garden City, NY, 1981.
25. Foukal, P., *Solar Astrophysics*, John Wiley & Sons, New York, 1990.
26. Markvart, T., Ed., *Solar Electricity*, John Wiley & Sons, Chichester, UK, 1994.
27. Shahan, Z., 13 charts on solar panel cost and growth trends, Clean Technica, September 4, 2014, http://cleantechnica.com/2014/09/04/solar-panel-cost-trends-10-charts/.
28. Growth of Photovoltaics, Worldwide, Wikipedia, https://en.wikipedia.org/wiki/Growth_of_photovoltaics#Deployment_by_country.
29. *International Marketing Data and Statistics*, 22nd Ed., Euromonitor PIC, London, 1998.
30. Maycock, P. D., The world PV market, production increases 36%, *Renewable Energy World*, July–August 2002, 147–161.
31. Earth Policy Institute, Eco-economy indicators: SOLAR POWER—Data, www.earthpolicy.org/Indicators/Solar/2007_data.htm.
32. Maycock, P. D., *PV News*, 22(3), March 2003.
33. International Energy Agency: PV Power Systems Programme, Solar Power International 08, San Diego, CA, October 2008.
34. Hagens, N., The energy return of (industrial) solar—Passive solar, PV, wind and hydro, The oil drum: Discussion about energy and our future, April 29, 2008, www.theoildrum.com/files/PV_table.PNG.
35. Fishbane, P. M., Gasiorowicz, S., and Thornton, S. T., *Physics for Scientists and Engineers*, 2nd Ed., Prentice Hall, Upper Saddle River, NJ, 1996.

SUGGESTED READING

Aubrecht, G. J., *Energy*, 2nd Ed., Prentice Hall, Upper Saddle River, NJ, 1995.
Gore, A., *Earth in the Balance: Ecology and the Human Spirit*, Roedale, New York, 1992.
Kerr, R. A., The next oil crisis looms large and perhaps close, *Science*, 281, 1998, 1128–1131.
Krupp, F. and Horn, M., *Earth: The Sequel*, W. W. Norton & Co., New York, 2008.

Lovins, A. B., Datta, E. K., Bustnes, O.-E., Koomey, J. G., and Glasgow, N. J., *Winning the Oil Endgame: Innovation for Profits, Jobs, and Security*, Rocky Mountain Institute, Snowmass, CO, 2004.
President Clinton announces million solar roofs by 2010, *FEMP Focus*, August/September 1997, 7.
Silberberg, M., *Chemistry: The Molecular Nature of Matter and Change*, Mosby, St. Louis, MO, 1996.
Starke, L., Ed., *Vital Signs 1998: The Environmental Trends That Are Shaping Our Future*, W. W. Norton, New York, 1998.

CHAPTER 2

The Sun

2.1 INTRODUCTION

Optimization of the performance of PV and other systems that convert sunlight into other useful forms of energy depends upon knowledge of the properties of sunlight. This chapter provides a synopsis of important solar phenomena, including the solar spectrum, atmospheric effects, solar radiation components, determination of Sun's position, measurement of solar parameters, and positioning of the solar collector. In this chapter, an attempt is made to quantify the obvious and, perhaps, some not-so-obvious observations: Why is it light during the day and dark at night? Why are there more daylight hours in summer than in winter? Why is the Sun higher in the sky in summer than in winter? Why are there more summer sunlight hours in northern latitudes than in latitudes closer to the equator? Why does less energy reach the surface of the Earth on cloudy days? Why is it better for a solar collector to face the Sun? What happens if a solar collector does not face the Sun directly? How much energy is available from the Sun? Why is the sky blue?

2.2 THE SOLAR SPECTRUM

The Sun provides the energy needed to sustain life in our solar system. In 1 hour, the Earth receives enough energy from the Sun to meet its energy needs for nearly a year. In other words, this is about 7500 times the input to the Earth's energy budget from all other sources. In order to maximize the utilization of this important energy resource, it is useful to understand some of the properties of this "ball of fire" in the sky.

The Sun is composed of a mixture of gases with a predominance of hydrogen. As the Sun converts hydrogen to helium in a massive thermonuclear fusion reaction, mass is converted to energy according to Einstein's famous formula, $E = mc^2$. As a result of this reaction, the surface of the Sun is maintained at a temperature of approximately 5800 K. This energy is radiated away from the Sun uniformly in all directions, in close agreement with Planck's blackbody radiation formula. The energy density per unit area, w_λ, as a function of wavelength, λ, is given by

$$w_\lambda = \frac{2\pi hc^2 \lambda^{-5}}{e^{(hc/\lambda kT)} - 1} \text{(W/m}^2\text{/unit wavelength in meters)}, \qquad (2.1)$$

where $h = 6.63 \times 10^{-34}$ W s^2 (Planck's constant), $c = 3.00 \times 10^8$ m/s (speed of light in a vacuum), $k = 1.38 \times 10^{-23}$ J/K (Boltzmann's constant), and T = absolute temperature of blackbody in K (Kelvin, where 0 K = −273.16°C).

Equation 2.1 yields the energy density at the *surface* of the Sun in W/m^2/unit wavelength in meters. By the time this energy has traveled 150 million km to the Earth, the total extraterrestrial energy density decreases to 1367 W/m^2 and is often referred to as the solar constant (see Problem 2.1) [1].

Figure 2.1 shows plots of Planck's blackbody radiation formula for several different temperatures, along with the extraterrestrial solar spectrum. Note that at lower temperatures, nearly all of the spectrum lies outside the visible range in the infrared range, whereas at 5800 K, the characteristic white color of the Sun is obtained due to the mix of wavelengths in the visible spectral range. At even higher temperatures, the color shifts toward blue, and at lower temperatures, the color shifts toward red. Incandescent lamp filaments, for example, typically are operated at temperatures of approximately 2700 K and, hence, emit an approximate blackbody spectrum characteristic of 2700 K. Depending on the color temperature of the light source, photographic equipment must be compensated to obtain true colors, unless appropriate filters are used. The extraterrestrial solar

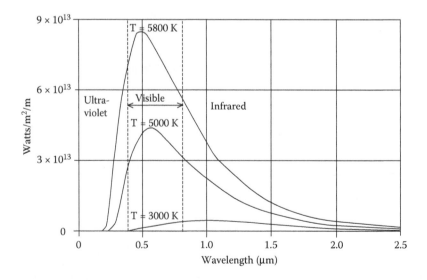

Figure 2.1 Blackbody radiation spectra for temperatures of 3000, 5000, and 5800 K.

spectrum indicates that the Sun can be reasonably approximated as a blackbody radiator.

Most sources of light are not perfect blackbody sources. Reasonable approximations of blackbody spectra are obtained from sources that emit light as a result of heating a filament to a high temperature. But to be a perfect radiator, the object must also be a perfect absorber of light, which is not the case for common light sources. Nonfilament light sources, such as gas discharge lamps, for example, emit either the discrete spectral characteristic of the specific gas or, alternatively, light containing many discrete spectral lines from emission from phosphorescent materials. Thus, fluorescent, high-pressure sodium and high-intensity discharge lamps have spectral characteristics of gas discharge systems. Light-emitting diodes (LEDs), depending upon design, also may have discrete spectra or a combination of spectral lines. These are important facts to recognize, since in the testing of PV systems, it is important to be able to produce standardized spectral testing conditions.

Knowledge of the spectral composition of the Sun is important for understanding the effects of the atmosphere on the radiation from the Sun and for understanding which materials should offer the best performance in the conversion of sunlight to electricity.

2.3 EFFECT OF ATMOSPHERE ON SUNLIGHT

As sunlight enters the Earth's atmosphere, some is absorbed, some is scattered, and some passes through unaffected by the molecules in the atmosphere and is either absorbed or reflected by objects at ground level.

Different molecules do different things. Water vapor, carbon dioxide, and ozone, for example, absorb significant amounts of sunlight at certain wavelengths. Ozone plays an important role by absorbing a significant amount of radiation in the ultraviolet region of the spectrum, while water vapor and carbon dioxide absorb primarily in the visible and infrared parts of the spectrum. These absorption lines are shown in Figure 2.2.

Absorbed sunlight increases the energy of the absorbing molecules, thus raising their temperature. Scattered sunlight is responsible for light entering north-facing windows when the Sun is in the south. Scattered sunlight, in fact, is what makes the sky blue. Without the atmosphere and its ability to scatter sunlight, the sky would appear black, such as it does on the moon. Direct sunlight consists of parallel rays, which are necessary if the light is to be focused. The reader has probably experimented with a magnifying glass and found that it is not possible to burn holes in paper when a cloud covers the Sun. This is because the diffuse light present under these conditions is coming from all directions and cannot be focused.

All of these components of sunlight have been given names of their own. Sunlight that reaches the Earth's surface without scattering is called *direct* or *beam* radiation. Scattered sunlight is called *diffuse* radiation. Sunlight that is reflected from the ground is called *albedo* radiation, and the sum of all three components of sunlight is called *global* radiation.

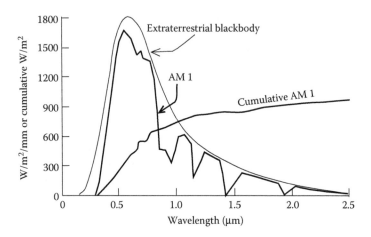

Figure 2.2 AM 1 solar spectrum after atmospheric absorption effects, including plots of extraterrestrial blackbody spectrum plus cumulative incident energy versus wavelength.

The amount of sunlight either absorbed or scattered depends on the length of path through the atmosphere. This path length is generally compared with a vertical path directly to sea level, which is designated as *air mass* = 1 (AM 1). Hence, the air mass at a higher altitude will be less than unity when the Sun is directly overhead and the air mass generally will be more than unity when the Sun is not directly overhead. In general, the air mass through which sunlight passes is proportional to the secant of the zenith angle, θ_z, which is the angle measured between the direct beam and the vertical. At AM 1, after absorption has been accounted for, the intensity of the global radiation is generally reduced from 1367 W/m² at the top of the atmosphere to just over 1000 W/m² at sea level. Hence, for an AM 1 path length, the intensity of sunlight is reduced to 70% of its original AM 0 value. In equation form, this observation can be expressed as, assuming that the absorption constant depends directly upon air mass,

$$I = 1367(0.7)^{AM}. \qquad (2.2)$$

This equation is, of course, obvious for AM 1. But does it hold for air masses different from unity? According to Meinel and Meinel [2], a better fit to observed data is given by

$$I = 1367(0.7)^{(AM)^{0.678}}. \qquad (2.3)$$

On the average, over the surface of the Earth, an amount of heat is reradiated into space at night that is just equal to the amount absorbed from the Sun during the day. As long as this steady-state condition persists, the average temperature of the Earth will remain constant. However, if for any reason the amount of heat absorbed is not equal to the amount reradiated, the planet will either cool down or heat up. This delicate balance can be upset by events such as volcanoes that fill the atmosphere with fine ash that reflects the sunlight away from the Earth, thus reducing the amount of incident sunlight. The balance can also be upset by gases such as carbon dioxide and methane, which are mostly transparent to short wavelength (visible) radiation, but more absorbing to long wavelength (infrared) radiation. Since incident sunlight is dominated by short wavelengths characteristic of the 5800 K Sun surface temperature, and since the reradiated energy is dominated by long wavelengths characteristic of the Earth surface temperature of approximately 300 K, these *greenhouse gases* tend to prevent the Earth from reradiating heat at night. In order to reach a new balance, it is necessary for the Earth to increase its temperature, since radiation is proportional to T^4, where T is the temperature in K.

The natural compensation mechanism is green plants on land and under water. Through the process of photosynthesis, they use sunlight and carbon dioxide to produce plant fiber and oxygen, which is released to the atmosphere. Hence, replacement of green plants with concrete and asphalt and adding carbon dioxide to the atmosphere by burning fossil fuels have a combined negative effect on the stability of the concentration of carbon dioxide in the atmosphere. Global warming (global climate change) has been discussed extensively in other literature, some of which is listed in the chapter references.

2.4 SUNLIGHT SPECIFICS

2.4.1 Introduction

Nearly everyone in the Northern Hemisphere has noticed that the Sun shines longer in the summer than in the winter. Nearly everyone also knows that the Sahara Desert receives more sunshine than does London. Another obvious observation is that cloudy places receive less sunlight than sunny places. It may be less evident, however, that the hours of sunlight over a year are the same for every point on the Earth, provided that only the hours between sunrise and sunset are counted, regardless of cloud cover. Those regions of the Earth closer to the poles that have long winter nights also have long summer days. However, since the Sun, on the average, is lower in the sky in the polar regions than in the tropics, sunlight must traverse greater air mass in the polar regions than in the tropics. As a result, polar sunlight carries less energy to the surface than tropical sunlight. Not surprisingly, the polar regions are colder than the tropics.

In this section, quantitative formulas will be presented that will enable the reader to determine exactly how long the Sun shines in any particular place on any particular day, and to determine how much sunlight can be expected, on the average, during any month at various locations. Means will also be presented for determining the position of the Sun at any time on any day at any location. Finally, the effects of varying the orientation of a PV array on the power and energy produced by the array will be discussed.

2.4.2 Definitions

Irradiance is the measure of the power density of sunlight and is measured in W/m². Irradiance is thus an instantaneous quantity and is often identified as the *intensity* of sunlight. The solar constant for the Earth is the irradiance received by the Earth from the Sun at the top of the atmosphere, that is, at AM 0, and is equal to 1367 W/m². After passing through the atmosphere with a path length of AM 1, the irradiance is reduced to approximately 1000 W/m² and has a modified spectral content due to atmospheric absorption. The irradiance for AM 1.5 is accepted as the standard calibration spectrum for PV cells.

Irradiation is the measure of energy density of sunlight and is measured in kWh/m². Since energy is power integrated over time, irradiation is the integral of irradiance. Normally, the time frame for integration is 1 day, which, of course, means during daylight hours.

Irradiation is often expressed as *peak sun hours* (psh). The psh is simply the length of time in hours at an irradiance level of 1 kW/m² needed to produce the daily irradiation obtained from integration of irradiance over all daylight hours. Figure 2.3 illustrates the result of this integration for an example day of sunshine with a few cloudy moments. Note that the figure plots irradiance versus time in order to determine irradiation.

Irradiance and irradiation both apply to all components of sunlight. Hence, at a given time, or for a given day, these quantities will depend on location, weather conditions, and time of year. They will also depend on whether the surface of interest is shaded by trees or buildings and whether the surface is horizontal or inclined. The daily irradiation is numerically equal to the daily psh.

In order to determine the amount of irradiation available at a given location at a given time for conversion to electricity, it is useful to develop several expressions for the irradiance on surfaces, depending on the angle between the surface and the incident beam. It is also interesting to be able to determine the number of hours of sunlight on a given day at a given location.

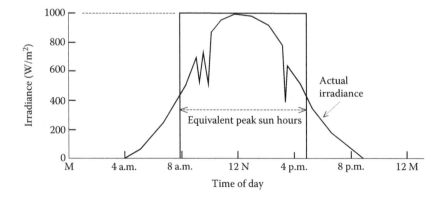

Figure 2.3 Determination of psh (irradiation) through integration of irradiance.

2.4.3 The Orbit and Rotation of the Earth

The Earth revolves around the Sun once per year in an elliptical orbit with the Sun at one of the foci, such that the distance from Sun to Earth is given by [1]

$$d = 1.5 \times 10^{11} \left\{ 1 + 0.017 \sin\left[\frac{360(n-93)}{365}\right] \right\} m, \qquad (2.4)$$

where n represents the day of the year with January 1 as Day 1. Since the deviation of the orbit from circular is so small, it is normally adequate to express this distance in terms of its mean value.

The Earth also rotates about its own polar axis once per day. The polar axis of the Earth is inclined by an angle of 23.45° to the plane of the Earth's orbit about the Sun. This inclination is what causes the Sun to be higher in the sky in the summer than in the winter. It is also the cause of longer summer sunlight hours and shorter winter sunlight hours. Figure 2.4 shows the Earth's orbit around the Sun with the inclined polar axis. Note that on the first day of Northern Hemisphere summer, the Sun appears vertically above the Tropic of Cancer, which is latitude 23.45°N of the equator. On the first day of winter, the Sun appears vertically above the Tropic of Capricorn, which is latitude 23.45°S of the equator. On the first day of spring and the first day of fall, the Sun is directly above the equator. From fall to spring, the Sun is south of the equator, and from spring to fall, the Sun is north of the equator. The angle of deviation of the Sun from directly above the equator is called the *declination*, δ. If angles north of the equator are considered as positive and angles south of the equator are considered negative, then at any given day of the year, n, the declination can be found from

$$\delta = 23.45° \sin\left[\frac{360(n-80)}{365}\right]. \qquad (2.5)$$

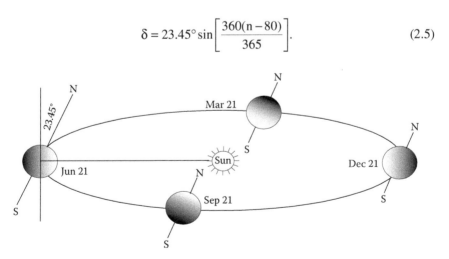

Figure 2.4 Orbit of the Earth and the declination at different times of the year.

This formula, of course, is only a good approximation, since the year is not exactly 365 days long and the first day of spring is not always the 80th day of the year. In any case, to determine the location of the Sun in the sky at any time of day at any time of year at any location on the planet, the declination is an important parameter.

It is also important to be able to determine the clock time at which solar noon occurs. Solar noon occurs at 12 noon clock time at only one longitude, L_1, within any time zone. At longitudes east of L_1, solar noon will occur before 12 noon, and at longitudes west of L_1, solar noon will occur after 12 noon. On a sunny day, solar noon can be determined as that time when a shadow points directly north or directly south, depending on the latitude. Note that in the tropics, part of the year the shadow will point north and the rest of the year the shadow will point south.

Fortunately, if the longitude is known, it is straightforward to determine the *approximate* relationship between clock noon and solar noon. Since there are 24 hours in a day and since the Earth rotates 360° during this period, this means that the Earth rotates at the rate of 15°/h. It is also convenient that longitude zero corresponds to clock noon at solar noon. As a result, solar noon occurs at clock noon at multiples of 15° east or west longitude. Furthermore, since it takes the Earth 60 minutes to rotate 15°, it is straightforward to interpolate to find solar noon at intermediate longitudes.

For example, at a longitude of 80°W, solar noon can be found by noting that 80° is between 75° and 90°, where solar noon occurs at clock noon *standard time*. Since 80° is west of 75°, when the Sun is directly south at 75°, the Sun will be east of south at 80°. The clock time at which the Sun will be south at 80° (solar noon for 80°) is thus found by interpolation to be

$$t = 12 + \frac{80-75}{15} \times 60 = 12 + 20 \text{ minutes} = 12{:}20 \text{ p.m.}$$

Note that this time is standard time relative to the time zone for which solar noon occurs at 75°W. If 90°W is used as the solar noon reference, then solar noon at 80° will occur 40 minutes before solar noon occurs at 90°. Note that the answer is still the same. At 80°W, solar noon occurs at 12:20 p.m. Eastern Standard Time or at 11:20 a.m. Central Standard Time.

One glitch in the solar noon argument involves those unique locations on the Earth such as Newfoundland, Canada or India, where there is only half-an-hour shift between adjacent time zones, or Alaska, where a single clock time zone covers nearly 30° of longitude.

Another glitch in determining solar noon results from the combination of the declination and the elliptical orbit of the Earth. During half the year, the Earth is moving closer to the Sun, and during the other half the year, the Earth is moving away from the Sun. This results in a variation in solar noon called the analemma [3]. One way of describing the analemma is to plot the position of the tip of a shadow from a fixed object at a specific time each day, each week, or each month. Whichever time period is used, the resulting plot over a year will look like the

THE SUN

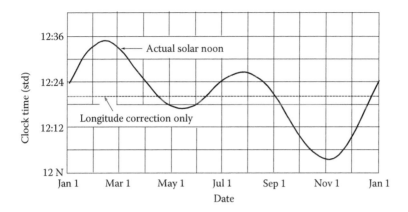

Figure 2.5 Daily variation of solar noon versus clock time in Boca Raton, FL.

number 8. The reader familiar with globes may recall seeing the analemma on the globe.

Another convenient way of describing the analemma is to plot the clock time at which solar noon occurs over the period of a year. This plot will refine the solar noon value determined from the longitude correction. Solar noon data can be obtained from http://aa.usno.navy.mil/data/docs [4]. Figure 2.5 shows a plot of solar noon versus clock time for Boca Raton, FL, which is located at a longitude of approximately 80°W. Note that the greatest variation in solar noon is approximately ±16 minutes from 12:20 p.m., the value calculated from the longitude correction formula.

2.4.4 Tracking the Sun

To completely specify the position of the Sun, it is necessary to specify three coordinates. If one assumes the distance from the Sun to the Earth to be constant, then the position of the Sun can be specified using two angles. Two common choices are the solar altitude and the azimuth.

The *solar altitude*, α, is the angle between the horizon and the incident solar beam in a plane determined by the zenith and the Sun, as shown in Figure 2.6.

The angular deviation of the Sun from directly south can be described by the *azimuth angle*, ψ, which measures the Sun's angular position east or west of south. The azimuth angle is zero at solar noon and increases toward the east. It is the angle between the intersection of the vertical plane determined by the observer and the Sun with the horizontal and the horizontal line facing directly south from the observer, assuming the path of the Sun to be south of the observer. *The reader should note that in many publications, the azimuth angle is referenced to north, such that solar noon appears at $\psi = 180°$.* In fact, compass angles are azimuth angles referenced to magnetic north.

Another useful, albeit redundant, angle in describing the position of the Sun is the angular displacement of the Sun from solar noon in the plane of apparent travel

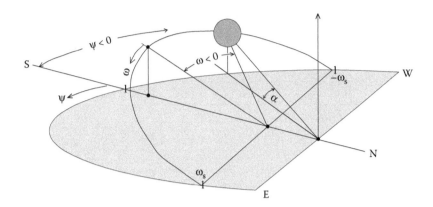

Figure 2.6 Sun angles, showing altitude, azimuth, and hour angle.

of the Sun. The *hour angle* is the difference between noon and the desired time of day in terms of a 360° rotation in 24 hours. In other words,

$$\omega = \frac{12-T}{24} \times 360° = 15(12-T)°, \tag{2.6}$$

where T is the time of day expressed with respect to solar midnight, on a 24-hour clock. For example, for T = 0 or 24 (midnight), $\omega = \pm 180°$ and for T = 9 a.m., $\omega = 45°$. By relating ω to the other angles previously discussed, it is possible to show [1] that the sunrise angle is given by

$$\omega_s = \cos^{-1}(-\tan\phi \tan\delta), \tag{2.7}$$

which, in turn, implies that the sunset angle is given by $-\omega_s$. This formula is useful because it enables one to determine the number of hours on a specific day at a specific latitude that the Sun is above the horizon. Converting the sunrise angle to hours from sunrise to solar noon, and then multiplying by two to include the hours from solar noon to sunset, yields the number of hours of daylight (DH) to be

$$DH = \frac{48}{360} \times \omega_s = \frac{\cos^{-1}(-\tan\phi\tan\delta)}{7.5} \text{ hour}. \tag{2.8}$$

Two very important relationships among α and ψ can be determined by the reader who enjoys trigonometry. If δ, ϕ, and ω are known, then the position of the Sun, in terms of α and ψ at this location at this date and time, can be determined from [1]

$$\sin\alpha = \sin\delta\sin\phi + \cos\delta\cos\phi\cos\omega \tag{2.9}$$

THE SUN

and

$$\cos \psi = \frac{\sin \alpha \sin \phi - \sin \delta}{\cos \alpha \cos \phi} \quad (2.10)$$

Note that in all above expressions, angles are measured in degrees.

It is interesting to look at the values of Equations 2.9 and 2.10 at solar noon, when $\omega = 0$. The results of Problem 2.9 show that at solar noon, $\theta_z = \phi - \delta$, where θ_z, the zenith angle, is the angle between the vertical and the Sun position. Thus, θ_z is the complement of α. During the year, the highest point of the Sun in the sky will be at $\theta_z = \phi - 23.45°$ and the lowest point of the solar noon Sun in the sky will be at $\theta_z = \phi + 23.45°$, provided that $\phi > 23.45°$. It is particularly interesting to note that if $\phi > (90° - 23.45° = 66.55°)$, then the lowest point of the Sun in the sky is below the horizon, meaning that the Sun does not rise or set that day. This, of course, is the situation in polar regions, which are subject to periods of 24 hours of darkness. These same regions, of course, are also subject to equal periods of 24 hours of Sun 6 months later. If $\phi < 23.45°$, θ_z will at some time during the summer be negative. This simply means that the Sun will appear north of directly overhead at solar noon. The relationships among θ_z, ϕ, and δ at solar noon are illustrated in Figure 2.7.

Since Equations 2.9 and 2.10 are somewhat difficult to visualize, it is convenient to plot α versus ψ for specific latitudes and days of the year. Figure 2.8 shows a series of plots of altitude versus azimuth at a latitude of 30°N. The curves show approximately how high in the sky the Sun will be at a certain time of day during a particular month, with the azimuth angle determined by the time of day.

Markvart [1] gives detailed formulas for Sun position and the components of global irradiance. The interested reader is encouraged to review them. Interestingly enough, when all of the mathematical work is finished, one ends up with an answer that does not account for cloud cover. The most reliable means of accounting for cloud cover is to make measurements over a long period of time in order to determine average figures. In fact, accurate prediction of the performance of PV systems also depends upon temperature.

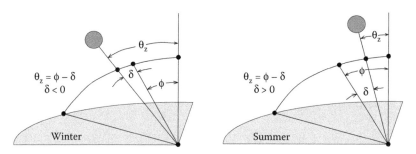

Figure 2.7 Relationships among zenith angle, latitude, and declination at solar noon in winter and summer.

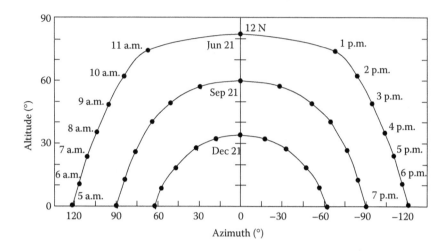

Figure 2.8 Plot of solar altitude versus azimuth for different months of the year at a latitude of 30°N.

Thus, if sunlight data are included as a result of measured meteorological data, which include irradiance and temperature over an hourly basis for a given location for the 8760 hours per year, reasonably reliable estimates can be made of PV system performance. Of course, measuring hourly data for a given year, say 2015, will not necessarily be a good predictor for other years. To achieve a better predictor for annual performance, experts have defined the *typical meteorological year* (TMY) by selecting data for each month from a different year. In later chapters, various performance models based upon TMY data will be discussed.

2.4.5 Measuring Sunlight

Since all the calculations and approximations in the world cannot yield exact predictions of the amount of sunlight that will fall on a given surface at a given angle at a given time at a given day in a given place, the design of PV power systems is dependent upon the use of data based on measurements averaged over a long time.

Depending upon whether it is desired to measure global, beam, diffuse, or albedo components of irradiance, or whether it is simply desired to measure when sunlight exceeds a certain brightness level, different types of instrument are used.

2.4.5.1 Precision Measurements

The *pyranometer* is designed to measure global radiation. It is normally mounted horizontally to collect general data for global radiation on a horizontal surface. However, it is also often mounted in the plane of a PV collector in order to measure the global radiation incident on the inclined surface, which is essential

THE SUN   35

when commissioning a system, that is, comparing actual system performance to predicted system performance.

The pyranometer is designed to respond to all wavelengths and, hence, it responds accurately to the total power in any incident spectrum.

The *black-and-white pyranometer* operates on the principle of differential heating of a series of black-and-white wedges. The temperatures of each wedge are measured with thermocouples that yield voltage differences dependent on the temperature differences. This instrument is somewhat less accurate than the precision unit.

The *normal incidence pyrheliometer* uses a long, narrow tube to collect beam radiation over a narrow beam solid angle, generally about 5.5°. Since the instrument is only sensitive to beam radiation, a tracker is needed if continuous readings are desired.

The pyranometer can be mounted on a *shadow band stand* to block out beam radiation so that it will respond only to the diffuse component. The stand is mounted so the path of the Sun will be directly above the band during daylight hours. Because δ changes from day to day and because the band blocks out only a few degrees, it is necessary to readjust the band every few days.

2.4.5.2 Less Precise Measurements

Many inexpensive instruments are also available for measuring light intensity, including instruments based on cadmium sulfide photocells and silicon photodiodes. These devices give good indications of relative intensity, but are not sensitive to the total solar spectrum and thus cannot be accurately calibrated to measure total energy. These devices also do not normally have lenses that capture incident radiation from all directions. Devices that capture solid angles from a few degrees to upwards of 90° are available.

2.5 CAPTURING SUNLIGHT

2.5.1 Maximizing Irradiation on the Collector

The designer of any system that collects sunlight must decide on a means of mounting the system. Perhaps the easiest mounting of most systems is to mount them horizontally. This orientation, of course, does not optimize collection, since the beam radiation component collected is proportional to the cosine of the angle between the incident beam and the normal to the plane of the collector, as shown in Figure 2.9. Depending on the ratio of diffuse to beam irradiance components, the fraction of available energy collected will be between cos γ and unity. Of course, in a highly diffuse environment, the beam irradiance will be only a small fraction of the global irradiance. Several alternatives to horizontal mounting exist. Since $\theta_z = \phi - \delta$ defines the position of the Sun at solar noon, if a collector plane is perpendicular to this angle, it will be perpendicular to the Sun at solar noon. This is the point at which the Sun is highest in the sky, resulting in its minimum path through the atmosphere

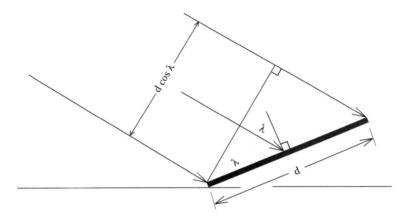

Figure 2.9 Two-dimensional illustration of effect of collector tilt on effective area presented to beam component of radiation.

and corresponding lowest air mass for the day. Since the Sun travels through an angle of 15° per hour, it will be close to perpendicular to the collector for a period of approximately 2 hours. Beyond this time, the intensity of the sunlight decreases due to the increase in air mass, and the angle between incident sunlight and the normal to the collector increases. These two factors cause the energy collected by the collector to decrease relatively rapidly during the hours before 10 a.m. and after 2 p.m. Figure 2.10 shows the approximate cumulative irradiation received by a south-facing collector tilted at the latitude angle in a region where the beam radiation component is significantly stronger than either the diffuse or the albedo components.

If the collector is mounted so it can track the Sun, then the incident irradiance is affected only by the increasing air mass as the Sun approaches the horizon. Figure 2.10 also shows the additional cumulative irradiation under direct beam conditions received

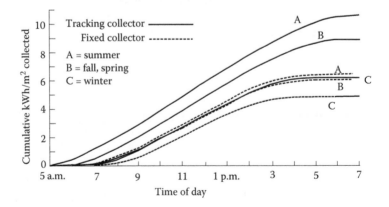

Figure 2.10 Cumulative daily irradiation received by fixed and tracking collectors for different seasons, direct beam contribution only.

by a tracking collector. Approximately 50% more energy can be collected in the summer in a dry climate such as that found in Phoenix, AZ, by using a tracking collector. During winter months, however, only about 20% more energy is collected using a tracker. In Seattle, WA, which receives somewhat more diffused sunlight than does Phoenix, a tracking collector will collect about 35% more in the summer but only 9% more energy compared to an optimized fixed collector in the winter [5]. Whether to use a two-axis tracking collector then becomes an important economic decision for the engineer, since a tracking mount is more costly than a fixed mount.

To make the mounting selection even more interesting, one can consider a single-axis tracker, which rotates about an axis perpendicular to θ_z. One can then also consider mountings that can be adjusted manually several times per day or, perhaps, several times per year. Each of these options will enable the collection of an amount of energy that lies somewhere between the optimized fixed collector and the two-axis tracking collector results.

Collector orientation may also be seasonally dependent. For example, a remote cabin, used only during summer months, will need its collector oriented for optimal summer collection. However, if the cabin is used in the winter as a ski lodge, or in the fall as a hunting base, then the collector may need to be optimized for one of these seasons. For optimal performance on any given day, a fixed collector should be mounted with its plane at an angle of $\phi - \delta$ with respect to the horizontal, as shown in Figure 2.11. This will cause the plane of the collector to be perpendicular to the Sun at solar noon.

For optimal seasonal performance, then, one simply chooses the average value of δ for the season. For summer in the Northern Hemisphere, δ varies sinusoidally between 0° on March 21 and 23.45° on June 21 and back to 0° on September 21. If this variation is plotted as half a sine wave with amplitude of 23.45°, those who have evaluated the average value of a sine wave in conjunction with the output of a rectifier circuit may recall that the average value of half a sine wave having amplitude A is $2A/\pi$. Hence, the average declination between March 21 and September 21 is

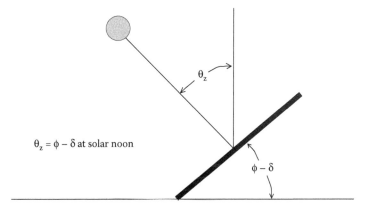

Figure 2.11 Optimizing the mounting angle of a fixed collector.

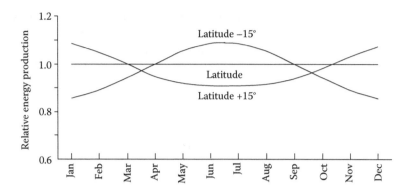

Figure 2.12 Monthly collector performance as a function of collector tilt angle.

14.93°. Similarly, the average declination for the period from September 21 to March 21 is −14.93°. Hence, for best average summer performance, a collector should be mounted at approximately ϕ − 15°, and for best average winter performance, it should be mounted at ϕ + 15°.

For optimum spring or fall performance or optimum annual performance, the collector should be mounted at ϕ°. Since normally it is acceptable to orient a collector for optimal seasonal or, perhaps, annual performance, it is a straightforward exercise to vary collector tilt angles in popular simulation programs, such as NREL SAM [5], to compare monthly, seasonal, or annual system performance for different collector tilt angles and different collector orientation directions. Typically, only a small percentage of optimal annual energy collection for a fixed collector is lost if the collector tilt is within ±15° of latitude and within ±45° of south-facing (north-facing in the southern hemisphere).

Figure 2.12 shows how monthly performance depends on collector tilt angle. Different locations will show different relative monthly performance, depending on local weather seasonal behavior.

2.5.2 Shading

Even a small amount of shade on a PV module can significantly reduce the module output current. It is thus of paramount importance to select a site for a PV system where the PV array will remain unshaded for as much of the day as possible. This is easy if there are no objects that might shade the array, but it is probably more likely that a site will have objects nearby that shade the array at some time of the day on some day of the year. The PV system designer must thus be able to use his or her knowledge of Sun position to determine the times at which a PV array might be shaded.

2.5.2.1 Field Measurement of Shading Objects

Figure 2.13 shows a device that incorporates plots of altitude versus azimuth for selected latitudes. The device is used at the proposed site to determine when the

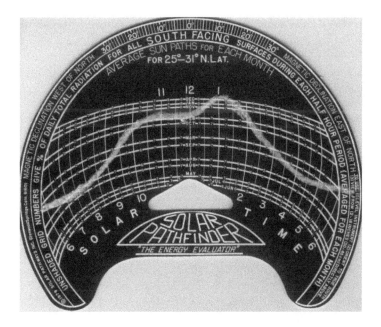

Figure 2.13 Solar Pathfinder showing region of shading. Shading occurs at points above the white line on the pattern. (From Florida Solar Energy Center.)

array will be shaded by observing the position of reflected objects on the screen of the device. By sketching the outlines of shading objects on the screen with the device at the proposed location of the collector, the user can then determine when the collector will be shaded. For example in Figure 2.13, the collector will be free of shade during May, June, July, and August. In September, the collector will be unshaded between about 9:15 a.m. and 3:15 p.m. Sun time, while in November, December, and January, the collector is shaded most of the day except for around 1 p.m. Sun time. This proposed collector location is thus acceptable if the collector is to be used only during the summer months.

The device also provides a measure of the fraction of available solar energy that will be collected under various shading conditions. For example, in September, 7% of the total daily insolation is collected between 11:00 and 11:30 a.m. on a south-facing surface, while in December, 8% of the total daily insolation is collected between 11:00 and 11:30 a.m. Sun time. For example in Figure 2.13, during March, about 67% of the daily available Sun will arrive at the collector during the unshaded period between 9:30 a.m. and 2:45 p.m. This instrument now also has software available that will analyze the data taken and predict solar harvest taking into account the observed shading [6].

The inclinometer is another device that can be used to determine potential shading problems. The inclinometer simply measures the angle between the horizontal at the height of the PV array and the top of objects that might present shading problems. This angle, along with an azimuth angle measured with a compass, corrected to true north, can then be compared with the altitude versus azimuth chart for the Sun for the location of the installation for different months of the year.

For the field person who is willing to spend a bit more on a measuring device, several photographic-based systems are also available that add some automation and a bit more precision to the field measurement of shade process [7].

2.5.2.2 Computational Methods of Determining Shading

Computational methods of determining shading range from solving Equations 2.9 and 2.10 for a specific site to using a variety of computer programs. Google Sketch-Up© is a popular freeware program that can be downloaded from the Web [8]. It can be used with Google Earth© to determine shading at existing sites. Autocad™ has a plug-in that does a very nice job of shading analysis on planned new construction drawings [9]. Any of these methods are useful for any mounting configuration.

As an example, consider a ground-mounted array as shown in Figure 2.14. Assume the array is located at 30°N latitude (ϕ) and that it is desired to space the rows such that rows do not shade each other between the hours of 8:30 a.m. and 3:30 p.m. Sun time at any time of the year.

If a shading program is not available, then the first step is to solve Equations 2.9 and 2.10 for the Sun position. Since the worst case will be on December 21, and since the angles will be symmetrical about south, one only needs to know δ, ϕ, and ω for 8:30 a.m. Sun time in order to solve for α and ψ. Solving for δ and ω yields

$$\delta = 23.45° \sin\left[\frac{360(n-80)}{365}\right] = 23.45° \sin\left[\frac{360(355-80)}{365}\right] = -23.45°$$

as expected, and

$$\omega = 15(12-T)° = 15(12-8.5) = 52.5°.$$

Thus, from Equation 2.9,

$$\sin\alpha = \sin(-23.45°)\sin 30° + \cos(-23.45°)\cos 30° \cos 52.5° = 0.2847$$

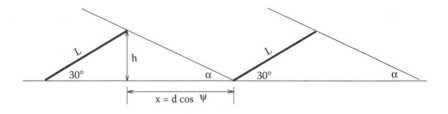

Figure 2.14 Determining the spacing between rows of PV modules.

and, from Equation 2.10,

$$\cos \psi = \frac{0.2847 \times \sin 30° - \sin(-23.45°)}{\cos(\sin^{-1} 0.2847) \cos 30°} = 0.6508.$$

Thus, $\alpha = 16.54°$ and $\psi = 49.4°$ east of south.

Next, the length of a shadow, d, can be found using a bit of trigonometry. If the length of the PV modules is L, then $h = L \sin 30°$, and $d = h/\tan \alpha = L \sin 30°/\tan \alpha$. So, for example, if L = 62 inch (157.4 cm), then d = 104 inch (264.2 cm). Since this shadow points 49.4° west of north, the projection of the shadow to the north will give a distance of 104 cos(49.4°) = 68 inch (172.7 cm), which will thus be the minimum row spacing.

This method can easily be adapted to an Excel spreadsheet by including the formulas for Sun position and then using the Sun position coordinates at specified hours on specified days along with known shading object heights and positions, to calculate the corresponding shadow length of the object. This method will be used in several examples in later chapters.

2.5.3 Special Orientation Considerations

It should be noted that all previous discussion about collector orientation has assumed fixed collectors to be facing directly south (or north if in the Southern Hemisphere). Sometimes it is not practical to orient a collector directly to the south and sometimes it may even be desirable to use a different orientation. In other situations, a horizontal mount may be preferable. The question is to what extent collection is sacrificed by a non-south-facing collector, or, alternatively, what may be gained with an alternate orientation.

2.5.3.1 Horizontal Mounting

In some PV installations, notably on floating buoys or watercraft, the orientation of the collector changes as the direction of travel of the watercraft changes. If a buoy has a single anchor, it will rotate. In these or similar cases, the collectors are normally mounted horizontally. Again, simulation software can readily estimate monthly or annual system performance.

The loss in performance with a horizontal mount depends mostly on the latitude of the installation and the season of the year, but also depends on the specific location and the ratio of beam to diffuse components of irradiance for that location. Since the optimal seasonal performance of collectors is obtained somewhere between $\phi \pm 15°$, if the installation is in the tropics, a relatively small amount of annual collection is lost with a horizontal mount.

2.5.3.2 Non-South-Facing Mounting

Sometimes it is not possible or convenient to install a collector facing directly south. If south-facing is the preferred orientation, simulations show that the collector can be facing up to 22.5° away from south with less than a 2% reduction in annual collection at latitudes up to 45°. In fact, the loss in monthly or annual solar harvest can be determined quite readily for any location and array orientation by using NREL SAM [5].

In other cases, maximum PV system output may be desirable at a time of day other than solar noon. For example, in many regions, peak utility electrical generation occurs between the hours of 3 and 6 p.m. If a PV system is connected to the utility grid, it may be desirable to maximize system output during the utility peak. For a fixed mounting, this would require having the collector face the Sun at the midpoint of this time period on a date near the middle of the period during which maximum collector output is desired. This orientation will produce peak output when the Sun path is at a higher air mass than at solar noon, resulting in a slightly lower peak output than would be obtained with a south-facing collector.

Figure 2.15 shows a comparison of PV output for south-facing and west-facing PV systems. Note the appreciable increase in system output later in the afternoon for the west-facing system. During utility peak time, the system power output is increased by 70% for the west-facing system, but the west-facing system develops 14.6% less total annual energy than the south-facing system. Although the total energy output of the west-facing system is lower than that of the south-facing system, the loss of total energy may be acceptable if the added power output during utility peak hours will offset the need to use more expensive peaking generation. This observation holds particularly if it will offset the need to install the peaking generation in the first place.

Computation of the desired orientation for maximum output in a direction other than south is straightforward:

1. Determine the latitude of the location.
2. Calculate the declination for the design day, using Equation 2.5.

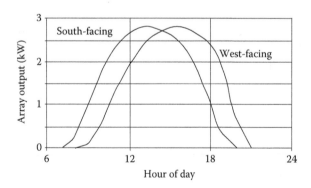

Figure 2.15 Comparison of power and energy output for south-facing and west-facing PV systems in Sacramento, CA, in summer. The south-facing system produced 22,417 Wh and the west-facing system produced 22,192 Wh. (Courtesy of Collier, D. E., *Interconnecting Small Photovoltaic Systems to Florida's Utility Grid*, A Technical Workshop for Florida's Utilities, Cocoa, FL, October 22, 1998.)

THE SUN

3. Determine the time of day (local time) for peak PV system output.
4. Convert the local time to solar time, based upon site longitude.
5. Calculate the hour angle using Equation 2.6.
6. Use Equations 2.9 and 2.10 to determine α and ψ.

EXAMPLE 2.1

Suppose it is desired to determine the position of the Sun over Atlanta, GA, at 2 p.m. on July 1. Starting with Step 1 of the six steps outlined above, the latitude of Atlanta is found to be 33°39′ = 33.65°N. Then, Equation 2.5 is used to determine the declination for July 1. Since July 1 is the 181st day of the year, Equation 2.5 gives

$$\delta = 23.45 \sin\left[\frac{360(181-80)}{365}\right] = 23.12°.$$

With 2 p.m. local time (Eastern Daylight Time) at the longitude of Atlanta (84°36′W = 84.60°W) as the time of day for peak system output, the solar time is found to be 1 p.m. (Eastern Standard Time) plus an additional

$$t = \frac{84.6 - 75}{15} \times 60 = 38.4 \text{ minutes,} \quad \text{or } 1:38 \text{ p.m.}$$

Next, substituting the solar time in Equation 2.8 yields the hour angle as

$$\omega = \frac{12 - 13.64}{24}(360°) = -24.60°.$$

The solar altitude is next found from Equation 2.9 with the result

$$\sin\alpha = \sin 23.12° \sin 33.65° + \cos 23.12° \cos 33.65° \cos(-24.6°) = 0.9161.$$

Next, the azimuth is found from Equation 2.10 to be

$$\cos\psi = \frac{0.9161 \sin 33.65° - \sin 23.12°}{\cos[\sin^{-1} 0.9161]\cos 33.65°} = 0.3445.$$

So, finally, $\alpha = \sin^{-1} 0.9161 = 66.36°$ and $\psi = \cos^{-1} 0.3445 = -69.85°$. Be sure to note that $\cos^{-1} 0.3445 = \pm 69.85°$. The minus sign is chosen because at 1:38 p.m. solar time the Sun is west of south.

PROBLEMS

2.1 Show that, for a surface temperature of 5800 K, the Sun will deliver 1367 W/m² to the Earth. This requires integration of the blackbody radiation formula over all wavelengths to determine the total available energy at the surface of the Sun in W/m². Numerical integration is recommended. Be careful to note the range of wavelengths that contribute the most to the spectrum. Then note that the energy density decreases as the square of the distance from the source, similar to the behavior of an electric field emanating from a point source. The diameter of the Sun is 1.393×10^9 m, and the mean distance from Sun to Earth is 1.5×10^{11} m.

2.2 If the diameter of the Sun is 1.393×10^9 m, and if the average density of the Sun is approximately $1.4 \times$ the density of water, and if 2×10^{19} kg/year of hydrogen is consumed by fusion, how long will it take for the Sun to consume 25% of its mass in the fusion process?

2.3 Calculate the zenith angles needed to produce AM 1.5 and AM 2.0 if AM 1.0 occurs at 0°.

2.4 Calculate the zenith angle at solar noon at a latitude of 40°N on May 1.

2.5 Calculate the number of hours the Sun was above the horizon on your birthday at your birthplace.

2.6 Calculate the irradiance of sunlight for AM 1.5 and for AM 2.0, assuming no cloud cover, using Equations 2.2 and 2.3. Then write a computer program that will plot irradiance versus AM for $1 \leq AM \leq 10$ for each equation.

2.7 Assume no cloud cover, and, hence, that the solar irradiance is predominantly a beam component. If a nontracking collector is perpendicular to the incident radiation at solar noon, estimate the irradiance on the collector at 1, 2, 3, 4, 5, and 6 hours past solar noon, taking into account air mass and collector orientation. Then estimate the total daily irradiation on the collector in kWh. Assume AM 1.0 at solar noon and a latitude of 20°N.

2.8 Write a computer program using MATLAB, Excel, or something similar that will plot solar altitude versus azimuth. Plot sets of curves similar to Figure 2.8 for the months of March, June, September, and December for Denver, CO; Mexico City, Mexico; and Fairbanks, AK.

2.9 Using Equations 2.9 and 2.10, show that at solar noon, $\theta_z = \phi - \delta$ and $\psi = 0$.

2.10 Write a computer program that will generate a plot of solar altitude versus azimuth for the 12 months of the year for your hometown. It would be nice if each curve would have time-of-day indicators.

2.11 Calculate the time of day at which solar noon occurs at your longitude. Then compare the north indicated by a compass with the north indicated by a shadow at solar noon and estimate the error in the compass reading or the error due to the analemma effect.

2.12 Noting the dependence of air mass on θ_z
   a. Generate a table that shows α, ψ, and I (using Equation 2.3) versus time for a latitude of your choice on a day of your choice.
   b. Assuming a tracking collector and no cloud cover, estimate the total energy available for collection during your chosen day.

2.13 Plot I versus AM and then plot I versus α.

2.14 Calculate the collector orientation that will produce maximum summer output between 2 p.m. and 5 p.m. in Tucson, AZ. Use July 21 as the assumed midpoint of summer and correct for the longitude of the site.

2.15 Calculate the collector orientation that will produce maximum summer output at 9 a.m. Daylight Savings Time in Minneapolis, MN, using July 21 as the assumed midpoint of summer and using a longitude correction.

2.16 A collector in Boca Raton, FL ($\phi = 26.4°N$, longitude = 80.1°W) is mounted on a roof with a 5:12 pitch, facing 30° south of west. Determine the time of day and days of the year that the direct beam radiation component is normal to the array.

THE SUN   45

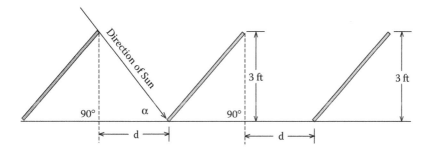

Figure P2.1 Rows of rack-mounted modules.

2.17 Determine the location (latitude and longitude) where, on May 29, sunrise is at 6:30 a.m. and sunset is at 8:08 p.m. Eastern Daylight Time. At what time does solar noon occur at this location?

2.18 Using Figure 2.13, make a table of unshaded collector times for each month of the year.

2.19 An array of collectors consists of three rows of south-facing collectors as shown in Figure P2.1. If the array is located at 40°N latitude, determine the spacing, d, between the rows needed to prevent shading of one row by another row.

2.20 The top of a tree is found to be at an angle of 15° between the horizontal and the corner of a collector and at an azimuth angle of −30°. If the site is at a latitude of 30°N, determine the months (if any) when the tree will shade the array at the point from which the measurements are taken.

2.21 A residence has a 5:12 roof pitch and is located at latitude 27°N and longitude 83°W, facing 15° west of south. If two rows of 66 cm × 142 cm collectors are to be mounted on the back roof so they will have a tilt of latitude, as shown in Figure P2.2
   a. How far apart will the rows need to be if collectors are to remain unshaded for 6 hours on December 21?
   b. Assuming Eastern Standard Time, over what time period on December 21 will the collectors be unshaded, assuming the collectors are facing 15° west of south.
   c. Assuming no other shading objects, over what time period will the collectors be unshaded on March 21?

2.22 A commercial building has a flat roof as shown in Figure P2.3. The roof has a parapet around it that is 3 ft high, and an air conditioner located as shown that is

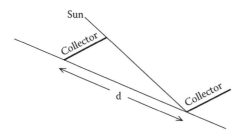

Figure P2.2 Rows of rack-mounted modules on a sloped roof.

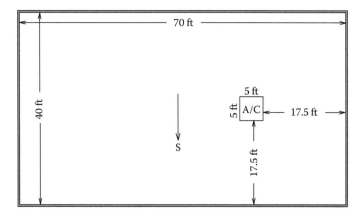

Figure P2.3 Roof layout for Problem 2.22.

4 ft above the roof. The building is located at 40°N. A PV system is to be installed using modules that measure 3 ft by 5 ft, in rows that will remain unshaded for at least 6 hours every day of the year. The modules are to be mounted at a tilt of 25°, in portrait mode, facing south. The bottom edges of the modules will be 6 inch above the roof. Determine the maximum number of modules that can be installed on the roof so that no module will be shaded during the 6-hour period. Pay attention to module rows shading other rows, shading from the air conditioner and shading from the parapet. Draw a diagram that shows your result.

2.23 Using NREL SAM [5] for Miami, FL; Sacramento, CA; Boston, MA; and Fairbanks, AK, create a table that shows the annual collection losses, in percentage, for fixed-tilt collectors, compared to maximum collection when the collector is tilted at latitude. Include tilt angles of latitude and latitude ±15°. If all that counts is total annual kWh, what generalizations can you offer with regard to array tilts?

REFERENCES

1. Markvart, T., Ed., *Solar Electricity*, John Wiley & Sons, Chichester, UK, 1994.
2. Meinel, A. B. and Meinel, M. P., *Applied Solar Energy: An Introduction*, Addison-Wesley, Reading, MA, 1976.
3. Discussion and explanation of analemma, complete with formulas and diagrams, www.analemma.com.
4. "Sun or Moon Rise/Set Table for One Year" tabulates sunrise, sunset, moonrise, and moonset on a daily basis for a year for a large number of cities, http://aa.usno.navy.mil/data/docs.
5. National Renewable Energy Laboratory, System advisory model, system advisor 2016, https://sam.nrel.gov/download.
6. Solar Pathfinder Assistant—Latest Version/Updates, www.solarpathfinder.com/spa-updates.html.

7. Wiley electronics Acme Solar Site Evaluation Tool information, http://www.we-llc.com/ASSET.html.
8. Google freeware program for shading analysis, http://sketchup.google.com.
9. A Web search for shading analysis programs will yield hundreds of results.
10. Collier, D. E., Photovoltaics in the Sacramento municipal utility district, In *Interconnecting Small Photovoltaic Systems to Florida's Utility Grid*, A Technical Workshop for Florida's Utilities, Cocoa, FL, October 22, 1998.

SUGGESTED READING

Bowen, M., *Thin Ice: Unlocking the Secrets of Climate in the World's Highest Mountains*, Henry Holt and Company, LLC, New York, 2005.

Howell, J. R., Bannerot, R. B., and Vliet, G. C., *Solar-Thermal Energy Systems: Analysis and Design*, McGraw-Hill, New York, 1982.

Hu, C. and White, R. M., *Solar Cells: From Basics to Advanced Systems*, McGraw-Hill, New York, 1983.

Hubbert, M. K., The energy resources of the earth, *Sci. Am.*, 225(3), 1971, 60–70.

McCluney, W. R., *Sun Position in Florida, Design Note*, Florida Solar Energy Center, Cocoa, FL, 1985 (FSEC-DN-4-83).

Nye, B., *Unstoppable: Harnessing Science to Change the World*, St. Martin's Press, New York, 2015.

Stand-Alone Photovoltaic Systems: A Handbook of Recommended Design Practices, Sandia National Laboratories, Albuquerque, NM, 1995.

Water Pumping: The Solar Alternative, Sandia National Laboratories, Albuquerque, NM, 1996.

CHAPTER 3

Introduction to PV Systems

3.1 INTRODUCTION

PV systems are designed around the PV cell. Since a typical PV cell produces less than 5 W at approximately 0.5 V_{dc}, cells must be connected in series-parallel configurations to produce enough power for high-power applications. Figure 3.1 shows how cells are configured into modules, and how modules are connected as arrays. Modules may have peak output powers ranging from a few watts, depending upon the intended application, to more than 400 W. Typical array output power is in the 100 W-to-kW range, although megawatt and gigawatt arrays are now becoming more commonplace.

Since PV arrays produce power only when illuminated, PV systems sometimes employ an energy storage mechanism so the captured electrical energy may be made available at a later time. Most commonly, the storage mechanism consists of rechargeable batteries, but it is also possible to employ more exotic storage mechanisms. In addition to energy storage, storage batteries also provide transient suppression, system voltage regulation (VR), and a source of current that can exceed PV array capabilities.

When a battery storage mechanism is employed, it is also common to incorporate a charge controller into the system, so the batteries can be prevented from reaching either an overcharged or overdischarged condition. It is also possible that some or all of the loads to be served by the system may be ac loads. If this is the case, an inverter will be needed to convert the dc from the PV array to ac. If a system incorporates a backup system to take over if the PV system does not produce adequate energy, then the system will need a controller to operate the backup system.

It is also possible that the PV system will be interconnected with the utility grid. Such systems may deliver excess PV energy to the grid or use the grid as a backup system in case of insufficient PV generation. These grid interconnected systems need to incorporate suitable interfacing circuitry, so the PV system will be disconnected from the grid in the event of grid failure. Figure 3.2 shows the components of several types of PV systems. This chapter will emphasize the characteristics of PV system components in order to pave the way for designing systems in the following chapters. The physics of PV cells, with an emphasis on the challenges to the cell innovator, are covered in Chapter 10, and current specific cell technologies are discussed in Chapter 11.

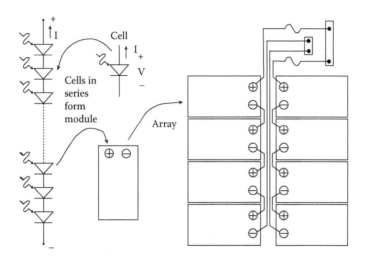

Figure 3.1 Cells, modules, and arrays.

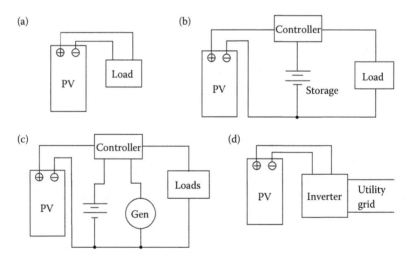

Figure 3.2 Examples of PV systems. (a) PV connected directly to load. (b) Controller and battery storage included. (c) System with battery storage and backup generator. (d) Grid-connected system.

3.2 THE PV CELL

The PV cell is a specially designed pn junction or Schottky barrier device. The well-known diode equation describes the operation of the shaded PV cell.

When the cell is illuminated, electron–hole pairs (EHPs) are produced by the interaction of the incident photons with the atoms of the cell. The electric field created by the cell junction causes the photon-generated EHPs to separate, with

INTRODUCTION TO PV SYSTEMS   51

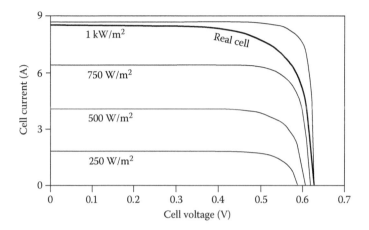

Figure 3.3 I–V characteristics of real and ideal PV cells under different illumination levels.

the electrons drifting into the n-region of the cell and the holes drifting into the p-region, provided that the EHPs are generated sufficiently close to the pn junction. This process will be discussed in detail in Chapter 10. For the purposes of this chapter, knowledge of the terminal properties of the PV cell is all that is needed.

Figure 3.3 shows the ideal I–V characteristics of a typical PV cell. Note that the amounts of current and voltage available from the cell depend upon the cell illumination level. In the ideal case, the I–V characteristic equation is

$$I = I_\ell - I_o(e^{(qV/kT)} - 1) \tag{3.1}$$

where I_ℓ is the component of cell current due to photons, $q = 1.6 \times 10^{-19}$ C, $k = 1.38 \times 10^{-23}$ J/K, and T is the cell temperature in K (0 K = $-273.16°$C). While the I–V characteristics of actual PV cells differ somewhat from this ideal version, Equation 3.1 provides a means of determining the ideal performance limits of PV cells.

Figure 3.3 shows that the PV cell has both a limiting voltage and a limiting current. Hence, the cell is not damaged by operating it under either open-circuit or short-circuit conditions. To determine the short-circuit current of a PV cell, simply set V = 0 in the exponent. This leads to $I_{SC} = I_\ell$. To a very good approximation, the cell current is directly proportional to the cell irradiance. Thus, if the cell current is known under standard test conditions, $G_o = 1$ kW/m² at AM 1.5, then the cell current at any other irradiance, G, is given by

$$I_\ell(G) = \frac{G}{G_o} I_\ell(G_o). \tag{3.2}$$

To determine the open-circuit voltage of the cell, the cell current is set to zero and Equation 3.1 is solved for V_{OC}, yielding the result

$$V_{OC} = \frac{kT}{q} \ln \frac{I_\ell + I_o}{I_o} \cong \frac{kT}{q} \ln \frac{I_\ell}{I_o}, \quad (3.3)$$

since normally $I_\ell \gg I_o$. For example, if the ratio of photocurrent to reverse saturation current is 10^{10}, using a thermal voltage (kT/q) of 26 mV yields $V_{OC} = 0.6$ V. Note that the open-circuit voltage is only logarithmically dependent on the cell illumination, while the short-circuit current is directly proportional to cell illumination.

Multiplying the cell current by the cell voltage yields the cell power, as shown in Figure 3.4. Note that at any given illumination level, there is one point on the cell I–V characteristic where the cell produces maximum power. Note also that the voltage at which the maximum power points occur is minimally dependent upon the cell illumination level.

If I_m represents the cell current at maximum power, and if V_m represents the cell voltage at maximum power, then the cell maximum power can be expressed as

$$P_{max} = I_m V_m = FF I_{SC} V_{OC}, \quad (3.4)$$

where FF is defined as the cell *fill factor*. The fill factor is a measure of the quality of the cell. Cells with large internal resistance will have smaller fill factors, while the ideal cell will have a fill factor of unity. Note that a unity fill factor suggests a rectangular cell I–V characteristic. Such a characteristic implies that the cell operates as either an ideal voltage source or as an ideal current source. Although a real cell does not have a rectangular characteristic, it is clear that it has a region where its operation approximates that of an ideal voltage source and another region where its operation approximates that of an ideal current source.

For the cell having an ideal I–V characteristic governed by Equation 3.1, with $V_{OC} = 0.596$ V and $I_{SC} = 9.0$ A, the fill factor will be approximately 0.825801. Typical fill factors for real PV cells, depending on the technology, may vary from 0.5 to 0.81.

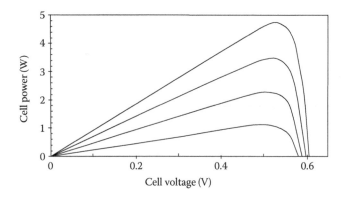

Figure 3.4 Power versus voltage for a PV cell for four illumination levels.

INTRODUCTION TO PV SYSTEMS

The secret to maximizing the fill factor is to maximize the ratio of photocurrent to reverse saturation current while minimizing series resistance and maximizing shunt resistance within the cell.

The cell power versus cell voltage curve is especially important when considering maximizing power transferred to the cell load. This topic will be investigated in more detail in Section 3.6, where methods of matching the load to the source are discussed.

The PV cell I–V curve is also temperature sensitive. A quick look at Equation 3.3 might suggest that the open-circuit voltage is directly proportional to the absolute temperature of the cell. A longer look, however, will reveal that the reverse saturation current is highly temperature dependent also. The net result, which will be covered in detail in Chapter 10, is that the open-circuit voltage of an ideal crystalline silicon PV cell decreases by approximately 2.3 mV/°C increase in temperature, which amounts to approximately 0.4%/°C. The short-circuit current, on the other hand, increases only slightly with temperature. As a result, the cell power also decreases by approximately 0.4%/°C. Figure 3.5 shows the temperature dependence of the PV cell power versus voltage characteristic.

It is important to remember that when a cell is illuminated, it will generally convert less than 22% of the irradiance into electricity. The balance is converted to heat, resulting in heating of the cell. As a result, the cell can be expected to operate above ambient temperature. If the cell is a part of a concentrating system, then it will heat even more, resulting in additional temperature degradation of cell performance.

The photocurrent developed in a PV cell is dependent on the intensity of the light incident on the cell. The photocurrent is also highly dependent on the wavelength of the incident light. In Chapter 2, it was noted that terrestrial sunlight approximates the spectrum of a 5800-K blackbody source. PV cells are made of materials for which conversion to electricity of this spectrum is as efficient as possible. Depending on the cell technology, some cells must be thicker than others to maximize absorption.

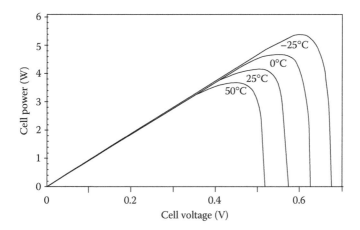

Figure 3.5 Temperature dependence of the power versus voltage curve for a PV cell.

Cells often have textured surfaces and may also be coated with an antireflective coating to minimize reflection of sunlight away from the cells.

3.3 THE PV MODULE

In order to obtain adequate output voltage, PV cells are generally connected in series to form a PV module. With silicon, single-cell open-circuit voltages are typically close to 0.6 V and maximum power voltages are close to 0.5 V at 25°C. Historically, when the majority of PV systems were stand-alone systems with battery backup, it was desirable to combine a sufficient number of cells in series to provide effective battery charging. Since nominal 12-V lead-acid battery charging voltages range from 14 to 16 V, allowing for operation of modules at elevated temperatures, 36-cell modules were the norm for many years. These modules operated close to their maximum power points when charging 12-V lead-acid batteries.

Recently, however, two things have changed the design parameters for PV modules. First of all, since 1999, more PV modules have been used in grid-connected systems worldwide than in stand-alone systems [1]. In a straight grid-connected system with no battery backup, it is now common to design the PV array such that the maximum open-circuit voltage is just under 600 V_{dc} for smaller systems and up to 1000 V_{dc} for larger systems. These arrays are connected directly to maximum power point tracking (MPPT) inverter inputs. Thus, many modern modules have 54–96 cells, and sometimes even more, with correspondingly higher open-circuit voltages and higher module power ratings.

The other change has been in the technology of charge controllers. When the first edition of this book was published, charge controllers connected the array directly to the batteries at the battery voltage. So the array was designed to operate close to the normal charging voltage of the batteries to approximate maximum power operation. Modern charge controllers, however, incorporate MPPT input circuitry, so the array can operate at a maximum power voltage that exceeds the battery voltage and still, via the charge controller, deliver the correct charging voltage to the batteries.

An important observation relating to the series connection of PV cells relates to shading of individual cells. If any one of the cells in a module should be shaded, the performance of that cell will be degraded. Since the cells are in series, this means that the cell may become forward biased if other unshaded modules are connected in parallel, resulting in heating of the cell. This phenomenon can cause premature cell failure. To protect the system against such failure, modules are generally protected with *bypass diodes*, as shown in Figure 3.6. If PV current cannot flow through one or more of the PV cells in the module, it will flow through the bypass diode instead.

When cells are mounted into modules, they are often covered with antireflective coating, then with a special laminate to prevent degradation of the cell contacts. The module housing is generally metal, which provides physical strength to the module. When the PV cells are mounted in the module, they can be characterized as having a *nominal operating cell temperature* (NOCT). The NOCT is the temperature the cells will reach when operated at open circuit in an ambient temperature of

INTRODUCTION TO PV SYSTEMS

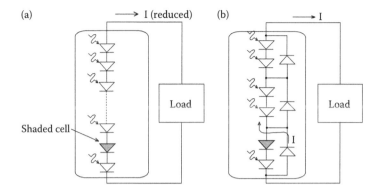

Figure 3.6 Use of bypass diodes to protect shaded cells. (a) Module without bypass diodes. (b) Module with bypass diodes.

20°C at AM 1.5 irradiance conditions, G = 0.8 kW/m², and a wind speed less than 1 m/s. For variations in ambient temperature and irradiance, the cell temperature (in °C) can be estimated quite accurately with the linear approximation that

$$T_C = T_A + \left(\frac{NOCT - 20}{0.8}\right)G. \tag{3.5}$$

The combined effects of irradiance and ambient temperature on cell performance merit careful consideration. Since the open-circuit voltage of a silicon cell decreases by approximately 2.3 mV/°C, the open-circuit voltage of a module will decrease by 2.3n mV/°C, where n is the number of series cells in the module.

Hence, for example, if a 36-cell module has a NOCT of 46°C with V_{OC} = 19.40 V, when G = 0.8 kW/m², then the cell temperature will rise to 62.5°C when the ambient temperature rises to 30°C and G increases to 1 kW/m². This 15°C increase in cell temperature will result in a decrease of the open-circuit voltage to 16.3 V, a 16% decrease. Furthermore, excessive temperature elevation may cause the cell to fail prematurely.

Finally, a word about module efficiency. It is important to note that the efficiency of a module will be determined by its weakest link. Since the cells are series connected, it is important that cells in the module be matched as closely as possible. If this is not the case, while some cells are operating at peak efficiency, others may not be optimized. As a result, the power output from the module will be less than the product of the number of cells and the maximum power of a single cell.

Figure 3.7 shows how individual cell operating characteristics must be combined to produce the composite operating characteristic of the module. Note that since all cells carry the same current, the voltages of individual cells must adjust accordingly to meet the current constraint imposed by the external load. The composite I–V characteristic for n cells in series is thus obtained by adding the individual cell voltages that result for each cell to deliver the required current. The maximum current

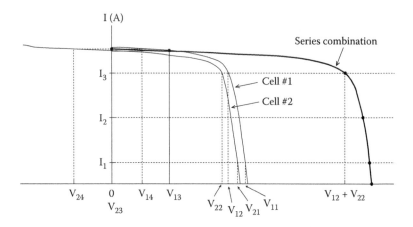

Figure 3.7 Determination of composite operating characteristic for a PV module.

available from the module is affected by the cell with the lowest current under specific load conditions at the operating irradiance. Hence, it is desirable for all cells in the module to have identical I versus V versus irradiance curves. It is not difficult to imagine circumstances where the power output of one cell may differ from the power output of other cells, especially if the module is operated in a location where flocks of birds find it to be a convenient rest stop.

To arrive at the composite I–V curve for two cells in series, simply add the corresponding voltages required to produce the required current, as shown in Figure 3.7. Note that if one cell has a larger I_{sc} than the other cell, as indicated, while the voltage across the combination is zero, the voltage of each cell will not necessarily be zero. In Figure 3.7, Cell 2 has a lower I_{sc} than Cell 1, and thus its voltage goes negative until its current equals the current of Cell 1. This means that Cell 1 is generating power and Cell 2 is dissipating this power.

When 30 or more cells are connected in series, if they do not have identical I–V curves, then when the module is short-circuited, some of the cells will be generating power while others will be dissipating power. The greater the mismatch among cells in the module, the greater the level of power dissipated in the weaker cells. If all cells are perfectly matched, no power is dissipated within the module under short-circuit conditions and the overall efficiency of the module will be the same as the efficiency of individual cells.

3.4 THE PV ARRAY

If higher voltages or currents that are available from a single module are required, modules must be connected into arrays. Series connections result in higher voltages, while parallel connections result in higher currents. Just as in the case for cells, when modules are connected in series, it is desirable to have each module's maximum power production occur at the same current. When modules are connected in parallel, it is desirable to have each module's maximum power production occur at the same voltage. Fortunately, modern quality control, along with individual module testing as the last manufacturing step, pretty much guarantees that all modules of the same module model will have nearly identical I–V characteristics over a wide temperature range. So as long as module models are not mixed, module voltages, currents, and powers will generally be well-balanced in an unshaded array. In later chapters, criteria will be developed for deciding how many modules to connect in series and how many modules to connect in parallel to optimize the performance of a PV array that provides power to the designer's choice of inverter or charge controller. Which, if any, of the positive or negative output leads of an array should be grounded will also be discussed later.

3.5 ENERGY STORAGE

3.5.1 Introduction

Most electrical engineers have some familiarity with batteries, especially the ones used in flashlights, toys, laptop computers, watches, calculators, cameras, automobiles, and golf carts. The extent of familiarity normally extends to the decision as to whether to purchase C–Zn batteries or whether to buy the more expensive, but longer lasting, alkaline cells. Remembering to get the correct size, whether a 9-V or a 1.5-V AA battery, is another of life's energy storage dilemmas. In this section, considerations in battery selection for PV systems are presented. Alternatives to battery energy storage are also presented, so the PV design engineer will be able to make educated choices, based on sound economic and engineering principles, as to which type of energy storage to incorporate into a specific PV system design.

Many different types of rechargeable batteries suitable for PV applications are currently available. Although several rather exotic technologies are now available, the lead-acid battery is still the most common for relatively economical storage of useful quantities of electrical energy, and will probably remain so for at least the next few years. However, unless improvements in energy density, cost, and lifetime are made with lead-acid technology, other promising technologies may surpass lead-acid technology. For example, Ni–Cd batteries have been in common use in applications that require sealed batteries capable of operating in any position, and still require high energy density. But they are more expensive per Joule stored than lead-acid units. Nickel metal hydride batteries were popular for a while, but at this time, it appears that several lithium technologies may soon make a major inroad in PV energy storage. And, for utility-sized battery backup, flow batteries are moving beyond the experimental stage.

But, then, batteries are not the only way to store energy. It is also possible to store energy by producing hydrogen, by pumping water uphill, by spinning a flywheel, or by charging a large chemical capacitor. This section explores the applications, advantages, and disadvantages of some of the storage methods currently in use as well as some of the new ideas that have been proposed.

3.5.2 The Lead-Acid Storage Battery

3.5.2.1 Chemistry of the Lead-Acid Cell

Figure 3.8 shows the basics of lead-acid cell operation. In simple terms, the battery consists of a lead cathode and a lead oxide (PbO_2) anode immersed in a sulfuric acid solution. The discharging reaction at the anode consists of the exchange of oxygen ions from the anode with sulfate ions of the electrolyte. At the cathode, the discharge involves sulfate ions from the electrolyte combining with lead ions to form lead sulfate. Removal of sulfate ions from the solution reduces the acidity of the electrolyte. To maintain charge neutrality, two electrons must enter the anode terminal and two electrons must leave the cathode terminal via the external circuit for each two sulfate ions that leave the electrolyte. This corresponds to a positive current leaving the anode terminal, which is consistent with the passive sign convention for power being delivered. Specifically, the chemical processes at the anode and cathode during discharge are represented by the respective equations

$$PbO_2 + 4H^+ + SO_4^{2-} + 2e^- \rightarrow PbSO_4 + 2H_2O \qquad (3.6)$$

and

$$Pb + SO_4^{2-} \rightarrow PbSO_4 + 2e^-. \qquad (3.7)$$

The oxygen ion, which enters the solution from the anode, combines with two hydrogen ions from the sulfuric acid electrolyte to form water.

When an external voltage source greater than the voltage produced by the reactions at the anode and cathode is applied across the battery terminals, the discharge

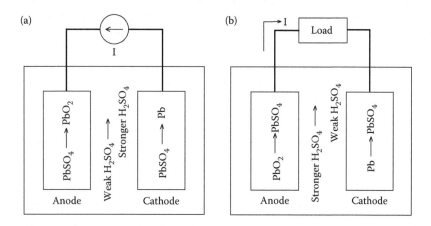

Figure 3.8 Charge and discharge of the lead-acid cell. (a) Charging process. (b) Discharging process.

process is reversed, and current flows into the anode rather than out of the anode, thus charging the battery. The chemical processes are then reversed at the anode and cathode to

$$PbSO_4 + 2H_2O \rightarrow PbO_2 + 4H^+ + SO_4^{2-} + 2e^- \quad (3.8)$$

and

$$PbSO_4 + 2e^- \rightarrow Pb + SO_4^{2-}. \quad (3.9)$$

To maintain charge balance, the two electrons shown in Equations 3.8 and 3.9 must leave the anode via the external connection and travel to the cathode. As the electrons leave the anode, they are replaced by an oxygen ion, which combines with a lead ion in the anode, which was created by a sulfate ion entering the electrolyte. As these two electrons are added to the cathode, a sulfate ion is liberated from the cathode to the solution. The two electrons combine with a lead ion, producing a lead atom. The sulfate ions liberated from the cathode and the anode to the solution increase the concentration of sulfuric acid in the electrolyte solution by balancing with hydrogen ions in the solution, while the oxygen ion left behind by the dissociated water molecule is left alone in solution. The positive potential difference between anode and cathode causes the negatively charged oxygen ion to migrate to the anode, where it becomes a part of the anode reaction, ultimately being converted to lead oxide at the anode, as previously indicated.

When the battery discharges, ultimately the surfaces of both electrodes are converted to lead sulfate. If excessive lead sulfate is allowed to build up on the electrodes, the effective surface area of the electrodes is reduced and cell performance may be affected. It is thus important to avoid fully discharging a lead-acid battery. During the charging process, some of the hydrogen ions combine with free electrons and are converted into gaseous hydrogen. At a certain point during charging, when the cathode is fully converted back to lead, there is no more sulfate at the cathode to maintain continuity in the charging current. At this point, if charging is continued, the electrons entering the cathode can no longer release sulfate ions, so the electrons continue into the electrolyte, combine with hydrogen ions, and produce hydrogen gas. This phenomenon is called gassing, and it is generally not desirable to charge a battery beyond this point, since the hydrogen gas can present a safety hazard. Occasional charging of a battery to the gassing stage so the gas bubbles up through the electrolyte can be useful, since the bubbling action tends to perform a cleaning action on the electrodes and a mixing action on the electrolyte.

The potential difference between electrodes in the lead-acid system is approximately 2.12 V when the cell is fully charged. As cells are connected in series, multiples of 2.12 V can be achieved. Most commonly, either three or six cells are connected in series, producing nominal voltages of 6 or 12 V.

3.5.2.2 Properties of the Lead-Acid Storage Battery

Ideally, the charging and discharging processes of the lead-acid system should be reversible. In reality, however, they are not. The temperature of operation, the rate of discharge, and the rate of charge all affect the performance of the battery. Since the electrical path of the battery presents ohmic resistance, some of the electrical energy intended for charging, that is, storage, is converted to heat. When hydrogen is lost, it also represents an energy loss. Typically, the charging process is about 95% efficient. The discharge process also results in some losses due to internal resistance of the battery, so only about 95% of the stored energy can be recovered. The overall efficiency of charging and discharging a lead-acid battery is thus about 90%, provided that the battery is not charged or discharged too fast.

Since battery losses to internal resistance are proportional to the square of the current, this means that high-current charging or high-current discharging will tend to result in higher internal losses and less overall performance efficiency. This effect is offset somewhat by the increase in temperature of the battery during charging or discharging at high rates due to the higher I^2R losses. It turns out that a warmer battery can hold more charge. It also turns out that if a battery is too warm for too long, its life expectancy is shortened, so charging and discharging rates need to be carefully observed so that they will not exceed rated values for a specific battery.

The amount of energy stored in a battery is commonly measured in ampere-hours (Ah). While Ah are technically not units of energy, but, rather, units of charge, the amount of charge in a battery is approximately proportional to the energy stored in the battery. If the battery voltage remains constant, then the energy stored in watt-hours (Wh) is simply the product of the charge in Ah and the voltage in V.

The capacity of a battery is often referred to as C. Thus, if a load is connected to a battery such that the battery will discharge in x hours, the discharge rate is referred to as C/x. The charging of a battery is measured in a similar fashion. Figure 3.9 indicates the effect of discharging rates on the relative amount of charge that can

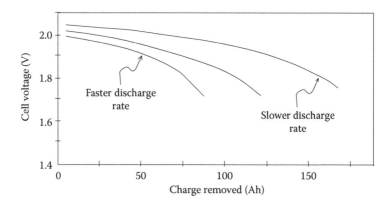

Figure 3.9 Effect of discharge rate on available energy from a lead-acid battery.

INTRODUCTION TO PV SYSTEMS

be obtained from a lead-acid battery. Note that higher discharge rates result in less charge being available as energy to a load. At higher charging rates, a smaller fraction of the charging energy is used for charging and a larger fraction is used to heat up the battery. The battery can be fully charged at higher charging rates, but it takes more energy at higher charging rates to obtain full charge.

These phenomena can be explained on the basis of the Thevenin equivalent circuit of the battery, which includes a resistance in series with the open-circuit cell voltages. Since the charging or discharging current must pass through the internal (Thevenin) resistance of the battery, a power loss equal to I^2R occurs. For constant-current charging or discharging, the charge delivered to or removed from the battery is given by $Q = It$. Since Q is proportional to energy storage, the energy stored is thus proportional to I, whereas the energy lost is proportional to I^2. Hence, at higher charging or discharging rates, a larger fraction of available charging energy is lost to resistive heating. This effect is shown in Figure 3.10.

Figure 3.10 shows the effect of discharge rates and temperature on the relative amount of charge that a battery can deliver. Again, slower discharge rates result in a higher overall amount of energy being delivered by the battery. This figure is especially notable to those readers who live in northern latitudes and find it more difficult to start their automobiles in subzero weather, especially if their batteries are about ready for replacement.

Depending upon specific composition of the electrodes, lead-acid batteries may be optimized for shallow discharge or for deep-discharge operation. The shallow discharge units typically have a small amount of calcium combined with the lead to impart greater strength to the otherwise pure lead. The plates can then be made thinner with greater area to produce higher starting currents. These units should not be discharged to less than 75% of their capacity. In automobile applications, these are satisfactory operating conditions, since the battery is needed primarily for operation of the starter motor until the engine starts. After this point, the alternator takes over, recharging the battery and operating the automobile electrical systems. The shallow

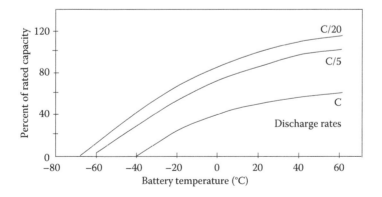

Figure 3.10 Effect of temperature and discharge rate on available energy from a lead-acid battery.

discharge units have a smaller quantity of lead and are correspondingly lighter and less expensive.

Deep-discharge lead-acid batteries use antimony to strengthen the lead and can be cycled down to 20% of their initial capacity. The plates are thicker, with less area and are hence designed for sustained lower level currents. These batteries are designed for use in golf carts, marine applications, electric forklifts, and PV systems. Although the deep-discharge batteries are designed for deep-discharge applications, their lifetime in cycles depends on the depth of discharge during normal operation. Figure 3.11 shows how the depth of discharge affects the number of operating cycles of a deep-discharge battery. The PV system designer must carefully consider the trade-off between using more batteries operating at shallower discharge rates to extend the overall life of the batteries versus using fewer batteries with deeper discharge rates and the correspondingly lower initial cost. Life-cycle costing of systems will be considered in Chapter 8.

Lead-acid batteries are available in vented (flooded) and nonvented (sealed) enclosures. There is a trade-off between lead–antimony and lead–calcium batteries when venting is considered. In certain lead–calcium batteries, minimal hydrogen and oxygen are lost during charging. This means minimal water is lost from the electrolyte. As a result, it is possible to seal off the cells of these batteries, making them essentially maintenance free. The trade-off, however, is that if these batteries are either purposely or inadvertently discharged to less than 75% of their maximum charge rating, their expected lifetime may be significantly shortened.

Lead–antimony electrodes, on the other hand, may be discharged to 20% of their maximum charge rating. This means that a 100-Ah lead–calcium battery has only 25 Ah available for use, while a 100-Ah lead–antimony battery has 80 Ah available for use, or more than three times the availability of the lead–calcium unit. However, the lead–antimony unit produces significantly more hydrogen and oxygen gas from dissociation of water in the electrolyte, and thus water must be added to the battery relatively often to prevent the electrolyte level from falling below the top of the electrodes.

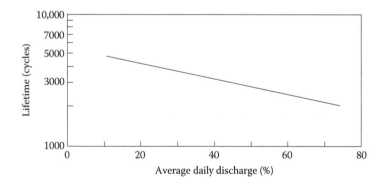

Figure 3.11 Lead-acid battery lifetime in cycles versus depth of discharge per cycle. (From *Maintenance and Operation of Stand-Alone Photovoltaic Systems*, Sandia National Laboratories, Albuquerque, NM, 1991.)

INTRODUCTION TO PV SYSTEMS

Water loss can be reduced somewhat by the use of cell caps that catalyze the recombination of hydrogen and oxygen back into water, which returns to the cell. Other cell caps have a flame-retardant structure that prevents any flame external to the battery from entering the cell. In any case, the bottom line is that nonsealed, deep-discharge batteries have open vents and require relatively frequent addition of water for maintenance, depending upon cycle time and depth.

An advantage of removable cell caps on batteries is that it is possible to measure the specific gravity of the cells of these batteries. Since one of the most accurate measures of the state of charge of a battery is the specific gravity of the electrolyte, this provides a convenient means of finding faulty battery cells. Figure 3.12 shows how the electrolyte specific gravity and voltage vary during charging and discharging of the battery. Note particularly the effect of the voltage drop across the internal cell resistance during charging and discharging. For applications where maintenance of batteries is inconvenient, maintenance free, sealed deep-cycle batteries exist, but are generally at least double the price of equivalent capacity nonsealed lead-acid units. These batteries are valve regulated to recombine gases, with additional treatment to immobilize the electrolyte. *Gel Cells* are sealed, lead-acid deep-cycle batteries that have silica gel added to the electrolyte. Absorbed glass mat (AGM) batteries have highly absorbent glass mat separators between plates to bind the electrolyte. In addition to being maintenance free, sealed deep-discharge units do not leak and have long life and low self-discharge. Charging requirements for sealed lead-acid batteries differ somewhat from nonsealed units. While nonsealed units are designed for occasional overcharge, sealed units should not be overcharged.

Once a battery type has been selected, a suitable enclosure must be selected or designed and appropriate safety precautions must be taken during installation, operation, and maintenance of lead-acid batteries. Two important references are the *National Electrical Code®* (*NEC*) [2] and the *IEEE Recommended Practice for Installation Design and Installation of Valve-Regulated Lead-Acid Storage Batteries for Stationary Applications* (IEEE 1187–2002 [3]). Annex C of IEEE 1187–2002

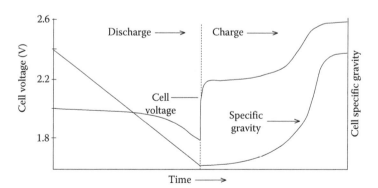

Figure 3.12 Variation of cell electrolyte specific gravity and cell voltage during charge and discharge at constant rate. (From *Maintenance and Operation of Stand-Alone Photovoltaic Systems*, Sandia National Laboratories, Albuquerque, NM, 1991.)

Table 3.1 Typical Deep-Cycle, Lead-Acid Battery Cell Properties versus State of Charge

State of Charge (%)	Specific Gravity	Cell Voltage (V)	Voltage of 12 V Battery	Freezing Point (°F)
100	1.265	2.12	12.70	−71
75	1.225	2.10	12.60	−35
50	1.190	2.08	12.45	−10
25	1.155	2.03	12.20	+3
0	1.120	1.95	11.70	+17

Source: Maintenance and Operation of Stand-Alone Photovoltaic Systems, Sandia National Laboratories, Albuquerque, NM, 1991.

provides information on the degree to which the life of lead-acid batteries is shortened if the batteries are operated at temperatures above 25°C. For example, at a temperature of 35°C, the battery lifetime will be shortened to 44% of its expected lifetime at 25°C. This means that it is important to provide adequate ventilation for batteries, not only to vent any hydrogen or other gases, but also to maintain a battery operating temperature close to 25°C. The serious PV system designer should be sure to have access to the *NEC* and IEEE 1187.

As a final note on lead-acid batteries, it is important to be aware of the temperatures at which the electrolytes of batteries will freeze. As the electrolyte becomes more acidic, the freezing temperature is lowered. Hence, fully charged batteries may be operated at low temperatures, while batteries in some state of discharge will need to be kept warmer. Table 3.1 [4] summarizes the relationships between state of charge, specific gravity, cell voltage, and electrolyte freezing point.

When a PV system is designed that uses lead-acid batteries for storage, the designer needs to determine proper charging rates for the batteries and also should determine the ventilation needs of the battery system. These considerations will be covered in Chapter 6.

3.5.3 Lithium-Ion Battery Technologies

Thanks to a renewed interest in electric and hybrid vehicles as well as the need for low-weight, high-energy-density storage for computing and communications devices, lithium technology has made significant advances over the past 20 years. Lithium-ion batteries are typically one-third the weight of a lead-acid unit with the same storage capacity. They are about half the volume of a comparable lead-acid unit, have significantly better low-temperature capacity, have a relatively steady voltage during discharge at a constant current, and have a lower internal impedance than comparable lead-acid units, which allows them to be charged and discharged at faster rates due to lower internal I^2R losses. They have minimal self-discharge, require almost no maintenance, and have nearly double the expected lifetime of their lead-acid counterparts.

Some lithium technologies have been found to have potentially unstable thermal performance that can result in the units catching on fire, and at present, the first

INTRODUCTION TO PV SYSTEMS

Table 3.2 History of Li Battery Commercialization

1991	Sony commercializes first Li-ion battery, used in cell phones and laptop computers
1996	Lithium polymer batteries become commercialized
1999	Lithium-phosphate batteries become commercialized
2006	Lessons learned from laptop computer fires: battery chemistry and manufacturing quality control are important
2007	Large-format lithium iron phosphate batteries become widely available for EV conversion

Source: Richmond, R., *Home Power Mag.*, Feb/Mar 2013.

cost of lithium-ion batteries exceeds the first cost of comparable lead-acid batteries. While lead-acid batteries are less sensitive to maintaining equal charge in every cell, lithium-ion batteries require accurate control of cell voltage and charge, which must be accomplished with a more sophisticated battery management system (BMS) that monitors the voltage and temperature of every cell. Table 3.2 [5] chronicles the commercialization of various lithium technologies, and Table 3.3 [5] compares the properties and uses of the more common lithium-ion technologies.

Presently, the most stable and most plausible technology for use in renewable energy storage applications is the lithium iron phosphate (LFP) cell. The reaction is described by

$$LiFePO_4 \leftrightarrow FePO_4 + Li^+ + e^- \qquad\qquad Li^+ + e^- + C \leftrightarrow LiC_6$$

Positive electrode   Electolyte $LiPF_6$ in   Negative electrode
ethylene carbonate/
dimethylcarbonate (DMC)   (3.10)

where discharging is in the right-to-left direction and charging is in the left-to-right direction.

This application is currently (2016) limited by the lack of compatible BMSs that properly interface with conventional PV charge controllers (pulse width modulated [PWM] or MPPT, see Sections 3.6 and 3.8). Otherwise, the characteristics of this technology are favorable for renewable energy use and will likely become more

Table 3.3 Comparison of Li Battery Technologies Available in 2016

Chemical Name	Material	Abbreviation	Applications
Lithium cobalt oxide	$LiCoO_2$	LCO	Cell phones, laptops, cameras
Lithium manganese oxide	$LiMn_2O_4$	LMO	Power tools, EVs, medical, hobbyist
Lithium iron phosphate	$LiFePO_4$	LFP	Power tools, EVs, medical, hobbyist
Li Ni Mn Co oxide	$LiNiMnCoO_2$	NMC	Power tools, EVs, medical, hobbyist
Li Ni Co Al oxide	$LiNiCoAlO_2$	NCA	EVs, grid storage
Lithium titanate	$Li_4Ti_5O_{12}$	LTO	EVs, grid storage

Source: Richmond, R., *Home Power Mag.*, Feb/Mar 2013.

popular over the near future as charge controlling algorithms are improved to maximize solar generation while simultaneously properly managing the individual cell charging of the LFP system. With the entry of electric vehicle manufacturers into this market, it is reasonable to assume the overlap of electric vehicle and renewable energy storage needs will continue to *drive* this market. LFP battery systems with BMS are presently available in sizes up to 200 Ah at 48 V, and it is reasonable to assume that larger units will appear in the near future. In fact, by the time this book is in press, they may already have appeared.

3.5.4 Nickel-Based Battery Systems [6]

3.5.4.1 Nickel–Cadmium Batteries

Ni–Cd batteries are more robust than lead-acid batteries. They can survive freezing and high temperatures, they can be fully discharged, and they are less affected by overcharging. As a result, in some applications, Ni–Cd batteries may be a better choice because their robust nature may enable the elimination of the system charge controller. If the batteries are to be used in a location where access for maintenance is difficult, the higher cost of these batteries can also often be justified.

The overall discharge reaction at the electrodes is given by

$$2NiOOH + 2H_2O + Cd \rightarrow 2Ni(OH)_2 + Cd(OH)_2. \tag{3.11}$$

The reaction is reversed in the charging direction. The voltage of the fully charged cell is 1.29 V. Unlike the lead-acid system where the specific gravity of the electrolyte changes measurably during discharge or charge, the KOH electrolyte of the Ni–Cd system changes very little during battery operation. In some batteries, LiOH is also added to the electrolyte to improve cycle life and also to improve higher temperature operation.

These batteries are in use in industrial, military, and space applications where the previously mentioned properties are important. Capacities range from small sizes up to more than 1200 Ah. Three different thicknesses are used for electrodes, depending on whether the battery is designed for high, medium, or low discharge rates.

Energy densities range from 20 Wh/kg to more than 50 Wh/kg, depending upon cell composition.

Unlike the lead-acid system, which loses capacity under conditions of heavy discharge, the Ni–Cd system can be discharged at rates up to C over a wide temperature range, while still providing more than 90% of its capacity to the load. This is partially attributable to the very low internal resistance of the cells, which, depending on cell area, can be less than a milliohm.

If Ni–Cd batteries are charged and then left unused, they will lose charge at the rate of approximately 2% per day for the first few days, but then stabilize to a relatively low loss rate. Over a 6-month period, the total loss is typically about 20%, depending on whether the battery is a high, medium, or low discharge rate battery. The higher the discharge rating of the battery, the greater the loss of charge over time. Loss of charge is also temperature dependent. The loss rate is greater at higher temperatures, but at a temperature of −20°C, there is almost no loss at all.

The lifetime of a Ni–Cd battery depends on how it is used, but is less dependent on depth of discharge than that of lead-acid batteries. A lifetime of at least 2000 cycles can be expected for a battery when it is not used extensively at elevated temperatures. As a result, under certain applications and operating conditions, a battery may last as long as 25 years, while under more frequent cycling, the lifetime may be reduced to 8 years. A Ni–Cd battery can last twice as long as its lead-acid counterpart.

Disadvantages of Ni–Cd batteries include difficulty in determining the state of charge of the batteries and the toxicity of the cadmium, which creates an environmental concern during production and disposal. They are also more expensive than most other rechargeable battery technologies.

As electric and hybrid automobiles become more popular, the demand for high-energy-density batteries will likely lead to the replacement of nickel–cadmium technology with lower-cost, higher-energy-density technologies.

3.5.4.2 Nickel–Zinc Batteries

The *nickel–zinc* battery is a combination of the Ni–Cd system and the Cu–Zn system, which provides some attractive features, including long life and a capacity advantage. The specific energy of the Ni–Zn system is double the specific energy of the Ni–Cd system. The overall discharge chemical reaction is given by

$$2NiOOH + Zn + 2H_2O \rightarrow 2Ni(OH)_2 + Zn(OH)_2. \qquad (3.12)$$

If the battery is allowed to overcharge, it dissociates the water in the KOH electrolyte, resulting in gassing and the release of hydrogen and oxygen. Hence, charging must be controlled to avoid overcharging. Fully charged cell voltage is approximately 1.73 V.

Ni–Zn battery systems can be made in sizes in excess of 200 Ah. At normal operating temperatures ($\cong 20°C$) and reasonable discharge rates ($\cong C/3$), the cell voltage remains quite stable at about 1.6 V over most of the discharge range.

3.5.4.3 Nickel–Metal Hydride Batteries

Another technology that is becoming very popular, particularly in smaller applications such as camcorders and laptop computers, is the *nickel–metal hydride* (NIMH) battery. This battery replaces the cadmium cathode with an environmentally benign metal hydride cathode, allowing for higher energy density at the cathode and a correspondingly longer lifetime or higher capacity, depending on the design goal. The anode is the same as in the Ni–Cd cell and KOH is used as the electrolyte. The overall discharge reaction is

$$MH + NiOOH \rightarrow M + Ni(OH)_2. \qquad (3.13)$$

The NIMH cell requires clever use of an oxygen recombination system to prevent loss of oxygen during the charging cycle. Electrodes are also carefully sized to ensure that the useful capacity of the battery is determined by the anode electrode to ensure that the cathode will moderate any overcharge or overdischarge condition.

Table 3.4 [7] compares existing commercialized battery storage technologies.

3.5.5 Emerging Battery Technologies

A number of additional battery types are currently under investigation for PV and similar uses, such as electric vehicles. A partial listing of these batteries includes zinc/silver oxide, metal/air, iron/air, zinc/air, aluminum/air, lithium/air, zinc/bromine, lithium–aluminum/iron sulfide, lithium–aluminum/iron disulfide, sodium/sulfur, and sodium/metal chloride. Perhaps, the technology to watch closely, however, is that of flow batteries.

Flow batteries have been around for over 40 years, but until recently, they have remained in the shadows. A number of technologies exist, each of which has its advantages and disadvantages. Attention has returned to flow technologies because they are capable of storing megawatt-hours of energy and are capable of releasing it at rates up to megawatts. Nonflow technologies can be optimized to release their energy quickly or slowly, that is, to provide high power, such as needed for starting an internal combustion engine, or to provide lower power, such as would be used in a PV application. These applications are generally classified as shallow discharge and deep discharge. Once the device has been configured, the charge and discharge limits are established. On the other hand, true redox flow batteries (RFBs) can be optimized for power and energy both, which can mean optimizing the unit to meet the load conditions.

In general, the main difference between flow batteries and previously discussed technologies is that in lead-acid, lithium-ion, nickel, and most other technologies, the energy is stored at the electrodes. In true RFBs, such as vanadium–vanadium and iron–chromium systems, the energy is stored in the electrolytes. In hybrid RFBs, such as zinc–bromine and zinc–chlorine systems, some energy is stored in the electrolytes and additional energy is stored at the electrodes. In order to store energy in electrolytes, separate electrolytes are needed for the anode side and the cathode side of the battery. The electrolytes are pumped from storage containers through the electrochemical stack, which houses the electrodes and a membrane that separates the two electrolytes, but allows ion transfer through the membrane as a part of the charging or discharging process. Disadvantages of flow batteries include lower volumetric energy storage and higher energy losses between charge and discharge than nonflow units, and some flow technologies use toxic chemicals.

Recently, work has been progressing on organic flow batteries that use nontoxic, organic electrolytes. Reasonable energy density has been achieved, and the researchers believe that this technology may result in a more cost-effective technology for utility-scale energy storage [8,9].

The young engineer can expect to see progress in these new technologies during his or her path toward becoming an older engineer, especially now that public

INTRODUCTION TO PV SYSTEMS

Table 3.4 Battery Technology Comparison

Specifications	Lead-Acid	NiCd	NiMH	Li-Ion Cobalt	Li-Ion Manganese	Li-Ion Phosphate
Specific energy density (Wh/kg)	30–50	45–80	60–120	150–190	100–135	90–120
Internal resistance (mΩ/V)	<8.3	17–33	33–50	21–42	6.6–20	7.6–15
Cycle life (80% discharge)	200–33	1000	300–500	500–1000	500–1000	1000–2000
Fast-charge time (h)	8–16	1 (typ)	2–4	2–4	1 or less	1 or less
Overcharge tolerance	High	Mod	Low	Low	Low	Low
Self-discharge/month (25°C)	5%–15%	20%	30%	<5%	<5%	<5%
Cell voltage	2.0	1.2	1.2	3.6	3.8	3.3
Charge cutoff voltage (V/cell)	2.4 (2.25 float)	Voltage signature	Voltage signature	4.2	4.2	3.6
Discharge cutoff (V/cell, 1C rate)	1.75	1	1	2.5–3.0	2.5–3.0	2.8
Peak load current[a]	5C	20C	5C	>3C	>3C	>30C
Peak load current (best result)[b]	0.2C	1C	0.5C	<1C	<10C	<10C
Charge temperature	−20 to 50°C	0–45°C	0–45°C	0–45°C	0–45°C	0–45°C
Discharge temperature	−20 to 50°C	−20 to 65°C	−20 to 65°C	−20 to 60°C	−20 to 60°C	−20 to 60°C
Maintenance requirement	3–6 months eq	30–60 days disch	60–90 days disch	None	None	None
Safety requirements	Thermally stable	Thermally stable, fuses common		Protection circuit mandatory		
Time durability (y)				>10	>10	>10
In use since	1881	1950	1990	1991	1996	1999
Toxicity	High	High	Low	Low	Low	Low

Source: batteryuniversity.com.
[a] Peak load current = maximum possible momentary discharge current, which could permanently damage a battery.
[b] Peak load current (best result) = the recommended operating current of the battery for best performance of the battery.

demand for electric vehicles, hybrid vehicles, and PV systems with battery storage is increasing.

In addition to battery storage, which is presently the most common method of energy storage for PV systems, a number of other energy storage methods are either in use or have been proposed. Many of these methods will also require a significant research effort to bring them into cost-effective, production-scale use.

3.5.6 Hydrogen Storage

Aside from using batteries to store energy produced by PV systems, it is also possible to use hydrogen as a storage medium [10]. The advantage of using hydrogen for energy storage is when it is desired to recover the stored energy, the hydrogen is reacted with oxygen to form water, which is an exothermic reaction that does not have any carbon byproducts. If the hydrogen is burned in air to produce steam to turn a turbine to generate electricity, water is still produced as the byproduct, with a relatively minimal amount of nitrogen oxides. In a fuel cell, hydrogen combines with oxygen to produce water, electricity, and heat. Of course, the hydrogen does not need to be used to produce electricity. It can also be used as an engine fuel to power a vehicle, such as an automobile, bus, or space shuttle. Indirectly, the result is a solar-powered vehicle that stores energy in hydrogen rather than in batteries. The engineering community has not yet developed a battery-powered, space shuttle booster engine.

Perhaps, the most important advantage of hydrogen storage is the energy density of hydrogen. One of the highest energy densities in conventional fuels is that of gasoline, which contains approximately 1,047,000 BTU/ft^3. To store the same amount of energy in high-capacity storage batteries would require approximately 120 batteries having approximately 2000 BTU/ft^3 storage capacity. The storage capacity of liquid hydrogen is close to 240,000 BTU/ft^3, and gaseous hydrogen can store approximately 47,000 BTU/ft^3. Since water cannot be decomposed into gasoline, and since gasoline burning produces CO_2, among other undesirable combustion products, clean burning hydrogen is a more attractive alternative for combustion. Aside from liquid and gaseous hydrogen storage, other means of obtaining relatively high energy density, including various hydrides, are also available.

Electrolysis of water is a common method of producing hydrogen. The output of a PV system is conveniently dc at a level consistent with that needed for relatively efficient electrolysis. Production of hydrogen by electrolysis may seem to be simply a matter of passing a current through water to produce hydrogen and oxygen, just as many readers have already done in chemistry lab. The challenge is to produce hydrogen efficiently with a system whose power output varies significantly whenever a cloud passes over the PV array.

Once the hydrogen is produced, it can be used on site or it can be transported to other locations. Hydrogen proponents have suggested that hydrogen be produced in tropical latitudes having greater annual insolation and shipped to regions having less sun. Thus, after the production of hydrogen, storage and transport of the gas also present engineering challenges.

INTRODUCTION TO PV SYSTEMS   71

Recent interest in fuel cells for use in automotive applications has once again piqued interest in the efficient and convenient production of hydrogen. The same technology that may emerge for automotive fuel cell applications may very well spin off into PV energy storage applications as well.

3.5.7 The Fuel Cell

Fuel cells provide a convenient means of reacting hydrogen and oxygen directly to produce electricity and water. They utilize essentially the inverse of the electrolysis process. The first fuel cell was built in 1839 by Sir William Grove [11]. The high cost of fuel cells, however, had kept them out of practical use until NASA decided to use them on space flights to provide power and water.

Fuel cells are being constructed in a wide range of power output capabilities. The smaller cells are sized to power video cameras and laptop computers. On a larger scale, The Southern California Gas Company had ten 200-kW plants online in 1993 and experimental units in the megawatt range are now in operation [11]. The fuel cell reaction is exothermic, so if cleverly applied, the cells can supply both heat and electricity to a building. By capturing the heat, the overall system efficiency is increased, provided that the heat can be put to good use, such as for district heating or other space or water heating application.

The alkaline fuel cell (AFC) contains a 30% solution of KOH as the electrolyte. The proton exchange (or polymer electrolyte) membrane fuel cell (PEMFC) incorporates a semipermeable membrane that will pass hydrogen ions, resulting in a proton exchange. The phosphoric acid fuel cell (PAFC) is based on a 103% phosphoric acid solution. The molten carbonate fuel cell (MCFC) is based on a eutectic mixture of molten lithium carbonate and potassium carbonate. The solid oxide fuel cell (SOFC) is based on a stabilized zirconium oxide electrolyte [12].

The low-temperature cells are approximately twice as efficient as conventional heat engines in the conversion of energy stored in hydrogen to electricity. This makes fuel cells a very attractive possibility for energy storage and conversion to electricity in an electric vehicle, since electric motors can be operated at very high efficiency, especially with modern solid-state controllers. The present disadvantage, of course, is that fuel cells remain quite costly and will require further development to reduce their cost. Fuel cells also require a source of hydrogen, so a means for providing fuel for vehicular applications of fuel cells will need to be created.

3.5.8 Other Storage Options

Perhaps, one of the simplest means of storing energy from the sun is to fill a water storage tank, so the water can be used after sundown. Maybe the use will be for drinking or irrigation, or maybe the water will be used to turn a turbine to generate electricity before it is used for some other purpose. It is not difficult to calculate the amount of potential energy stored in a gallon of water raised to a height of 10 ft above ground.

Other possible storage mechanisms include compressed air, flywheels, superconducting magnets, and chemical capacitors. All of these can be shown to be useful for certain end uses, but, in general, the cost per kWh stored is quite high at present.

3.6 PV SYSTEM LOADS

The importance of operating PV modules near their maximum power points has already been discussed. Maximum power operation is a challenging problem, since it requires that the system load be capable of using all power available from the PV system at all times. Not only does this mean that the system load needs to be maximum when PV system output is maximum, but it means the system load must adjust itself rather quickly at the onset or dispersion of cloud cover. In any case, the I–V characteristic of the ideal load will intersect the locus of maximum power points on the I–V characteristics of the PV array for varying illumination levels. The I–V characteristic for this ideal load is shown in Figure 3.13 along with the I–V characteristics of several other common PV system loads.

If the intersection of the I–V characteristic of a load with that of a PV source departs significantly from the maximum power point of the PV source, it may be desirable to employ an electronic maximum power point tracker (MPPT) between the array and the load. MPPTs generally employ pulse width modulation techniques to switch from an input dc voltage to an output dc voltage at a different level, similar to a switching dc power supply. The MPPT employs a feedback loop to sense the PV array output power and change the array output voltage accordingly until the output power is maximized.

Figure 3.14 shows how the MPPT electronically moves the operating point along the maximum power hyperbola (IV = constant) associated with the PV array until it intersects the load I–V characteristic. The price paid for this MPPT is not just the dollar price, since there is also some power loss in the MPPT. It is thus necessary to justify that the expense of the MPPT will be recovered by the value of the

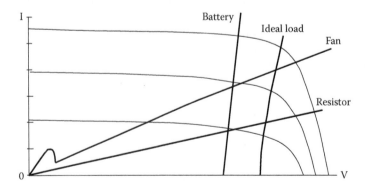

Figure 3.13 I–V characteristics for several common loads along with ideal load I–V characteristic for maximum power operation of a PV system.

INTRODUCTION TO PV SYSTEMS

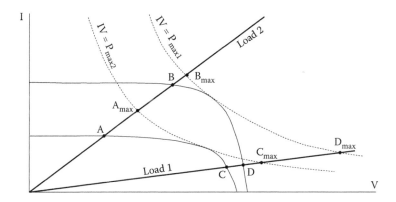

Figure 3.14 Operation of the MPPT.

additional energy made available by the device. In Chapter 8, this will be shown to be a straightforward analysis.

Another consideration for loads on PV systems is the trade-off between lifetime operating cost of a load and initial cost of the load. Sometimes a more efficient load is available at a higher initial cost and sometimes a more efficient load is available at a lower initial cost. If the initial cost of a load is lower, and there are no hidden maintenance costs, then it is a no-brainer to use the more efficient load, unless aesthetic or functional considerations should rule it out.

On the other hand, if the initial cost of a more efficient load is higher, then it is important for the PV system designer to perform an economic analysis of both system options, to determine which load is a better choice. In particular, the choice between relatively inexpensive first-cost incandescent lamps and more expensive but more efficient fluorescent lamps involves an interesting cost analysis. Now that LED lamps are becoming commonplace, but at somewhat higher initial cost and correspondingly longer lifetimes, these devices add yet another interesting dimension to the choice of lighting sources. These considerations will be accounted for in Chapter 8.

3.7 PV SYSTEM AVAILABILITY: TRADITIONAL CONCERNS AND NEW CONCERNS

3.7.1 Traditional Concerns as Applied to Stand-Alone Systems

Critical loads are defined as loads for which power is required at least 99% of the time. Noncritical loads require power for at least 95% of the time. It is important to recognize that these definitions involve a statistical distribution of downtime over the expected lifetime of the system. In other words, a critical system may have availability during any given month or year of its operation that may be less than 99%, while it will have availability during other years in excess of 99%. The 99% figure is considered to be an average over the lifetime of the system.

The reader is probably familiar with causes of downtime on conventional grid-connected systems. Sometimes it is a failure of an electrical generator. Sometimes it is a system overload that requires parts of the system to be shut down selectively to prevent generator overload. Sometimes a tree falls and breaks a power line. About every 100 years, an ice storm of the century causes long-term power outages for millions of utility customers, and sometimes a short circuit occurs and causes a circuit breaker to trip. Even hurricanes have been known to cause an occasional power outage. Clearly, many other situations may occur that will result in the loss of power.

PV systems are subject to similar failure modes. Loose or corroded connections, battery failure, controller failure, and module failure represent a few of the things that might go wrong in a PV system. However, a good preventive maintenance program can keep these failure modes at a minimum. The warrantees on most major PV system equipment are now 10 years or longer.

PV systems do have a factor that affects system performance to which conventional systems are not subjected—unpredictable cloud cover. Unless a PV system is grid-connected, so grid power will be available at night and when it is cloudy, it is necessary to provide battery backup not only for hours of nighttime operation, but also for those days when the sun either does not shine at all or when it shines too little to make sufficient electricity to meet the required daily need. Most readers are aware that different geographical locations have different seasons when cloudy days are more common. The reader in the north will be familiar with the winter weeks when the sun almost never peeks through the cloud cover. The reader in the south will be familiar with the rainy summer days when little sun is available. The bottom line is that for different geographical locations, different amounts of battery backup are required for critical and noncritical loads if the PV system is not connected to the grid.

Figure 3.15 shows the statistical distribution of downtimes due to weather over a 23-year system lifetime for critical and noncritical systems. Note that a 95% available system is allowed 438 hours of downtime per year, but that for 3.5 years of the

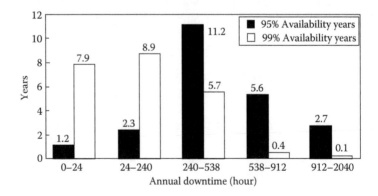

Figure 3.15 Statistical distribution of annual downtime for critical and noncritical PV systems over a 23-year system lifetime. (From *Stand-Alone Photovoltaic Systems: A Handbook of Recommended Design Practices*, Sandia National Laboratories, Albuquerque, NM, 1995.)

INTRODUCTION TO PV SYSTEMS

23 years, the system will be down for less than 240 hours and for 8.3 years the system will be down for somewhere between 538 and 2040 hours per year. The critical system is allowed to be down for 88 hours per year, but will be down for less than 24 hours per year for 7.9 years and will be down for more than 240 hours per year for 6.2 years out of the 23-year lifetime.

Figure 3.16 shows a comparison of necessary days of battery backup for critical and *noncritical* operation as a function of available peak sun hours (psh), assuming a stand-alone system. An Internet search will generally result in tables of psh for various geographic locations, months, and array tilts and orientations. These tables, however, do not yield psh for all possible tilts and orientations, so an alternative method of determining psh is to use the PVWatts model found in NREL SAM for the desired location and array tilt and orientation.

By selecting an array tilt and orientation and then running SAM for a 1-kW PV array at that location, the simulation will present a monthly ac kWh summary for the array. Noting that this result includes a 1.1 dc to ac conversion ratio of the system, the monthly dc kWh can be determined by multiplying the monthly ac kWh result by 1.1 to obtain the monthly dc array kWh output.

The next step is to recognize that if the monthly psh is defined as the number of hours per month the sun would need to shine at peak intensity (1 kWh/m^2) to generate the simulated dc monthly kWh output, then simply dividing by the number of days in the month will give the average daily psh for that month at the specific simulation conditions. As an example, suppose the SAM PVWatts simulation results in an estimated 110 kWh(ac)/kW production for the month of January. The equivalent daily psh will thus be $110 \times (1.1/31) = 3.9$.

Days of battery backup are also known as storage days or *days of autonomy*. Using Excel, second-order polynomial curve fits can be obtained for the data in Figure 3.16. For those familiar with least mean square curve fitting to data, it will

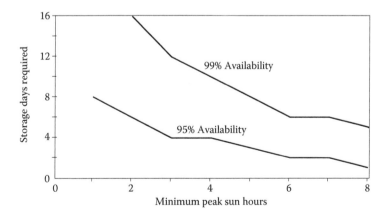

Figure 3.16 Necessary days of autonomy for critical and noncritical PV systems operation versus minimum available psh. (From *Stand-Alone Photovoltaic Systems: A Handbook of Recommended Design Practices*, Sandia National Laboratories, Albuquerque, NM, 1995.)

be recognized that an R^2 value of 1.0 indicates a perfect fit of the curve to the data, while an R^2 value that approaches zero indicates random scatter of the data. The best fits to the data of Figure 3.16 yield the following equations for estimating necessary storage days, based on minimum average psh over the year, T_{min}, as determined from insolation data for the listed sites.

$$D_{crit} = 0.2976T_{min}^2 - 4.7262T_{min} + 24 \quad (R^2 = 0.9914) \quad (3.14)$$

for critical applications, and

$$D_{non} = 0.1071T_{min}^2 - 1.869T_{min} + 9.4286 \quad (R^2 = 0.9683) \quad (3.15)$$

for *noncritical* applications, provided that $T_{min} > 1$ hour. These equations should be used only if the critical and *noncritical* storage times have not already been determined for a site, since site-specific cloud cover or sunlight availability may differ from the averages assumed in arriving at Equations 3.14 and 3.15.

Extending the availability of a PV system from 95% to 99% may at first appear to be a simple, linear extension of the 95% system. Such a linear extension would involve only an additional 4% in cost. However, this is far from the case, and should obviously be so, considering that essentially no systems can provide 100% availability at any cost. Figure 3.17 shows the sharp increase in system cost as the availability of the system approaches 100%.

The cost of extending the availability of a system toward 100% depends on the ratio of available sun during the worst time of the year to available sun during the

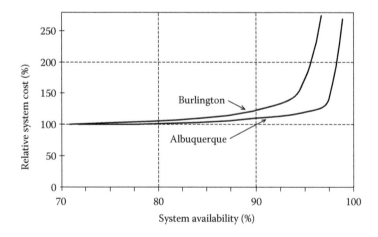

Figure 3.17 Relative costs versus availability for PV systems in Burlington, VT and Albuquerque, NM. (From *Stand-Alone Photovoltaic Systems: A Handbook of Recommended Design Practices*, Sandia National Laboratories, Albuquerque, NM, 1995.)

INTRODUCTION TO PV SYSTEMS 77

best time of the year. The extreme of this situation is represented by the polar latitudes, where no sun at all is available during the winter. Thus, near-100% availability would require enough batteries to store enough energy to meet all the no-sun load needs, plus a sufficient number of PV modules to charge up the batteries. This situation is further complicated by the likelihood that the winter loads would be larger than the summer loads. Clearly, a system with 180 days of autonomy should be more costly than a system with 10 days.

In the less extreme situation, such as found in Seattle, to provide the winter system needs, the system must be overdesigned with respect to the summer needs, resulting in excess power from the array during the summer. There is a good chance that this excess availability will be wasted unless a creative engineer figures out a way to put it to use. One way to do so, of course, is to connect it to the local electrical grid if the grid is available. However, in remote locations, the system may be stand-alone and the creativity of the engineer will be challenged.

3.7.2 New Concerns about PV System Availability

In the United States, for the past few years, more grid-connected renewable capacity has been installed annually than any other generation technology. In 2014, new wind capacity (5319 MW) accounted for 27.4% of total new energy capacity and new solar (3776 MW) accounted for 19.4% of new capacity. In 2015, solar added 2598 MW to the grid and solar plus wind comprised 62.4% of all new generation capacity added [13]. Solar is no longer a power source of interest to hobbyists only.

Because of this rapid growth in small, medium, and large systems, some utilities are now expressing concern over how PV generation may impact the rest of the grid. For example, PV only produces power during the day, but, for the most part, nighttime electrical needs must be provided by the nonrenewable utility sources, except for nighttime wind and storage of renewable energy.

A somewhat less obvious concern regards the potential rapid variation of PV output during the day from any one PV system. If a 10-kW residential grid-connected PV system is clouded over in a matter of 5 seconds, reducing its output from, say, 8 to 3 kW, this will have about the same effect on the grid as an electric water heater or an electric clothes dryer turning on. When the sun returns in a few minutes, if the PV output returns to 8 kW, this will be equivalent to the appliance turning off. For a single appliance or PV system, the grid can readily handle the disturbance. But for hundreds of similar systems in the same neighborhood, the grid impact of changing daytime cloud cover presents an interesting transient power flow analysis challenge, especially if the clouds are small and not all systems change power output simultaneously.

Another concern is the effects of cloud cover changes on utility-scale PV systems. An increase or decrease of 10 MW over a few seconds resulting from changing cloud cover presents an even more serious challenge to the utility in terms of transient power flows. Major utility disruptions have generally involved the reaction of the grid to the loss of feeder capacity. If the loss of major PV generation occurs, can the remaining utility generation or storage make up for the PV loss?

And what if the utility is operating near peak capacity and a conventional generation facility goes off-line? In Section 3.8, IEEE 1547 will be discussed. This standard requires a PV system to shut down if utility power is lost. There are now some concerns that PV should be allowed to continue to operate to help stabilize the grid if utility generation is lost. All of the preceding concerns will become of greater interest as PV generation continues to become more cost-effective. Predictably, massive energy storage will play an important role in smoothing out the bumps on this road. In fact, these and other efficiency considerations will be involved in the continued evolution of the smart grid.

3.8 ASSOCIATED SYSTEM ELECTRONIC COMPONENTS

3.8.1 Introduction

This section introduces the basic electronic components of PV systems. These components include charge controllers, MPPTs, linear current boosters (LCBs), and inverters. All of these components handle relatively large amounts of power and are thus classified under the realm of power electronics. In each case, a simplistic explanation of the operation of the component will be given, with an emphasis on system performance requirements and how to best achieve them. Readers are encouraged to extend their understanding of these systems by consulting a power electronics text.

3.8.2 Charge Controllers

In nearly all systems with battery storage, a charge controller is an essential component. The charge controller must shut down the PV array when the battery is fully charged and may also shut down the load when the battery reaches a prescribed state of discharge. When the "battery" is really a system of batteries connected in series and parallel as needed to meet system needs, the control process becomes somewhat more of a challenge. The controller should be adjustable to ensure optimal battery system performance under various charging, discharging, and temperature conditions for lead-acid battery systems and may require more sophisticated individual cell charge control for lithium-ion battery systems.

Battery terminal voltage under various conditions of charge, discharge, and temperature has been presented in Figures 3.9 and 3.10 and Table 3.1. These results can be used to determine a Thevenin equivalent circuit for the battery system as shown in Figure 3.18. The key is that during charging, the battery terminal voltage, V_T, will exceed the battery cell voltage, V_B, since $V_T = V_B + IR_B$. During discharge, $V_T < V_B$, since under discharge conditions, $V_T = V_B - IR_B$ as the current direction is now reversed. The battery cell voltage is simply the battery open-circuit voltage.

The requirements for charging and discharging are made more complicated by the fact that the Thevenin equivalent circuit for the battery system is temperature dependent both for the open-circuit voltage and for the resistance. As temperature decreases, open-circuit voltage decreases [14] and resistance increases. Furthermore,

INTRODUCTION TO PV SYSTEMS

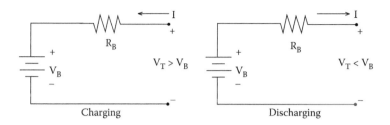

Figure 3.18 Thevenin equivalent circuit for a battery under charging and discharging conditions.

the Thevenin equivalent circuit for an old battery is different from that of a new battery of the same type. Hence, in order for a charge controller to handle all of these parameters, it should incorporate several important features. Depending upon the specific application, it may be possible to omit one or more of the following features.

3.8.2.1 Charging Considerations

First, consider the charging part of the process. Assume that the battery is fully charged when the terminal voltage reaches 15 V with a specific charging current. Assume also that when the terminal voltage reaches 15 V, the array will be disconnected somehow from the batteries and that when the terminal voltage falls below 15 V, the array will be reconnected. Now note that when the array is disconnected from the terminals, the terminal voltage will drop below 15 V, since there is no further voltage drop across the battery internal resistance. The controller thus assumes that the battery is not yet charged and the battery is once again connected to the PV array, which causes the terminal voltage to exceed 15 V, which causes the array to be disconnected. This oscillatory process may continue until ultimately the battery becomes overcharged or until additional circuitry in the controller senses the oscillation and decreases the charging current or voltage. It should be noted that if the charging current pulses on and off, in fact, its average value will decrease to the duty cycle of the pulsing times the peak value of the current pulses.

Figure 3.19 shows how the terminal voltage of a battery depends on the charge or discharge rate and the state of charge for a typical battery. Note, for example, that if the battery is charged at a C/5 rate, full charge will be reached at a terminal voltage of 16 V, whereas if the battery is charged at C/20, then the battery will reach full charge at a terminal voltage of 14.1 V. If the charging current is then reduced to zero, the terminal voltage will drop to below 13 V.

One way to eliminate overcharging resulting from the oscillatory process would be to reduce the turnoff set point of the controller. This, however, may result in insufficient charging of the battery. Another method is to introduce hysteresis into the circuit, as shown in Figure 3.20, so that the array will not reconnect to the batteries until the batteries have discharged somewhat. The reader who has been wondering what to do with the regenerative comparator circuit that was presented in an electronics course now may have a better idea of a use for this circuit.

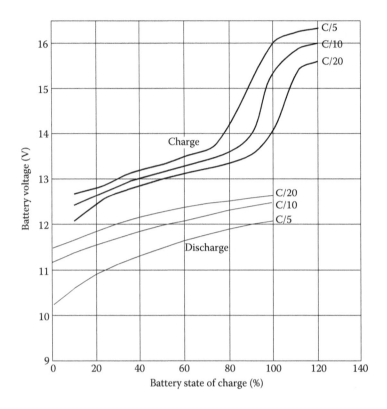

Figure 3.19 Terminal voltage as a function of charging or discharging rate and state of charge. (From *Maintenance and Operation of Stand-Alone Photovoltaic Systems*, Sandia National Laboratories, Albuquerque, NM, 1991.)

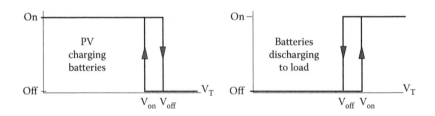

Figure 3.20 Hysteresis loops in charge controller using regenerative comparator for voltage sensing.

A more careful examination of Figure 3.19 suggests that an even better charging algorithm might be to initially charge at a relatively high rate, such as C/5. When the terminal voltage reaches about 15 V, indicating approximately 85% of full charge, the charging rate is then decreased, taking temperature into account, until the battery ultimately reaches 100% charge at a very low charging rate and a correspondingly

INTRODUCTION TO PV SYSTEMS

lower voltage. This method is employed in many of the charge controllers currently being marketed for use with PV systems.

Figure 3.21 shows the regions of charging associated with the algorithm suggested in the previous paragraph. Initially, the charge controller acts as a current source. If the charging mechanism is a PV array, then presumably full array current will be used for charging. This is the bulk stage. When the charging voltage reaches a preset level, the *bulk voltage*, the charging mode is switched to constant voltage, during which the charging current decreases nearly linearly. This is called the *absorption* stage. Note that sometimes the bulk voltage is referred to as the absorption voltage, since it is the voltage at which constant voltage charging begins. The absorption mode is continued for a time preprogrammed into the controller, after which the charging voltage is decreased to the *float* voltage. The float voltage is then maintained by the charge controller. The float voltage must be set to a level that will not result in damage to the battery.

In fact, since battery temperature affects battery terminal voltage and state of charge, modern charge controllers incorporate battery temperature sensor probes that provide temperature information to the controller that results in automatic adjustment of charging set points for the charging modes. Since battery cell voltage drops slightly with temperature decrease, but battery internal resistance increases with temperature decrease, the battery charging voltage is increased as temperature decreases to overcome the increased resistance and maintain the desired charging current levels.

A further mode that is available in modern chargers is the equalization mode. The equalization mode involves application of a voltage higher than the bulk voltage for a relatively short time after the batteries are fully charged. This interval of overcharging causes gassing, which mixes the electrolyte as a result of the turbulence caused by the escaping gases. This mixing helps prevent sulfate buildup on the plates and brings all individual cells to a full state of charge.

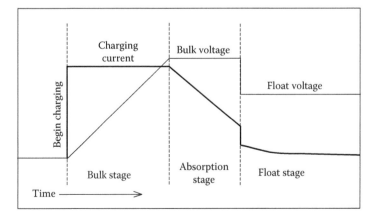

Figure 3.21 Three-stage battery charge control.

Only unsealed or vented batteries need equalization. For specific equalization recommendations, manufacturers' literature on the battery should be consulted. Some charge controllers allow for automatic equalization every month or so, but often it is also possible to set the controllers for manual equalization. Electrolyte levels should be checked before and after equalization since the gassing process results from decomposition of the electrolyte.

The battery disconnect may result in the array's being short-circuited, open-circuited, or, perhaps, connected to an auxiliary load that will use excess array energy. If the array is short-circuited to disconnect it from the batteries, the controller is called a shunt controller. Open circuiting the array is done by a series controller. One advantage of the shunt controller is that it maintains a constant battery terminal voltage at an acceptable level by bypassing enough charging current to achieve this result. The disadvantage is the amount of power that must be dissipated by the shunt and the heat sinking necessary to remove this heat from the shunt device.

3.8.2.2 Discharging Considerations

Now consider the discharge part of the cycle. Assume the battery terminal voltage drops below the prescribed minimum level. If the controller disconnects the load, the battery terminal voltage will rise above the minimum and the load will turn on again, and once again an oscillatory condition may exist. Thus, once again, an application for hysteresis is identified, and another regenerative comparator circuit is justified for the output of the controller.

At this point, all that remains is to make the set points of the charging regenerative comparator temperature sensitive with the correct temperature correction coefficient and the controller is complete. Of course, if the controller is designed to reduce charging current in order to bring the batteries up to exactly full charge before shutting down, and if the controller selectively shuts down loads to ensure the battery is optimally discharged, then the overall system efficiency will be improved over the strictly hysteresis-controlled system.

New designs continue to emerge for controllers as engineers continue their quest for the optimal design. Ideally, a charge controller will make full use of the output power of the PV array, charge the batteries completely, and stop the discharge of the batteries at exactly the prescribed set point, without using any power itself.

3.8.3 Maximum Power Point Trackers and Linear Current Boosters

Electronic MPPTs have already been mentioned in Section 3.6. LCBs are special-purpose MPPTs designed for matching the PV array characteristic to the characteristic of dc motors designed for daytime operation, such as in pumping applications. In particular, a pump motor must overcome a relatively large starting torque. If a good match between array characteristic and pump characteristic is not made, it may result in the pump operating under locked rotor conditions and may result in shortening of the life of the pump motor due to input electrical energy being converted to heat rather than to mechanical output.

INTRODUCTION TO PV SYSTEMS 83

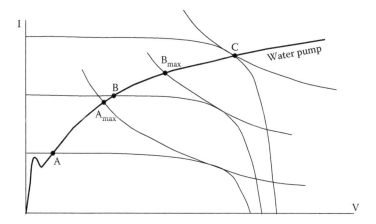

Figure 3.22 Pump and PV I–V characteristics, showing the need for use of LCB.

Figure 3.22 shows a typical pump I–V characteristic, along with a set of PV array I–V curves for different illumination levels. The fact that the pump characteristic is relatively far from the array characteristic maximum power point for lower illumination levels shows why an LCB can enable the pump to deliver up to 20% more fluid. The LCB input voltage and current track V_m and I_m of the PV array. The LCB output voltage and current levels maintain the same power level as the input, except for relatively small conversion losses, but at reduced voltage and increased current levels to satisfy the pump motor characteristic. The fact that the LCB increases current to the load accounts for the name of the device. In effect, the LCB acts as an electronic transformer for dc currents and voltages.

MPPTs and LCBs are generally adaptations of dc-to-dc switching voltage regulators, as indicated in Section 3.6. Coupling to the load for maximum power transfer may require providing either a higher voltage at a lower current or a lower voltage at a higher current. Either a buck–boost or a boost–buck conversion scheme is commonly used in conjunction with load voltage and current sensors tied into a feedback loop using a microcontroller to vary the switching times on the switching device to produce optimal output voltage.

The LCB is used in special cases where only a boost of current is needed. This means a decrease in voltage will accompany the current boost in order to keep output power equal to input power. Since only a decrease in voltage is required, no boost is needed in the converter. Hence, a simple buck converter with associated tracking and control electronics will meet the design requirements of the device. In fact, this is also the case for most MPPTs. They typically convert a higher array voltage and lower array current to a lower voltage and a higher current, with minimal loss of power in the conversion process. The efficiency of good MPPT devices is typically 98% or better.

Figure 3.23a shows a simplified diagram of a buck converter circuit. When the MOSFET is switched on, current from the PV array can only flow through the inductor into the parallel RC combination, where the capacitor voltage increases. When the

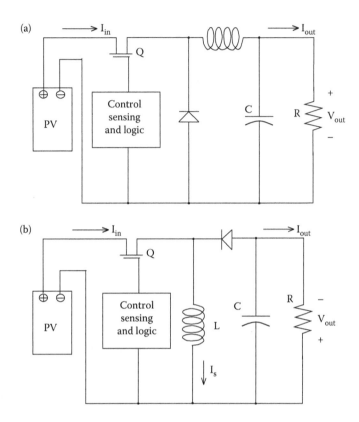

Figure 3.23 Maximum power point tracker and linear current booster. (a) LCB using buck converter. (b) MPPT using buck-boost converter.

MOSFET is off, current must remain flowing in the inductor, so the inductor current is now supplied by the capacitor through the diode, causing the capacitor to discharge. The extent to which the capacitor charges or discharges depends upon the duty cycle of the MOSFET. If the MOSFET is on continuously, the capacitor will charge to the array voltage (V_{in}). If the MOSFET is not on at all, the capacitor will not charge at all. In general, the output voltage and current of an ideal buck converter are given by

$$V_{out} = DV_{in} \tag{3.16}$$

and

$$I_{out} = I_{in}/D, \tag{3.17}$$

where D is the duty cycle of the MOSFET, expressed as a fraction ($0 < D < 1$). Note that the polarity of the output voltage is the same as the polarity of the input voltage, so there is no problem keeping the same grounded conductor at the input and the output of the converter.

INTRODUCTION TO PV SYSTEMS 85

Figure 3.23b shows the basic elements of a buck–boost converter that might be used in an MPPT. Note that in this system, the polarity of the output voltage is opposite the polarity of the input voltage. This can be a problem if a common negative (grounded) conductor is required for the PV circuit and the output of the converter. In this case, the circuit needs to be somewhat more sophisticated. On the other hand, if, for example, the PV input required a grounded positive conductor and the controller output required a grounded negative conductor, this switch of grounds can be achieved with this design.

The output voltage and current of the ideal buck–boost converter are given by

$$V_{out} = \frac{D}{1-D} V_{in} \qquad (3.18)$$

and

$$I_{out} = \frac{1-D}{D} I_{in}, \qquad (3.19)$$

where, again, D is the duty cycle of the MOSFET. Note that if $D < 0.5$, $V_{out} < V_{in}$, and if $D > 0.5$, $V_{out} > V_{in}$. Hence, the MPPT is capable of either increasing or decreasing its output voltage in order to track an array maximum power point. Of course, if the MPPT output voltage decreases, then its output current will increase, and vice versa, such that, in the ideal case, the output power will equal the input power.

During the time Q is on, energy is stored in the inductor. When Q turns off, the inductor current must continue to flow, so it then flows through R and C and the diode, charging C to V_{out}, since the capacitor voltage equals the output voltage. When Q turns on again, the diode becomes reverse biased, and current is built up again in the inductor while the capacitor discharges through the resistor. Finally, when Q turns off, the cycle repeats itself. The values of L and C and the switching frequency determine the amount of ripple in the output voltage. Since no energy is lost in ideal inductors and capacitors, and since Q and the diode approximate ideal switches, essentially all power extracted from V_{in} must be transferred to the load. Of course, in reality, these components will have some losses, and the efficiency of the MPPT will be less than 100%. However, a well-designed MPPT will have an overall efficiency greater than 95%, with many units currently being marketed having advertised efficiencies that are close to 98% [15,16].

Another application of the MPPT is to ensure optimal charging of batteries. The MPPT charge controller electronically tracks the PV array maximum power point to ensure that maximum charging current is delivered to the battery bank. The result is a charge controller with high input voltage, low input current, lower output voltage, and higher output current. Charge controllers without MPPT input circuitry connect the PV array directly to the battery voltage, while MPPT charge controllers operate the array at its maximum power point for charging the batteries at the proper charging voltage, as shown in Figure 3.24.

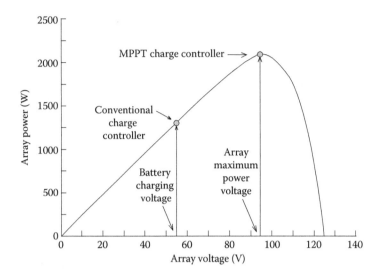

Figure 3.24 Comparison of PV array operating points for conventional and MPPT charge controllers.

By designing the array to have a higher maximum power voltage, smaller wiring can be used between the array and charge controller, with lower power loss in the wiring and lower wiring cost. Note that with the PV array of Figure 3.24, the MPPT charge controller will deliver about 2100 W to the batteries, while the conventional controller will deliver only about 1300 W to the batteries.

In terms of charging current, the MPPT charge controller delivers approximately $2100/56 = 37.5$ A, while the conventional charge controller delivers only $1300/56 = 23.2$ A. In fairness to the conventional charge controller, however, one should admit that normally with a conventional charge controller, the array would be designed with its maximum power point closer to 60 V, so the battery charging voltage will be closer to the array maximum power point voltage.

At the time of this writing, essentially all charge controllers and inverters used in grid-connected PV systems used MPPT. The array-to-battery efficiency of good MPPT charge controllers is typically close to 98%. Problem 3.14 explores this concept further.

3.8.4 Inverters

Depending on the requirements of the load, a number of different types of inverters are available. Selection of the proper inverter for a particular application depends on the waveform requirements of the load and on the efficiency of the inverter. Inverter selection will also depend on whether the inverter will be a part of a grid-connected system or a stand-alone system. Even though warrantees are now typically 10 years on PV inverters, many opportunities still exist for the design engineer to improve on inverters.

INTRODUCTION TO PV SYSTEMS

Table 3.5 Summary of Inverter Performance Parameters

Parameter	Square Wave	Modified Sine Wave	Pulse Width Modulated	Sine Wave[a]
Output power range (W)	Up to 1,000,000	Up to 6000	Up to 800,000	Up to 1,500,000
Surge capacity (multiple of rated output power)	Up to 20×	Up to 4×	Up to 2.5×	Up to 4×
Typical efficiency over output range	70%–98%	>95%	>90%	>97%
Harmonic distortion	Up to 40%	>5%	<5%	<5%

Source: *Maintenance and Operation of Stand-Alone Photovoltaic Systems*, Sandia National Laboratories, Albuquerque, NM, 1991.
[a] Multilevel H-bridge or similar technology to yield utility grade sine wave output.

Table 3.5 [17,18] summarizes inverters presently available. Inverter performance is generally characterized in terms of the rated power output, the surge capacity, the efficiency, and the harmonic distortion. Since maximum efficiency may be achieved near rated output, it is important to consider the efficiency versus output power curve for the inverter when selecting the inverter. Certain loads have significant starting currents, so it is important to provide adequate surge current capacity in the inverter to meet the load surge requirements. Other loads will either overheat or introduce unwanted noise if the harmonic distortion of their power supply is not below a specific level.

In general, the square wave inverter is the least expensive and is relatively efficient, but has limitations on its applications. It has the best surge capacity but the highest harmonic distortion. The modified sine inverter is more complicated, but still relatively efficient. The PWM inverter has higher cost, high efficiency, and minimal distortion. The pure sine inverter, with digitally generated waveform, has the least distortion, and can have efficiencies in excess of 97%.

3.8.4.1 Square Wave Inverters

The simplest inverters are square wave inverters. These inverters employ solid-state switches connected as either astable multivibrators or as externally controlled switches. Use of astable multivibrators enables the inverter to be used in a self-contained PV system, whereas externally synchronized switches are used when the inverter is to be synchronized with an external ac power source. Figure 3.25 shows how a single source can be switched alternatively in a positive direction, then in a negative direction to produce a square wave. This configuration is called an "H-bridge."

An important feature of Figure 3.25 is that *there is no common negative terminal between dc input and ac output*. If a common ground is desired between the dc input and the ac output, it is necessary to couple V_o via a transformer to the inverter output. As long as the transformer primary and secondary windings are isolated, this enables the connection of one side of the secondary winding to a common dc/ac

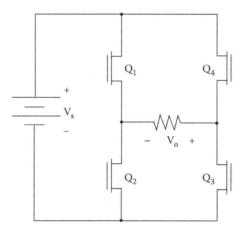

Figure 3.25 Converting a dc source to a square wave.

ground. Furthermore, it allows the selection of a transformer to step the inverter bridge circuit output voltage either up or down. The transformer can also assist in filtering the output waveform to remove harmonics.

The disadvantage of transformer coupling is in the cost of the transformer, the weight of the transformer, and the losses encountered in the transformer. In the last few years, with the continued improvement in the design and control of multilevel H-bridges, inverter design has been transitioning away from transformer coupling. This requires that the dc input to the bridge must be floating, which is accomplished by using an ungrounded array. This practice will be further discussed in Chapter 4.

The key to efficient switching is to have current flowing from the array at all times while maintaining 0 V across the switching element when it passes current, and zero current through the switching element when it has voltage across it. This can be approximated reasonably well with insulated gate bipolar transistors, power MOSFETs, or silicon-controlled rectifiers. Once the dc is converted into a square wave, its amplitude normally needs to be increased to produce a 120-V rms ac waveform. Since the rms value of a square wave is simply the amplitude of a square wave, a system with a 12-V_{dc} input will need a transformer with a 10:1 turns ratio.

It is important, however, to realize that the transformer must be designed with a sufficient number of turns so that the time constant determined by the magnetizing inductance of the transformer and the source resistance will be long enough to maintain the square wave. Too few turns on the transformer will cause the output waveform to droop as shown in Figure 3.26. When specifying the transformer for this application, it is easy to assume that transformers are ideal and to simply try to use a 120/12-V transformer backwards. The 120/12-V transformer, however, probably would have been designed for use with a sinusoidal signal, and thus would not have enough turns to handle the square wave effectively. A square wave inverter with a good transformer and efficient switching can operate at efficiencies in the 90% range.

INTRODUCTION TO PV SYSTEMS

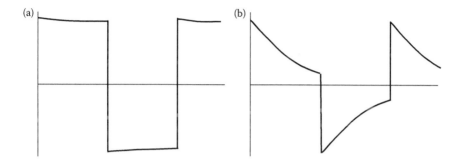

Figure 3.26 Output waveforms for a square wave inverter with adequate and with inadequate transformer turns. (a) Adequate. (b) Inadequate.

3.8.4.2 Modified Sine Wave Inverters

For a number of applications, a square wave is inadequate for meeting the harmonic distortion requirements of the load. For example, since square waves have significant harmonic content, and since hysteresis and eddy current losses in magnetic materials increase significantly with increase in frequency, square wave excitation may cause some motors or magnetic fluorescent ballasts to overheat.

Square wave harmonics can also introduce noise into a system. Thus, before selecting an inverter, it is important to verify that the proposed load will operate with square wave excitation.

If square wave excitation is not suitable for a load, it is possible that a modified sine wave will work. A number of methods are available to convert the output of a dc source into some approximation of a sine wave. One such method involves using a multilevel H-bridge as shown in Figure 3.27a to generate a waveform such as the one shown in Figure 3.27b. The idea is to time the individual voltage levels so that harmonic distortion will be minimized and that the rms value of the output voltage will remain constant in the event that the input dc voltage should vary.

The sequence for closing switches to obtain portions of the waveform in Figure 3.27b is shown on the waveform diagram. The actual switches may be bipolar transistors, MOS transistors, SCRs, or insulated gate bipolar transistors. As an example of how the various voltage levels in V_o are obtained, consider the interval T_1, where, according to the output voltage waveform, $V_o = 0.5V_{dc}$. If S_1 is closed, diode D_1 becomes reverse biased and appears as an open circuit. If S_2 is closed, then the positive terminal of V_o is connected to $0.5V_{dc}$. Closing S_7 provides a path to ground from the negative terminal of V_o through forward biased D_4. Hence, if voltage drops across diodes and electronic switches are neglected, $V_o = 0.5V_{dc}$. Next, consider the interval where V_o is shown as $-V_{dc}$. During this interval, when S_3 and S_4 are closed, D_3 becomes reverse biased and appears as an open circuit and the positive terminal of V_o is thus connected to $-0.5V_{dc}$. Closing S_5 and S_6 reverse biases D_2 and connects the negative terminal of V_o to $+0.5V_{dc}$. Hence, $V_o = -V_{dc}$. It is left as an exercise for the reader to verify the remaining switching combinations.

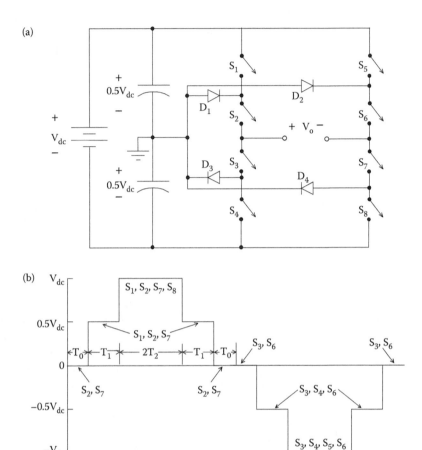

Figure 3.27 Multilevel H-bridge modified sine wave inverter. (a) Switching configuration. (b) Output voltage waveform.

Noting the symmetry of V_o, and recognizing that $T_0 + T_1 + T_2 = 0.25\,T$, where T is the period of the waveform, it is possible to determine the rms value of V_o from the first quarter cycle by solving

$$V_{rms} = \left[\frac{4}{T}\int_0^{T/4} V_o^2(t)dt\right]^{1/2} = \left[\frac{4}{T}\left\{\int_0^{T_0} 0\,dt + \int_{T_0}^{T_0+T_1}(0.5V_{dc})^2 dt + \int_{T_0+T_1}^{T_0+T_1+T_2} V_{dc}^2 dt\right\}\right]^{1/2}$$

$$= V_{dc}\left[\frac{T_1}{T} + \frac{4T_2}{T}\right]^{1/2}. \tag{3.20}$$

As an example, assume that $V_{dc} = 160$ V, and it is desired to produce $V_{rms} = 120$ V by keeping $T_0 = T_1$ and solving for T_2. The result is that $T_0 = T_1 = T/16$ and $T_2 = T/8$.

Problem 3.18 offers the reader a chance to solve for T_0, T_1, and T_2 for different values of V_{dc}. Note that since $T_1 + 4T_2 < T$, if $V_{dc} < V_{rms}$, there is no solution for T_0, T_1, and T_2.

H-bridges can be designed with $4n + 1$ levels, where n is an integer. The more levels, the closer V_o can be made to approximate a sine wave. In addition, the voltage drop across any individual switch is reduced as the number of levels is increased. Problem 3.21 involves the design of a nine-level H-bridge.

The design challenge in modern inverters is to maximize the range of input voltages over which the output of the multilevel H-bridge maintains a constant rms value with minimal distortion. The next design constraint, of course, is to ensure that the inverter input will track the PV array maximum power point and, finally, the inverter should have an efficiency close to 100%. Modern inverters generally have MPPT input voltage ranges of nearly 300 V between maximum and minimum tracking voltages.

Many computer uninterruptible power supplies produce a five-level modified sine wave, so it is safe to say that this waveform is acceptable for use in providing backup ac power to a computer. However, IEEE Standard 1547–2003 [19] requires that any source connected to the utility line must have less than 5% total harmonic distortion (THD). For those who haven't memorized the formula for THD from their electronics textbook, recall that THD is the percentage ratio of the sum of the rms values of all the harmonics above the fundamental frequency to the rms value of the fundamental frequency. It is unlikely that a five-level modified sine wave inverter will meet the 5% THD rule, especially if V_{dc} is allowed to vary while keeping V_{rms} constant. This is the reason for the existence of pure sine inverters. Using higher level H-bridges to approximate pure sine waves is one method. Another method is to use pulse width modulation techniques.

3.8.4.3 PWM Inverters

The PWM inverter produces a waveform that has an average value at any instant equivalent to the level of a selected wave at that instant. PWM inverters are perhaps among the most versatile of the family of inverters. They are similar to the PWM described in the discussion of the MPPT with the exception that the MPPT PWM signal is designed to have a constant average value to produce a regulated dc output, while the inverter PWM signal is designed to have a time-dependent average value that can have any arbitrary waveform at any arbitrary frequency at any arbitrary amplitude. For use in PV applications, it is generally desirable to have a sinusoidal waveform with a predictable amplitude and frequency.

Figure 3.28 shows how a PWM waveform can produce waveforms of differing amplitudes and frequencies by controlling the on-and-off time of a pulse waveform. The waveform is controlled by controlling the relative duty cycle of successive pulses. The amplitude is controlled by controlling the overall duty cycle, and the frequency is determined by controlling the repetition time for the pulse sequence.

By switching the pulse between a positive level and a negative level, it is possible to construct waveforms with zero average value, which is particularly important

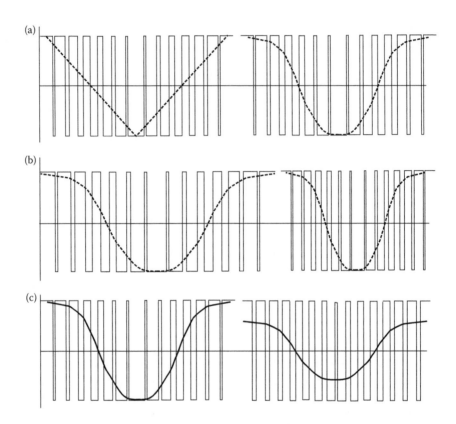

Figure 3.28 PWM control of waveform, frequency, and amplitude. (a) Controlling waveform. (b) Controlling frequency. (c) Controlling amplitude.

when driving loads for which a dc component in the excitation may cause losses in the load. This includes applying dc to an ac motor or to a transformer. Application of dc to such loads can result in significant I^2R heating with possible failure of the winding insulation and subsequent catastrophic failure of the motor or the transformer, including the possibility of fire. For this reason, elimination of any dc component in the output of an inverter is important regardless of the type of inverter.

PWM inverters are particularly useful when used as ac motor controllers, since the speed of an ac motor can be controlled by adjustment of the frequency of the motor excitation. It must be recalled, however, that the flux developed in the motor is inversely proportional to the excitation frequency, and that every motor has a saturation flux above which the motor draws significant current without further flux increase. Hence, as a motor excitation frequency is decreased, the peak value of the applied voltage must also decrease proportionally to keep the motor out of saturation. A PWM controller coupled with MPPT capability can be a very efficient means of maximizing the efficiency of a PV pumping system.

Figure 3.29 shows one means of generating a PWM waveform. The process begins with a triangle wave at the frequency of the pulse waveform applied to one

INTRODUCTION TO PV SYSTEMS   93

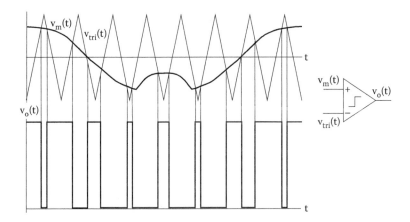

Figure 3.29 Generation of a PWM waveform.

input of a voltage comparator. Next, the desired waveform is applied as a modulating waveform at the other input of the comparator. Whenever the modulating signal exceeds the triangle waveform, the comparator output goes high, and whenever the modulating signal is less than the triangle waveform, the comparator output goes low. The comparator output is then used to switch the output pulse on and off. The result is an output PWM waveform that has a moving average value proportional to the modulating signal. The average value of the PWM output is obtained in a manner similar to the detection process used to pick off the modulation signal of an amplitude modulated communications signal. A filter needs to be incorporated that will not allow the voltage (or current) of the load to change instantaneously, but will allow it to change quickly enough to follow the moving average of the PWM waveform. This can be achieved with inductive and/or capacitive filtering, depending on whether it is desired to smooth the current, the voltage, or both.

The harmonic content of the two-level PWM waveform can be significantly reduced through the use of a three-level PWM waveform. A three-level PWM waveform can be generated by incorporating a second comparator into the circuit, so that one comparator will control the positive-going pulse when the modulating signal is positive and the other comparator will control the negative-going pulse when the modulating signal is negative. The challenge here, however, is to switch the PV input so the array operates at maximum power voltage if the inverter is powered directly from the array. If a storage mechanism is used for array current, so the array is continuously supplying dc power to the storage mechanism, then the storage mechanism can be switched on and off to meet the load requirements without interrupting the power flow from the array. Alternatively, if the array voltage is switched between V_{OC} and some voltage less than V_{mp}, then the duty cycle of the connection to the array can be varied such that the average value of the array voltage seen by the inverter is V_{mp}. Figure 3.30 shows a three-level PWM inverter output signal derived from dc input from battery storage.

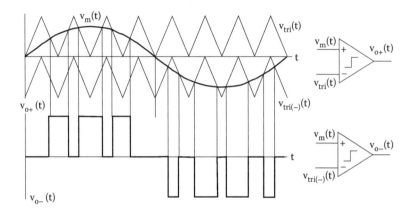

Figure 3.30 Three-level PWM inverter configuration and output signal.

Figure 3.31 shows a PWM inverter circuit using a full-bridge switching network for the dc source, including MPPT. This is an adaptation of the full-bridge that was shown for the generation of a square wave. The inverter can be fixed frequency, line commutated (synchronized), or variable frequency, depending on the source of the modulation signal. The controller generates the triangle waveform and the modulating waveform, v_m, with the amplitude and/or frequency of v_m determined by the PV array output voltage and current as sensed by the MPPT. Switches Q_1 and Q_3 of the bridge are switched on during the positive pulse excursions and Q_2 and Q_4 are switched on during the negative excursions. This provides a continuous power drain on the PV array, thus enabling MPPT.

Keeping the switching elements as close to ideal as possible, along with minimal loss in the filtering elements, can yield very high overall efficiencies for these units. When a switching element has either zero voltage or zero current, it does not dissipate power. During a switching transition, however, switching elements have both voltage

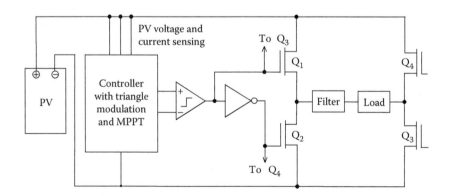

Figure 3.31 Full-bridge PWM converter showing controller and PV connections.

and current and, hence, dissipate power. The trade-off in a PWM inverter, then, is to have the frequency as high as possible to give adequate waveform reproduction, but not so high as to increase switching losses to unacceptable levels. Most manufacturers now advertise PWM inverters with efficiencies above 90% over a wide range of inverter output power.

3.8.4.4 Transformerless Inverters

A closer look at the multilevel H-bridge inverter schematic (Figure 3.27a) shows that there is no common connection between the input and output that can be grounded without shorting out at least two of the switches. This means that if a common ground is desired between input and output, the output must be isolated from the switching network output with a transformer connected across V_O. The transformer primary winding is connected across V_O and then either the transformer can have a grounded center-tapped winding to generate a dual-voltage output, such as 120/240 V, or the secondary can have just two terminals, one of which can share the ground connection with the input. Other convenient features of a transformer is that it tends to smooth out the high frequency components of the inverter output and is capable of stepping the output voltage either up or down, depending upon the dc input voltage of the inverter and the corresponding rms value of the output of the multilevel H-bridge.

But there are inconveniences to having a transformer, not the least of which is the cost of the transformer as well as the inconvenience of the additional weight of the transformer. Perhaps even more important is the observation that the transformer losses are close to half the total losses of an inverter. So, for example, a 10,000-W inverter with a transformer that operates at 96% efficiency will lose 400 W at full output, whereas a transformerless inverter of the same size will lose only about 200 W. So the challenge is to figure out how to eliminate the transformer.

Fortunately, *NEC* 690.35 already allows for ungrounded PV source circuits and output circuits, provided that they meet the conditions of Sections A–G. Section G allows PV arrays to be ungrounded provided that the inverter is listed for use with ungrounded dc input conductors. Note that even though the current-carrying conductors of the array are ungrounded, all the rest of the metal parts of the array (module frames, module mounts, etc.) must be separately grounded with equipment grounding conductors.

There remains a challenge for the transformerless inverter designer to provide an acceptable range of operating voltages and currents. Fortunately, this challenge has been met with typical input voltage ranges of 125–600 V and MPPT tracking range of 175–480 V and output voltage of either 208 or 240 V_{ac}.

3.8.4.5 Other Desirable Inverter Features

Modern inverters for PV applications often incorporate very sophisticated features to optimize inverter performance. In fact, the instruction and installation manuals for some of these inverters are nearly as long as this textbook [20,21].

For stand-alone inverters and for grid-connected inverters with stand-alone option, a "search" mode is often incorporated into the design. In the search mode, the inverter uses minimal energy to keep its electronics operational when no loads are connected. In the search mode, approximately once every second, the inverter sends out an ac voltage pulse a few cycles long. If the inverter senses that no current flows as a result of the pulse, it concludes that no load is connected and waits to send another pulse. On the other hand, if current flows at a magnitude greater than a preprogrammed value, the inverter then recognizes that a load has been connected and supplies a continuous output voltage. If the inverter is not producing an output voltage on a continuous basis, power losses in the inverter are minimized.

In some cases, however, an inverter in the search mode can be confused. For example, if the inverter is set to provide power as long as more than 25 W is connected, it can be confused by a 20-W incandescent light bulb. The reason is that when the lamp filament is cold, it has a lower resistance and a surge of current flows until the filament is heated. Hence, a 20-W incandescent light bulb will appear to be a higher wattage load when it first turns on, and then will drop back to 20 W when the filament is hot. This, then, will cause the inverter to turn off and resume the search mode, where it will find the same light bulb and will turn it on again until it gets hot, resulting in a light that flashes on and off. Other confusing loads include the remote receivers in electronic equipment such as televisions, DVD, and blue-ray players. When they are on, they draw very small amounts of power, so when an inverter is in the search mode, it is quite possible that the remote receivers will shut down.

For stand-alone inverters, it is necessary that the output appear as a voltage source, preferably as close to ideal as possible. For utility-interactive inverters, however, it is more convenient to let the grid voltage fulfill the voltage source role and to have the inverter act as a current source that feeds current into the utility grid. Utility-interactive inverters with battery backup that are capable of supplying power to standby loads if the utility grid is disconnected must be capable of switching over from grid-synchronized current source to internally synchronized voltage source to power standby loads.

Utility-interactive inverters must be designed so that if the utility goes down, the inverter output to the utility also shuts down. This safety feature is relatively easy to design into an inverter if no other distributed electrical source is connected to the utility. However, if another nonutility source is connected to the line, it is important that the PV system inverter will not recognize this source as a utility source and continue to supply power to the line. This is called "islanding" and will be discussed in detail in Chapter 4.

Many utility-interactive inverters that have a standby backup feature also incorporate a battery charger in the inverter so that the utility can be used to charge the batteries in the event that the PV system has not fully charged the batteries. This feature only works, of course, when the utility grid is energized. Fortunately, this is generally most of the time.

Another convenient feature found in many inverters is a generator start option. This allows for a separate ac generator to be used as a backup to the PV system in either a utility-interactive system with backup or a stand-alone system. When the battery level of charge drops to a prescribed level, the inverter will send a starting signal to the generator. If the inverter has a real-time clock, the time of day for generator starting can also be controlled, so the generator will not come on just before the sun is about to charge the batteries.

Essentially all straight grid-connected inverters that are connected directly to a PV array incorporate MPPT circuitry at their dc input so that the PV array will deliver power at its maximum power level to the inverter. Ground fault detection and interruption (GFDI) and arc fault circuit interrupter (AFCI) are often incorporated integral to the dc input circuitry of straight grid-connected inverters. Both will be discussed in more detail in Section 3.10.2.

3.9 GENERATORS

3.9.1 Introduction

In some cases, where either a large discrepancy between seasonal loads exists or where seasonal sun availability varies greatly, a stand-alone system designed completely around PV components will result in the deployment of a large PV array to meet the needs of one season. Meanwhile, during other seasons, much of the energy available from the array is not used. This is similar to the problem of meeting critical system needs with PV, where the cost generally increases rapidly as the system availability exceeds 95%. In such cases, it is often more cost-effective to employ an alternate source of electricity to be available when the PV array is not meeting system needs. While it is conceivable that the backup source for a stand-alone system may be wind or other renewable source, it is more common to employ a gasoline, diesel, or propane generator as a system backup. In a grid-connected system, the utility grid provides any needed backup power, unless, of course, the grid goes down.

With relatively low acquisition costs of small and portable generators, it may appear that it would make better economic sense to simply use the generator without the PV array. However, life-cycle cost analysis, as is discussed in Chapter 8, often shows the use of a mix of PV and conventional generation to be more economical than an engine-powered generator. Furthermore, the fact that PV generation is quiet and clean adds further appeal to using a maximum practical amount of PV generation in the system.

When it makes economic sense to use a generator, it is important for the engineer to have some knowledge of generator options. The next section deals with some of the considerations in selecting a generator.

3.9.2 Types and Sizes of Generators

While most small electrical generators use a gasoline engine as the mechanical prime mover, methane-, propane-, and diesel-powered engines are also available.

Table 3.6 Common Generator Sizes, Features, and Approximate Costs

Max kW	Engine HP	Duty Rating	Voltage Regulation	Noise Level (dBA)	≈kWh/gallon at Rated Load	Approx Cost (USD)
2.5	5.5	Economy	Yes	74	≈4.0	510
5.0	8.0	Economy	Yes	77	≈4.0	670
5.5	8.0	Deluxe	Yes	82	≈5.4	1425
8.0	16.0	Deluxe	Yes	76.5	≈4.6	1890
10	20	Deluxe	Yes	76	≈4.8	2475
5.5	10	Industrial	Yes	77	≈5.6	2455
6.0	10	Industrial	Yes	77	≈6.9	2475
13	19	Industrial	Yes	81	≈7.2	5140

Source: http://www.electricgeneratorsdirect.com/power/honda-inverter-generators.html?gclid=ClyRh_HW-ssCFcUmhgodt-0KrA.

A number of factors will enter the selection process, including initial cost, power requirements, fuel availability, and maintenance requirements.

While it is possible to obtain dc generators for use in battery charging applications, dc generators are generally not recommended because of maintenance requirements. They have brushes and commutators that require frequent replacement and cleaning. Many years ago, automotive batteries were charged with dc generators, but now all automotive batteries are charged with alternators, which are simply ac generators that are connected to rectifiers to convert their output to dc. Similarly, ac generators are used in PV systems, even if the systems are dc systems, to minimize maintenance costs.

Table 3.6 [22] lists a few common generator sizes and approximate costs on the basis of recent data. The smaller units are normally used as portable units, while larger sizes are intended for use as fixed units. Rated output is generally about 90% of maximum output.

3.9.3 Generator Operating Characteristics

Factors other than peak power output will enter the selection process if a generator is carefully chosen. Generator specifications also include rotation speed, efficiency, fuel type, altitude effects, waveform harmonic content, frequency stability, noise levels, type of starting, and overload characteristics.

3.9.3.1 Rotation Speed

Normally, 60-Hz generators operate at either 3600 or 1800 rpm. Corresponding speeds for 50-Hz generators are 3000 and 1500 rpm. The 3600 rpm units are two-pole machines and are of simpler construction, resulting in lower acquisition cost. The 1800-rpm machines are four-pole machines and are somewhat more expensive, but more common in the larger sizes or heavy duty units.

INTRODUCTION TO PV SYSTEMS

In general, the higher the rpm, the more wear and tear on the bearings, which means more frequent maintenance requirements. Two-pole generators are thus most convenient for use in relatively light-duty applications that require less than 400 hours per year of operation. Four-pole generators are recommended when more than 400 hours of operation per year are anticipated.

To increase efficiency at low load levels, some generator manufacturers [23] have incorporated inverters into some of their generators. The advantage of this design is that the load on the inverter determines the amount of mechanical power needed to meet the electrical load requirements. Since the interface between rotating generator and inverter output involves conversion of the generator ac output to dc at the inverter input, operation of the generator at a lower speed with a correspondingly lower ac output frequency is acceptable, since the rectifier will rectify at any frequency. As a result, the generator is controlled to operate at variable engine speeds up to maximum engine horsepower. This significantly reduces fuel consumption at low load levels and, as a result, increases the overall efficiency of the generator. In addition, these models tend to run more quietly than conventional fossil fuel generators.

3.9.3.2 Efficiency versus Electrical Load

Electrical and mechanical losses are present in all generators. However, the greatest losses in a generator system are attributable to the prime mover engine. Since the prime mover and electrical generator will each generally have a particular load at which they will operate at maximum efficiency, manufacturers endeavor to carefully match the two components to produce maximum efficiency at somewhere between 80% and 90% of rated full electrical load. Figure 3.32 shows approximate plots of efficiency versus percentage of rated electrical load and kWh per gallon versus percentage of rated electrical load for typical small generators compared to small inverter/generator models [20,21].

As generator size increases, the overall maximum efficiency also increases. Figure 3.33 shows maximum efficiency versus generator size [20]. Whether or not

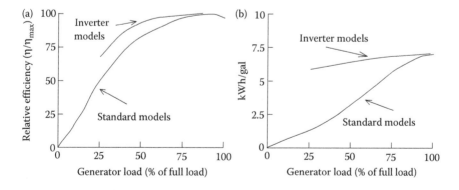

Figure 3.32 Two methods of characterizing generator efficiency. (a) Overall efficiency versus rated load. (b) Fuel consumption versus rated load.

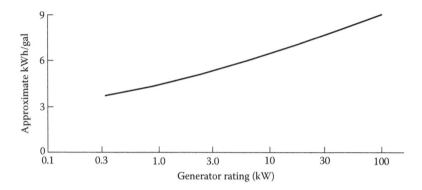

Figure 3.33 Approximate efficiency versus generator size at rated output.

to use a larger, more expensive generator, operating at higher efficiency for shorter intervals, presents an interesting design challenge to the design engineer.

Maximum generator size is limited in most cases by maximum allowable charging rates for the system batteries, assuming the generator is incorporated into a system with batteries. For systems with highly variable electrical loads, it is particularly inefficient to incorporate a generator unless battery storage is provided to present a nearly constant load on the generator or a generator/inverter is used rather than a conventional generator. In either case, it is more efficient to charge batteries at maximum generator efficiency and then turn off the generator.

3.9.3.3 Fuel Types

Choice of fuel for a generator will depend on several factors. Although diesel generators tend to be the most efficient, diesel fuel is sometimes more difficult to obtain and is more likely to become contaminated with water or other impurities than other fuel choices. In cold temperatures, diesel engines tend to be more difficult to start, so they need to be used under adequately controlled temperature conditions if they are to be started and stopped. Diesel is probably not the best choice for remote locations for these reasons, especially if the site is unattended and reliability is important.

Gasoline is probably the easiest fuel to obtain and will be the fuel of choice for many applications, since gasoline engines can be started and run under relatively adverse conditions. In some situations, however, when either natural gas or propane is used on site for other purposes, such as cooking or space heating, it is convenient to also use it as a fuel for backup electrical generation. This eliminates the need to keep track of more than one supply of fuel, and, in standby operations, such as during utility outages, it is often very inconvenient to obtain replacement supplies of gasoline. However, as will be noted in the next section, natural gas efficiency decreases more rapidly with increase in altitude than gasoline or propane efficiency, so this observation may need to be taken into account.

INTRODUCTION TO PV SYSTEMS 101

3.9.3.4 Altitude Effects

Many engineers have observed that their automobiles do not respond as well at high altitudes as they do at lower altitudes. As altitude increases, less air is available for the combustion process, and combustion engines thus degrade in performance with increase in altitude.

The design engineer should first check manufacturers' specifications for altitude derating on generators. If manufacturers' specifications do not list derating factors, then it is reasonable to derate gasoline, diesel, and propane generators by 3% per 1000 ft of altitude above sea level and to derate natural gas engines by 5% per 1000 ft of altitude above sea level.

3.9.3.5 Waveform Harmonic Content

Generally the output waveform of an electrical generator is adequate for nearly all applications. For battery charging, almost any waveform is satisfactory, depending on whether the battery charger contains a transformer.

If the charger contains a transformer, there is a possibility that excessive dc or harmonic content in the generator waveform may damage the transformer. If the generator is connected directly to the PV system ac loads at any time, then it is a good idea to be sure the generator output waveform is sufficiently "clean" to meet the requirements of the loads. Normally, the only loads with possible sensitivity to waveform quality will be electronic systems, magnetic fluorescent light ballasts, and some motors.

3.9.3.6 Frequency Stability

Again, depending on load requirements and whether the generator is connected to ac loads with critical frequency requirements, the frequency stability of a generator may need to be taken into account. It is generally desirable to maintain frequency fluctuations at less than ±0.5 Hz for ac loads, but this degree of frequency stability is not necessary for many PV system loads. Conventional generators often have specified frequency stability of 5%.

3.9.3.7 Amplitude Stability

It is generally desirable to have the generator output voltage amplitude remain within 5% of its no-load value. This implies a VR figure of 5%. Depending on whether the load is inductive or capacitive, the load voltage under full-load conditions may actually exceed the no-load voltage. The VR of the generator is defined by

$$VR = \frac{V_{nl} - V_{fl}}{V_{fl}}, \qquad (3.21)$$

where V_{nl} and V_{fl} are the no-load and full-load voltages, respectively. Hence, if $V_{fl} > V_{nl}$, VR will be negative.

Again, VR requirements will depend on the specific application. If the PV system is designed to operate the generator at maximum efficiency, the load will remain nearly constant and the output amplitude and frequency will remain quite stable.

3.9.3.8 Noise Level

Some generators are noisy and others are less noisy. Local ordinances should be checked, but normally the noise level demands of the user will be more stringent than local ordinances. National parks, for example, have relatively strict noise regulations.

3.9.3.9 Type of Starting

If the generator is to be controlled by the system controller, then it will need to be equipped with electric starting. Otherwise, it may be satisfactory to use a generator with manual starting.

3.9.3.10 Overload Characteristics

A synchronous generator can suffer serious consequences if it is overloaded. Overloads can occur with the application of excessive steady-state loads, but also can occur as the result of transient loads, such as motor starting.

For example, if the generator is to operate an electric motor, the starting current of the motor can place the generator in an overload condition that can cause the generator to slow down, resulting in reduced generator output. This, in turn, can result in insufficient starting current for the motor and slow the generator even more. This can result in damage to the prime mover and possibly to the generator armature winding and lead to an expensive repair bill.

3.9.3.11 Power Factor Considerations

The generator must be able to meet the real power requirements of the load as well as the reactive power requirements of the load. Since I^2R copper losses in the generator armature winding depend on the magnitude of the armature current, generators are rated in kVA rather than in watts. It is also possible that the generator will have a minimum power factor rating under full-load conditions. If so, the minimum power factor anticipated for the load must also be considered when the generator is selected.

3.9.4 Generator Maintenance

Table 3.7 shows cost and maintenance information on 3600-rpm gasoline generators, 1800-rpm gasoline generators, and diesel generators. Note that gas includes gasoline, natural gas, and propane. Note also that "small" is defined as sizes less than 100 kW. Diesel generators up to 5 MW are not uncommon, but are not typically intended for home or recreational vehicle use.

INTRODUCTION TO PV SYSTEMS

Table 3.7 Cost and Maintenance Information on Small Electrical Generators

Type	Size Range (kW)	Application	Cost ($/W)	Oil Ch (h)	Tune-Up (h)	Engine Rebuild (h)
3600 rpm gas	1–20	Light use	0.50	25	300	2000–5000
1800 rpm gas	5–20	Heavy use	0.75	50	300	2000–5000
Diesel	3–100	Industrial	1.00	125–750	500–1500	6000

Source: *Stand-Alone Photovoltaic Systems: A Handbook of Recommended Design Practices*, Sandia National Laboratories, Albuquerque, NM, 1995.

Not included in the table, but important to the reliable performance of the generator, is the need to "exercise" all generators to ensure that they will operate properly when needed. Just as an automobile or a lawn mower may resist starting after a prolonged period of nonuse, the internal combustion engines that power the generator behave similarly. It is important to follow manufacturer's recommendations for exercising any generator. Some may require weekly exercise and others may require monthly exercise. Sometimes, especially with larger generators, the generator will have a controller that will automatically start and run the generator for a predetermined time at preset intervals, such as 20 minutes, once a week. When this is the case, it is important to monitor fuel levels, especially if the generator is used infrequently.

3.9.5 Generator Selection

Selection of a generator for a system may involve all the previous considerations. If the generator is not expected to be used very much, a light duty or portable unit may suffice. If it is to be an essential, reliable system component, capable of running for long periods and for a significant number of hours per year, then a heavy duty unit is preferable. The interesting challenge for the engineer is to determine the dividing line between light duty and heavy duty. After this choice is made, then all the other items previously listed may be considered. Generally, the unit with the lowest life-cycle cost will be the best unit for the application, provided that all operating parameters have been incorporated into the analysis.

3.10 BALANCE OF SYSTEM COMPONENTS

3.10.1 Introduction

Aside from the major components of a PV system, that is, PV modules, charge controllers, batteries, inverters, and, possibly, generators, there are numerous other smaller/less expensive components necessary to complete a PV system installation. They include overcurrent protection devices, disconnects, surge protectors, array mounts, receptacles, GFDI devices, arc fault detection, provisions for rapid shutdown, wiring, connectors, wiring enclosures, inverter bypass switches, source circuit combiner boxes, grounding connections, and battery cables and battery containers. In some cases, the connected loads are considered to be part of the balance of system (BOS) and in other cases the loads are not considered to be a part of the system. The distinction is generally made when the system is installed to operate a specific load, as opposed to being installed to be one contributor to the operation of any load.

Certain BOS components are regulated by codes or standards. Array mounts, for example, must meet any wind loading requirements of applicable building codes. Battery compartments are covered in the *NEC*.

In certain environments, BOS components may need to be resistant to corrosion from exposure to salt air or may need to be appropriate for other environmental considerations. If a PV system is part of a building integrated structure, then a number of other codes and standards may become applicable.

The *NEC* [2] specifies the requirements for choosing most of the electrical BOS components. *NEC* requirements will be described and used in detail in the next few chapters. ASCE-7 is essential for guidance in the design of array mounts and structural components of a PV system. Structural design is covered in Chapter 5.

3.10.2 Switches, Circuit Breakers, Fuses, and Receptacles

All switches, circuit breakers, and fuses used in the dc sections of PV systems must be rated for use with dc. Switches, circuit breakers, and fuses used in ac circuits must be rated for ac use. Switches have both current and voltage ratings. If a switch is used to control a motor, it must be rated to handle the horsepower of the motor at the operating voltage of the motor.

Circuit breakers must be sized in accordance with *NEC* requirements. For example, although the maximum current ratings for #14, #12, and #10 THHN wire are 25, 30, and 40 A, respectively, the maximum fuse or circuit breaker sizes allowed for use with these wire sizes are 15, 20, and 30 A, respectively. Larger wire sizes may be fused at their rated ampacities (i.e., I_{max}).

For motors, it may be necessary to install a fuse or a circuit breaker with a rating that exceeds the circuit ampacity in order to accommodate the starting current of a motor. When this is the case, the motor must have a form of overload protection that will disconnect the motor if the motor current exceeds approximately 125% of its rated running current, depending on the size and type of motor. Details of wiring for motors and motor controllers are covered in great detail in *NEC* Article 430.

Different voltages require different receptacles. While it is not very likely that 12 V_{dc} will damage a piece of 120 V_{ac} equipment except possibly a motor or transformer, it is almost certainly true that 120 V_{ac} will damage a piece of 12 V_{dc} equipment. It is thus necessary to use different attachment cap and receptacle configurations for different voltages.

Since the *NEC* will be referenced in nearly all the design examples in later chapters, specific applications of *NEC* requirements will become more clearly evident as the examples are developed. Further discussion of overcurrent protection,

INTRODUCTION TO PV SYSTEMS 105

disconnects, and wire sizing in accordance with *NEC* requirements will be covered in detail in these examples.

3.10.3 Ground Fault, Arc Fault, Surge, and Lightning Protection

3.10.3.1 Grounding versus Grounded and Ground Fault Protection

Presumably, current will leave the PV array via the positive conductor and the same amount of current will return to the array via the negative conductor. This will be the case, provided that no alternate return paths are present.

The *NEC* requires that metallic frames and other metal parts of PV systems be connected to ground. The conductors used for this purpose are called *grounding conductors*. In older systems, but not as much in newer systems, the positive or negative conductor (but not both) is also connected to ground at some point along the system. If so, then the conductor connected to ground is called the *grounded conductor*. Since the grounded conductor is connected to ground at only one point, current will flow in the grounded conductor, but will not flow in any of the grounding conductors, since there is no closed circuit in which the current can flow when the system is operating properly.

However, if for some reason the ungrounded conductor were connected to ground, then there would be an alternate closed path in which current would be able to flow through the grounding conductors. Such a condition is classified as a ground fault. If this is the case, then Kirchhoff's current law requires that the current in the ungrounded conductor will equal the sum of the currents in the grounded and grounding conductors. The net result is that no longer will the currents in the positive and negative conductors be equal. The greatest danger in dc ground fault currents in a PV system is the means by which they are established. If the ground fault results from a loose connection, the connection may begin to arc and become a fire hazard.

NEC 690.5 requires ground fault protection to be provided (GFDI) to disconnect the array in the event of a difference in current between the positive and negative dc array conductors leading to the controller/inverter/loads, unless the system is a small system with all dc output circuitry isolated from any buildings.

3.10.3.2 Arc Fault Protection

NEC 690.11 requires that all dc circuits operating at a maximum system voltage of 80 V or greater be protected by a listed dc arc fault circuit interrupter (AFCI). Arc faults can occur as series faults, such as arcing across a broken wire, or as parallel faults, such as arcing between two conductors. The AFCI device must trip in either situation. The AFCI device provides additional protection over and above the protection provided by a GFDI. While GFDI devices sense a difference in current between positive and negative (supply and return) conductors of approximately 0.5 A for small systems, a difference in supply and return current is not necessary to activate an AFCI device. The AFCI device electronically detects the electronic signature of an arc and responds by opening the circuit and providing a visual indication that the device has detected a fault.

3.10.3.3 Surge Protection

Another important component used to provide protection to the array and inverter is the surge protector. Surge protectors are similar to Zener diodes in that they are made of material that is essentially insulating until a predetermined voltage appears across the material. At this point, avalanche breakdown occurs and the surge protector acts as a current shunt. Metal oxide varistor (MOV) types of surge protector respond in nanoseconds and can bypass many joules of surge energy. However, a disadvantage of MOV surge protectors is that they draw a small current at all times and generally the failure mode of a MOV device is a short circuit. SiO varistor (SOV) surge protectors are an improvement in terms of the MOV disadvantages. While it is not required by the *NEC*, it is good practice to incorporate listed dc surge protection into system design at a location close to the common system ground point in order to protect system electronic components. Additional ac surge protection may be installed near the service entrance to prevent any backfed surges from power line disturbances.

Additional protection from lightning strikes can be obtained with lightning rod systems. Contrary to popular belief, lightning rods do not attract lightning. Rather, they present a sharp point (or points) that are connected to ground. If a potential difference should appear between the rod and the surrounding atmosphere, a very high electric field builds up around the point to the extent that a corona discharge takes place, equalizing the charge between the air and the ground. Corona discharge is also known as glow discharge. Lightning is known as arc discharge. Arc discharge is violent, but glow discharge is essentially harmless since the charge is equalized over a much longer period of time.

3.10.4 Inverter Bypass Switches and Source Circuit Combiner Boxes

3.10.4.1 Inverter Bypass Switches

Inverter bypass switches are used in battery backup systems to bypass the inverter in the event that maintenance is needed on the inverter and an alternate connection between grid and standby loads is required. Inverter bypass switches will be used in Chapter 6.

3.10.4.2 Source Circuit Combiner Boxes

Source circuit combiner boxes are used to combine the outputs of individual source circuits of a PV array into a single PV output circuit. The combiner boxes include either fuses or circuit breakers in each source circuit to protect the source circuit from backfed current from other source circuits in the event that a problem should occur in a source circuit that either lowers the current or voltage in the circuit.

INTRODUCTION TO PV SYSTEMS

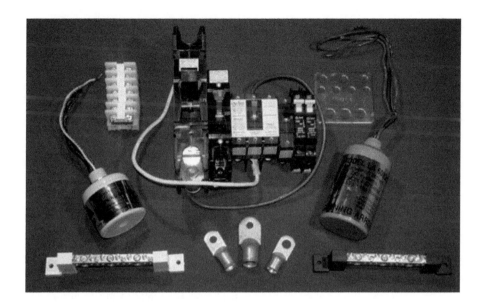

Figure 3.34 Examples of BOS components.

At the time of this writing, PV systems that are directly grid connected with no battery backup generally operate at array voltages greater than 150 V. Also, at the time of this writing, no economical dc-rated circuit breakers are available with voltage ratings greater than 150 V. As a result, dc-rated fuses are almost always used in source circuit combiner boxes designed for operating at voltages between 150 and 600 V_{dc}, and dc-rated circuit breakers are generally used in source circuit combiner boxes that are designed to operate at voltages below 150 V. The exception is inverters larger than 100 kW, where high-current, high-voltage fuses are also expensive, such that circuit breakers that can be reset are sometimes considered to be more cost-effective. Figure 3.34 shows a collection of BOS components.

3.10.5 Grounding Devices

All noncurrent-carrying metallic parts of the system, such as array mounts, PV module frames, equipment enclosures, junction boxes, and conduit must be grounded. Grounded means that these items must be at ground potential. The *NEC* has established the sizes of conductors and the types of connectors that must be used for grounding equipment. As system designs are completed in later chapters, careful attention will be paid to the selection of proper grounding equipment.

3.10.6 Rapid Shutdown

In a typical PV system prior to 2014, for rooftop PV systems, it was common to use open wiring with conductors rated for the purpose for wiring from module-to-module and then to transition from open wiring at the array to conductors in metallic conduit to protect the conductors, via a junction box near the array. The conduit would then be routed, often through an attic, to the inverter or to a dc disconnect switch located close to the inverter. These systems were code compliant. But it was realized that, albeit very unlikely that any problem would occur with the wiring inside metallic conduit installed, in an emergency situation where the system had to be shut down, the dc conductors from rooftop to inverter would still be live, posing a potential danger to emergency responders.

The 2014 *NEC*, in 690.12, now requires provisions for rapid shutdown of systems installed in or on buildings. More specifically, array conductors must be shut down within 10 ft of the array or within 5 ft of entering a building, whichever wire run is shorter. Because of large capacitors at inverter inputs, when a dc circuit is shut down, it may take a while for the capacitors to discharge. Thus, rapid shutdown has been defined to take into account the time it takes for the system voltage to drop below 30 V and the dc VA to drop below 240 VA. The VA requirements adds an additional safety element, since otherwise, at 30 V, if the system current is still in excess of 8 A, the VA requirement would not be met.

The challenge to the engineer/designer is to meet the intent of this requirement. For example, having a switch in the attic located 4 ft from the array does not meet the intent for rapid shutdown if someone has to climb into the attic to turn off the switch. *NEC* 690.12 is sufficiently flexibly stated that it invites the industry and designers to find creative solutions to achieving the goals of this section. As a result, rapid shutdown will also likely become a rapidly developing set of options as industry strives to achieve cost-effective, meaningful solutions. Remote operation of rapid shutdown devices located close to the array surely meets the intent of the section, provided that the remote operation is sufficiently reliable that it will meet listing requirements.

PROBLEMS

3.1 If $I_\ell = 2$ A, $I_o = 10^{-10}$ A, and $T = 300$ K for a PV cell, determine the maximum power point of the cell by differentiating the expression obtained by multiplying Equation 3.1 by the cell voltage.

3.2 Determine the range of operating voltages for which a 60-cell module will have power output within 90% of maximum power. You may assume $I_o = 10^{-9}$ A and $V_{OC} = 0.600$ V at an operating temperature of 300 K. Assume all cells in the module to be identical.

3.3 A PV module is found to operate at a temperature of 60°C under conditions of $T_A = 30$°C and $G = 980$ W/m². Determine the NOCT of the module.

3.4 Plot V_m versus T for $-25 < T < +75$°C for a 36-cell Si module for which each cell has $V_m = 0.5$ V at 25°C. Use a typical value for $\Delta V_m/\Delta T$.

3.5 Plot P_m versus T for the module of Problem 3.4 if $I_m = 5.85$ A at 25°C. Use a typical value for $\Delta I_m/\Delta T$.

3.6 Two 36-cell PV modules are connected in series. One is shaded and one is fully illuminated, such that the I–V characteristics of each module are as shown in Figure P3.1.

INTRODUCTION TO PV SYSTEMS

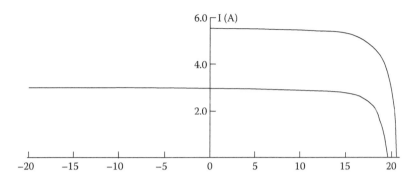

Figure P3.1 I–V characteristics of two modules, one shaded and one fully illuminated.

    a. If the output of the two series modules is shorted, estimate the power dissipated in the shaded module.
    b. If the two modules are equipped with bypass diodes across each 12 series cells, estimate the power dissipated in the shaded module.

3.7 Rewrite the chemical equations of the Ni–Cd and NIMH systems showing the half-reactions at the anodes and at the cathodes. Determine the charge-transfer mechanism.

3.8 How many gallons would have to be pumped into a tank raised 10 ft above the ground in order to be able to recover 1 kWh of electricity at a conversion efficiency of 100%, assuming the water is allowed to fall the entire 10 ft to operate an electrical generator?

3.9 Determine the capacitance of a capacitor needed to store 1 kWh if it is charged to 500 V.

3.10 Determine the inductance needed to store 1 kWh if the inductor is carrying a reasonable current, to be determined by the problem solver.

3.11 Determine the size and rotation speed of a flywheel designed to store 1 kWh of energy. Note that there is no unique solution to this problem. Use reasonable assumptions.

3.12 Describe two additional charging algorithms that will result in a fully charged battery and minimal waste of PV array energy.

3.13 If an MPPT operates at 98% efficiency with a PV array that has the characteristics shown in Figure P3.2, using the load curves as shown in the figure
    a. Determine the additional power available to each load when the array operates at 1000 W/m².
    b. For each load, determine the charge controller input current and output current at 1000 W/m².
    c. For each load, how long would the system need to operate at 1000 W/m² in order to recover the cost of the MPPT if the MPPT cost $500 and the PV-generated electricity has a value of $.40/kWh?
    d. Repeat Parts a–c if the array operates at 500 W/m².

3.14 A battery charge controller incorporates an MPPT to optimize charging of the batteries. Assume the maximum power voltage of a PV array to be 99 V and the bulk charge voltage level of a 48-V vented lead-acid battery bank to be 56 V. If the conversion efficiency of the MPPT is 98%, estimate the percentage increase in charge delivered to

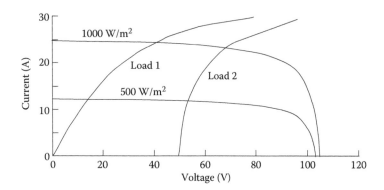

Figure P3.2 I–V characteristics for PV array and two loads.

the battery under these conditions over that which would be delivered by a controller that causes the array to operate at 56 V rather than 96 V. You may assume the PV array consists of three modules connected in series, V_{mp} is independent of irradiance levels, and each module has $V_{OC} = 42$ V, $I_{SC} = 6.25$ A, $V_m = 33$ V, and $I_m = 5.76$ A.

3.15 For the circuit of Figure 3.25, explain the on–off sequencing of the MOSFET switches to produce a symmetrical square wave output. Sketch the result.

3.16 Determine whether the dc source in Figure 3.25 remains on constantly during the switching. If not, explore modifications to the circuitry to ensure that the dc source is delivering constant dc current.

3.17 Show how a dc–dc converter can be used as a part of an inverter designed to have a square wave output of 120 V with a 12-V_{dc} input. The idea is to design the inverter without a transformer. Sketch a block diagram, showing some components to clearly express your design.

3.18 For the five-level H-bridge of Figure 3.27, determine T_2 if $T_1 = T_0$, $V_{rms} = 120$ V, and
   a. $V_{dc} = 150$ V
   b. $V_{dc} = 155$ V
   c. $V_{dc} = 160$ V
   d. $V_{dc} = 165$ V

   Note that Excel or MATLAB can be very useful for this problem.

3.19 For the five-level H-bridge of Figure 3.27, determine a set of values of T_0, T_1, and T_2 that will minimize the THD of the waveform if $V_{dc} = 160$ V and $V_{rms} = 120$ V. A convenient tool is to construct the waveform in PSPICE using series pulse voltage sources and then to run a fast Fourier transform of the resulting waveform.

3.20 For the five-level H-bridge of Figure 3.27, determine the maximum value for V_{dc} if the design requires $T_2 > 0.025$ T and $T_0 > 0.025$ T.

3.21 Referring to Figure 3.27a, design a nine-level H-bridge, using four capacitors and an appropriate number of series switches in each string. Show the waveform and the switching needed to obtain each of the nine levels of the output voltage.

3.22 For the nine-level H-bridge output waveform, derive a formula for the rms value of the output voltage.

3.23 Construct the nine-level H-bridge output waveform using series pulse voltage sources in PSPICE and perform a fast Fourier transform on the waveform to explore the harmonic distortion of the waveform. Keep V_{dc} and V_{rms} constant, while varying

INTRODUCTION TO PV SYSTEMS

the duration of the different levels of the waveform in order to minimize the THD of the waveform.

3.24 Make a list of loads that might confuse an inverter that is in the "sleep" mode. Explain why each load causes a problem.

3.25 The following data are given for a series of gasoline-powered electrical generators:

Rated output (W)	1500	2300	3000	4500
Fuel tank size (gallon)	2.9	2.9	4.5	4.5
Run time/tank (h)	9	9	8.3	5.6

Calculate the kWh/gallon for each of these generators under rated load conditions.

REFERENCES

1. RISE Information Portal, The future of grid-connected systems, www.rise.org.au/info/Applic/Gridconnect/indix.html (accessed November 2008).
2. *NFPA 70 National Electrical Code*, 2014 Ed., National Fire Protection Association, Quincy, MA, 2013.
3. *IEEE Standard 1187-2002, IEEE Recommended Practice for Installation Design and Installation of Valve-Regulated Lead-Acid Storage Batteries for Stationary Applications*, Institute of Electrical and Electronics Engineers, Inc., New York, 2002.
4. *Maintenance and Operation of Stand-Alone Photovoltaic Systems*, Sandia National Laboratories, Albuquerque, NM, 1991.
5. Richmond, R., *Home Power Mag.*, Feb/Mar 2013.
6. Linden, D. Ed., *Handbook of Batteries*, 2nd Ed., McGraw-Hill, New York, 1994.
7. Batteryuniversity.com.
8. Energy Storage Association, Washington, DC.
9. Aziz, M. J., Materials Science Group, Harvard School of Engineering and Applied Science.
10. Hancock, O. G., Jr., Hydrogen storage: The hydrogen alternative, *Photovoltaics Int.*, 1986.
11. Skerrett, P. J., Fuel UPD, *Pop. Sci.*, 1993.
12. Roland, B., Nitsch, J., and Wendt, H., *Hydrogen and Fuel Cells—The Clean Energy System*, Elsevier Sequoia, Oxford, UK, 1992.
13. U.S. Department of Energy, Federal Energy Regulatory Commission, Office of Energy Projects, *Energy Infrastructure Update*, December 2015.
14. Buchanan, J.R., Lead-acid batteries intro, www.buchanan1.net (accessed April 2016).
15. *UL 1741: 2005, Standard for Static Inverters and Charge Controllers for Use in Independent Power Production Systems*, Underwriters Laboratories, Inc., Northbrook, IL, May 1999.
16. *ANSI C84.1-1995, Electric Power Systems and Equipment—Voltage Ratings (60 Hertz)*, American National Standards Institute, New York, 1995.
17. For information on a wide selection of lower power inverters, see http://www.powerinverters.com.
18. For information on utility scale inverters, see http://www.eaton.com/.
19. IEEE 1547-2003, *Standard for Interconnecting Distributed Resources with Electric Power Systems*, IEEE Standards Coordinating Committee 21, Photovoltaics, 2003.

20. SMA America, Inc., *Solar Inverter Sunny Boy 5000US, 6000US,7000US Installation Guide, V2.2*, 2008, www.sma-america.com.
21. Outback Power Systems, FX Series Inverter/Charger FX/VFX/GTFX/GVFX/Mobile Installation Manual, 2008, www.outbackpower.com.
22. Welcome to Generator Joe home page, www.generatorjoe.net/store.asp (a wealth of information on a wide variety of fossil fuel generators).
23. For information on generators with internal inverters, see http://www.electricgeneratorsdirect.compower/honda-inverter-generators.html?gclid=CIyRh_HW-ssCFcUmhgodt-0KrA.
24. *Stand-Alone Photovoltaic Systems: A Handbook of Recommended Design Practices*, Sandia National Laboratories, Albuquerque, NM, 1995.

SUGGESTED READING

ASCE Standard 7–10, Minimum Design Loads for Buildings and Other Structures, American Society of Civil Engineers, Reston, VA, 2010.

Hoogers, G. Ed., *Fuel Cell Technology Handbook*, CRC Press, Boca Raton, FL, 2002.

Hu, C. and White, R., *Solar Cells: From Basics to Advanced Systems*, McGraw-Hill, New York, 1983.

Krein, P. T., *Elements of Power Electronics*, Oxford University Press, New York, 1998.

Markvart, T. Ed., *Solar Electricity*, John Wiley & Sons, Chichester, UK, 1994.

Roden, M. S. and Carpenter, G. L., *Electronic Design: From Concept to Reality*, Discovery Press, Burbank, CA, 1997.

Sedra, A. S. and Smith, K. C., *Microelectronic Circuits*, 4th Ed., Oxford University Press, New York, 1998.

Silberberg, M., *Chemistry: The Molecular Nature of Matter and Change*, Mosby, St. Louis, 1996.

Skvarenina, T. L. Ed., *The Power Electronics Handbook*, CRC Press, Boca Raton, FL, 2002.

Vithayathil, J., *Power Electronics: Principles and Applications*, McGraw-Hill, New York, 1995.

CHAPTER 4

Grid-Connected Utility-Interactive Photovoltaic Systems

4.1 INTRODUCTION

As the cost of PV systems continues to decrease, utility-interactive systems are now at grid parity in many locations, meaning that the cost of PV-generated kWh is less than or equal to the cost of kWh purchased from the grid. Furthermore, increases in consumer concern over global climate change are resulting in an interest in reducing CO_2 emissions by replacing fossil fuel-generated electricity with PV-generated electrical energy. This feedback loop, coupled with the increased demand for stand-alone systems, as well as the financial incentives offered by a number of utilities, states, and countries, has resulted in a healthy market for PV systems. And this increased demand has enabled PV module and balance of system (BOS) component manufacturers to scale up manufacturing facilities to take advantage of economies of scale to further reduce system costs.

In addition to cost reductions, the increased demand for PV systems has led to significant improvements in the reliability of PV system components, designs, and installations. So not only are PV systems decreasing in cost, but they are also increasing in reliability.

By 1999, the technical issues associated with connecting PV systems to the utility grid had essentially been solved. In 2000, IEEE adopted Standard 929-2000 [1]. Any PV system meeting the performance criteria of IEEE Standard 929, using power conditioning units (inverters) listed under UL 1741 and installed in accordance with the current *National Electrical Code*® (*NEC*), automatically met all established technical performance criteria. Since then, IEEE has adopted further refinements to Standard 929-2000, which are reflected in IEEE 1547-2003 [2]. Presently, a goal has been set to complete a full revision of IEEE 1547-2003 by 2018 that will address the rapid deployment of grid-connected PV and other renewable energy systems.

Grid-connected utility-interactive PV systems are capable of supplying power to the utility grid. There are essentially two classes of grid-connected systems: straight grid-connected with no battery backup and grid-connected with battery backup. Grid-connected systems with battery backup can be further separated into dc-coupled systems and ac-coupled systems, and in some cases, battery-backup systems do not sell back to the grid, but only purchase power from the grid for the purpose of charging the system batteries.

This chapter will focus on the design of straight grid-connected PV systems. Straight grid-connected systems are the most straightforward systems to understand and are currently the most commonly installed systems. Battery-backup systems will be covered in Chapter 6, and stand-alone PV systems will be covered in Chapter 7.

Before beginning the design process, it will be helpful to take a closer look at some of the codes and standards that govern the design and installation of PV systems.

4.2 APPLICABLE CODES AND STANDARDS

A number of codes and standards have been created to ensure the safety of electrical systems. These codes and standards also generally address the efficiency and reliability of systems. Since a PV system is capable of generating sufficiently high voltages to present a potential electrical safety hazard, PV systems are included in these codes. Perhaps the two most common codes and standards that deal with PV systems are the *NEC* [3] and IEEE 1547-2003 [2]. The PV engineer should be familiar with both. But the list does not end here. PV systems have a mechanical component as well as an electrical component, and, hence, a number of mechanical and structural codes also apply. If the PV system is to be integrated into the construction of a building, the list of codes and standards becomes even longer, now explicitly including fire codes.

Table 4.1 [1–10] provides a partial listing of the many codes and standards that may be applicable to any particular PV installation.

4.2.1 The *National Electrical Code*

4.2.1.1 Introduction

The *NEC* [3] is published by the National Fire Protection Association and is updated approximately every 3 years. This text will use the 2014 edition of the *NEC* as a reference. The *NEC* consists of a collection of articles that apply to considerations such as wiring methods, grounding, motor circuits, and nearly every conceivable topic in which electrical safety and efficient utilization is a consideration, including Article 690, which deals specifically with PV systems.

The *NEC* specifies the sizes and types of switches, fuses and wire to be used, and specifies where these items must be located in the system. These components are necessary, not only to protect the end user, but also to protect the maintenance technician. As specific PV system design examples are discussed, compliance with the *NEC* will be incorporated into the design process discussion. In particular, proper wire sizing to limit voltage drop in connecting wires to acceptable limits, proper use of switches, circuit breakers and fuses, types of insulation and conductors, types

GRID-CONNECTED UTILITY-INTERACTIVE PHOTOVOLTAIC SYSTEMS

Table 4.1 Partial Listing of Codes and Standards That May Apply to PV Systems

Reference #	Title/Contents
NEC 2014	*National Electrical Code*/Wiring methods (comprehensive)
IEEE 937	IEEE Recommended Practice for Installation and Maintenance of Lead-Acid Batteries for Photovoltaic Systems
IEEE 1013	IEEE Recommended Practice for Sizing Lead-Acid Batteries for Photovoltaic Systems
IEEE 1187	Recommended Practice for Design and Installation of Valve-Regulated Lead-Acid (VRLA) Storage Batteries for Stationary Applications
IEEE 1361	Recommended Practice for Determining Performance Characteristics and Suitability of Batteries in Photovoltaic Systems
IEEE 1526	Recommended Practice for Testing the Performance of Stand-Alone Photovoltaic Systems
IEEE 1547	IEEE Standard for Interconnecting Distributed Resources with Electric Power Systems
IEEE 1561	Guide for Optimizing the Performance and Life of Lead-Acid Batteries in Remote Hybrid Power Systems
IEEE 1562	Guide for Array and Battery Sizing in Stand-Alone PV Systems
IEEE 1661	Guide for Test and Evaluation of Lead-Acid Batteries Used in PV Hybrid Systems
IEC TC-82	A compendium of 25 standards relating to the electrical and mechanical performance testing and measurement of PV systems
ISO 9001	An international quality standard, composed of 20 segments, dealing with all aspects of design, manufacturing, and delivery of service
UL 1741	Standard for Static Inverters and Charge Controllers for Use in Photovoltaic Power Systems
ANSI Z97.1	Relates to safety relating to potential glass breakage
ASCE 7-10	Minimum Design Loads for Buildings and Other Structures
ASTM	A compendium of tests and standards that may apply to BIPV systems

Source: IEEE 929-2000, IEEE Recommended Practice for Utility Interface of Residential and Intermediate Photovoltaic (PV) Systems, IEEE Standards Coordinating Committee 21, Photovoltaics, 2000; IEEE 1547-2003, IEEE Standard for Connecting Distributed Resources to Electric Power Systems, IEEE Standards Coordinating Committee 21, *Fuel Cells, Photovoltaics, Distributed Generation and Energy Storage*, 2003; *NFPA 70 National Electrical Code*, 2014 Ed., National Fire Protection Association, Quincy, MA, 2013; IEEE Standards Coordinating Committee 21 Standards and Solar America Board for Codes and Standards National and international Standards Panel, January 16, 2008, http://www.solarabcs.org/about/publications/meeting_presentations_minutes/2008/01/pdfs/16Jan08-IEEE-SCC21.pdf; IEEE Standard 1187-2013, https://standards.ieee.org/findstds/standard/1187-2013.html; ISO Standard 9001, http://www.solarabcs.org/about/publications/meeting_presentations_minutes/2012/09/pdfs/16-IEC-Barikmo-14Sept 2012.pdf; Catalog of all ISO standards, http://www.iso.org/iso/catalogue_detail?csnumber=46486; American National Standards Institute Standard Z97, http://www.ansiz97.com/standard/; *ASCE Standard 7-10, Minimum Design Loads for Buildings and Other Structures*, American Society of Civil Engineers, Reston, VA, 2010; ASTM International incorporates over 12,000 standards that operate globally. www.astm.org.

of electrical conduit, electrical safety devices for detecting and interrupting ground faults and arc faults, and proper grounding will be considered.

Normally, copper wire will be used in PV system wiring. Aluminum wire is allowed by the *NEC*, but generally it is used only over longer distances for carrying higher currents when the cost of using copper would be prohibitive. For example, since copper is significantly heavier than aluminum, the insulators on power poles would need to be considerably stronger if copper wire were used instead of aluminum. Aluminum has a lower conductivity than copper, it oxidizes faster than copper, and unless terminals are tightened properly, it is more likely to loosen in a connector. Furthermore, aluminum oxide is not readily visually distinguishable from aluminum.

Since aluminum oxide is a good insulator, special care must be taken when terminating aluminum to ensure that the exposed aluminum is free of oxidation. If aluminum wire is used in a system, devices with terminations approved for use with aluminum must be used.

The intent of this section is to highlight several *NEC* requirements. The reader who goes beyond the walls of the classroom into the design of PV systems should invest in the latest edition of the *NEC* to be sure to have all the latest requirements on items such as ground fault detection and interruption (GFDI), source- and output-circuit maximum voltage ratings, acceptable conductor placement, rapid disconnect methods, and storage batteries.

4.2.1.2 Voltage Drop and Wire Sizing

Table 4.2 shows the dc resistance per 1000 ft (304.8 m) and rated current (ampacity) for copper conductors with insulation rated at 90°C (194°F). The ampacities are for no more than three current-carrying conductors in conduit at a temperature of 30°C (86°F) or less. If more than three current-carrying conductors are in a conduit, or if the conductors are operated in an ambient that exceeds 30°C (86°F), their ampacity must be derated. Ampacities for single conductors in free air are higher than those listed in Table 4.2, since heat is not trapped inside the conductor by crowding with other conductors or conduit. While many different types of insulation exist, as shown in *NEC* Article 310, type THWN-2 is a high-temperature (90°C), moisture- and oil-resistant, thermoplastic insulation that is commonly in use for wiring protected by conduit. Type THHN insulation is almost always also rated as THWN-2.

The *NEC* requires that the total voltage drop in feeder and branch circuits be less than 5%, with the drop in either feeder or branch circuits limited to no more than 3%. A feeder circuit is a circuit that provides power to an electrical distribution panel. In a common residential electrical service, the feeder circuit is the wiring

Table 4.2 Properties of Copper Conductors with 90°C Insulation

Size (AWG)	18	16	14	12	10	8	6	4
dc (Ω/kft)	7.77	4.89	3.07	1.93	1.21	0.764	0.491	0.308
I_{max} (A)	14	18	25	30	40	55	75	95
Size	3	2	1	0	00	000	0000	250 kcm
dc (Ω/kft)	0.245	0.194	0.154	0.122	0.0967	0.0766	0.0608	0.0515
I_{max} (A)	110	130	150	170	195	225	260	290

Source: NFPA 70 *National Electrical Code*, 2014 Ed., National Fire Protection Association, Quincy, MA, 2013.

Note: Single conductor wire sizes 18–8. Stranded for larger sizes.

between the electric meter and the circuit breaker panel to which branch circuits are connected. The branch circuits are the circuits that provide power to the individual electrical loads, such as lighting, refrigerators, dishwashers, air conditioners, and so on. The PV equivalence of branch circuits can be considered to be the PV source circuits that connect the PV array to the power conditioning equipment, and the PV equivalence of feeder circuits can be considered to be the PV inverter output circuits that connect to the utility. In any case, good PV system design generally requires voltage drop in any PV circuits to be less than 2%.

Many tables exist in various design manuals that list the maximum distance a certain size conductor can be run with a given current and still not produce excessive voltage drop. For the engineer with a calculator, however, all one needs to do is to recognize that a circuit consists of wire in both directions, so that a load located 50 ft (15.24 m) from a voltage source will need 100 ft (30.48 m) of wire to carry the current to and from the load. If d is the distance from source to load in ft, and V_s is the source voltage, Ohm's law can be used to calculate the percentage voltage drop in the wire via

$$\%VD = 100\frac{I}{V_s}\left(\frac{\Omega}{kft}\right)\left(\frac{2d}{1000}\right) = \frac{0.2Id}{V_s}\left(\frac{\Omega}{kft}\right), \quad (4.1)$$

as long as the circuit is a dc circuit or a single-phase, two-wire ac circuit. Three-phase voltage drop calculations will be introduced in Section 4.6.

EXAMPLE 4.1

A 20-W, 12-V dc LED lamp is located 50 ft (15.24 m) from a 12-V battery. Specify the wire size needed to keep the voltage drop between battery and lamp under 2%.

Solution

First solve Equation 4.1 for (Ω/kft) to obtain

$$\frac{\Omega}{kft} = \frac{(\%VD)V_s}{0.2Id}.$$

Then, determine the load current from P = IV, assuming the load voltage to be essentially equal to the supply voltage. Substituting the known values on the right-hand side yields Ω/kft = (2 × 12)/(0.2 × 1.67 × 50) = 1.437. Table 4.2 shows that #12 (3.31 mm²) wire has too much resistance, so it is necessary to use #10 (5.261 mm²). The actual voltage drop with #10 (5.261 mm²) wire can now be found from Equation 4.1 to be 1.68%.

Note that although #10 wire will carry 40 A, the current in the circuit is limited to 1.67 A due to the voltage drop limitation. It is very important to be aware of the need for larger wire in low-voltage systems. Equation 4.1 and Table 4.2 will find considerable use in examples to follow in this and later chapters. Note also that if the wire resistivity is given in Ω/km, then Ω/kft = (Ω/km)/3.28.

Because of the problem with voltage drop at low voltages, and the correspondingly larger wire sizes necessary, it is generally desirable to operate PV systems that deliver any significant amounts of power over any reasonable distances at voltages higher than 12 V. If the previous 20-W load were connected to a 24-V system, the load current would be halved, and the load voltage would be doubled. This allows the resistance of the wiring to be four times higher, or as much as 5.648 Ω/kft, which means that #16 (1.31 mm^2) wire is now adequate. From a total cost standpoint, however, the cost of the additional 12-V battery will exceed the difference in cost between #10 wire and #16 wire. Presumably, the cost of the PV modules will be the same, since the amount of power required has not changed.

For concealed wiring, the minimum wire size is #14 (2.08 mm^2). Smaller wire sizes are normally used only for portable cords or for attaching single loads.

The *NEC* is an important source of information for the PV design engineer, since it clearly defines acceptable PV system design practice. Article 690 deals exclusively with PV systems, but refers to other articles such as Article 240 on overcurrent devices, Article 250 on grounding, and Article 310 on conductor ampacities. Other parts of the *NEC* also apply to specific installations or installation methods. Table 4.3 summarizes the sections of *NEC* Article 690 and lists some of the other *NEC* articles that apply to PV system installations.

Table 4.3 **Summary of Contents of *NEC* Article 690**

Section	Contents	*NEC* Cross-References
I	*General:* scope, definitions, general req, ground fault protection, ac modules	Articles 240, 705
II	*Circuit requirements:* maximum voltage, circuit sizing and current, overcurrent protection, stand-alone systems, arc fault circuit protection, rapid shutdown	Articles 110, 210, 240
III	*Disconnecting means:* structures supplied by a PV system, disconnection of PV equipment, fuses, disconnect types, installation and service	Article 705
IV	*Wiring methods:* methods permitted, component interconnections, connectors, access to boxes, ungrounded PV systems	Articles 110, 250, 310, 400
V	*Grounding:* system grounding, point of system grounding connection, equipment grounding, size of equipment grounding conductor, grounding electrode system, continuity of equipment grounding systems, continuity of PV source- and output-circuit grounded conductors, equipment bonding jumpers	Article 250
VI	*Marking:* modules, ac modules, dc PV power source, point of common connection, PV systems with energy storage, identification of power sources	Article 110
VII	*Connection to other sources:* load disconnect, identified interactive equipment, loss of interactive system power, unbalanced interconnections, point of connection	Article 705
VIII	*Storage batteries:* installation, charge control, battery interconnections	Articles 110, 400, 480
IX	*Systems over 1000 V*: general, listing, definitions	Articles 300, 490

The design examples that follow in this chapter as well as in Chapters 6 and 7 will introduce specific sections of the *NEC* that apply to the specific designs as the need arises.

4.2.2 IEEE Standard 1547-2003

4.2.2.1 Introduction

Prior to the adoption of IEEE Standard 1547-2003 [2], grid-connected PV inverters were required to comply with IEEE Standard 929. IEEE Standard 929 was developed specifically for the purpose of addressing concerns of utilities regarding quality of power delivered to the grid and the need to disconnect the PV system from the utility grid in the event of utility power failure. IEEE Standard 1547 was developed "to provide a uniform standard for interconnection of distributed resources with electric power systems," by providing requirements for performance, operation, testing, safety considerations, and maintenance of the equipment. Thus, in addition to covering electronic inverters as used in PV systems, IEEE Standard 1547 also covers distributed resources that use synchronous machines and induction machines, such as wind and low-head hydropower. If more than one type of distributed resource, up to an aggregate capacity of 10 MVA, is connected at a common point of utility connection (PUC), then all must meet the requirements of this standard at the connection point.

While the PV system designer only needs to specify that the inverter must be listed to UL 1741, which is based on IEEE 1547, to be assured that it will meet all utility interconnect requirements, it is interesting to explore some of the requirements of IEEE 1547 in order to appreciate the amount of creative engineering that was involved in developing the standard and the subsequent amount of creative engineering that is necessary to meet the standard with an inverter design. In fact, the engineer who loves to speak in acronyms should probably obtain a copy of IEEE 1547 in order to add another dozen or so acronyms to his or her vocabulary.

4.2.2.2 Specific Requirements

IEEE 1547 is essentially a functional standard as opposed to a prescriptive standard. It specifies functions that must be performed by a distributed generation system rather than specify how the function must be implemented. It is up to the designer to figure out how to satisfy the functional requirements. Section 4.1 of the standard delineates the general requirements of voltage regulation, integration of grounding, synchronization with the grid, distributed secondary spot networks, disconnect from the grid when the grid is de-energized, monitoring, isolation, electromagnetic interference (EMI), withstanding surges, voltage rating of interconnecting device, and response to abnormal grid conditions. Abnormal grid conditions include over-voltage, under-voltage, over-frequency, under-frequency, and reconnection to the grid after the abnormal condition is cleared. The following discussion will apply to PV inverter compliance with IEEE 1547.

Table 4.4 Required Clearing Times for PV Inverters under Abnormal Grid Voltage Conditions

Voltage range in % of base voltage	<50%	50%–88%	110%–120%	≥120%
Clearing time (s)	0.16	2.0	1.0	0.16

Source: IEEE 1547-2003, IEEE Standard for Connecting Distributed Resources to Electric Power Systems, IEEE Standards Coordinating Committee 21, *Fuel Cells, Photovoltaics, Distributed Generation and Energy Storage*, 2003.

Table 4.4 defines abnormal voltage conditions and lists the necessary response times for the PV inverter, as listed in Table 1 of IEEE 1547-2003. Since PV inverters may have output voltages of 120, 208, 240, 277, or 480 V, the table lists abnormal limits as a percentage of the "normal" voltage. The listed clearing times are maximum limits for inverters rated at less than 30 kW and are default clearing times for larger inverters.

Table 4.5 defines clearing times for abnormal grid frequencies for inverters that are rated either more than 30 kW or less than 30 kW. The clearing times on the smaller units can either be fixed or be field adjustable. The clearing times on the larger units must be field adjustable. As in the case of voltage clearing times, the times for the smaller units are maximum values, while the times for the larger units are default values.

If a PV inverter shuts down due to a voltage or frequency disturbance, it must monitor the voltage and frequency of the utility for 5 minutes upon restoration of the grid. If voltage and frequency remain stable for this time period, then the inverter will restart. This same situation, of course, applies to the initial start-up of the inverter. When the ac side of the inverter is switched on, the inverter will not deliver any power for 5 minutes. For the installer who turns on the switch and waits for the inverter to start, this is the longest 5 minutes of his or her day.

IEEE 1547 also defines necessary power quality requirements for PV inverters. The maximum dc component of an inverter output must be less than 0.5% of rated inverter output current. The inverter must not cause objectionable flicker for other grid users, generally defined as not causing their lights to flicker. The final power

Table 4.5 Required Clearing Times for PV Inverters under Abnormal Grid Frequency Conditions

Inverter Size (kW)	Frequency Range (Hz)	Clearing Time (s)
≤30	>60.5	0.16
	<59.3	0.16
>30	>60.5	0.16
	<(59.8–57.0) (adjustable)	Adjustable 0.16–300
	<57.0	0.16

Source: IEEE 1547-2003, IEEE Standard for Connecting Distributed Resources to Electric Power Systems, IEEE Standards Coordinating Committee 21, *Fuel Cells, Photovoltaics, Distributed Generation and Energy Storage*, 2003.

Table 4.6 Maximum Allowable Harmonic Amplitudes for PV Inverter Output in Percentage of Maximum Load Current

Harmonic range	n < 11	11 ≤ n < 17	17 ≤ n < 23	23 ≤ n < 35	35 ≤ n
% of rated current	4.0	2.0	1.5	0.6	0.3

Source: IEEE 1547-2003, IEEE Standard for Connecting Distributed Resources to Electric Power Systems, IEEE Standards Coordinating Committee 21, *Fuel Cells, Photovoltaics, Distributed Generation and Energy Storage*, 2003.

quality measure is the accuracy of the sine wave output, defined in terms of the harmonic content of the waveform. Table 4.6 is a summary of the maximum allowed harmonic amplitudes. Note that the overall harmonic distortion must not exceed 5%. The table entries apply to odd harmonics. Even harmonics must be less than 25% of the odd harmonic limits.

The final requirement of IEEE 1547 is that if an unintentional island should occur in which a collection of inverters is connected to the grid and grid power is lost, all of the inverters must be able to distinguish the grid from other inverters such that all inverters in the island are shut down within 2 seconds of loss of grid power.

Since the adoption of IEEE 1547-2003, several supplements have been adopted. IEEE Standard 1547.1 (2005) specifies tests to be made to ensure interconnects between distributed generation systems and the grid meet the requirements of IEEE 1547. IEEE Standard 1547.2 (2008) provides technical descriptions, schematics, application guidance, and interconnection examples to enhance the use of IEEE 1547. IEEE Standard 1547.3 (2007) defines an Information Exchange Interface and provides an Information Exchange Agreement template. IEEE Standard 1547.4 (2011) provides support for connecting distributed generation systems in microgrids such that the microgrid will remain energized and isolated from the main grid if the main grid is down. IEEE Standard 1547.6 (2011) gives an overview of distribution secondary grid networks where grid reliability is enhanced by geographically separated generation serving the customer.

By 2013, significant presence of distributed generation on utility distribution grids in some areas had resulted in utility concern over the effect of this increased presence on distribution network stability. IEEE Standard 1547.7 (2013) addresses criteria, scope, and extent for engineering studies of the impact of distributed generation on the stability of the utility grid. IEEE Standard 1547.8 is under review at the time of this writing. The purpose of this standard is to "lead to the development of advanced hardware and software and help streamline their implementation acceptance." In other words, this standard encourages the development of smart grid technologies that will ensure grid stability as the penetration of distributed generation resources increases.

Amendment 1 (May, 2014) of this standard recognizes that there may be an instance where, under the condition of intentional islanding, it is desirable to have inverters remain connected and provide grid voltage regulation, voltage ride-through, and frequency ride-through in coordination with the operators of the distribution grid. One such example might be a military base that is connected to the grid, but also has PV systems in operation. If the grid goes down, it would still be desirable to maintain as much power supply as possible for critical base operations. Additionally, in some areas where the grid is served by limited generation capacity, as customer peak load is approached, the available utility generation may be significantly challenged, such that customer voltages and frequencies may decrease as primary generation approaches maximum capacity. Normally, IEEE 1547 would require the distributed generation to disconnect under these conditions. Amendment 1 allows the coordination of distributed generation with utility generation such that the distributed generation can remain connected under these conditions, since disconnection of the distributed generation would result in an even greater demand on the utility generation and possibly result in the utility generator overload protection relay shutting down the utility completely.

It is relatively straightforward to detect islanding, since PV inverters that are synchronized with the grid voltage act as nearly ideal current sources. Figure 4.1 shows how a typical inverter will generate a current waveform that is slightly off grid frequency. As long as the grid voltage is present, the inverter adjusts the current waveform frequency at zero crossings of the grid voltage. If the grid is lost, then the inverter current drifts in frequency until it exceeds either the upper or the lower frequency limit. This algorithm, developed at Sandia National Laboratories, is called the Sandia Frequency Shift (SFS) [11,12]. A similar algorithm, the Sandia Voltage Shift (SVS) [11,12], is used to detect the output voltage of the inverter going out of range, and shutting down the inverter until the grid is restored.

While IEEE 929 required that the voltage at the inverter output should not be more than 5% higher than the voltage at the PUC, IEEE 1547 no longer contains this requirement. IEEE 1547 is only concerned with conditions at the PUC. To this end, IEEE 1547 prescribes test procedures to verify compliance. These procedures are used by Underwriters Laboratories when they perform these tests prior to listing an inverter to UL 1741. Thus, any inverter listed under UL 1741 has been tested to meet the criteria established in IEEE 1547.

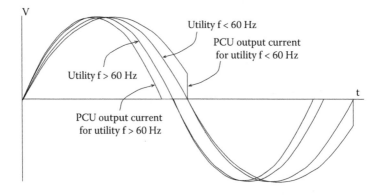

Figure 4.1 Comparison of utility voltage and PCU current for under-frequency and over-frequency utility voltage.

4.2.2.3 Comparison of PV Inverters to Mechanically Rotating Generators

When considering the connection of a PV source to the grid, it is important to distinguish between the electrical characteristics of an inverter and a conventional rotating generator. First of all, most utility-interactive inverters are best modeled as dependent current sources, while rotating generators appear as voltage sources. In the event of a short-circuit fault, a rotating generator can deliver a very large current, limited only by the ability of the prime mover to keep the generator rotating. Any energy stored as rotational energy can be dissipated into a short circuit as electrical energy.

On the other hand, if a short circuit occurs at the output of an inverter, say, at the PUC, little more current than full-load value will flow *from* the inverter. On the other hand, if the output circuitry of the inverter becomes a short circuit, then a high current can flow to the inverter from the utility. The inverter output-circuit breaker is thus required to protect the inverter from the world, as opposed to protecting the world from the inverter. Proper overcurrent protection at the output of an inverter can minimize inverter damage in the event of a short circuit within the inverter itself.

Because the inverter acts as a current source, it is easier to ensure that the inverter will meet the standards for utility interconnection. The reason is that the utility is close to being an ideal voltage source. Hence, the inverter can sense the utility voltage and frequency and inject current only if the voltage and frequency fall within prescribed limits. This same circuitry can be used to ensure that the current is injected in phase with the utility voltage. This assures a high power factor for the inverter output. The sensing circuitry has high impedance inputs and can remain connected to the utility at all times in order to monitor the voltage and frequency stability of the utility.

It should be noted that the voltage values apply to the PUC, also known as the point of common coupling (PCC) for the inverter. If the inverter is located some distance from the PCC, there may be voltage drop on the line between the inverter and the PCC. If so, compensation can be made at the inverter output, since the inverter output voltage in these cases will be higher than the voltage at the PCC in order for the inverter to deliver power to the PCC.

4.2.2.4 Islanding Analysis

The greatest concern over islanding occurs when more than one inverter is in operation within the island. In this case, it is possible that they will support a feedback situation in which each inverter, sensing the combined output of the other inverters, thinks the output of the other inverters constitutes the grid. This possibility is enhanced under worst-case load conditions. It is thus useful to consider what might constitute worst-case load conditions.

Worst-case load conditions occur if, when the island is created by the utility fault, the island voltage does not change quickly to steady-state fault value. This will occur under two conditions: resonance at the utility frequency and island motor loads with low damping, such as grinding wheels that continue to rotate even though power is removed from the motor.

If a circuit is at resonance with a relatively high Q, then it will continue to oscillate at its natural resonant frequency until the oscillation is ultimately damped out. As long as a motor continues to rotate, it will act as a generator and return its back electromotive force (emf) to the grid. Except for those used in wind turbines, induction motors are normally inefficient generators, but synchronous motors are very efficient. The generation frequency, of course, depends on the rotation speed, so as the motor slows down, the generated frequency changes.

Circuit analysis textbooks generally prove that any parallel RLC circuit will be underdamped if it has a Q > 0.5. Equation 4.2 represents the response of an underdamped parallel RLC circuit.

$$v(t) = V_m e^{-\alpha t} \cos(\omega_d t + \phi), \tag{4.2}$$

where V_m is the initial amplitude, $\alpha = 1/2RC$, $\omega_o = 1/\sqrt{LC}$, $\omega_d = \sqrt{\omega_o^2 - \alpha^2}$, and ϕ is a phase angle that depends upon initial energy storage in the inductor and capacitor. The Q of the circuit can be found from

$$Q = \frac{\omega_o}{2\alpha} = \omega_o RC. \tag{4.3}$$

It is interesting to use Equations 4.2 and 4.3 to calculate how long it will take for the amplitude of the voltage to decrease to the inverter trip limit as a function of Q.

The two important parameters of the equation are the exponentially decaying amplitude part, $V_m e^{-\alpha t}$, and the natural frequency, ω_d. Assuming the energized grid has a natural resonant frequency outside the trip limits, the inverter will sense the frequency departure and disconnect within the prescribed time. If the natural resonant frequency is within the trip limits, then the inverter will need to disconnect when the voltage falls outside the trip limits. For the case of the motor/flywheel load, again, as soon as the motor speed drops enough to move the motor output frequency outside the trip limits, the inverter will trip. On the other hand, if the inverter keeps the motor running, it may never trip. Given that neither the inverter nor the motor will normally have any provisions for stabilizing the frequency or rotation, this occurrence is extremely unlikely.

It is useful to determine the number of cycles required for the amplitude of the island load voltage to decrease to 50% of its initial value. The results are obtained by setting $V_m e^{-\alpha t} = 0.5\, V_m$. Solving for t yields the result

$$t = \frac{\ln 2}{\alpha}. \tag{4.4}$$

GRID-CONNECTED UTILITY-INTERACTIVE PHOTOVOLTAIC SYSTEMS

Table 4.7 Number of Cycles for the Voltage to Reach Half the Original Amplitude in an Underdamped, Decaying, Parallel RLC Circuit as a Function of the Q of the Circuit

Q	1	2	3	4	5	6	7	8	9	10
N	0.19	0.43	0.65	0.88	1.10	1.32	1.54	1.76	1.98	2.20

But it can be shown (see Problem 4.2) that $\alpha = \pi f_o/Q$. Thus, the time for the voltage amplitude to fall to half its starting value is given by

$$t = \frac{Q \ln 2}{\pi f_o}. \tag{4.5}$$

It also can be shown that (Problem 4.3) the relationship between resonant frequency, ω_o, and natural resonant frequency, ω_d, is

$$f_o = \frac{\omega_o}{2\pi} = \frac{\omega_d}{2\pi\sqrt{1-\frac{1}{4Q^2}}} = \frac{f_d}{\sqrt{1-\frac{1}{4Q^2}}}. \tag{4.6}$$

So, finally, solving for t in terms of the period of the natural frequency,

$$t = \frac{Q \ln 2}{\pi f_d}\sqrt{1-\frac{1}{4Q^2}} = \left[0.2206\sqrt{Q^2 - \frac{1}{4}}\right] T_d. \tag{4.7}$$

Since T_d represents one cycle at the natural resonant frequency, the coefficient of T_d represents the number of cycles, N, that it takes for the voltage amplitude to decay to half its initial value. Table 4.7 tabulates the number of cycles needed for a signal to decay to half its original amplitude as a function of the Q of the circuit.

The previous analysis was based on the assumption that the utility island load had only initial stored energy. If energy continues to be added to the load in a synchronous manner, the load will continue to oscillate in a manner not very different from an electronic class C amplifier, depending upon the conduction angle of the current source. The time to decay is thus prolonged, perhaps indefinitely, depending upon the match between the load and the inverter output. So this is the worst-case condition that must be overcome and this is why the SVS and SFS, as previously discussed, were developed.

4.2.3 Other Issues

4.2.3.1 Aesthetics

Although not a part of Article 690, *NEC* 110.12 requires that "electrical equipment shall be installed in a neat and workmanlike manner." This addresses the importance of a professional appearance for an installation and helps to remove any potential objections to the aesthetics of the installation. The aesthetics of a PV installation are of particular concern to architects and building owners who want to make the PV system look like it is an intended part of a structure as opposed to looking like an add-on. The PV community has responded with building integrated PV (BIPV) products such as shingles, tiles, windows, and laminates, as well as framed modules with "more aesthetic appeal" than other framed modules, to address the wishes of the design community.

4.2.3.2 Electromagnetic Interference

The U.S. Federal Communications Commission requires that any electronic device with an electronic clock that runs at a frequency greater than 9 kHz must meet its standards for EMI [13]. EMI occurs in two forms: conducted and radiated. Conducted EMI is coupled directly through connections to the power line. Radiated EMI consists of radio frequency signals transmitted by the device directly to the surroundings. Both types of EMI, if sufficiently strong, can interfere with other electronic devices.

Since all modern utility-interactive inverters are microprocessor controlled, all have internal clocks that can be expected to run at frequencies in the MHz range. Hence, all are subject to the radiative and conductive emission standards as set forth in Federal Communications Commission (FCC) Part 15 of the Code of Federal Regulations. Each unit must have a label that indicates compliance.

Conducted emission will normally not be a problem because of the low-pass filters needed at the output of an inverter to average out the synthesized sinusoidal signal. High-frequency currents that might otherwise be conducted to the utility interconnection are thus suppressed by this filter.

Radiated emission is caused by clock and other high-frequency currents circulating in sufficiently long conductors on the printed circuit boards of the inverter. The long, sometimes looped, conductors act as antennas and radiate the clock pulse signal. These same conductors act as receiving antennas for signals from other devices. Since the clock is not sinusoidal, it contains many harmonics that spread across the spectrum. An important part of suppression of radiated emission is the inverter cover. Hence, it is important to avoid operating the inverter without its cover for the prevention of electrical shock, but also for FCC compliance.

4.2.3.3 Surge Protection

PV system surge protection is not directly addressed in the *NEC* or in IEEE 1547, but surge protection in general is covered in *NEC* article 280. Surges may appear on either the utility line or on the PV line. Both are subject to lightning strikes, and the utility line is particularly susceptible to transients from events such as motor starting and stopping, which may cause sharp inductive voltage spikes. A good PV system design will provide MOV or SOV surge protection on both the dc side and the ac side of the inverter to protect the sensitive pn junctions in inverter electronic components.

Surge protectors act much like bilateral zener diodes with response times measured in nanoseconds. They can bypass hundreds of joules of energy that might otherwise enter the inverter or the PV modules and damage the components.

Lightning protection is a wise choice for systems installed in areas with a high incidence of lightning. The idea is to dissipate any uneven charge balance between ground and the atmosphere prior to the buildup of sufficient voltage to cause arc discharge.

4.2.3.4 Structural Considerations

The primary consideration in the structural design of utility-interactive PV systems is safety. This includes the safety of individuals as well as the protection of structures and other property that could be damaged as the result of mechanical failure.

Building codes are adopted and enforced by local jurisdictions to help ensure a safe environment in and around buildings. A major difference exists between the installations of utility-interactive versus stand-alone PV systems in that codes are much more likely to be enforced for the former. This is primarily because of the need to electrically connect the interactive system to the grid, which can result in multiple inspections: (a) by the electric utility service provider and (b) by local code officials, which may include both electrical and building inspections. Stand-alone PV systems need only be inspected as required by local building officials.

As will be discussed in Chapter 5, large area arrays may present a significant wind load on a structure. They are also subject to corrosion and degradation by ultraviolet sunlight components. Furthermore, they may be subjected to other conditions such as rain, snow, ice, and earthquakes. While most arrays are tested for mechanical performance, and while most array mounts have been pre-engineered to withstand the worst of loading conditions, it is still important for the engineer to determine that the specific array mounting is adequate with respect to existing structural parameters and local weather conditions. The American Society of Civil Engineers (ASCE) standard procedures and formulas for computing the various types of mechanical forces should be followed. In addition to the *NEC*, the PV engineer should keep abreast of the periodic updates to *Minimum Design Loads for Buildings and Other Structures* (SEI/ASCE 7-10) [9].

Of the various types of forces mentioned above, aerodynamic wind loading generally presents the most concern. At most locations around the world, the force effects due to wind loading are much higher than the other forces acting on the structure. For example, both dead- and live-weight loads added by a PV array to a building are usually less than 5 pounds per square foot (psf) (239 pa). In contrast, the computed wind loads are typically between 24 and 55 psf (1149 and 2633 pa), but sometimes greater than 80 psf (3830 pa), depending on location and the specific array mounting configuration. Earthquake zones also require special attention as discussed in ASCE 7 [9].

In addition to the above considerations, the design engineer must select and configure the array mounting materials such that corrosion and ultraviolet degradation are minimized. Also, if the array is mounted on a building, the structural integrity of the building must not be degraded, and building penetrations must be properly sealed such that the building remains watertight over the life of the PV array. Structural considerations and calculations are covered in detail in Chapter 5.

4.3 DESIGN CONSIDERATIONS FOR STRAIGHT GRID-CONNECTED PV SYSTEMS

4.3.1 Determining System Energy Output

Depending upon the technology, PV modules in common use may have sunlight-to-electrical conversion efficiencies ranging from 6% to over 20%. Assuming 15% overall module efficiency, this would mean that for irradiance of 1 kW/m^2, each square meter of PV area would generate 150 W if the modules are operating at their maximum power points. Hence, to generate 10 kW would require 66.7 m^2 (678 ft^2) of roof space, assuming the array to be roof mounted. For most dwellings, this area comes close to being the maximum area that may be south facing, especially if room is provided for access to the modules for maintenance, to comply with fire code(s) and to minimize wind loading. Hence, 10 kW is a reasonably practical size limit for residential PV systems, and systems under 10 kW are considered to be small systems.

The next question is how many kWh can the system be expected to generate? The answer depends upon the location of the system and on the overall conversion efficiency from dc array output to ac inverter output. While some areas receive upwards of 7 average daily peak sun hours over the year, other areas receive less than 4. A 10-kW PV array can thus be expected to generate somewhere between 40 and 70 kWh/day, or 1200 and 2100 kWh/month for an annual average array output energy. Loss mechanisms, however, generally result in an overall conversion efficiency between 65% and 80%, depending on whether the system incorporates battery backup and whether a maximum power point tracking charge controller is used for systems with battery backup. Figure 4.2 shows loss mechanisms that occur between the generation of electrical power by the individual modules and final delivery of power to the utility grid.

Historically, it was difficult for manufacturers to produce identical PV modules, even of the same model number. It was not uncommon for modules of the same model number to have a tolerance in rated output power of ±10% or more. Currently, however, improvements in manufacturing quality control, combined with I–V testing of each individual module, enable manufacturers to market modules that are guaranteed to produce at least their rated power, and possibly up to 5 W in excess of rated power. Any module testing at 5 W more than rated power gets sorted out and marketed with a 5 W higher rated power. As a result, at this point in history, minimal array degradation results from module mismatch, whereas up to about 2005, it was prudent to include a 10% mismatch factor when estimating array performance.

Module output does degrade as modules age, but manufacturers generally warrant annual power decrease to be less than 0.5%. Thus, a 25-year-old module should

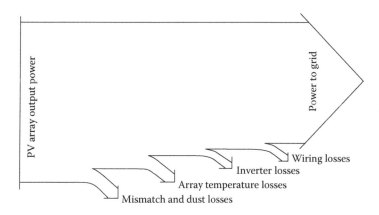

Figure 4.2 Loss mechanisms between module output and inverter output.

still produce at least 87.5% of its original rating. Other than mismatch and aging, the only other significant degradation factor is dust or dirt on the modules, which can be minimized by cleaning the modules as necessary.

Returning to the question of annual kWh delivered to the grid from the array, if the loss mechanisms are accounted for, it is seen that, depending upon the specific losses and on the specific irradiation data for the site, a 10-kW array can be expected to deliver somewhere between 800 and 1800 kWh/month to the utility grid or to PV system owner loads.

A number of computer programs are available for calculating monthly and annual PV system performance. Some are more detailed than others, such as including more worldwide weather and sunlight data, and have a price tag. Others are free (i.e., the government already paid for them). Perhaps the most popular free program for PV system performance analysis is the System Advisor Model (SAM) which can be downloaded from the U.S. National Renewable Energy Laboratory (NREL) [14]. By inputting location and array orientation information along with specific system component information (module models and inverter models), estimated monthly and annual system performance and savings can be obtained. A simplified version of this simulation program is based on the PVWatts model that does not incorporate performance specifications of modules and inverters.

Since the average residential electrical consumption is in the 1000-kWh/month range, depending upon location once again, it appears that installation of a 10-kW PV system on every rooftop can produce sufficient kWh to exceed the demands of the dwellings upon which the PV systems are installed. Of course, the PV systems will be generating electrical energy only while the sun is shining, so if all electricity used is to be PV generated, obviously massive storage capability will be required. One way to reduce the storage needs is to reduce nighttime electrical consumption. This, of course, is contrary to most current utility marketing schemes, since it is economically beneficial to them to operate existing large nuclear and fossil generation at nearly constant capacity all the time. This requires nighttime electrical use to distribute their load more evenly.

One major advantage of distributed electrical generation on rooftops is the elimination of power plant siting concerns. Most consumers tend to be happy to install PV systems on their roofs, whether they own the system themselves or whether the system is owned by someone else. Customer surveys consistently have shown customer support of cleaner generation, even if it has a premium price. Now that in many instances, PV-generated electricity carries a price tag below utility rates, customer interest is now no longer limited to those who are willing to pay extra for their systems. Furthermore, if the choice is between a PV system on their roof versus a coal or nuclear plant nearby, the PV system becomes an even more attractive choice for most consumers. For those who perceive PV systems as aesthetically unappealing, it must be remembered that beauty is in the eye of the beholder.

4.3.2 Array Installation

The preferred orientation of a PV array will normally be south facing (in the Northern Hemisphere), mounted in a fixed position, and tilted at 90% of latitude. This positioning will usually result in maximum annual kWh generation by the array. However, generally little is lost if the array is within 20° of south with a tilt within ±15° of latitude. A check with NREL SAM will verify this statement for any specific location.

Depending upon the duration of utility peak load and the difference between utility peak load and utility base load, it may be desirable to mount the array with a different exposure. Although the alternate mounting may result in generation at preferred utility times, the annual kWh production will most likely be reduced. Thus, unless the utility has a preferred rate for peak electricity, it may not be worth it to the system owner to orient the array to coincide with utility peaking. In some cases, a compromise can be had, especially when the roof of the dwelling has more west-facing exposure than south-facing exposure. If the array has multiple exposures, separate inverters will be required for each exposure unless a single inverter has multiple maximum power point tracking (MPPT) inputs. Again, NREL SAM can be used to estimate the performance of any orientation or orientations of the array.

4.3.3 Inverter Selection and Mounting

It should be remembered that inverters are not necessarily noise-free. In a manner similar to magnetic fluorescent light ballasts and other magnetic loads, the inverters with transformers may produce a 120-Hz hum or they may have an internal cooling fan. They should thus be mounted in as practical a location as possible from an electrical and mechanical perspective, but should also not be mounted where noise will be an annoyance. Generally, mounting the inverter outdoors or in a garage or carport is acceptable. If it is mounted near an air conditioner compressor or a swimming pool pump, the inverter noise will be pretty much masked by the other noise. If a transformerless inverter is used, noise is less of a problem.

When selecting an inverter, it is important to remember to specify a utility-interactive unit, since there are a wide variety of units available. Some are straight utility interactive, some can be used as utility interactive as well as stand-alone, and some are intended for stand-alone use only. While a wide range of performance specifications may be available for any unit, it is important to identify those parameters that are most significant. Features to consider may include any or all of the following:

IEEE 1547 and UL 1741: If the inverter has a UL 1741 listing [15], then it has been tested to IEEE Standard 1547 and consequently complies with *NEC* requirements. For small systems, all voltage and frequency trip points are factory preset and are not field adjustable.

Power rating: Maximum inverter output power is normally specified at an ambient temperature of 25°C (77°F). Surge power rating is less important for a utility-interactive inverter, since any surges encountered in the load can generally be met by the utility. For a stand-alone inverter, surge capability is more important. Utility-interactive units with battery backup act as stand-alone units if the utility disconnects, so in this case, the surge capacity becomes relevant.

Peak efficiency: It is important to note the input power over which the stated peak efficiency is obtained. Since the utility-interactive inverter always is loaded by the utility connection, if it is designed with maximum power point tracking, it will deliver maximum power to the grid over a wide range of PV input power and will operate close to peak efficiency over most of the maximum power point tracking range. Modern straight grid-connected inverters should have peak efficiencies in excess of 96% over most of the operating range.

Maximum power point tracking range: Look for the widest possible range of input voltages over which the unit is capable of tracking maximum power for greatest flexibility in PV array design. Some inverters have multiple MPPT inputs.

No-load power consumption: No-load power consumption should be very low, although normally a straight grid-connected inverter will operate at maximum power output during the day since the grid always provides a load for all available PV power, but during nighttime hours, a straight grid-connected inverter should consume only a small amount of power to keep its control circuitry energized.

Nighttime losses: If the unit operates on power supplied by the PV system, then there will be no nighttime losses. If the unit has a stand-alone mode of operation, then one can normally expect it to have a "sleep" mode that consumes very little power if no loads are connected. Straight grid-connected units with power ratings below 10 kW will generally have power losses of a few watts.

Warranty: Many quality inverters carry a 10-year warranty, sometimes extendable. Check what the warranty covers.

Code compliance: UL 1741 listing indicates compliance with IEEE 1547. It is possible to purchase inverters that incorporate additional *NEC* compliance components, such as ground fault protection, arc fault protection, input and output disconnects, fusible combiners for multiple string inputs (so strings will not need to be fused in a remote or difficult-to-access location), and weatherproof packaging.

Data monitoring: Most grid-connected inverters provide optional data monitoring capability by digitizing system parameters and making them available on an output bus. Some monitoring is available at the inverter and some units provide for remote data logging via telephone dial-up, Internet, or other communication means.

Rapid shutdown: The 2014 *NEC* requires that PV systems mounted on buildings have a means of rapid shutdown, which may be incorporated into inverter design, especially the requirement for quick discharge of capacitors.

4.3.4 Other Installation Considerations

Prior to commencing the installation, since the system will very likely be inspected by both the utility and the local municipality, it is important to first complete all the necessary permitting paperwork. This may offer an excellent opportunity for the engineer to perform an educational public service by answering any questions either authority may have about the installation.

Persons not familiar with IEEE 1547, UL 1741 and Article 690 of the *NEC* may be reluctant to accept the straightforward installation allowed by these codes and standards. It is possible that the most time-consuming part of the installation may be convincing the inspectors that the installation will be safe, competent, and code compliant. Fortunately, electrical inspectors tend to be well-versed on the *NEC* and are generally interested in new technology.

It is also conceivable that a zoning restriction on installation of rooftop solar systems on the street side of a building may exist. The permitting process may then involve a trip or two to zoning boards or town council meetings, perhaps with an attorney or a delegation of solar enthusiasts.

4.4 DESIGN OF A SYSTEM BASED ON DESIRED ANNUAL SYSTEM PERFORMANCE

4.4.1 Array Sizing

PV system design may be targeted at the achievement of a variety of objectives. In this example, it will be assumed that the objective is to provide a specified percentage of the electrical consumption of an occupancy. In this case, the building will be a residence in Oklahoma City, OK that has an average monthly electrical consumption of 1000 kWh. The design goal is to provide 50% of the annual electrical consumption with a grid-connected PV system, which amounts to an average of 500 kWh/month. Since it is desired to provide 50% of the annual consumption, this amounts to 6000 kWh/year.

The dwelling has 600 ft^2 (55.7 m^2) of unobstructed, unshaded, south-facing roof that has a 5:12 slope. In other words, for every 12 horizontal units, the roof rises 5 vertical units. So the roof slope is $\tan^{-1}(5/12) = 22.6°$.

While a completely formal analysis would involve estimating all the factors that affect dc–ac conversion efficiency, along with looking up the peak sun hours for each month for Oklahoma City, for this example, NREL SAM [14] will be used instead. By opening up SAM, and choosing the PVWatts with no financials option and selecting Oklahoma City as the location, with a south-facing roof with a 22.6° slope, it is possible to iterate array sizes in order to determine the approximate array size needed to produce the annual 6000 kWh.

Perhaps the easiest way to proceed is to start with an array size of 1 kW, using all of the default settings. If this is done, the annual ac kWh production is found to be 1532 kWh. Thus, to generate 6000 kWh annually, it would appear the array size

GRID-CONNECTED UTILITY-INTERACTIVE PHOTOVOLTAIC SYSTEMS 133

will need to be 6000/1532 = 3.916449086 kW. Just to confirm this, entering 3.92 kW as the array size yields 6005 kWh/year.

Before moving on to select an inverter and modules, the reader should experiment with the entries in the SAM dialog boxes to explore the variations in monthly and annual system output for various values of array tilt and array azimuth. The first thing to notice is that up to 20° either side of south does not make much difference on annual system performance. Then it is also interesting to note that any tilt within about ±15° of latitude (35.4°) also does not make much difference on annual results. At a tilt of 35.4°, annual performance is estimated to be 6084 kWh, a 1.0% increase. Thus, mounting parallel to the roof is clearly the best option from structural, cost, and aesthetic perspectives. Since the system is grid-connected, normally there will not be concern over monthly performance, as long as the desired annual performance is achieved.

Another very interesting simulation to perform with this array is to look at the hourly data over the year. This is easily done by selecting time series to obtain simulations of array output each of the 8760 hours of the year. When time series is selected, numerous parameters can be displayed, one or two at a time, as indicated on the right-hand side of the screen. For this example, choose *ac inverter power* at the bottom of the list and then zoom in on the highest power peak shown over the year. Note that peak power at any hour never exceeds 3.6 kW. This number is important when selecting an inverter for the system. In fact, returning to the SAM System Design page, under System Parameters, the program shows a rated inverter size of 3.56 kW, which is consistent with the simulation results.

It should be remembered that the data used in SAM simulations are based on a typical meteorological year (TMY), which means that the actual system output on any single day will probably not be as predicted by the model, since the model is based on months of past years. But on a monthly basis, the prediction for monthly output is pretty good, and the prediction for yearly output is even better, since it essentially averages all the daily/hourly data over an entire year.

4.4.2 Inverter Selection

Since the maximum inverter output has been determined to be 3.56 kW (ac), it now makes sense to find an inverter or two that are rated to handle this amount of output power. When searching for a suitable inverter, it must be remembered that it will be unlikely to find an inverter rated exactly at the maximum estimated inverter output power. In fact, it may not be possible to find a module combination that will produce exactly the rated array power. Table 4.8 shows the input and output

Table 4.8   Input and Output Performance Characteristics of Two Different Inverters

Inverter #	Rated ac Power (W)	Max Array Power (W)	Max dc V_{in} (V)	V_{in} MPPT Range (V)	I_{in} Max (A)	V_{ac} Out (V)	I_{ac} Out Max
1	3840	4400	600	$125 < V_m < 500$	24	240	16
2	3800	4400	600	$230 < V_m < 500$	17.8	240	15.8

characteristics of two inverters that can handle the calculated array output, provided that the array output does not exceed the maximum allowed inverter input voltage, nor does the array output fall below the lowest maximum power point tracking voltage level. Before selecting the inverter, since the main difference between the specifications of the two inverters is the MPPT range, it will be useful to explore PV array operating voltage and current options.

Other considerations in inverter selection may include whether or not the inverter includes ac and/or dc disconnects, source-circuit combining inputs, and, of course, the need for the inverter to be listed to UL 1741. Then there are factors such as peak efficiency, CEC (California Energy Commission) rated efficiency, inverter size, weight, and weather rating (outdoor or indoor), whether the inverter incorporates GFDI and/or arc fault circuit interruption (AFCI), and provision for rapid shutdown, which may influence the selection decision.

The next step in the system design is to select a module or modules and then see if the modules can be configured to meet the array power requirements and inverter input limits.

4.4.3 Module Selection

Assume that after searching the Internet for suitable modules, four possibilities have been found, having the performance characteristics as shown in Table 4.9. If ambient temperatures range from −20°C to +38°C (−4°F to 100°F), then Equation 3.5 can be used to determine the operating temperature range of the module. For the low-temperature limit, solve $T_C = T_A + (NOCT - 20)G/0.8 = T_A = -20°C$. Note that this assumes G = 0, which would be close to the case at sunrise. The assumption is that the module achieves its lowest temperature just after sunrise. For the high-temperature limit, solve $T_C = 38 + (47 - 20)1/0.8 = 72°C$ (162°F) (or 71°C if NOCT = 46°C). These cell/module temperature limits are now used to determine the maximum and minimum number of modules that can be used in a single source circuit. The calculation of the maximum open circuit voltage of the modules is required by *NEC* 690.7 to ensure that the maximum array open circuit voltage will not exceed 600 V, but if the selected inverter has a maximum input voltage less than 600 V, then the array voltage must not exceed this value or the inverter may be damaged. The determination of the minimum module operating voltage, V_m, at high temperatures is needed to ensure that the array output voltage will not fall below the minimum allowable value for the inverter maximum power point tracking range.

Table 4.9 Electrical Characteristics of Four Different PV Modules

Module #	V_{oc} (V)	I_{sc} (A)	V_m (V)	I_m (A)	$\Delta V_{oc}/\Delta T$ (%/°C)	$\Delta V_m/\Delta T$ (%/°C)	NOCT (°C)
A	48.3	5.9	39.8	5.5	−0.292	−0.39	47
B	48.8	6.43	40.5	6.05	−0.254	−0.324	46
C	39.5	9.71	31.2	9.07	−0.30	−0.45	47
D	46.9	9	39.3	8.4	−0.324	−0.432	46

GRID-CONNECTED UTILITY-INTERACTIVE PHOTOVOLTAIC SYSTEMS 135

Since the module temperature coefficients are expressed in %/°C departure from Standard Test Conditions (STC) at 25°C, it is useful to tabulate the maximum module V_{OC} values and the minimum module V_m values for a single module of each type. When writing the formulas for these determinations, it is important to keep all the minus signs in order so that the resulting values make sense. The formulas become

$$V_{OC}(\max) = V_{OC}(25°C)\left[1 + (T_{\min} - 25)\frac{\Delta V_{OC}}{\Delta T}\right] \quad (4.8)$$

and

$$V_m(\min) = V_m(25°C)\left[1 + (T_{\max} - 25)\frac{\Delta V_m}{\Delta T}\right], \quad (4.9)$$

where temperatures are expressed in °C and the temperature coefficients are expressed in %/°C, which, in turn, is expressed as a decimal fraction.

Expressing the temperature coefficients as decimal fractions and substituting the known maximum and minimum module temperatures yield the values listed in Table 4.10 for each of the four selected modules. Using the values in Table 4.10, it is now possible to determine the maximum and minimum number of modules of each type that can be used in source circuits for each inverter.

Both inverters allow a maximum input voltage of 600 V, so dividing 600 by the maximum module V_{OC} and then *rounding down* gives the maximum number of modules that can be connected in series and still keep the maximum series voltage below 600 V. For example, for Module 1, 600/54.6 = 10.98, which rounds down to 10.

Inverter 1 has a minimum MPPT voltage of 125 V, so dividing 125 by the minimum module V_m and then *rounding up* gives the minimum modules allowed per source circuit. For example, for Module 2, 125/34.3 = 6.70, which rounds up to 7.

Since Inverter 2 has a minimum MPPT voltage of 230 V, the minimum number of modules per source circuit is obtained by dividing 230 by the minimum module voltage and rounding up.

Table 4.11 shows the allowable numbers of modules for each inverter, for one and two source circuits of each module, which will not exceed the rated array power for either inverter. Note that when more than one source circuit is used, all should contain equal numbers of modules unless more sophisticated systems are used.

Table 4.10 Tabulation of V_{oc}(max) and V_m(min) for the Modules of Table 4.9

Module	V_{oc}(max)	V_m (min)
A	54.6	32.5
B	54.4	34.3
C	44.8	24.6
D	53.7	31.3

Table 4.11 Summary of Allowed Numbers of Modules That Fall within Inverter Input Voltage Limits

Inverter	Allowed # of Module A	Allowed # of Module B	Allowed # of Module C	Allowed # of Module D
Inverter # 1 with:				
One source circuit	4, 5, 6, 7, 8, 9, or 10	4, 5, 6, 7, 8, 9, 10, or 11	6, 7, 8, 9, 10, 11, 12, or 13	4, 5, 6,7, 8, 9, 10, or 11
Two source circuits	8, 10, 12, 14, 16, 18, or 20	8, 10, 12, 14, 16, or 18	12 or 14	8, 10, 12, or 14
Inverter #2 with:				
One source circuit	8, 9, or 10	7, 8, 9, 10, or 11	10, 11, 12, or 13	8, 9, 10, or 11
Two source circuits	16, 18, or 20	14, 16, or 18	N/A	N/A

Note: Inverter input power limits are not considered so that some of the module combinations may not meet inverter array power restrictions.

The final step in module selection is to determine the number of modules that most closely match the required array power.

Since the minimum required array power is 3920 W, one can conclude that it will require 17.81818 of Module A, 16.0 of Module B, 14.0 of Module C, or 11.87879 of Module D. Noting that fractional modules are not available, the conclusion is that it will require 18 of Module A, 16 of Module B, 14 of Module C, or 12 of Module D to achieve the design goal.

Comparing these numbers with the allowed numbers of modules as shown in Table 4.11, one can conclude that it is acceptable to use 9 of Module A in each of two source circuits, regardless of which inverter is used. For Module B, two 8-module source circuits also will work for either inverter.

For Module C, because of the lower V_{OC}, it will take two source circuits of seven modules each provided that Inverter 1 is used. For Inverter 2, it is not possible to use either a single 14-module source circuit or two 7-module source circuits. For Module D, once again Inverter 1 is the only inverter that will work. For this module and inverter, two source circuits of six modules each will be needed to achieve the design goal.

So, for the possible module and inverter combinations, the array power options are either $18 \times 220 = 3960$ W if Module A is used, 3920 W if Module B is used, 3920 W if Module C is used, and 3960 W if Module D is used.

So all of the allowable module/inverter combinations are reasonable choices and other considerations such as price, warranty, or which modules best fit on the roof may be taken into account.

4.4.4 Balance of System

Assuming that Inverter 1 and Module C meet further requirements of cost, availability, warranty, and other features that may be desired by the owner, the rest of the system can now be designed. Selection of the array mount will be saved for the next chapter, so this means that all that is left is to select module extension wire, a junction box, wire and conduit to connect between the array junction box and the inverter, dc and ac disconnects if they are not already incorporated into the inverter, dc ground fault and arc fault detection and interruption if they are not already incorporated into the inverter, rapid array dc shutdown provisions, and wiring from the ac side of the inverter to the PUC.

4.4.4.1 Wiring from Array to Rooftop Junction Box

The modules come with permanently attached positive and negative leads with #12 (3.31 mm^2) copper conductors with type PV insulation. The leads are 39.37 inch (1 m) long, which means that each positive-to-negative connection involves 6.6 ft (2 m) of #12 copper wire. This length is long enough for the positive lead of one module to connect to the negative lead of the adjacent module, or, for that matter, for alternate modules to be connected to each other.

If the 14 modules can be placed in two rows of 7, as shown in Figure 4.3, then each row of modules can be connected to each other in series and to a junction box, without any additional extension wire. In some cases, however, it may be necessary to install the modules in a single row of 14, or some other combination of modules and rows, depending upon the particular roof shape. If this should be the case, it may be necessary to use one or two additional pieces of PV extension wire to connect to the module wiring to extend to the rooftop junction box.

For the module configuration of Figure 4.3, there is a total of 6.6 ft (2 m) of wire per module, so that the total length of open wiring between modules and junction box for each source circuit is 46.2 ft (14 m). This information will be needed in calculating the voltage drop between the array and the inverter.

As an aside, this is a good time to note that the loop of wire among the modules acts as an antenna for electromagnetic disturbances. Since the voltage generated in

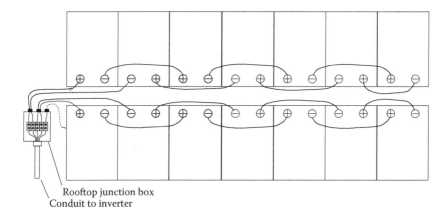

Figure 4.3 Possible rooftop arrangement of PV modules showing transition from open wiring to wiring in conduit via rooftop junction box.

a loop from a time-varying magnetic flux density is proportional to the area of the loop, it is important to *minimize the area of the loop of wire to minimize coupling to nearby lightning strikes*. A secondary precaution to protect the equipment is to install a surge arrestor at the point where the array conductors enter the inverter. This will be shown on the final system circuit diagram (see Figure 4.5 later in this chapter).

Note also that a grounding conductor is shown in Figure 4.3. The *NEC* requires that listed grounding techniques must be used to ground all metallic components of the PV module frames and the array mounting structure. The size of the grounding conductor is determined by the requirements of *NEC* Section 690.45, which indicates that a system with GFDI is to comply with *NEC* Section 250.122. Since essentially all straight grid-connected inverters that comply with UL-1741 have internal GFDI provisions, including the selected inverter for this design, this means that *NEC* 250.122 will apply.

Since *NEC* 250.122 specifies the size of the grounding conductor in terms of the circuit breaker or fuse that protects the circuit conductors, it will thus first be necessary to determine the sizes of the circuit conductors and the corresponding fuse size.

4.4.4.2 Wire and Conduit from Rooftop Junction Box to Inverter

To determine the proper wire size for connecting between the rooftop junction box and the inverter, the distance from the junction box to the inverter must be known and the highest anticipated ambient temperature must also be specified. Typically in a residential PV installation, the distance between the rooftop junction box and inverter will be less than 100 ft (30.48 m), unless the dwelling is unusually large. For this example, a distance of 60 ft (18.3 m) will be assumed. Also, when the maximum array temperature was determined, it was based on a maximum ambient temperature of 100°F (38°C).

The wire size must be based on satisfying two constraints: ampacity and voltage drop. Table 4.2 shows the ampacities of a range of conductor sizes at a temperature of 30°C (86°F). These ampacities are excerpts from *NEC* Table 310.15(B)(16), which applies to conductors in electrical conduit. If the conductors are exposed to ambient temperatures higher than 30°C, then they must be derated according to the correction factors of several tables in *NEC* 310.15.

NEC 690.8(A) defines maximum current for (1) PV source circuits, (2) PV output circuits, (3) inverter output circuits, (4) stand-alone inverter input circuits, and (5) dc–dc converter output circuits. For PV source circuits, the maximum current is defined as 125% of the rated module short-circuit current. This takes into account cloud focusing, which has been observed to increase module output current by as much as 25%.

NEC 690.8(B) requires that the maximum current as defined in *NEC* 690.8(A) be applied to the criteria listed in *NEC* 690.8(B)(1) and *NEC* 690.8(B)(2), such that the larger of the two results from these sections will be used to determine the required conductor ampacity at 30°C.

NEC 690.8(B)(1) involves increasing the maximum current of *NEC* 690.8(A) by 25% before applying any correction factors for ambient temperature or conduit fill.

This results in two successive applications of 25% increases in the module short-circuit current, which is the familiar 156% multiplier of previous editions of the *NEC*. In this particular case, with $I_{SC} = 9.71$ A, the 156% result is 15.15 A as the required 30°C wire ampacity *if this section applies*.

NEC 690.8(B)(2) involves applying the adjustment factors for ambient temperature and conduit fill to the maximum current of *NEC* 690.8(A). Table 4.12 summarizes the adjustment factors found in *NEC* 310.15 for ambient temperature correction of conductors with insulation rated at 90°C (194°F), and Table 4.13 summarizes ambient temperature correction factors from *NEC* 310.15 for conduit run across rooftops. Table 4.14 summarizes derating factors based upon conduit fill from *NEC* 310.15.

For this example, we have determined that the highest expected ambient temperature is 100°F (37.8°C). Applying the rooftop correction factor assuming the conduit to be mounted 1 inch (2.54 cm) above the rooftop adds another 22°C, resulting in an ambient temperature correction for 59.8°C, which is, from Table 4.12, 0.71.

Since there will be two current-carrying conductors from each source circuit in the conduit plus a noncurrent-carrying grounding conductor, the derating factor for four current-carrying conductors in conduit must be applied. From Table 4.14, this factor is 0.8. Thus, the overall combined derating for conductor temperature and conduit fill will be $0.71 \times 0.8 = 0.568$. Dividing the maximum current from Section 8(A) by this combined correction factor results in a required ampacity of $1.25 I_{SC}/0.568 = 21.37$ A. Since 21.37 A > 15.15 A, the required 30°C ampacity must

Table 4.12 Summary of Ambient Temperature Correction Factors for Wire with 90°C Insulation

Temp (°C)	21–25	26–30	31–35	36–40	41–45
Correction	1.04	1.00	0.96	0.91	0.87
Temp (°C)	46–50	51–55	56–60	61–65	66–70
Correction	0.82	0.76	0.71	0.58	0.41

Source: NFPA 70 National Electrical Code, 2014 Ed., National Fire Protection Association, Quincy, MA, 2013.

Table 4.13 Ambient Temperature Correction Factors for Conduit Run across Rooftops

Height of conduit above roof (in.)	0–0.5	0.5–3.5	3.5–12	>12
Add to ambient (°C)	33	22	17	14

Source: NFPA 70 National Electrical Code, 2014 Ed., National Fire Protection Association, Quincy, MA, 2013.

Table 4.14 Derating Factors Based on Conduit Fill

Number of current-carrying conductors	0–3	4–6	7–9	10–20	21–30
Derating factor	1.0	0.8	0.7	0.50	0.45

Source: NFPA 70 National Electrical Code, 2014 Ed., National Fire Protection Association, Quincy, MA, 2013.

be no less than 21.37 A. Referring back to Table 4.2, #14 wire with 90°C insulation is rated at 25 A at 30°C ambient temperature. Thus, provided that #14 wire does not result in excessive voltage drop between junction box and inverter, it will serve the purpose. So the next step is to check the total percentage voltage drop (%VD) between the array and the inverter.

When using Table 4.14, it is important to identify those conductors that are considered to be current-carrying versus those that are not considered to be current carrying. In particular, *grounding* conductors are not considered to be current-carrying. Grounding conductors must be distinguished from *grounded* conductors. Grounding conductors are for the specific purpose of establishing ground potential on all metal parts of an electrical system. Grounded conductors are current-carrying conductors that are connected to the grounding conductor *at only a single point*. This way the grounded conductor is maintained at a potential close to ground, but no current can flow from the grounded conductor to the grounding conductor because the single connection point does not provide for a return path for current flow. In a properly wired system, the voltage drop in a circuit will not exceed 3%, and will usually be less than 2%. In this case, if half the voltage drop is on the ungrounded conductor and half is on the grounded conductor, then the maximum voltage above ground potential of the grounded conductor will be about 1% of the source voltage.

To keep track of all the ampacity and derating factors that influence the selection of the source-circuit and output-circuit conductors, it is useful to construct a small table, or, more conveniently, a spreadsheet.

The next step is to calculate the voltage drop in the source circuit. For the voltage drop calculation, normally V_m and I_m of the source circuit are used. In some instances, it may be preferable to use the value of V_m at an elevated temperature, but for a small PV system, this is usually not necessary.

There are two components to the voltage drop calculation. The first is the voltage drop in the open module wiring and the second is the voltage drop in the wiring between rooftop junction box and the inverter. Equation 4.1 along with Table 4.2 can now be used to determine the voltage drops.

For the module open wiring, since there is a total of 46.2 ft (14.1 m) in the complete loop of each source circuit, this is equivalent to a one-way distance of 23.1 ft (7.04 m). So, using 23.1 ft for d in Equation 4.1 and using $I_m = 9.07$ A and $V_m = 7 \times 31.2 = 218.4$ V gives

$$\%VD = 100 \frac{I}{V_s}\left(\frac{\Omega}{kft}\right)\left(\frac{2d}{1000}\right) = \frac{0.2 \times 9.07 \times 23.1}{218.4}(1.93) = 0.370.$$

Next, the voltage drop between rooftop junction box and inverter is calculated. For this calculation, two approaches can be used. The first approach is simply to use the smallest allowable wire, which, in this case, has been shown to be #14, and calculate the %VD for this wire run. An alternate approach is to solve the %VD formula for Ω/kft to determine the minimum allowable wire size and then compare this result with the ampacity-based calculation for minimum wire size. Since the

%VD formula has been used to determine the %VD for the module leads, this shows that for a maximum of 2% VD overall, the remaining allowable %VD is 1.63%. The minimum wire size can thus be determined from

$$\frac{\Omega}{\text{kft}} = \frac{(\%VD)V_S}{0.2Id} = \frac{1.63 \times 218.4}{0.2 \times 9.07 \times 60} = 3.27.$$

Since #14 wire has $\Omega/\text{kft} = 3.07$, this shows that #14 (2.08 mm^2) is acceptable, but that the overall voltage drop will be close to 2%. Because of the difference in cost between #12 (3.31 mm^2) and #14 for this amount of wire, perhaps it might be preferred to use #12 instead of #14. Problem 4.7 gives the reader an opportunity to check on the %VD if #12 THWN-2 is used between the rooftop junction box and the inverter instead of #14.

Allowed conduit wire fill is given in *NEC* Chapter 9 and supplemented by the tables in *NEC* Informative Annex C. In general, if one wire is run through conduit, the cross-sectional area of the wire may not exceed 53% of the cross-sectional area of the conduit. If two wires are run through a conduit, then the combined cross-sectional area of the wires may not exceed 31% of the cross-sectional area of the conduit, and if three or more wires are run through conduit, then the total conductor cross-sectional area may not exceed 40% of the conduit cross-sectional area. The practical reason for these percentages is the difficulty of pulling the wire through the conduit. In any case, because there are so many different wire and insulation types with different cross-sectional areas, even for the same wire size, and since there are also many different types of conduit, there are numerous variations in how many wires can be run in a given size and type of conduit. The serious design engineer will generally tabulate these numbers on a spreadsheet to expedite the calculation of conduit size, especially when different wire sizes are run in a single conduit.

For the present design, the total of four #12 THWN-2 current-carrying conductors and another #10 or #12 THWN-2 ground wire will fit into any type of half-inch (metric #16) conduit.

4.4.4.3 Rapid Shutdown

The system as designed to this point includes an array and a means of transferring the power generated by the array on the roof to a dc disconnect located at the inverter, which is located elsewhere, such as on the side of the building or in a garage, utility room, or basement. The significance of this situation is that whenever the sun is shining, the array is generating voltage, and this voltage cannot be shut down until it reaches the dc disconnect.

In previous editions of the *NEC*, this situation was recognized as a potential hazard, which resulted in the requirement that the energized source-circuit conductors be run in conduit from rooftop to inverter/dc disconnect. The 2014 edition of *NEC* now includes Section 690.12, which requires rapid shutdown of PV systems on buildings. Specifically, *NEC* 690.12 has five subsections that address control of specific conductors as follows:

1. If conductors remain within a building for more than 5 ft (1.5 m) or extend more than 10 ft (3 m) from a PV array, then they must be controlled.
2. Controlled conductors are allowed 10 seconds to discharge to a voltage less than 30 V and volt-amperes less than 240 VA.
3. The voltage and power listed in (2) are to be measured between any two conductors and between any one conductor and ground.
4. The rapid shutdown method needs to be labeled conveniently for emergency responders.
5. Rapid shutdown equipment needs to be listed and identified.

It should be noted that since the rapid shutdown concept is newly adopted, the *NEC* allows a significant amount of flexibility in the implementation of the requirement. As a result, it is expected that many creative ideas will result in a variety of methods to achieve the intent of this requirement.

For the purpose of this design example, rather than suggesting anything specific, it will simply be noted that rapid shutdown must be incorporated. A quick look on the Internet for rapid shutdown of PV systems will no doubt reveal new solutions on a regular basis. Several good articles have been published in trade journals that indicate early adopted methods of meeting the intent of this *NEC* provision as well as ways of getting it wrong [16].

4.4.4.4 Ground Fault and Arc Fault Protection

As safety features are recognized as desirable, industry is generally responsive in the development of desired features as long as they have reassurance that there will be a market for the product. This has been the case with ground fault and arc fault protection, both of which at this point have been incorporated as part of the input circuitry of inverters, with the exception of certain battery-backup systems where these features tend to be incorporated into charge controllers. Two of the features of the selected inverter for this design are internal ground fault and arc fault protection that meet *NEC* requirements. Ground fault and arc fault protection are discussed in Chapter 3, Section 10.

4.4.4.5 DC and AC Disconnects and Overcurrent Protection

As luck would have it, the selected inverter comes equipped with a dc disconnect, so the wires from the array are connected to the inverter array input terminals and the dc disconnect is already there. But the inverter does not have an ac disconnect, so one will need to be selected. Several considerations may influence the choice of ac disconnect.

The first consideration is the inverter rated ac output voltage and current, which is, for Inverter 1, 16 A at 240 V.

This gives several useful pieces of information, such as the minimum wire size to connect the inverter output to the PUC, the size of the circuit breaker to be used at the PUC if the PUC is to be in a circuit breaker panel, the minimum size of the circuit breaker panel busbars, the size of the ac disconnect, and whether or not the ac disconnect must be fused. The distance from the inverter to the PUC is also needed to confirm that the voltage drop in the selected wire from inverter to PUC is not too large.

Noting that #12 (3.31 mm^2) wire is rated at 30 A, but can only have 20 A of overcurrent protection per *NEC*, the first attempt at sizing the wire suggests to try #12. The ac disconnect need only be rated at 20 A, so, in principle, a two-pole 20-A snap switch could be used. The only reasons for not using a simple switch would be if the utility were to require an accessible, lockable disconnect with an external handle or if the PUC were to be at the line side of the main disconnect or if the inverter is within sight of the PUC, which would allow the circuit breaker at the PUC to also serve as the ac disconnect for the system. If the inverter is not within sight of the PUC, then it will need a separate disconnect for maintenance purposes to be located within sight of (usually next to) the inverter.

The *NEC* does not require a lockable disconnect, but some utilities require a lockable disconnect for redundant assurance that the inverter can be shut down if utility power is lost or if the utility has justifiable reason to believe the inverter output does not meet utility specifications.

If the PUC ends up at the line side of a main disconnect, either because *NEC* 690.64 (which references *NEC* 705.12) is not satisfied or because there is no space in the distribution panel for another two-pole, 20-A circuit breaker, then a 20-A fused or circuit breaker disconnect will be needed near the PUC.

For this example, it will be assumed that there is room in the distribution panel for another two-pole, 20-A circuit breaker as the PUC. It will also be assumed that the inverter will be located 6 ft (1.83 m) from the PUC. The reader is encouraged to use the voltage drop formula to determine the voltage drop for 16 A in #12 wire over a distance of 6 ft. If it is less than 2%, then #12 is adequate. If it is more than 2%, then larger wire will need to be used, but it will still have 20 A of overcurrent protection.

4.4.4.6 Point of Utility Connection

NEC 705.12(D)(3)(b) states that the sum of 125% of the output current ratings of all inverters that supply power to an electric panel plus the rating of the overcurrent protection device protecting the busbar shall not exceed the rating of the panel busbars. However, if the PUC and utility connections are at opposite ends of the distribution panel busbars, and the panel is labeled so that no one will be tempted to move the PUC circuit breaker, it states that the sum of 125% of the output rating of the inverter(s) plus the rating of the overcurrent protection circuit breaker feeding power to the panel must be less than 120% of the busbar rating.

The reason for this rule is that if a busbar is fed from both ends, Kirchhoff's current law requires that for each circuit breaker connected downstream from a feed, if there is a load on the circuit breaker, then the current to that load is no longer carried beyond that point on the busbar. Figure 4.4 shows how feeding the opposite end of a busbar with the PUC helps to distribute current more evenly on the busbar.

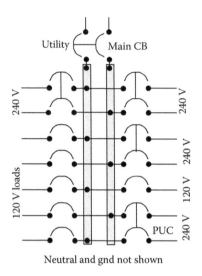

Figure 4.4 PV PUC at distribution panel.

The 20% rule implies that the busbar to which the 16-A inverter output is attached must have a rating of at least 100 A, presumably fed from the utility with a 100-A circuit breaker, since 120% of 100 A is 120 A and 100 A + 20 A = 120 A. Note that in this case, 125% of the inverter output is conveniently equal to the size of the circuit breaker (20 A) at the PUC. Since almost all circuit breaker panels are rated at 100 A or more, the 20-A circuit breaker can be attached to almost any 120/240 V panel that has space for it. In fact, depending upon whether the inverter has a setting for an output voltage of 208 V, the circuit breaker could also be connected to a 120/208 V panel as long as the increased inverter output current at 208 V is accounted for. In this case, however, with a 3800-W rated inverter output, the inverter output current would be 18.3 A and 125% of this value is 22.8 A, which would require a 25-A PUC circuit breaker. Since 22.8 A + 100 A exceeds 120% of the busbar rating of a 100-A distribution panel, a 100-A distribution panel used with a 120/208 service would be too small. However, it is extremely rare for a single-family residence to be served with 120/208 V. This is more common in apartments and condominiums.

When the PUC circuit breaker is switched on and the inverter output disconnect is switched on, the inverter output automatically will assume the phase of the utility voltage.

4.4.4.7 Final System Electrical Schematic Diagram

Figure 4.5 shows the final electrical schematic diagram of the system. Note that although the diagram shows #12 conductors between the rooftop junction box and the inverter, it is likely that Problem 4.7 will show that #14 will also be acceptable for this circuit. The 20-A circuit breaker at the PUC determines the size of the equipment grounding conductor that is required. *NEC* 250.122 indicates that for wiring protected

GRID-CONNECTED UTILITY-INTERACTIVE PHOTOVOLTAIC SYSTEMS

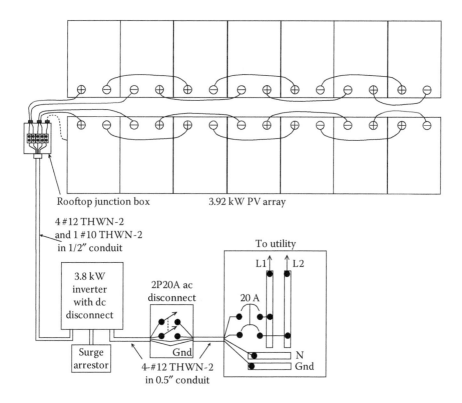

Figure 4.5 Electrical schematic diagram of 3.92-kW PV system.

by a 20-A circuit breaker, a #12 grounding conductor is adequate. Note that a #10 wire is used for array grounding. The reason for this is simply that a #10 wire has greater physical strength, and, since it will be a bare wire, the larger size should survive the outdoor environment better than a smaller conductor. The inverter installation manual will show where all the wire connections are to be made within the inverter enclosure.

Conduit sizes are determined from tables in Chapter 9 of *NEC* and from additional tables in Annex C of *NEC*.

The only other *NEC* requirement of note is that *NEC* 690.31(A) requires that if source-circuit or output-circuit conductors are run inside a structure, then the conductors must be guarded or installed in metallic conduit.

4.5 DESIGN OF A SYSTEM BASED UPON AVAILABLE ROOF SPACE

4.5.1 Array Selection

A common challenge in PV system design is to fit as much power onto the roof as possible. With a roof that is "just the right size and just the right shape," that is, rectangular, this may not seem to be much of a challenge. But in many areas of the country, finding a south-facing, rectangular roof is like finding a needle in a haystack. Odd roof shapes, shading, and roof protrusions all influence the placement of PV modules. In this example, the roof of Figure 4.6 will be used.

Roof diagrams may come in two forms. One form is a diagram showing the actual measurements of the roof, as in Figure 4.6. The other form is a plan view, where the projection of the roof onto the horizontal is shown. Plan views, believe it or not, are what are found on sets of plans for new construction. The plan view is also what is seen from satellite photos, such as on Google Earth. Plan sets (but not Google Earth) also generally indicate the slope of the roof in terms of rise/run ratios, that is, the ratio of vertical distance to horizontal distance. The slope of the roof is then simply

$$\theta = \tan^{-1}\left(\frac{y}{x}\right), \qquad (4.10)$$

where y is the vertical rise and x is the horizontal run of a right triangle. The horizontal dimension, x, is the projection of the actual roof dimension, r, in the direction of the slope. The actual dimension in the direction of the slope is the hypotenuse of the right triangle formed by x, y, and r. Thus, if x is known from the plan view, and the ratio y/x is known, then r can be found from

$$r = \frac{x}{\cos\theta}. \qquad (4.11)$$

While Equations 4.10 and 4.11 are not exactly rocket science, if the slope is not taken into account in the plan view by the designer, then the installer may end up with some surprises when laying out the array on the actual roof. The roof dimension perpendicular to the slope direction on the plan view is the same as the actual dimension.

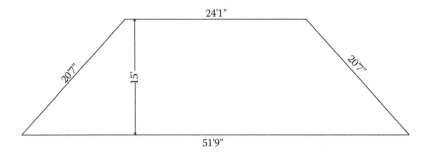

Figure 4.6 A roof in need of a PV array.

GRID-CONNECTED UTILITY-INTERACTIVE PHOTOVOLTAIC SYSTEMS 147

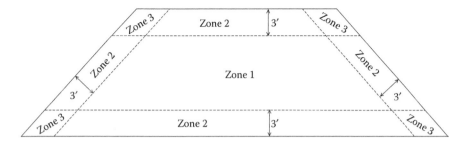

Figure 4.7 Identification of roof zones.

Chapter 5 will show that in hurricane-prone regions, there is good reason to keep modules at least 3 ft (0.91 m) from the edges or corners of a roof to minimize wind loading. However, in most regions where wind speeds are not expected to exceed 90 mph, as opposed to >125 mph hurricane winds, the designer need not be concerned about the modules ending up too close to the edges of the roof except for compliance with fire codes. Just for practice, this example will assume that the installation will be in a hurricane-prone region and the modules will be kept at least 3 ft away from the edge of the roof. When this is done, generally an added bonus of fire code compliance is also accomplished.

So the first step in determining an array layout is to mark the 3-ft (0.91 m) boundary on the roof of Figure 4.6, as shown in Figure 4.7. Figure 4.7 also shows the three wind load zones of the roof, as defined in ASCE 7-10 [9]. Keeping the modules 3 ft away from the edges and corners of the roof means keeping them in Roof Zone 1.

Since Figure 4.6 shows the actual dimensions of the roof, then the array layout can be accomplished by drawing the roof to scale and then drawing the proposed modules to scale. The proposed modules are then fitted onto the roof in what looks like a logical configuration to the designer, within any constraints that may be imposed by shading, protrusions, wind loading, or other loads such as snow loads. In fact, there may even be a constraint imposed by the location of the roof trusses. If the designer is willing to try more than one module type, each with different dimensions, then scale model drawings can be made of each module and scaled drawings are made for each possible module so that they can be fitted to the roof. Figure 4.8 shows three different modules drawn to scale showing the module dimensions.

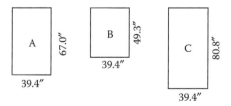

Figure 4.8 Three possible choices for modules.

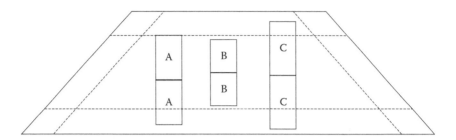

Figure 4.9 Checking module fit in Zone 1.

Figure 4.9 shows an attempt to create two rows of each type of module with the modules in portrait position, keeping the rows within Roof Zone 1. Note that for this roof, only Module B can be mounted in two portrait rows and still be kept completely within Zone 1. So Module B will be used in this design, even though a slight intrusion into Zone 2 is sometimes acceptable, as will be seen in Chapter 5. Figure 4.10 shows that 16 of the B-variety of module can fit within Roof Zone 1. So the next step is to look at the electrical characteristics of Module B.

The data sheet for Module B shows $P_m = 250$ W, $V_{OC} = 50.93$ V, $I_{SC} = 6.2$ A, $V_m = 42.8$ V, $I_m = 5.84$ A, $\Delta V_{OC}/\Delta T = -0.291\%/°C$, and $\Delta V_m/\Delta T = -0.39\%/°C$. The nominal operating cell temperature (NOCT) of the module is 46°C.

With $V_{OC} = 50.93$ V, it should be obvious that it is not possible to connect all 16 modules into a single source circuit, since $16 \times 50.93 \gg 600$ V. So the next logical possibility is two source circuits of eight modules each. Assume that the system will be installed in hurricane-prone South Florida, where the ambient temperature might be expected to fluctuate between −2°C and 37°C (28°F and 99°F). The maximum open circuit voltage for an eight-module source circuit, using Equation 4.8, is

$$V_{OC}(\max) = V_{OC}(25°C)\left[1+(T_{\min}-25)\frac{\Delta V_{OC}}{\Delta T}\right] = 8\times 50.93[1+(-2-25)(-0.00291)]$$
$$= 439.4 \text{ V} < 600 \text{ V}.$$

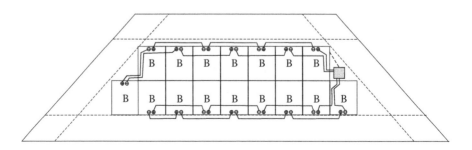

Figure 4.10 Portrait layout of Module B in high-wind region.

GRID-CONNECTED UTILITY-INTERACTIVE PHOTOVOLTAIC SYSTEMS 149

At the high-temperature end, the maximum module operating temperature can be found from Equation 3.5 to be

$$T_C = T_A + (NOCT - 20)\frac{G}{G_0} = 37 + (46 - 20)\frac{1}{0.8} = 69.5°C.$$

Thus, the minimum value of V_m can be found from Equation 4.9 to be

$$V_m(min) = V_m(25°C)\left[1 + (T_{max} - 25)\frac{\Delta V_m}{\Delta T}\right]$$
$$= 8 \times 42.8[1 + (69.5 - 25)(-0.0039)] = 282.9 \text{ V}.$$

4.5.2 Inverter Selection

If an inverter is available for which these high and low voltages fall within the inverter input voltage limits and for which the array power meets the inverter array power limits, then the inverter will work in the system. Noting that 16 modules at 250 W/module gives a 4000-W array power, and that the inverter will need to have a maximum dc input rating of 600 V, with a minimum MPPT voltage of less than 283 V, inverter characteristics can now be checked. It will also be nice if the inverter will have GFDI and AFCI at the input, a dc disconnect, and, perhaps, some provision for rapid shutdown.

These particular specifications might bring to mind the inverter used in the last example, and, indeed, that inverter (3800 W) will conveniently work for this example as well. In fact, both inverters considered for the previous example will work, since, in this case, the lowest maximum power input voltage is 230 V for the second inverter. So this time, inverter selection may be determined by another consideration, such as price, warranty, reputation of manufacturer, or any other desirable feature of whichever inverter is selected. The remaining calculations will be based on the use of Inverter 2 of Table 4.8 for no reason other than dealing with slightly different numbers for the following calculations.

4.5.3 Balance of System

In this example, there are two source circuits that must be run, either to a combiner box or directly to the inverter. It will be assumed that a rooftop junction box will be used for a transition from open wiring of the modules to wiring in conduit from rooftop junction box to the inverter. It will also be assumed that the inverter will be mounted to the wall of the occupancy within a few feet of the main disconnect of the occupancy and that the distance from the rooftop junction box to the inverter is 50 ft (15.2 m). It will also be assumed that the roof has a 4:12 pitch (18.43°).

4.5.3.1 Wiring from Array to Rooftop Junction Box

Figure 4.10 shows the proposed location of the rooftop junction box and an accompanying wiring layout. Again, modules have 39.37-inch (1 m) positive and negative leads of #12 PV wire, so an eight-module source circuit will have 52.5 ft (16 m) of module wiring. In the configuration of Figure 4.10, assume that the top row of modules plus the module in the bottom row furthest from the junction box will be one source circuit and the remaining eight modules in the bottom row will be the other source circuit. In this case, assuming that the modules are installed with the module junction boxes as shown, the leads can be connected as shown in the figure. One might imagine the challenge of connecting and strapping the module leads to the mounting frame if the module junction boxes were at the top of the bottom row and the bottom of the top row. However, the proposed wiring layout shown may involve using some short extension cables, so for estimating voltage drop in the module open wiring, a total round-trip of 60 ft (18.3 m) is a reasonable figure to use.

When the grounding conductor for the module frames and array mount frames is included, there will be five conductors running from the rooftop junction box to the inverter. If the conduit between rooftop junction box and inverter runs inside the structure, then the inside section of conduit must be metallic and it must be properly grounded.

This time an alternate, but equivalent method will be used to determine the necessary ampacity of the conductors from rooftop junction box to inverter. The array conductors are #12, and I_{SC} for each source circuit is known to be 6.2 A, so at this point, the size of the conductors in the conduit can be determined. To determine the highest ambient temperature to which the conductors in conduit are exposed, it will be assumed that the conduit is supported 1.5 inch above the roof for a short distance before it penetrates into the attic of the structure. Referring to Table 4.13, an additional 22°C must be added to the maximum ambient temperature (37°C) to correct for the conduit over the roof. So the highest wire temperature can be expected to be 59°C. The ambient temperature derating factor for the 90°C wire is 0.71, according to Table 4.12.

Next, an additional derating will need to be applied since there are two source circuits, which have a total of four current-carrying wires for the conduit. Thus, a derating factor of 0.8 must also be applied to account for the four current-carrying conductors in the conduit.

Next, 156% of I_{SC} = 9.67 A. If the derating factors are applied to 125% of I_{SC}, this results in an ampacity of 6.2 × ((1.25/0.71)/0.8) = 13.64 A, so the 30°C ampacity of the conductors from rooftop junction box to inverter must be greater than 13.64 A, since 13.64 A > 9.67 A. Thus, if #12 wire is used, with a 30°C ampacity of 30 A, this will be more than adequate as long as this wire size meets the voltage drop requirements. So now the last step is to check the voltage drop in the wiring between source circuits and inverter.

The total equivalent one-way distance between array and inverter is the sum of the equivalent one-way length of the open wiring of the modules (30 ft) plus the one-way length from rooftop junction box to inverter, which happens to be 84 ft (25.6 m). Using Equation 4.1 with I_m = 5.84 A and V_m = 342 V, to evaluate the voltage drop if #12 wire is used the entire distance, gives

$$\%VD = 0.2Id\frac{\Omega/kft}{Vs} = 0.2 \times 5.84 \times (84+30) \times (1.93 \div 342) = 0.751.$$

The worst-case %VD occurs when the modules operate at their maximum temperature, causing V_m of the source circuits to drop to 286 V. The resulting %VD would then be found by replacing 342 by 286 to get 0.90%, which is still well-below the allowed 2%. So the wiring from rooftop junction box to inverter can be #12 THWN-2 and easily meet all ampacity and voltage drop constraints.

4.5.3.2 Rapid Shutdown, Ground Fault, and Arc Fault Protection

The inverter selected combines the inputs of the two source circuits, provides internal dc ground fault and arc fault detection and interruption, and has convenient monitoring and communications capabilities. Rapid shutdown is, however, still to be accomplished. Since the concept and requirement of rapid shutdown is emerging as this is being written, rather than at this point specifying a particular method, it is interesting to explore what needs to be done.

One method might be to install a listed switch at the rooftop junction box. If it is switched off, then no current flows from array to inverter. However, it is possible that capacitors at the inverter input may take a long time to discharge, especially if the inverter ac output is switched off, which will likely happen if the house main is shut down. If the inverter has no load, then the input capacitors need to discharge into whatever equivalent parallel resistance may be present. If this takes too long, then the rapid shutdown criteria is not achieved, since the wiring from load side of the rooftop switch is open circuited at the array, but still connected to the inverter input, unless the inverter dc disconnect is switched off. This suggests that there be a communication link, either wireless or low voltage, between the rooftop switch and the inverter dc disconnect such that if either is shut down, the other will also be shut down. With this thought in mind, at this point it will assumed that such a system will be used.

4.5.3.3 DC and AC Disconnects and Overcurrent Protection

The inverter comes with an integrated dc disconnect, but an external ac disconnect will be required. The maximum continuous inverter input current rating is 17.8 A. When both source circuits are combined, they add up to a dc output short-circuit current of 12.4 A; however, this is not necessary because normally the inverter maximum input current from the array will be 11.68 A at full sun, and 125% of that with cloud focusing, which is still less than the rated input current.

With two input circuits, they do not necessarily need to be fused unless recommended by the module manufacturer. If one source circuit is rated to carry the full current of the other source circuit in the reverse direction, then no fuse is needed. However, generally the dc source circuits at the inverter input will be protected with fuses rated at approximately 156% of I_{SC}, or, in this case, 10 A.

The rated inverter output current at 240 V is 15.8 A. This requires wiring rated at 125% of this value, or 19.9 A, which suggests using #12 (3.31 mm^2) wire fused at 20 A, provided that voltage drop between inverter and PUC is acceptably low. If the inverter is not within sight of the PUC, a separate two-pole, 20-A, 250-V, ac disconnect switch will be needed near the inverter. The disconnect switch does not need to be fused as long as the PUC is connected to a circuit breaker that meets the requirements of the inverter manufacturer. So the wiring from inverter to PUC will consist of three #12 THWN-2 conductors. Note that two of these are phase conductors and one is the grounding conductor. A neutral conductor is not required with this inverter. So, once again, these three conductors easily fit in 0.5-inch (metric 16) conduit. In fact, it is acceptable that, depending upon the location of the inverter, it may be possible to run nonmetallic cable instead of conduit, as long as the cable is properly protected from physical damage.

4.5.3.4 Point of Utility Connection

This time, it is assumed that there is no room for an additional two-pole circuit breaker in the distribution panel. This means that the PUC will need to be at the line side of the main disconnect, unless 125% of the inverter output current added to the main disconnect rating is less than the rating of the distribution panel busbars (*NEC* 705.12(D)(2)). It is also necessary that the feeder conductors between the PUC and the distribution panel be rated to carry this same current, which might require installing larger feeder conductors, which may not be possible if the conduit is too small. When this method is used for the PUC, whether the PUC is located on the line side or the load side of the building main disconnect, it is necessary to follow *NEC* 240.21(B) on tap conductors. The procedure is to install a fused disconnect within 10 ft (3.1 m) of the main disconnect for the occupancy, so *NEC* 240.21(B) will apply. The issue here is that if #12 conductors are connected to the line side of the main disconnect, they will not have overcurrent protection between the fused disconnect and the main disconnect. This is permissible, provided that the fused disconnect is located within 10 ft (3.04 m) of the main disconnect, the conductors between the two disconnects are protected by conduit, and the ampacity of the main circuit breaker is no more than 10 times that of the proposed 20-A fuses. Furthermore, regardless of how tempting it may be, it is not acceptable to fasten the #12 conductors under the same lugs as the service conductors in the main disconnect. Either the lugs must be changed to accommodate both conductors or listed tap assemblies must be used on the service entrance conductors.

4.5.3.5 Estimating System Annual Performance

Now that the system design is essentially complete, before proceeding with the final electrical schematic diagram, the PVWatts version of SAM will be consulted to determine the expected system performance. Since the installation is in South Florida, West Palm Beach will be chosen as the site for the NREL SAM PVWatts analysis. Entering the pertinent data on roof slope (18.43°), roof orientation (210°), and array size (4 kW), with the default dc–ac conversion factor, produces an estimate of 5596 kWh/year, with estimated maximum production of 535 kWh in the month of May. If the array had been facing directly south, annual kWh production would be 5683 kWh.

GRID-CONNECTED UTILITY-INTERACTIVE PHOTOVOLTAIC SYSTEMS 153

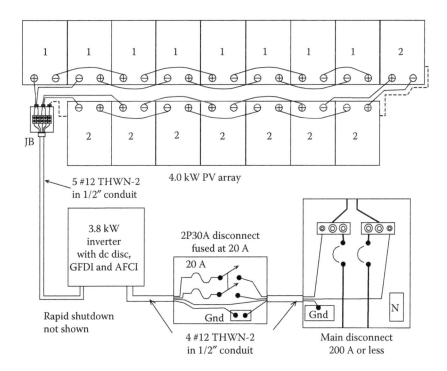

Figure 4.11 Electrical schematic diagram of 4.0-kW PV system.

4.5.3.6 Final System Electrical Schematic Diagram

Figure 4.11 shows the final system electrical schematic diagram. Note that this diagram is very similar to the previous system with the exception of the PUC at the main disconnect rather than at the distribution panel.

4.5.4 Extension of Design to Lower Wind Speed Region

Before leaving this example, it is interesting to explore how modules can be fit onto the roof if the modules are allowed to spill over into Roof Zones 2 and 3. This is generally the case for lower wind speeds, as will be determined in more detail in the next chapter.

Figure 4.12 shows that 31 modules can be conveniently mounted onto the roof. But there is a problem with this number of modules, since 31 is a prime number and a single 31-module source circuit will have all of its voltage parameters outside allowable *NEC* and inverter limits. And, of course, it is unlikely that fire code regulations would permit this amount of roof coverage, anyway. In any case, a better design would involve at least one less module and an inverter in the 7000 W range. Completion of this design is left as an exercise for the reader in Problem 4.8.

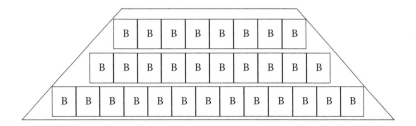

Figure 4.12 Lower wind speed array possibility with Module B.

4.6 DESIGN OF A MICROINVERTER-BASED SYSTEM

4.6.1 Introduction

Toward the beginning of the 1990s, the concept of ac modules was introduced by Ascension Technology, Inc. [17]. The idea was to include a small inverter for each PV module, with the inverter attached to the module itself. The inverters were designed to be straight grid-connected inverters that met the requirements for grid connection. At the time, however, the inverter was generally considered to be the least reliable system component, so the PV community met the idea with an element of skepticism. Even though the *NEC* included a section on ac modules, very few systems were installed.

Proponents of the idea persisted, however, with the argument that if each module had its own inverter, overall efficiency of the entire system would be improved because array mismatch would be eliminated. Furthermore, if one module were shaded, the rest would still operate at maximum efficiency and failure of one module or one inverter would not affect the operation of the rest of the system. In addition, if the array required modules to have a number of different orientations, since each module already had its own inverter, individual module orientation did not affect the performance of any other modules in the system. In any case, after about a decade of neglect, the ac module concept has made a comeback in the form of microinverters [18]. This time, however, rather than mounting the inverter directly to the module, the inverter is mounted close to (generally underneath) the module. Module choice is then governed by matching module voltage, current, and power ratings to the inverter ratings. The dc connecting leads of the module connect directly to the dc input of the inverter, and the 208 or 240 V ac outputs of the individual inverters are daisy-chained, that is, connected as current sources in parallel, with the maximum number of inverters being limited by the maximum load that can be handled by a circuit breaker, consistent with *NEC* requirements for the use of flexible cords. In addition, each inverter has its own built-in data monitoring capability so that a master system monitor can be used to observe the individual performance of each inverter as well as the overall system performance.

An important feature of microinverters is that they comply with UL 1741. This means that if the ac power is shut down at the PUC, the outputs of all microinverters in the system are immediately shut down, thus making the system automatically compliant with *NEC* 690.12 Rapid Shutdown requirements, since the PV system output is shut down at each individual module. Furthermore, since each individual module maximum output voltage is nearly always less than 80 V, especially those modules that are compatible with existing microinverters, no arc fault protection is required for the source circuit, that is, the module output.

Some newer models have provisions for operation with smart grids, including features such as fixed power-factor voltage and frequency ride-through, adjustable output power factor, remote updating to respond to changing grid requirements, configuration for variable grid profiles, and compatibility with battery management systems and storage systems.

4.6.2 System Design

Since the inverters come separate from the modules, it is necessary to determine whether an inverter will operate under the output conditions of a particular module. The inverters still have a defined MPPT range and maximum operating voltages and powers. Since $V_m < 60$ V for most modules, microinverters must incorporate boost converters rather than buck converters or, perhaps, buck–boost converters. At the time of this writing, additional modules are being incorporated into the operating range of microinverters. And, of course, new modules are being introduced to the marketplace at a rapid rate, as well.

For the purposes of this example, it will be assumed that the installation will be in Albany, NY on a structure that has a roof with which the architect took pride in incorporating six different slopes and orientations between east and west, not counting three more sections that face north or close to north. A 260-W module with $V_{OC} = 38.4$ V, $I_{SC} = 8.94$ A, $V_m = 31.4$ V, $I_m = 8.37$ A, $\Delta V_{OC}/\Delta T = -0.31\%/K$, and $\Delta V_m/\Delta T = -0.43\%/K$ and NOCT = 46°C has been selected. A suitable microinverter to match this module has $V_{dc}(max) = 48$ V, MPPT voltage range of 27–37 V, and maximum module power = 285 W. The rated inverter output voltage is 240 V and the rated output current is 960 mA, with maximum output power of 230 W. The maximum number of inverters per branch (i.e., connected in parallel to feed a 2P20A circuit breaker) is 17.

It will be assumed that the roof has room for 14 modules, to be located as indicated in Table 4.15.

Since the inverters are designed to be mounted to the module mounting rails, each module will have its inverter mounted essentially beneath the module. The connections are straightforward, since the two-module leads connect directly to the inverter.

Inverters are interconnected as current sources in parallel. Each inverter is connected to a cable supplied by the inverter manufacturer that includes an integrated equipment grounding conductor for the inverter. Thus, a single open-wire cable can be connected from inverter to inverter to a rooftop junction box. Between junction box and distribution panel, since this is an ac circuit, conventional ac wiring techniques are acceptable. Wire in conduit is still allowed, but it is also possible to use

Table 4.15 Summary of Module Placements on Dwelling Roof for Microinverter-Based PV System

Section #	Azimuth with South Ref = 180°	Slope	# Modules on This Section	Annual kWh per NREL SAM
1	90°	45°	2	523
2	135°	45°	2	631
3	180°	60°	3	948
4	210°	45°	3	982
5	240°	60°	2	558
6	270°	45°	2	523
			Total annual kWh	4165

cable, such as Type NM (Romex) to feed a circuit breaker in the distribution panel. The sizing of wire and circuit breakers on the ac side of the modules is determined by the inverter maximum output characteristics, whether or not the inverter will ever achieve maximum rated output. Thus, the maximum current to a circuit breaker will be the sum of the output currents of the 14 inverters, or 14×0.96 A = 13.4 A. The wire and PUC circuit breaker must thus be rated at 125% of this value, or at least 16.77 A. Thus, a 2P20A circuit breaker is adequate. Cabling supplied by the manufacturer will meet the system ampacity requirements as well as all UL and *NEC* requirements, since the cabling is custom-designed to attach to the connectors on the inverters. It should be noted that since the inverters include communications with downstream equipment via the ac output, that cables must include a proper termination at the end of the cable at the first inverter to prevent reflection of data in the data stream, resulting in faulty communication.

To determine the expected annual and/or monthly and/or hourly output of the overall system, the PVWatts version of SAM can be used, or, for that matter, the detailed version can also be used if the inverter and module manufacturers and model numbers are known. Whichever version is used, it will require a separate simulation for each subarray orientation. Table 4.15 also shows the annual results for each of the six subarrays as obtained using the PVWatts version.

Observing that the total rating of the 14 modules of the array is 3.64 kW, it is interesting to compare these results with the expected annual kWh production of a 3.64 kW array oriented south facing with a latitude tilt. The multiple-orientation microinverter system is expected to produce 4165 kWh/year, while the optimized south-facing system is expected to produce 4738 kWh/year. This represents a 12% annual loss of production because of the suboptimal orientation of each of the subarrays. Considering the range of orientations of the six subarrays, the microinverter solution represents a reasonable PV solution, given the difficult roof configuration.

4.6.3 Bells and Whistles (i.e., Monitoring Possibilities)

Another interesting feature of the microinverter-based system is the monitoring option. Each inverter has its own address and communicates performance data via

GRID-CONNECTED UTILITY-INTERACTIVE PHOTOVOLTAIC SYSTEMS 157

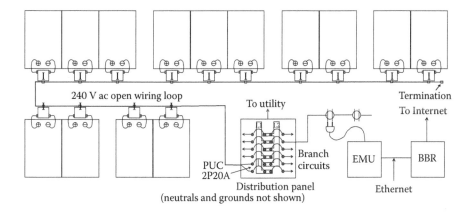

Figure 4.13 Schematic diagram of 3220-W microinverter-based PV system.

power line carrier connection to a communications gateway. The gateway enables performance monitoring of individual module/inverter systems and also can be connected via a broadband router to the Internet. This particular inverter is also an advanced smart grid ready unit. Figure 4.13 shows the schematic diagram of the total system.

4.7 DESIGN OF A NOMINAL 20 kW SYSTEM THAT FEEDS A THREE-PHASE DISTRIBUTION PANEL

4.7.1 Introduction

Thus far, the systems that have been designed have been relatively small systems that connect into single-phase distribution systems. For the reader who has been wondering when an example of a larger system would be presented, the wait is over.

In general, any PV system between 10 and 500 kW is considered to be a medium-sized system. While systems at the lower end of this range may connect into a 208, 240, or 277 V (U.S. versions) single-phase ac supply, as system sizes increase, it becomes more likely that they will be connected into either 208, 240, or 480 V three-phase power distribution systems. These connections can be made with multiple inverters having single-phase outputs or with single, larger, inverters with three-phase outputs or with multiple inverters, each with three-phase outputs.

In this example, the system will be designed around a single 20-kW inverter that feeds a 277/480-V three-phase power distribution system. The system will be located near Denver, CO. The latitude of Denver is 39.8°.

4.7.2 Inverter

The transformerless inverter has two MPPT inputs, each of which can accommodate three source circuits with an input voltage range of 320–1000 V. The rated input voltage is 600 V, so the array should be designed for $V_m = 600$ V for optimal system performance. Each dc input is limited to a maximum of 33 A. Because the inverter is transformerless, input circuits must be ungrounded, with overcurrent protection in positive and negative leads. This topology, however, results in a maximum inverter efficiency of 98.4%.

The inverter is equipped with internal GFDI, internal arc fault protection, a dc disconnect, and optional dc surge arrestor. Accessories include an RS-485 interface, a power control module, and a multifunction relay. The ac output voltage range is 180–280 V, which means, since it is a three-phase inverter, it can be configured for either 120/208 V or 277/480 V grid connections and is designed to interact with a smart grid. For example, the output power factor can be controlled by the grid and can range from purely capacitive to purely inductive, depending upon the needs of the grid.

4.7.3 Modules

Assuming a maximum of six source circuits, with the goal of achieving $V_m = 600$ V at the most likely operating temperature, it makes sense to first estimate a likely operating temperature of the modules. Denver climate data are available on a number of websites. U.S. climate data [19] reports an average annual high temperature of 64°F (17.8°C) and an average annual temperature of 50.15°F (10.1°C). In recent history, the lowest temperature in Denver has not been below −20°F (−28.9°C), and in recent history, the highest temperature in Denver has not exceeded 105°F (40.6°C). It is interesting to note that the coldest temperature occurred in 1875 and the hottest days on record occurred in 2012 [20].

Since generally the maximum array power can be at least 120% of the maximum rated inverter power, simply because it is a rare occasion that an array will operate at 100% of its rating, either because of elevated temperatures or reduced sunlight, it makes sense to look for a module that can be incorporated into an array size of approximately 24 kW, provided that inverter input current and voltage limitations are met.

One commercial 72-cell module is rated at $P = 350$ W, $V_{OC} = 48$ V, $I_{SC} = 9.82$ A, $V_m = 38.4$ V, $I_m = 9.17$ A, $\Delta V_{OC}/\Delta T = -0.304\%/°C$, and $\Delta V_m/\Delta T = -0.47\%/°C$, and the module is rated at 1000 V. If this module is used, then, at a temperature of 60°F (15.6°C), and $G = 800$ W/m², the cell temperature will be approximately $15.6 + (46 - 20) = 41.6°C$ on an average day. At this cell temperature, V_m drops to 35.4 V, which is the value that will be used to determine whether an integral number of this module model will reach a voltage close to 600 V. Dividing 600 by 35.4 results in 16.95. Thus, with this module, using 17 in series will result in $V_m = 602$ V for a 17-module source circuit on an average day.

The maximum value of V_{OC}, obtained at $T = -28.2°C$, is 55.9 V. Thus, for 17 modules in series, $V_{OC} = 950.3$ V, which is within the 1000 V limit. The minimum value of V_m, obtained at $T = 41.6°C$, is 30.3 V, so that for 17 modules in series, $V_m = 515$ V. Thus, this particular module nicely meets the desired V_m design point and also satisfies the inverter input requirements at maximum and minimum array temperatures.

Next, the STC power rating of a 17-module source circuit is 5950 W. Thus, four of these source circuits will produce 23,800 W, which is close enough to the target value of 24,000 W. Since I_{SC} of each source circuit is 9.82 A, if two source circuits are attached to each of the two MPPT inputs, then the input current at each input is 19.64 A, which is well within the 33-A inverter limit. So, a satisfactory array has been defined and the design can proceed to determination of dc wiring requirements.

4.7.4 System DC Wiring

Before wire lengths can be determined, it is necessary to determine the array layout. It will be assumed that the building roof has adequate east–west and north–south dimensions to allow four rows of 17 modules, adequately spaced so that no row shades any other row. Figure 4.14 shows the determination of row spacing for the array, showing necessary spacing between rows to ensure that the modules remain unshaded between 9 a.m. and 3 p.m. every day. Note that the modules are mounted in portrait position, and that the length of a module is 78.5 inch (1.993 m).

Solving Equations 2.9 and 2.10 for Denver on December 21 shows that the altitude of the sun at 9 a.m. and 3 p.m. sun time is 14° and the azimuth is 42° east of south at 9 a.m. and 42° west of south at 3 p.m. Thus, the sun angles that define the module spacing are established. The assumption is that there are no other objects on the roof, such as fans, air conditioners, plumbing vents, drains, skylights, lightning rods, or fire hydrants that will either shade the array or require different module placement. In a real installation, all of these possibilities need to be accounted for.

Figure 4.14 shows the length and direction of the shadow at 9 a.m. sun time on December 21. To obtain this value, it is first necessary to decide upon a tilt for the array, which, in this case, has been chosen to be 30°. This angle is chosen for three reasons: (1) it is within less than 10° of the 39.8° Denver latitude, (2) it allows for

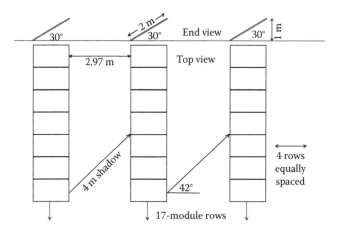

Figure 4.14 End view and top view of PV array row layout showing incident sunlight at 14° and array tilt at 30°.

reasonable row spacing without rows shading each other, and (3) it is sufficient for snow to slide off the modules relatively quickly.

The height of the top side of the module above the array mount is found from h = Lsinθ, where L is the module length (2 m) and θ is the array tilt angle. The shadow length at 9 a.m. is thus found by observing that the horizontal component of the shadow is given by D = h/tanϕ, where ϕ = 18°, the 9 a.m. sun altitude, and h is the result of the previous calculation. Next, since the azimuth of the sun at 9 a.m. is 42° east of south, Figure 4.14 shows the sun positioned at this angle. With the azimuth angle and the length of the shadow known, it is now possible to use simple trigonometry to determine the spacing between the rows. It is left as an exercise for the reader to determine that the dimensions indicated on Figure 4.14 are correct (Problem 4.9).

Next, assuming that a rooftop junction box is installed at the west end of each row of 17 modules, it is possible to estimate the length of open wiring between modules and junction box for each row. The length of the negative module lead is 1.0 m (3.281 ft) and the length of the positive module lead is 1.2 m (3.94 ft). This means the modules can be wired as shown in Figure 4.15, which might bring to mind the wiring of modules in Figure 4.11. Note that using this wiring method, no additional extensions are required, so the total length of module wiring is simply 17 × (1 + 1.2) = 37.4 m (123 ft).

The %VD of this loop can now be calculated. Interestingly enough, the %VD for a source circuit of these or any module will be the same, regardless of the number of modules. The reason is because as each module is added, the lead length increases, but also V_m increases, such that for two modules, the lead length will be 2L and the total V_m will be $2V_m$ of a single module. Thus, in the %VD formula, the number of modules appears in the numerator and the denominator, such that they cancel out. In this case, using #12 (3.31 mm²) module leads with a combined length of 2.2 m (7.22 ft) and V_m = 38.4 V yields %VD = (0.2 × 9.17 × 3.61 × 1.93)/38.4 = 0.333. This leaves 1.67% VD remaining for the extension leads from source circuit to inverter.

The inverters will be located one floor down, a distance of 50 ft (15.24 m) from the ends of the two center rows and an additional 17 ft (5.18 m) from the ends of the two outer rows. Thus, if #12 wiring rated at 1000 V is used, the %VD for the two outer rows will be (0.2 × 9.17 × 67 × 1.93)/(38.4 × 17) = 0.363% and the %VD for the two center rows will be (0.2 × 9.17 × 50 × 1.93)/(38.4 × 17) = 0.271%. Thus, the combined voltage drop between modules and inverter is <1% for all four source circuits. Of course, it will still be necessary to check the ampacity of #12 wire against temperature and conduit fill deratings.

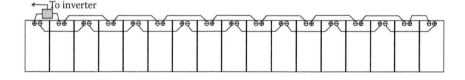

Figure 4.15 Determination of extension cable lengths from scale drawing of module row.

To check on ampacity of the source-circuit wiring, first note that 156% of I_{SC} = 15.32 A. Since the array has nice symmetry, the source circuits from the outer rows will be routed via the source circuits from the two inner rows using a single conduit from each of the two inner row junction boxes to the inverter. Assuming that each rooftop junction box contains terminal strips with 90°C rating, and that four current-carrying conductors, two for each source circuit, leave each junction box, and that the highest temperature of the junction box will be 50°C, the derating factors from Tables 4.12 and 4.14 are 0.82 and 0.8. Applying these factors to 125% of I_{SC} results in $9.82 \times ((1.25/0.82)/0.8) = 18.71$ A. Thus, the ampacity of the wiring from the inner row junction boxes to the inverter will need to be the larger of 15.32 and 18.71 A. In either case, #12 wire with a 30°C ampacity of 30 A is more than adequate. In fact, #14 has a 30°C ampacity of 25 A, so it also has adequate ampacity, provided that the source circuits can be fused at 15 A. Since source-circuit fuses are generally rated at 156% of I_{SC}, 15 A fuses are ever so slightly small. On the other hand, since each inverter input is connected to only two source circuits, and the modules in each source circuit have a maximum reverse current of 25 A, no fuses will be needed, since the maximum reverse current will be the maximum current of the other source circuit, which, if 156% of I_{SC} is used, is 15.32 A. Again, it must be remembered that wiring insulation must be rated at 1000 V for all dc circuits.

At this point, the remaining consideration for the dc-side of the inverter is rapid shutdown, since the system is on a building roof. At this point, this is mentioned as a reminder to the designer, who may find that the inverter may be able to communicate with remote switches on the roof that are within 10 ft (3.1 m) of the array to shut down the array if the inverter is shut down. The inverter manufacturer can also supply an accessory connection unit that can be mounted on the roof. The unit has source-circuit combiners, a dc disconnect, and two output circuits. And, for that matter, it is also possible to mount the inverter on the roof within 10 ft of the array to provide rapid shutdown. In this case, if the inverter PUC is turned off, the inverter shuts down and no array current flows.

The inverter provides arc fault and ground fault protection internally.

4.7.5 System AC Wiring

4.7.5.1 Wire and Overcurrent Protection Sizing

The ac wiring of the system will depend on the supply voltage to which the inverter is connected. It will first be assumed that the inverter will be connected to a 400-A, 277/480-V, Y-connected ac distribution panel, located 150 ft from the inverter. It will also be assumed that the distribution panel has a spare space for a three-pole circuit breaker. But it will first be necessary to determine whether 125% of the rated output current of the inverter will exceed 20% of the 400-A panel busbar rating, or 80 A.

Since the rated output current of the inverter is 24 A, 125% of 24 A is 30 A. So a three-pole 30-A circuit breaker will be needed for connecting the inverter to the distribution panel. In fact, this inverter could be connected to a 150-A, three-phase distribution panel and still be code compliant.

The previous calculations have also established the wire sizing for the ac side of the inverter. The inverter output wiring must have an ampacity of at least 30 A, so #10, with an ampacity of 40 A, may be used along with a three-pole 30-A circuit breaker.

4.7.5.2 Voltage Drop Calculations

Since this is a balanced three-phase source, the formula for voltage drop changes. Figure 4.16 shows a balanced, Y-connected, three-phase source and load, along with indications of all voltages and currents. It is assumed that the lengths of each phase and neutral conductor are the same.

The first observation is that if the magnitudes of I_1, I_2, and I_3 are the same, then $I_N = 0$, since the three-phase currents are each 120° out of phase. Thus, the voltage drop in the wiring to and from each of the voltage sources is simply IR_W, since there will be no voltage drop in the neutral. Note that this is only half as much as would be the case if the same current were to flow in the neutral as in a phase conductor.

So in terms of the line-to-neutral voltage, V, the %VD can be expressed as

$$\%VD = \frac{100IR}{V} = \frac{100I(\Omega/\text{kft})L}{1000V} = \frac{I(\Omega/\text{kft})L}{10V}. \quad (4.12)$$

Now, since the line-to-line voltage, V_L, is $\sqrt{3}V_P = \sqrt{3}V$, Equation 4.12 can be expressed in terms of the line-to-line voltage as

$$\%VD = \frac{100IR\sqrt{3}}{V_L} = \frac{100I\sqrt{3}(\Omega/\text{kft})L}{1000V_L} = \frac{I\sqrt{3}(\Omega/\text{kft})L}{10V_L}. \quad (4.13)$$

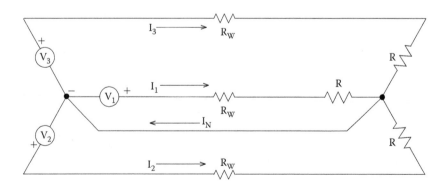

Figure 4.16 Determination of voltage drop in balanced three-phase system.

GRID-CONNECTED UTILITY-INTERACTIVE PHOTOVOLTAIC SYSTEMS

Thus, substituting $\Omega/\text{kft} = 1.21$ for #10 wire, $V_L = 480$ V, $I = 24$ A, and $L = 150$ ft into Equation 4.13 yields %VD = 1.57. Since %VD < 2, #10 is an acceptable wire size.

4.7.6 Annual System Performance Estimate

Using the PVWatts version of NREL SAM to simulate system performance results in an estimate of 36,942 kWh/year for the 30° tilt. For the optimal 39° tilt, the result is 37,214 kWh/year, an increase of 0.7%.

4.8 DESIGN OF A NOMINAL 500-KW SYSTEM

4.8.1 Introduction

It is now time to design a larger system so the concept of subarrays can be introduced. Large systems are sometimes composed of multiples of small systems, such as the system of the last example. Combining the outputs of 25 of the previous 20-kW design will result in a 500-kW system. A 1-MW system can be built around 50 of the previous designs. All that is needed is to figure out the best way to combine the ac outputs of each system and how to implement the PUC. With larger systems, sometimes the output will be fed directly to the utility distribution system rather than to a building system. In these cases, it will be necessary to transform the PV system output to distribution voltage levels, which are typically 7.6/13.2 kV.

The advantage of using multiple smaller (20–30 kW) inverters is the availability of multiple MPPT inputs to the inverters and the ability to locate the inverters close to the subarrays that supply each inverter. Another advantage of using multiple small inverters is the lower cost of ac components, such as transformers, switches, circuit breakers, and fuses. Furthermore, with inverter level performance monitoring, if one inverter out of 500 in a 10-MW system should fail, it will only reduce system output by 0.2%, but still will be located by the monitoring system, provided that the monitoring system is working properly.

On some occasions, however, it is possible that larger building blocks can be used for large systems. For example, for a 25-MW PV system, it may be preferable to build it with fifty 500-kW inverters. If the inverters can be conveniently located among the array, dc wiring can still be kept reasonable. The following system will be designed around a 540-kW inverter and a ground-mounted array.

4.8.2 Inverter

Since this system is intended to be a nominally 500-kW system, it will use a 540-kW inverter. Again, this does not mean that a 500-kW system must be designed with a 540-kW inverter. It just happens that a 540-kW inverter with attractive design features is available. The larger inverter will be used in order to introduce the reader to an additional layer of dc input components and the corresponding design procedure to incorporate very high dc currents into the design. Since very high dc or ac currents, if subjected to fault conditions, can result in extremely dangerous arcing conditions, arc flash safety and engineering computations will be introduced. The array will be mounted on racks at ground level. Rack mounts will be further discussed in the next chapter.

The choice of inverter may depend on price, availability, weight, physical dimensions, reputation, or any number of other considerations. It will be assumed that the 540-kW, UL 1741-listed inverter to be used in this example is characterized by $V_{OC}(max) = 1000$ V, $440 < V_m < 800$ V, $I_{IN}(max) = 1280$ A, $V_{OUT} = 300$ V, and $I_{OUT}(max) = 1040$ A. The unit can be configured with up to 10 combining inputs fused between 200 and 400 A each. The unit has a CEC rated efficiency of 97.5%. It has built-in load-break dc disconnect switch and ac circuit breaker and has a utility adjustable power factor between 0 and ±1 and internal GFDI. In this case, since the entire system will be located outside, no arc fault protection is required. The weight of the unit is 2240 kg (4938 lb).

4.8.3 Modules and Array

This example will use the modules of the last example, since they are intended for use in commercial installations. This does not necessarily mean that lower power modules cannot be used in commercial installations, since megawatt systems have been designed around modules having power ratings less than 100 W. The main reason for using the modules of the last example is because it is already known that 17 modules per source circuit will produce an output that falls within the input voltage range of many inverters. In addition, larger modules generally reduce installation time by reducing the number of attachment points that need to be secured.

4.8.4 Configuring the Array

4.8.4.1 Sizing the Array

The design goal is to configure an array that will exceed the inverter rating by about 15%, since the inverter is rated for *output* power, and its dc-to-ac conversion efficiency is rated at 97.5%. Furthermore, since the array will very seldom, if at all, operate at full-rated output, a reasonable array power to set as a goal is $540 \times 1.15 = 621$ kW. It will be assumed that the location of the installation will have the same latitude and ambient temperature conditions as in the last example, so the expected range of array output voltage for 17-module source circuits during normal system operation will be between 515 and 800 V, which is within the MPPT range of the inverter.

Noting that a 17-module source circuit will have a rated power of 5950 W, all that is now needed is to determine how many source circuits it will take to produce approximately 621 kW. To begin, divide 621 kW by 5950 to find the approximate number of source circuits. The result is 104.3697479.

So now what should be done with 104.3697479? The answer is that when possible, it is preferred to be able to combine source circuits such that a balanced set of output currents can be obtained, consistent with the number of available recombining inputs on the inverter. So a set of source-circuit combiner boxes need to be chosen that will hopefully each have the same output current. These combiner boxes will define the subarrays of the system that will feed the recombining inputs of the inverter.

For example, if 100 source circuits were to be used, then the 100 source circuits could be split into 10 groups of 10 so that 10 source-circuit combiner boxes could be used, each combining 10 source circuits. With 10 source circuits as input, the expected operating output current of the combiners will be $10I_m = 91.7$ A. But 100 source circuits will only produce 595 kW, which is less than the desired array size. So it will be useful to look for other combinations. If it is not possible to balance the array perfectly, then one or two groups of combined source circuits may end up having fewer or more source circuits than the other groups. Before proceeding to narrow down the number of source circuits, it is first useful to look at the *design* current per source circuit, that is, 156% of I_{SC}, which is 15.32 A. So if 10 source circuits are combined, then the *design* output current will be 153.2 A, which is marginally acceptable for a 150-A fuse. Since the inverter has only 10 inputs, this means that with 104 source circuits, at least one combiner will need to handle more than 10 source circuits. So, at this point, a look at available external combiner boxes could be useful.

4.8.4.2 Combiner Boxes

Combiner boxes exist with a number of configurations, including AFCI combiners, disconnect combiners, contactor combiners, and higher voltage (i.e., 1000 or 1500 V units) combiners. AFCI combiners are useful on buildings. Disconnect combiners, whether or not used on a building, are useful when it is desirable to isolate a group of source circuits from the remaining source circuits. Contactor combiners are useful on buildings when used with controllers for rapid shutdown, and higher voltage combiners are obviously useful with higher voltage arrays. Most of the other features are also available in higher voltage configurations.

The number and current ratings of inputs, type of overcurrent protection for each source circuit and output ratings, and environmental ratings are additional considerations in specifying combiner boxes.

One manufacturer advertises disconnect combiner boxes that range in size from 6 to 36 inputs, rated at 600 or 1000 V, with 100-, 200-, 250-, 320-, and 400-A integrated disconnects [21]. They are available with 90°C terminals so the wire connected to these terminals does not need to be considered as 75°C wire, which would be the case if 90°C wire were used and the terminals were rated at 75°C, and can be configured for grounded, floating, and bipolar arrays.

Thus, it appears that a significant amount of flexibility is possible in the choice of combiner boxes. Other considerations are whether to keep all the combiner boxes in close proximity to each other or to distribute them among the array rows. Wire sizing, ease of construction, and BOS component costs can be affected by the location of combiner boxes, so before proceeding further with selection and location of combiner boxes, it is now time to develop an array layout, keeping in mind the need to include no more than 10 combiner boxes for the approximately 104 source circuits.

4.8.4.3 Array Layout

Since this will be a ground-mounted system, it will be assumed that the area will be essentially flat, with no surprises beneath the ground where array footings will be located. Presumably, a geotechnical engineer will have submitted a report on soil information so the array mount can be appropriately designed. Wind load, of course, will be an additional consideration. At this point, a layout will be developed that will keep the system as compact as possible without sacrificing performance. The design criteria will be to produce close to maximum kWh/acre, which means that the tilt of the array will be kept at 20° to reduce the row spacing that would be required if the array were to be tilted at the latitude (approximately 39°). This will allow more modules per acre. As long as the increase in modules per acre exceeds the decrease in individual module annual kWh production, the annual kWh/acre will be increased.

Figure 4.17 shows half of a proposed layout that can be adapted to 104, 105, 106, or 107 source circuits. Note that each uninterrupted row contains six source circuits, except for one or two rows in the center of the array that will need to be shorter in order to allow space for the inverter, switched recombiner, and transformer (not shown). Depending upon the physical size of utility transformer that will be needed to transform the 300-V, delta-connected inverter output to distribution-level voltage, it may be desirable to eliminate an additional source circuit in the area of shaded modules, similar to what has been done in the center row.

Module sizes are close to 1 m × 2 m (3.28 ft × 6.56 ft). The modules are mounted in portrait configuration with 102 modules laid out to conveniently accommodate six source circuits on each of 16 module racks of the overall array. Two racks at the center of the array are shortened to make room for the system inverter and associated components by the elimination of one source circuit from each. Since the modules are mounted at an angle of 20°, the length of the plan view of the modules in the tilted direction becomes 3.76 m (12.34 ft).

The system is shown as close to scale as is possible considering the limitation in page size. Row spacing is based upon keeping all modules unshaded by other modules or other system components between the hours of 9 a.m. and 3 p.m. on December 21. Equations 2.9 and 2.10 were used to determine the azimuth and altitude of the sun at these times. Shadow length and direction were determined from the calculated azimuth and altitude and are shown on Figure 4.17. Problem 4.11 offers the reader a chance to verify these calculations for a latitude of 39°N.

The diagram shows possible locations of 9 source-circuit combiner boxes. Eight of the nine boxes will combine 12 circuits, and Combiner 5 will combine 10 or 11 source circuits, depending upon whether the number of source circuits is reduced to 106.

GRID-CONNECTED UTILITY-INTERACTIVE PHOTOVOLTAIC SYSTEMS

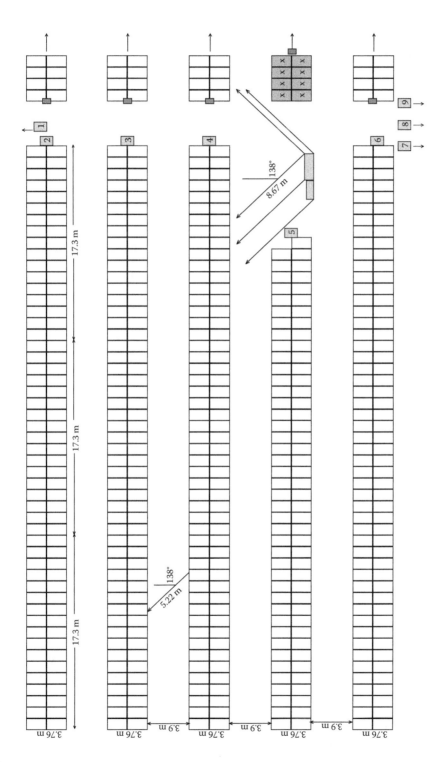

Figure 4.17 Array layout for 560-kW system showing locations of source-circuit combiner boxes and inverter.

Note that the original design goal was to use 104 source circuits. To confirm that using 106 or 107 source circuits is acceptable, the question that needs to be answered is how much total annual kWh production will be lost due to the array producing more total dc power than the inverter can handle. This question can be answered by simulating the system with the 540-kW inverter and the extra source circuits. If the expected system output continues to increase linearly as source circuits are added, then minimal losses due to inverter overload are expected to occur.

As a further check, a time series plot can be obtained for the simulation results. The time series produces an hour-by-hour estimation of system output, based upon a TMY at the installation site, over the 8760 hours of the year. By zooming in on daily estimated production curves, it can be verified that minimal kWh loss occurs, as indicated by inverter output power curves being limited at 540 kW, at input array ratings up to 120% of the inverter rating, or, in this case, 648 kW. This exercise is left for the reader as Problem 4.12.

The remainder of this design will be based upon using 106 source circuits for no other reason than it will result in a very nice, symmetrical system that occupies a 65 m × 108 m (213 ft × 354 ft) rectangular space. This means it will use eight 12-circuit combiner boxes and one 10-circuit combiner box. This, of course, is not the only possible array layout. It is simply an example of the thought process that might be used to develop an array configuration.

4.8.5 Wire Sizing and Voltage Drop Calculations

The system layout as shown in Figure 4.17 is intended to be friendly to everyone—to the installer because all wiring is based upon integral numbers of source circuits to each rack along with reasonable access space for installing all system components, to the owner because it is designed for minimal cost and maximum performance, to the designer because of the symmetry that enables repetitive calculations, to the inspector because of the neatness, and to the teams responsible for commissioning and maintenance of the system. All wire sizing will begin with the calculation of *NEC*-required wire ampacities and then voltage drop calculations will be made to be sure that no total %VD from source circuit to inverter input will exceed 2%.

4.8.5.1 Source-Circuit Calculations

Source circuits will be wired with type PV insulated wiring with insulation rated at a minimum of 1000 V. The module leads are #12 copper. It has already been shown that the ampacity of the source-circuit wiring must be at least 14.3 A. As wiring is run through conduit, it may require deratings for conduit fill and for ambient temperature. In this example, the only wiring of source circuits in conduit will be between the junction boxes on the right (east) subarrays and the combiner boxes at the ends of the left (west) subarrays. All other wiring up to the combiner boxes and junction boxes will be open wiring, consisting of the module leads as connected and

Table 4.16 Tabulation of Source Circuit (SC) Lengths and Voltage Drops[a] for the Source Circuits of the 560 kW System

Location	Outer SC	Wire Size	%VD	Middle SC	Wire Size	%VD	Inner SC	Wire Size	%VD
6 sc rack west	118	#10	0.79	59	#10	0.58	3	#12	0.38
6 sc rack east	138	#10	0.87	79	#10	0.65	23	#10	0.44
5 sc rack west	88	#12	0.88	29	#12	0.53	3	#12	0.38
5 sc rack east	170	#10	0.99	111	#10	0.77	85	#10	0.67

[a] Includes 0.36% VD from source circuit module leads.

extension cables between the end of each source circuit and the combiner box corresponding to the subarray to which the source circuit is assigned. Wiring must also include properly sized equipment grounding conductors. In general, each combiner box will contain six source circuits from each half of the array from racks that are across from each other. Minimum size of extension cables will be #12, even though technically #14 has adequate ampacity to meet the source-circuit ampacity requirements. Table 4.16 summarizes the lengths of extension cables for the outer, middle and inner source circuits on each rack, along with proposed wire sizes and resulting %VD.

The source-circuit wiring in conduit will consist of 12 #10 THWN-2 conductors along with a ground wire. The derating factor for 12 current-carrying conductors in a conduit is 0.5 and the derating factor for ambient temperatures between 31°C and 35°C is 0.96. Since all wiring crossing in conduit between the east and west subarrays will be #10, if 12 conductors are in a single conduit, the wire will be derated to $40 \times 0.5 \times 0.96 = 19.2$ A, which is greater than the design current of 15.3 A.

Since the voltage at the combiner output is the same for each source circuit that is combined, it is good design practice to keep the %VD from each source circuit to the combiner box as close as possible. This may require using larger wire sizes for longer wire runs. Table 4.17 shows an attempt to achieve a reasonable balance of

Table 4.17 Tabulation of Subarray Circuit Lengths, Wire Sizes and Voltage Drops for the 9-PV Output Circuits of the 560 kW System

Combiner #	d, ft	Wire Size	%VD	Max Total %VD[a]
1	125.3	2/0	0.44	1.31
2	100.1	2/0	0.35	1.22
3	74.8	2/0	0.26	1.13
4	45.9	2/0	0.16	1.03
5	19.7	1/0	0.07	1.06
6	32.8	2/0	0.12	0.98
7	61.7	2/0	0.22	1.09
8	87.0	2/0	0.31	1.17
9	112.2	2/0	0.40	1.26

[a] Includes all voltage drops between modules and recombiner.

%VD for the outer, middle, and inner source circuits on the west and east subarrays, with the highest %VD = 0.99, the lowest %VD = 0.38, and the average %VD = 0.66. Thus, the %VD of all source circuits is within 0.33% of the average.

4.8.5.2 Combiner to Recombiner Wiring (PV Output Circuits)

The recombiner is located next to the inverter at the center of the array. The next step is to determine appropriate wire sizes to connect from the nine combiner boxes to the recombiner.

Table 4.17 summarizes the distances, wire sizes, and %VD for each of these wire runs. All %VD calculations are based upon $V_m = 602$ V and $I_m = 110$ A for 12-circuit combiners and $I_m = 91.7$ A for the 10-circuit combiner. Wire sizing is based upon 184 A for 12-circuit combiners and 153 A for the 10-circuit combiner.

Using the minimum wire size based upon ampacity requirements results in the use of #2/0 (67.43 mm^2) copper for wiring the 12-circuit combiners to the recombiner, with 200-A fuses at the recombiner. For the 10-circuit combiner, #1/0 (53.49 mm^2) copper is adequate, fused at 175 A. Use of these conductors results in maximum %VD = 1.31% between the most distant outer source circuits and recombiner box for each of the nine combiner boxes, which is an acceptable result, since a maximum of 2% is considered to be acceptable.

For this wiring selection, it is assumed that each combiner box will have a separate conduit run to the recombiner box, since it is impractical to install more than five large conductors in a single conduit.

The wiring design for the dc side of the inverter is now complete except for the sizing of the equipment grounding conductors. *NEC* Table 250.122 specifies the minimum equipment grounding wire sizes based upon the size of the overcurrent protection device that protects the circuit of which the equipment grounding wire is a part. Thus, since each source circuit is protected by a 20-A fuse, the allowable size for the equipment grounding conductor is #12 (3.31 mm^2) copper. However, since this conductor is exposed, but protected from damage, it is common practice to use a bare #10 (5.261 mm^2) solid copper wire to ground the module frames as well as the racking, with the equipment grounding conductor terminated in the combiner boxes.

Because this is a ground-mounted system, ground rods are driven at each combiner box and a grounding electrode conductor sized according to *NEC* Table 250.66 is used. Since 2/0 (67.43 mm^2) copper is used between the combiner boxes and recombiner (or, for Combiner 5, 1/0 [53.49 mm^2] copper), *NEC* Table 250.66 requires a #4 (21.15 mm^2) copper grounding electrode conductor for the 2/0 circuits and a #6 (13.3 mm^2) copper grounding electrode conductor for the 1/0 circuit. However, since the grounding electrode at the combiner boxes is a ground rod, *NEC* 250.66(A) permits #6 grounding electrode conductors for all combiner boxes.

Finally, since the conductors between all combiner boxes except #5 and the recombiner are #2/0 (67.43 mm^2), protected with 200-A fuses, the required equipment grounding conductor to be run along with the #2/0 conductors between combiner and recombiner will be #6 (13.3 mm^2) copper per *NEC* Table 250.122.

4.8.5.3 Disconnects, GFDI, and Overcurrent Protection on the DC Side of the Inverter

The sizes of overcurrent protection have already been determined for source circuits and for combiner box output circuits. In this particular design, 200-A load-break dc switches are already incorporated in each combiner box and in the recombiner box. In addition, the inverter itself includes a dc load-break disconnect and an ac circuit breaker at the output as well as GFDI protection at the input. The recombiner is directly attached to the inverter via busbars.

4.8.6 AC Wire Sizing, Disconnects, and Overcurrent Protection

The ac output of the inverter is 300 V at 1040 A, so wire sizing, disconnects, overcurrent protection, and the PUC must take these values into account. Since this system is a ground-mounted system intended to feed directly into the utility distribution system, which will be assumed to be a 7620/13,200 V system, a transformer will be needed to step up the voltage from 300 V delta to 13,200 V delta. This transformer will logically be located close to the inverter to minimize the run of large cable from inverter to transformer. Note that at 13,200 V, the line current is reduced to 23.6 A. Typically, high-voltage cable minimum size is #2 (33.62 mm^2), so this cable can be run a long distance with minimal %VD (see Problem 4.13).

If the recombiner, inverter, and transformer are mounted on a single concrete pad, the pad must be maintained at equipotential with ground by thermal bonding of a wire mesh within the concrete to a large copper conductor to be approved by the local jurisdiction and utility, usually 2/0 (67.43 mm^2) or 3/0 (85.01 mm^2) that encircles the mounting pad. There will be multiple connections between the mesh and the buried copper conductor. The medium voltage wiring between transformer and utility should be coordinated with the local utility.

With the inverter next to the transformer, the inverter ac disconnect (circuit breaker) will serve as the disconnect between inverter and transformer primary. The disconnect for the transformer secondary will likely be at the point of connection to the utility distribution system. Although the inverter will have utility-grade monitoring of system output and other system parameters, the utility may elect to install a revenue meter on the primary or secondary side of the transformer as well.

Wiring between inverter and transformer will need to be rated at 125% of the inverter output, or 1300 A. It should be noted that the inverter is rated for full power output up to an altitude of 1000 m, but must be derated at altitudes between 1000 and 2000 m (3280 and 6562 ft). It will thus be assumed that the inverter will be at an altitude below 1000 m, which rules out Denver, CO. In any case, however, to accommodate 1300 A will require either busbars or parallel conductors. If the transformer can be integrated into the overall inverter system, busbars are the likely candidate, but if it is located several meters away from the inverter, which may be preferred anyway for safety reasons, the choice of parallel conductors might be three sets of three 500 kcm (253 mm^2) copper in three parallel conduits or four sets of three 350 kcm (177 mm^2) copper in each of four conduits, along with appropriately sized equipment grounding conductors in each conduit. For the equipment grounding conductors, *NEC* Table 250.122 requires the equivalent of a 4/0 (107.2 mm^2) equipment grounding conductor. If the three-conduit solution is chosen, then the parallel equivalent ampacity of 4/0 is still needed, provided that the parallel wires are no smaller than 1/0 (53.49 mm^2). Thus, wiring will also include 1/0 equipment grounding conductors in each of the three conduits between inverter and transformer.

Finally, what about the neutral? Since the inverter output is delta-connected, the inverter will have no neutral. In some cases, however, the utility may require a noncurrent-carrying neutral to be run to the meter socket as a zero voltage reference for metering purposes. This neutral will be the same size as the equipment grounding conductor unless the utility will approve a smaller conductor(s) and will be derived at the transformer primary grounding connection (low-voltage side).

4.8.7 Arc Flash Calculations

4.8.7.1 Introduction

At voltages exceeding 120 V, it is possible to strike an arc between two conductors of different potentials, as any electrician who has lost part of the head of a screwdriver in a live switch enclosure can attest. For that matter, any time a switch is switched off, interrupting current, at least a small arc, is created just at the instant the contacts break. If the source of the arc is capable of delivering high currents at high voltages, the arc can be substantial and possibly can be sustained and sufficient to be life and limb threatening. The arc itself can reach temperatures as high as three times the surface temperature of the sun [22]. If the arc is sustained and if the distance between arcing components is large enough, substantial power can be contained in the arc that will produce intense heat that will radiate away from the arc. If a person is too close, significant, and, possibly, fatal, burns may result.

Arc flash can occur as a result of human error when working on live components or it can also result from equipment failure, such as switching off a switch that has corroded contacts as a result of not being used over a long period. It is thus important to take precautions for possible equipment failure when switching a large switch either off or on.

The most common arc flash incidents involve high energy ac circuits and/or components, primarily because there are a lot more of them. But dc arc flash can be a significant concern at the combiner level, where, in this case, currents between 100 and 1000 A at voltages of 600 V may be present, or at the recombiner level, where dc currents exceeding 1000 A may be present. Imagine standing in front of an oven with a 60,000 W heating element. It certainly would not take long to notice that the oven is on. Now imagine this 60,000 W being contained in a relatively small space at a temperature of nearly 20,000°C (36,000°F). Planck's radiation law, as discussed in Chapter 2, shows that the radiated power density is proportional to T^4. So, one need not even come in contact with the arc to sustain significant injury. The question is how far away one must be to avoid second-degree burns.

IEEE Standard 1584-2002, Guide for Performing Arc Flash Hazard Calculations, provides computational methods, based upon empirical studies, to determine the level of energy and the distance required to avoid second-degree burns for ac arc flash [23]. To determine the key quantities of interest, one must first estimate the arc flash current, the incident energy, and the flash boundary. To make these estimates, one needs estimates of maximum fault current, clearing time of the protective device, and the expected distance between a worker and the energized object. It is also important to distinguish ac voltage sources, such as utilities or inverters from dc current sources, such as PV source circuits or PV output circuits.

4.8.7.2 AC Voltage Sources

When the driving source for the arc is an ac voltage source, the IEEE Standard presents two empirical formulas for estimating the arc current: one for voltages less than 1 kV and one for voltages above 1 kV. For voltages less than 1 kV,

$$\log I_a = K + (0.662 + 0.5588\,V - 0.00304G)\log I_{bf} + 0.0966\,V + 0.000526G, \quad (4.14)$$

where I_a is the arc current in kA, $K = -0.153$ in open air and -0.097 in enclosures, I_{bf} is the three-phase bolted fault current in kA, V is the system nominal line-to-line voltage in kV, and G is the gap between arcing points in mm.

For voltages above 1 kV,

$$\log I_a = 0.00402 + 0.983 \log I_{bf}. \quad (4.15)$$

Upon calculation of I_a, it is now possible to estimate the incident energy of the arc. The incident energy is the energy received by a surface at a certain distance from the arc. The first step in calculating the incident energy of the arc is to determine the normalized incident energy of the arc, E_n, which can be determined from

$$\log E_n = K_1 + K_2 + 1.081 \times \log I_a + 0.0011G, \quad (4.16)$$

where E_n is the incident energy normalized in time and distance in cal/cm², $K_1 = -0.792$ for open air and -0.555 for enclosures, $K_2 = 0$ for ungrounded systems and -0.113 for grounded systems, and G is the gap between arcing points in mm.

Then, for voltages up to 15 kV, the incident energy of the arc can be determined from

$$E = C_f E_n \left(\frac{t}{0.2}\right)\left(\frac{610}{D}\right)^x, \quad (4.17)$$

where E is the incident energy of the arc in cal/cm², $C_f = 1.5$ for systems below 1 kV and 1.0 for systems above 1 kV, t is the arcing time in seconds before the fault is cleared by the protection device, D is the distance in mm between the arcing point and the person, and x is the distance factor that depends upon voltage and equipment type and is tabulated for typical combinations in Table 4 of IEEE Standard 1584-2002.

Next, since the threshold incident energy for second-degree burns is 1.2 cal/cm^2, Equation 4.17 can be solved for D by substituting 1.2 for E and using the remaining known quantities.

4.8.7.3 DC Current Sources

The arc flash computation procedure is somewhat different when the source of the arc is a high-current PV combiner or recombiner output circuit. In these cases, current is limited by the intensity of the sunlight on the array and power is limited by the maximum power point of the source circuits. For an arc fault situation, since the current is limited, no fuses are expected to blow unless multiple source circuits or output circuits backfeed through a single overcurrent protective device, although if the arc builds up enough heat, it may melt and vaporize components near the arc. Thus, clearing time for an overcurrent protection device is not necessarily a useful parameter. Instead, exposure time is generally assumed to be the time it takes to move out of range of the fault, which is commonly set at 2 seconds. To justify this number, the reader might try pretending this page has suddenly reached a temperature of 20,000°C and imagine how long it might take to let go of or pull back from the book.

In a source-circuit/output-circuit configuration, available arc power at the combiner box will depend upon whether the combiner box disconnect is closed or open. At the recombiner box, the maximum available arc power is the same if all recombiner disconnects are closed and all recombiner fuses are in place, whether or not the recombiner disconnect is closed. There are thus three separate calculations to be made when both combiners and recombiners are used.

Shapiro and Radibratovic present an excellent summary of work on PV dc arc flashes [24]. Their February/March 2014 article cites an article by David Smith in which he concluded that "no consensus standard exists for calculating arc energies in dc systems." Otherwise, no empirical observations are included in the analysis, such as might involve a conductive or convective heat-transfer process. Over the next few years, it is anticipated that the following results will be refined as empirical data are gathered by the IEEE and a consensus standard is adopted.

The Shapiro and Radibratovic's article presents their proposed method of calculating PV dc arc flash incident energy based upon the assumption that the maximum power available from a PV source circuit will be 125% of P_m of the source circuit, to account for possible cloud focusing and the corresponding increase in source-circuit power.

They also account for a directional component of the arc on the assumption that the arc will be radiating from inside an enclosure. After application of these assumptions as well as conversion factors for units, their resulting formula for determining incident energy from an arc flash event at the output of a source-circuit combiner box with the combiner box switch closed is

$$IE_{enc} = \frac{142.5 P_{arc}}{D^2} \text{ cal/cm}^2, \qquad (4.18)$$

where IE_{enc} is the incident energy if the arc source is in an enclosure, P_{arc} is the total MPPT power in kW developed at STC by all source circuits connected to the combiner box, D is the distance between the arc and the worker in cm, and 142.5 is a conversion factor constant that incorporates the 125% of STC factor for arc power, 2 seconds for clearing time, a factor of $3/(4\pi)$ for enclosure focusing of arc energy, and conversion factors from kW to cal.

For the 12 source-circuit combiners of the 560-kW design example, each source circuit has $P_m = 5.95$ kW, so the value of P_{arc} to use in Equation 4.18 will be 71.4 kW. At a working distance of 18 inch (45.7 cm), $IE_{enc} = 4.87$ cal/cm^2. Thus, since the threshold for second-degree burns is 1.2 cal/cm^2, protective clothing would need to be worn for working at this distance.

If 1.2 cal/cm^2 is substituted into Equation 4.18 for IE_{enc}, then the arc flash boundary can be determined by solving Equation 4.18 for D with the result that D = 92 cm (36 inch).

It is also possible that an arc can be generated at a combiner box if the combiner box disconnect is open. The reason is that if all the disconnects in the recombiner box are closed, with all recombiner fuses in place, and arcing begins between the load side of the combiner disconnect switch and ground (or the ungrounded side of the array), the available arc current will be limited by the output-circuit fuse at the recombiner. In this case, eight of the nine output circuits are fused at 200 A and the other is fused at 175 A. This would allow a backfeed fault current of 1575 or 1600 A to feed back to the faulted combiner disconnect, depending upon which disconnect is faulted, which would be interrupted in a timely fashion by the 200- or 175-A fuse, as applicable.

The formula reported by Shapiro and Radibratovic to apply to this situation is

$$IE_b = 1.67(IE_{enc}) = \frac{1.67 \times 142.5 P_{arc}}{D^2} = \frac{238 P_{arc}}{D^2}, \qquad (4.19)$$

where IE_b is the incident energy at a distance D from the combiner box due to backfeed from the recombiner if the combiner disconnect is open and faulted.

The 1.67 factor results from the overcurrent protection being rated at 156% of I_{SC} of the combiner box output circuit and then rounding the overcurrent protection size up to the nearest standard value for overcurrent protection sizes of 800 A or less and the fact that I_m, which is normally about 90% of I_{SC}, will be flowing in the circuit.

Thus, if both sources are contributing to the arc energy at the combiner box, the total arc incident energy will be the sum of the array side and the recombiner side contributions.

Finally, for the arc flash energy available at the recombiner output, if all the combiner output switches are closed, it will be necessary to use the total energy produced by the array as the arc energy. This will be the case whether or not the recombiner disconnect is closed. Thus, if P_{arc} in Equation 4.18 is replaced by P_{array}, which is the total STC rated array power, then the incident energy at a distance D from the recombiner output, IE_{re}, can be estimated by

$$IE_{re} = \frac{142.5 P_{array}}{D^2} \text{ cal/cm}^2. \quad (4.20)$$

Again, it must be remembered that the formulas reported in this dc section are preliminary, yet to be refined by the IEEE.

4.9 SYSTEM COMMISSIONING

Up to this point, an important part of PV engineering has been taken for granted. Presumably, all components used in a PV system have been factory tested and can be assumed to operate as specified. Thus, once the system has been designed and installed, it has been assumed that it will work as predicted.

While this is mostly true for small systems with small numbers of components, when systems have thousands of modules, hundreds of source circuits, and a lot more wire and enclosures, it is much easier to miss something during system installation. Thus, for larger systems, it is good practice to have a third party conduct a thorough commissioning procedure on the system. As a minimum, system commissioning should include

- A visual check of all components and connections
- Verification of proper polarity of all source circuits
- Measurement of V_{OC}, V_m, I_{SC}, and I_m for all source circuits and conversion to STC values using correction for cell temperature and irradiance at the time of measurement
- A high-voltage insulation check on all wiring to be sure no connections or insulation are faulty
- A check to verify that all module frames and mounting frames are adequately grounded
- Inverter performance testing
- Measurement of I–V curves of each source circuit (optional)

The visual check should reveal any loose clamps or hardware, any open module connections, any missing junction box covers, whether expansion joints have been incorporated into long conduit runs, and whether array wiring is neatly and securely fastened to the modules or array mount or cable tray(s). It should also be a check on missing fuses and whether all switches are in the open position.

Verification of proper polarity is very important and easily accomplished by voltage measurements within the combiner boxes with the fuses open. It is important to take proper safety precautions to minimize any chance of injury from arc flash when making any of these or the following measurements on live equipment.

Measurement of source-circuit voltage and current parameters will indicate whether all modules are functioning properly and whether there are any loose or high impedance connections between any modules. All source-circuit electrical parameters, when normalized to STC, should be within a few percent of each other. In making these measurements, typically a specialized instrument can be used that will automatically cycle through the desired measurements for each source circuit and store the results in memory in a CSV format that can be imported into an Excel file. The important point to remember when performing these measurements is that when modules are shaded by clouds, their current drops instantaneously, but their cell temperature drops gradually. When the sun appears again, the current increases instantaneously, but the temperature rises gradually. It is thus important, in order to achieve reliable STC values, to wait for the irradiance and cell temperature to stabilize before recording voltage measurements. These measurements should normally be performed between 10 a.m. and 2 p.m. sun time or, at least, when irradiance levels are higher than 800 W/m^2. To minimize risk of arc flash injury when making these measurements, it is helpful to have only one source-circuit fuse connected at a time when measuring I_m or V_m. When V_{OC} and I_{SC} are measured, all fuses should be open.

The high-voltage insulation check will reveal any potential short circuits or even high impedance connections to ground. This check is done with a megger, which is an ohmmeter with a high-voltage internal dc source. At least one commercially available instrument incorporates an insulation test in addition to the voltage and current tests.

When performing this test, usually on an individual source-circuit basis from the combiner box, it will likely be necessary to disconnect the grounded conductor connection between combiner and recombiner, since the grounded conductor is grounded via the GFDI fuse in the inverter. If this conductor is not disconnected, the megger test will show a short circuit between the source-circuit conductors and ground. An interesting nuance of performing this test during early morning hours is that sometimes moisture will condense on the internal connections of the combiner box, causing a resistance to ground on the order of 10 kΩ, when normally this resistance should exceed 1 MΩ. Repeating the test later in the day or after drying out the enclosure will reveal whether moisture is the problem.

Since this test is not irradiance-dependent, except for waiting for moisture to clear, it can be performed when irradiance is less than 800 W/m^2.

The equipment grounding check can also be performed outside the 800-W/m^2 window. This check involves connecting a long wire to the grounding terminal of the combiner box and then running this wire to each of the modules of the array, where an ohmmeter is used to check the resistance between the module frame and the mounting frame and the combiner box ground. Since the wire between combiner box and module may be several hundred feet long, its resistance must be subtracted from the measured resistance value. The result should be less than 5 Ω for each measurement taken. If it is higher, then an immediate check can be made for any loose mechanical or electrical connections at the module or frame.

The inverter performance check is made last. In order to perform this check, all source-circuit fuses need to be inserted, all combiner box switches need to be on, all recombiner fuses must be in place, and all recombiner switches must be on. At this point, the inverter can be started in accordance with manufacturer's instructions, which usually involves turning on the dc and ac disconnects and then waiting the 5 minutes required by UL 1741 for something to happen. Generally, the inverter will have a display that shows input and output voltage and input and output current. If the inverter incorporates the possibility of power factor adjustment, either manually or automatically, this also needs to be noted. Ideally, this check will be an efficiency check of the inverter, which is determined by the ratio of output power to input power. In fact, this measurement is more difficult than one might imagine, simply because of catastrophic subtraction errors. For example, if the inverter efficiency is supposed to be 98%, and there is 0.5% uncertainty in measuring each of input power and output power, then it is possible that the *computed* efficiency, based upon these measurements, will be lower than 98% or higher than 100%. Thus, an error analysis should be made when the uncertainties in these quantities are known. It is good practice to measure the input power and output power at least five times and then take the average. This results in a lower uncertainty of inverter performance. If any of the inverter checks fail, it is important to have the instruction book and the manufacturer's phone number handy.

If the owner or designer specifies I–V checks as part of the commissioning, instrumentation is available for these checks. Strange I–V checks, having more than one maximum power point or some other unusual or unexpected behavior, point to a source circuit that needs additional attention. Generally, the I–V tester manual will have suggestions of causes of unusual I–V curves.

4.10 SYSTEM PERFORMANCE MONITORING

It is not at all unusual for a PV system owner to wonder how well the system is functioning. One method is to compare this year's electric bills with last year's. Another way is to go out to the inverter each day and read what the display has to offer, if there is a display. But it is becoming more and more commonplace for system owners to want to be able to connect system performance data to the Internet so they can monitor the performance of their systems from wherever they may be. A system owner in Daytona Beach, FL compared "a PV system without monitoring as the equivalent of an automobile without a speedometer."

Monitoring systems can monitor a wide range of variables, such as source-circuit current and voltage, inverter input power, inverter output power, inverter self-diagnostics, weather data such as temperature, irradiance and irradiation, power purchased from the utility, and power sold to the utility. A wide range of communication protocols are available, including hard-wired data communication, radio frequency wireless data communication, and powerline carrier data communication. Most inverters monitor input and output voltage, current, and power in order to accomplish functional tasks of the inverter, such as maximum power point tracking and self-diagnostics. So the quantities of interest are digitized and made available to display devices, either on the inverter or remote to the inverter.

Web-based monitoring systems are becoming very popular. In general, PV system data and sometimes weather data are collected, digitized, and then transferred to a controller/display unit and then to a broadband router as shown in Figure 4.13. Some manufacturers provide their own communications equipment, others make data available at a port to be used with third-party hardware and software, and some do both.

As the number of performance-based systems increases, the need for utility-grade monitoring also increases. In some cases, third-party monitoring is required by the performance-based contract. As a result, equipment manufacturers are rapidly developing data monitoring capability for their own products as well as providing for third-party monitoring of performance as well as fault diagnostics. As will be discussed in Chapter 8, performance-based contracts simply guarantee a certain level of energy production by the PV system. If the system does not meet the performance objectives, the system installer must pay the owner the difference. With this type of contract, the installer will want to be the first to know if there is a problem with a system.

PROBLEMS

4.1 Refer to Table 4.2 or to the *NEC* in order to
   a. Determine the wire size needed to limit voltage drop to 3% for a 100-W, 24-V load at a distance of 75 ft from the voltage source.
   b. Using the wire size determined in Part a, determine the actual voltage drop for the wiring.

4.2 An inverter has a balanced three-phase, 277/480-V output and is installed a distance, d, feet from the PUC.
   a. Derive an expression for the voltage drop between inverter and PUC in terms of d.
   b. If the maximum inverter output current is 14.4 A, and the inverter is located 200 ft from the PUC, determine a wire size that will have adequate ampacity if the wire has 90°C insulation and will have a voltage drop of less than 2% between inverter and PUC.

4.3 Show that $\alpha = \pi f_o/Q$ for a parallel RLC circuit.

4.4 Show that $f_o = \dfrac{\omega_d}{2\pi\sqrt{1-(1/4Q^2)}}$ for a parallel RLC circuit.

4.5 Design a resonant (RLC) load with Q = 5 that will dissipate 1000 W at 120 V (rms) at 60 Hz. Then use PSPICE or your favorite network analysis program to simulate the SFS and SVS algorithms by connecting a sinusoidal current source to the load and varying the frequency above and below 60 Hz and observing the total response (transient plus forced) as the excitation frequency moves farther away from 60 Hz.

4.6 Using Module 1 as described in Section 4.4.3, and an inverter that has $V_{in}(max) = 600$ V, $V_m(min) = 250$ V, $I_{in}(max) = 30$ A and is rated at 7000-W output power, with a maximum array power of 8750 W and provisions for combining four inputs, determine three configurations of the modules that will provide between 7000 and 8000 W to the inverter. Assume the module operating temperature range will be between −20°C and +65°C.

4.7 For the 14-module array used in Section 4.4.3, consisting of two 7-module source circuits that use Module C,
   a. Determine the %VD between rooftop junction box and inverter if #12 THWN-2 wire is used between the two points.
   b. Verify that #12 THWN-2 has adequate ampacity if its operating temperature is 60°C.
4.8 For the module configuration in Figure 4.12, assume a latitude of 30°N with $T_{min} = -20°C$ and $T_{max} = +70°C$.
   a. Determine a suitable arrangement of source circuits. If necessary, eliminate one or more modules.
   b. Select an inverter that will have input and output characteristics that will match your array of Part a.
   c. Complete the design of the rest of the system, including a schematic diagram of the entire system.
4.9 For the PV array of Section 4.7, shown in Figure 4.14
   a. Verify the shadow length at 9 a.m. on December 21.
   b. Account for the azimuth angle and determine the minimum row spacing based on the shadow direction at 9 a.m. on December 21.
   c. Using the row spacing shown in Figure 4.14, determine the time interval during which the array will be unshaded on November 21.
4.10 A microinverter system consists of 42 microinverters rated at 230 W with 208 V outputs. The system is to be tied into a commercial 225-A, 120/208-V distributon panel with a single three-pole circuit breaker in a balanced fashion.
   a. Determine the maximum number of units that can be connected in a single branch circuit with a maximum of 16-A output.
   b. Show that the system can be implemented with three separate microinverter branch circuits.
   c. Show in a circuit diagram how the outputs of the three separate systems can be combined to provide a balanced, three-phase output. Then calculate the size of the three-pole circuit breaker needed for connecting to the distribution panel.
   d. Check the PUC rules to verify that the PUC input current does not exceed 120% of the busbar rating, assuming the distribution panel is fed with a 225-A main circuit breaker.
4.11 For the PV array of Figure 4.17, verify the values of altitude, azimuth, and shadow length and direction to confirm that the stated row spacing is acceptable.
4.12 For the system of Section 4.8, use NREL SAM, detailed version, to estimate the array size where the ac annual kWh output is no longer proportional to the array size. In order to run the detailed version, you will need to enter a specific module and a specific inverter. Ideallly, the module will be a SolarWorld SW350XL and the inverter will be a Conext Core XC540. However, if your version of SAM does not have the SW350XL in its module library, you may substitute a SW295 Mono module and use strings of 20 rather than 17. This gives approximately the same string voltage and power as the 17-module string of the 350-W module. Plot your result as a graph of annual kWh versus array size to illustrate where the graph becomes sublinear.
4.13 Assume a balanced three-phase inverter output to a medium voltage transformer that will supply a balanced, 13,200-V delta-connected output of 26 A to the utility distribution system. If #2 Cu cable is used between the transformer secondary

GRID-CONNECTED UTILITY-INTERACTIVE PHOTOVOLTAIC SYSTEMS 181

and the power lines, how far can the cable be run without exceeding a voltage drop of 2%?

4.14 The output of a 540-kW inverter is rated at 300 V at a maximum current of 1040 A. It is connected to the utility line via a pad-mounted, 500-kVA transformer that has a rated fault current of 32,000 A. If the gap between arcing points, G, is 13 mm,
  a. Calculate the arc current in the inverter output wiring compartment.
  b. Estimate the incident energy of the arc normalized in time and distance.
  c. Estimate the incident energy of the arc at a distance of 500 mm from the arc. You may assume the arc will be cleared in 0.2 seconds and the distance factor, x, per IEEE 1584-2002 is 2.
  d. Estimate the distance from the arc at which second-degree burns will occur (arc flash boundary).

4.15 If the 540-kW PV system in Section 4.8 had been configured with seven 15-circuit combiner boxes for a total of 105 source circuits, calculate the arc flash boundary at the following locations. You may assume the combiner boxes are fused with 225-A fuses in the recombiner.
  a. The output of a combiner box with all source circuits and combiner box switched on.
  b. The output of a combiner box if only one source circuit and the combiner box are switched on.
  c. The output of a combiner box if the output of the box is switched off.
  d. The output of the recombiner.

4.16 A source circuit consists of 12, 280 W (at STC), 60-cell modules, each of which has $I_{mp} = 8.84$ A, $V_{mp} = 31.7$ V, $I_{SC} = 9.45$ A, $V_{OC} = 39.2$ V, $\Delta V_{OC}/\Delta T = -0.29\%/°C$, $\Delta V_{mp}/\Delta T = -0.42\%/°C$, and $\Delta I_{SC}/\Delta T = 0.04\%/°C$. The following measurements are made on the source circuit under non-STC conditions with a cell temperature of 50°C and an irradiance level of 800 W/m² normal to the array: $V_{OC} = 430$ V, $I_{SC} = 7.62$ A, $V_{mp} = 335$ V, and $I_{mp} = 7.12$ A. While current and voltage measurements have better than 1% accuracy, irradiance measurements are within ±5% and cell temperature measurements are within ±3°C.
  a. Calculate the expected values of the four parameters under test conditions. You may neglect the temperature dependence of the currents and the irradiance dependence of the voltages.
  b. Explain why it is reasonable to neglect the temperature dependence of the currents and the irradiance dependence of the voltages. You might want to check some of the formulas in Chapter 3.
  c. Conclude whether or not, based on the field measurements, the source circuit is operating within expectations.

DESIGN PROBLEMS

4.17 Specify all the components for a nominal 6000-W ground-mounted utility-interactive PV system. Assume the code in effect is the most recent version of the *NEC*. Use a utility connection point on the utility side of the meter.

4.18 Specify all the components and show the design for a nominal 5000-W residential rooftop-mounted, utility-interactive PV system, based on the most recent version of the *NEC*. The system will be located in Atlanta, GA and will be standoff mounted on a roof with a 30° tilt. Connect the system on the customer side of the meter,

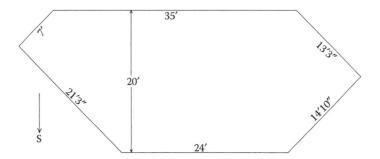

Figure P4.1 Roof dimensions for roof of Problem 4.19.

specifying the allowable sizes of main circuit breaker and distribution panel busbars if the PUC is to be in the distribution panel. Use NREL SAM to estimate the monthly and annual performance of the system.

4.19 For the roof shown in Figure P4.1, assume there are no obstructions in the area shown and make scale drawings that show
   a. How a set of modules could be mounted within Roof Zone 1 (in this case, more than 3 ft from the edge) to maximize the array power. Select whatever module you feel will best accomplish the design goal. Be sure the modules can be configured to meet the input voltage requirements of an inverter that has $V_{OC}(max) = 600$ V and $250 < V_m < 520$ V as the MPPT range. Specify the approximate power rating needed for the inverter.
   b. How a set of modules could be mounted on the entire available area of the roof, without extending beyond the roof. Select whatever module you feel will best accomplish the design goal. Be sure the modules can be configured to meet the input voltage requirements of an inverter that has $V_{OC}(max) = 600$ V and $250 < V_m < 520$ V as the MPPT range. Specify the approximate power rating needed for the inverter.

4.20 a. The south-facing portion of the roof shown in plan view in Figure P4.2 has a 5:12 slope. Select a module that has a rating of at least 250 W and lay out to scale an array configuration on the south roof. Select a suitable inverter for use with the system.
   b. On the east-facing and west-facing portions of the roof, all of which also have 5:12 slopes, configure modules that use microinverters to fill as much roof as possible.
   c. Assume your configurations of Parts a and b are in Sacramento, CA. Use NREL SAM to estimate total PV production on a monthly and annual basis of the combined set of arrays.

4.21 Design a 36-kW commercial roof-mounted PV system that will feed a 120/208-V, three-phase, 400-A distribution panel on the customer side of the meter. Comment on the bus capacity of the distribution panel and the allowable size of the main circuit breaker on the distribution panel feeder circuit.

4.22 A 120-kW utility-interactive system is to be installed as a parking lot canopy. It is to feed the grid with a three-phase, 277/480-V balanced output. Design a system, showing all necessary electrical components, including modules, inverter(s), wire

GRID-CONNECTED UTILITY-INTERACTIVE PHOTOVOLTAIC SYSTEMS

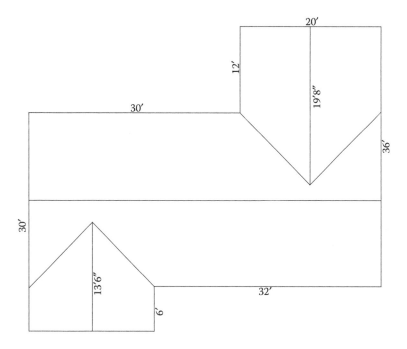

Figure P4.2 Roof dimensions for roof of Problem 4.20.

sizes, and fuse/circuit breaker sizes and locations. Recommend a horizontal layout of modules that will provide shading for cars, and estimate how many parking spots will be covered. Comment on your choice of whether to use a single inverter or multiple inverters.

REFERENCES

1. IEEE 929-2000, IEEE recommended practice for utility interface of residential and intermediate photovoltaic (PV) systems, IEEE Standards Coordinating Committee 21, *Photovoltaics*, 2000.
2. IEEE 1547-2003, IEEE standard for connecting distributed resources to electric power systems, IEEE Standards Coordinating Committee 21, *Fuel Cells, Photovoltaics, Distributed Generation and Energy Storage*, 2003.
3. *NFPA 70 National Electrical Code*, 2014 Ed., National Fire Protection Association, Quincy, MA, 2013.
4. IEEE Standards Coordinating Committee 21 Standards and Solar America Board for Codes and Standards National and international Standards Panel, January 16, 2008, http://www.solarabcs.org/about/publications/meeting_presentations_minutes/2008/01/pdfs/16Jan08-IEEE-SCC21.pdf.
5. IEEE Standard 1187-2013, https://standards.ieee.org/findstds/standard/1187-2013.html.
6. ISO Standard 9001, http://www.solarabcs.org/about/publications/meeting_presentations_minutes/2012/09/pdfs/16-IEC-Barikmo-14Sept2012.pdf.

7. Catalog of all ISO standards, http://www.iso.org/iso/catalogue_detail?csnumber=46486.
8. American National Standards Institute Standard Z97, http://www.ansiz97.com/standard/.
9. *ASCE Standard 7-10, Minimum Design Loads for Buildings and Other Structures*, American Society of Civil Engineers, Reston, VA, 2010.
10. ASTM International incorporates over 12,000 standards that operate globally. www.astm.org.
11. Kern, G. A., Bonn, R. H., Ginn, J., and Gonzalez, S., Results of Sandia National Laboratories grid-tied inverter testing, In *Proc. 2nd World Conf and Exhibition on PV Solar Energy Conversion*, Vienna, Austria, July 1998.
12. Kern, G. A., New techniques for islanding protection, *Interconnecting Small PV Systems to Florida's Utility Grid, A Workshop for Florida Utilities*, Cocoa, FL, October 22, 1998.
13. CFR 47, Federal Communications Commission, Rules and Regulations, Radio Frequency Devices, Part 15, 1994.
14. System Advisor 2016 (SAM 2016.3.14), National Renewable Energy Laboratory, Golden, CO, 2016.
15. *UL 1741, Inverters, Converters, Controllers and Interconnection System Equipment for Use with Distributed Energy Resources*, 2nd Ed., Underwriters Laboratories, Inc., Northbrook, IL, 2010.
16. Brooks, B., Rapid shutdown for PV systems: Understanding NEC 690.12, *SolarPro Mag.*, (8.1), Jan/Feb 2015.
17. Kern, G., *SunSine™ 300: Manufacture of an AC Photovoltaic Module, Final Report*, Phases I and II, 25 July 1995–30 June 1998, NREL/SR-520-26085, March 1999, www.nrel.gov/docs/fy99osti/26085.pdf.
18. For information on Enphase microinverter systems, see www.enphaseenergy.com.
19. For Denver temperature information, see www.usclimatedata.com/climate/Denver/Colorado/united-states/usco0105.
20. For Denver temperature extremes, see http://www.thorntonweather.com/noaa/hotcold.php.
21. Solar BOS markets an extensive line of combiner and recombiner boxes, http://www.solarbos.com/products/combiner-boxes/disconnect-combiners.
22. El-Sharkawi, M., *Electric Safety Practice and Standards*, CRC Press, Taylor & Francis Group, Boca Raton, 2014.
23. IEEE Standard 1584-2002, IEEE Guide for Performing Arc Flash Hazard Calculations, IEEE Standards Association.
24. Shapiro, F. R. and Radibratovic, B., Calculating DC arc-flash hazards in PV systems, *SolarPro Mag.*, (7.2), Feb/Mar 2014, http://solarprofessional.com/articles/design-installation/calculating-dc-arc-flash-hazards-in-pv-systems.

SUGGESTED READING

NFPA 70E: Standard for Electrical Safety in the Workplace®, 2015 Ed., National Fire Protection Association, Quincy, MA, 2014.

Smith, D., *Arc Flash Hazards on Photovoltaic Arrays*, Colorado State University, Department of Electrical and Computer Engineering, Fort Collins, CO, 2013.

CHAPTER 5

Mechanical Considerations

5.1 INTRODUCTION

Because the primary function of a PV system is to convert sunlight to electricity, often the role and importance of the mechanical aspects of the system are overlooked. Most PV modules are designed to last 20 years or longer. It is important that the other components in the system, including mechanical components, have lifetimes equivalent to those for the PV modules. It is also important that the mechanical design requirements of the system be consistent with the performance requirements as well as with the operational requirements of the system. The mechanical design of PV systems cuts across a variety of disciplines, most notably civil and mechanical engineering and, to a lesser extent, materials science, aeronautical engineering, and architecture. More specifically, mechanical design involves the following:

- Determining the mechanical forces acting on the system.
- Selecting, sizing, and configuring structural members to support these forces with an adequate margin of safety.
- Selecting and configuring materials that will not degrade or deteriorate unacceptably over the life of the system.
- Locating, orienting, and mounting the PV array so that it has adequate access to the sun's radiation, produces the required electrical output, and operates over acceptable PV cell temperature ranges.
- Designing an array support structure that is aesthetically appropriate for the site and application and provides for ease of installation and maintenance.

Each of these elements of the mechanical system will be discussed in more detail throughout this chapter.

5.2 IMPORTANT PROPERTIES OF MATERIALS

5.2.1 Introduction

Before discussing the mechanical design process, it is useful to review certain properties of materials, especially the non-PV materials that are important components of PV systems. The properties of materials can be grouped into four general categories: electrical properties, mechanical properties, chemical properties, and thermal properties [1].

Electrical properties of semiconductor materials were introduced in Chapter 3 and are discussed in more detail in Chapters 10 and 11.

The mechanical properties most important to PV systems are those associated with the strength of the structural members, including PV modules, in response to static and dynamic forces.

The most important chemical properties are those related to material degradation and deterioration due to corrosion and exposure to ultraviolet radiation. Also of interest are the rates of chemical degradation and the associated reduction in lifetimes of various materials, such as those used for weather sealing and insulation, which are caused by repeated exposure to high operating temperatures.

The thermal properties of most concern involve thermal expansion and contraction and the resulting thermal stresses, particularly on long runs of PVC conduit.

5.2.2 Stress and Strain

The PV system, in particular the PV array and its structural support members, is subjected to a variety of mechanical forces—both static and dynamic, such as wind, snow, and earthquakes as well as hailstones and other possible projectiles. These forces result in stress and strain in the structural system.

Stress is defined as force per unit area. Uniform normal stress occurs when a force P is normal to and distributed uniformly over a cross-sectional area A. It can be calculated using the simple equation:

$$S = \frac{\text{force}}{\text{area}} = \frac{P}{A}. \tag{5.1}$$

In addition to stress, another important concept in discussing the strength of materials is *strain*, ε. Whereas stress is a measure of force intensity (i.e., force per unit area), strain is a measure of deformation per unit of length and can be defined by the equation

$$\varepsilon = \frac{\text{elongation}}{\text{length}} = \frac{\delta}{L}, \tag{5.2}$$

where δ = elongation and L = length. The cause-and-effect relationship between stress and strain should be intuitive to many. The intensity of the force, that is, stress, causes the structural member to deform, that is, strain. However, engineers need a more useful and quantitative relationship between stress and strain.

Figure 5.1 shows a bar of length L and cross-sectional area A. Force P acts uniformly over the cross-sectional area, putting the bar in tension. The experiment is simple: gradually increase the force P from small to larger values until the bar

MECHANICAL CONSIDERATIONS

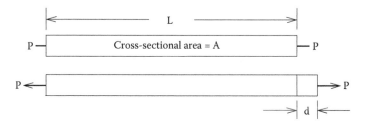

Figure 5.1 Simple pull test of a bar showing the resulting elongation.

ruptures. During the experiment, the force and the elongation will be measured continuously. Remembering that stress is simply force per unit area and strain is elongation per unit length, stress can be plotted versus strain. The resulting graph from this experiment for a typical material such as steel is shown in Figure 5.2, which should be familiar to anyone who may have studied strength of materials.

Note that the stress and resulting strain are directly proportional to each other up to a limit, appropriately named the *proportional limit*. If the force P is increased to produce stresses above the proportional limit, then the stress–strain relationship is no longer linear. However, the bar may still be elastic to a slightly higher stress called the *elastic limit*. If the force P is removed at any point up to and including the elastic limit, then the bar will return to its original dimensions. If the bar is stressed beyond the elastic limit, permanent deformation occurs. In addition to the elastic limit, the *yield point* is defined as the point on the stress–strain curve corresponding to a specified permanent deformation (usually when the elongation per unit length equals 0.002). The yield point is used because it is easier to determine than the elastic limit for some materials. The stress corresponding to the yield point is defined as the *yield strength*.

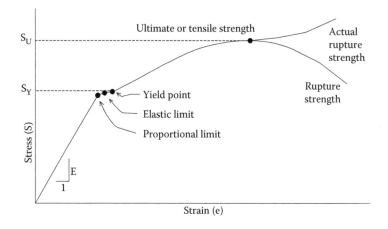

Figure 5.2 Stress–strain curve for steel.

Other key locations on the stress–strain curve include the maximum stress point on the graph, called the *ultimate tensile strength*, or, simply, the *tensile strength*, and two *rupture strengths*. The lower value for rupture strength results from using the original cross-sectional area A, while the higher value at the end of the dashed curve results from using the reduced area measured after rupture. Note that as material is stretched, its cross-sectional area decreases, resulting in higher stress for the same applied force.

In addition to the proportional limit, elastic limit, yield strength, ultimate strength, and rupture strength, another important parameter obtained from the stress–strain curve is the slope of the straight-line portion of the graph from the origin to the proportional limit. This slope is defined as the *modulus of elasticity* (also known as *Young's modulus*) and is usually represented by the letter E. It represents the ratio of stress to strain in the linear portion of the graph, which is where most structural materials are designed to operate. In equation form, it can be expressed as

$$E = \frac{\text{stress}}{\text{strain}} = \frac{S}{\varepsilon}. \tag{5.3}$$

A better name for E would be modulus of stiffness. The higher the value of E, the less the material deforms for a given stress. For a spring, the spring constant is analogous to the modulus of elasticity. The higher the spring constant, the harder you have to pull it to stretch it a given amount (i.e., the stiffer it is). Equation 5.3 is known as *Hooke's law*. For most materials, the modulus of elasticity is the same for both tension and compression.

Referring to Figure 5.2 once again, the area under the stress–strain curve has special physical significance in that it is proportional to the total energy required to rupture the bar. This energy is referred to as the *toughness* of the material.

For the simple case of one-dimensional, normal stress that has been discussed thus far, a useful formula for the elongation of the bar can be obtained by substituting for S (Equation 5.1) and for ε (Equation 5.2) into Equation 5.3, yielding

$$\delta = \frac{PL}{AE}. \tag{5.4}$$

This equation shows that the amount of elongation depends on the force applied, P, the dimensions of the bar (length L and cross-sectional area A), and the bar's material, as represented by its stiffness E.

The normal stresses considered up to this point were one-dimensional and were the result of normal forces acting uniformly over the cross-sectional area. In addition to normal forces, other forces act along or parallel to the area resisting the forces. These forces are called *shear forces*. Shear forces per unit area are called *shear*, or *tangential, stresses*. Figure 5.3 shows examples of structural members and fasteners subjected to forces that produce shear stresses.

In general, both normal and shear stresses are present in structural members. Like the modulus of elasticity for normal stresses, there is an analogous modulus of elasticity in shear, but it is more commonly referred to as the *modulus of rigidity*

MECHANICAL CONSIDERATIONS

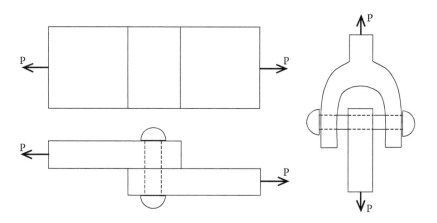

Figure 5.3 Two examples of shear loading.

and is represented by G. Table 5.1 [2] presents the modulus of elasticity, E, and the modulus of rigidity, G, for common structural materials.

The normal and shear stresses discussed up to this point are important and occur frequently in structural support members for PV systems. In computing these stresses, forces are assumed to be uniformly distributed over the cross-sectional areas resisting the forces. However, other forces may act on the structure and produce stresses that vary over the cross section. Examples include

- *Bending (or flexural) stresses*: As an example, imagine a person standing on the end of a diving board. The top surface of the diving board would be in tension, the bottom in compression, and the normal stresses along the mid-horizontal plane would be zero.
- *Torsional shear stresses*: Imagine the torque applied to a drive shaft of an automobile. The shear stresses across the section of the shaft vary from zero at the center to a maximum at the outer radius.
- *Combined stresses consisting of axial loading, bending, and torsion*: All three of these types of loading may occur in various members of the PV array and support structure.

Table 5.1 Moduli of Elasticity and Rigidity for Structural Materials

Metal	E (psi)	G (psi)
Stainless steel, 18–8	27.6×10^6	10.6×10^6
All other steels	$(28.6–30.0) \times 10^6$	$(11.0–11.9) \times 10^6$
Cast iron	$(13.5–21.0) \times 10^6$	$(5.2–8.2) \times 10^6$
Aluminum alloys, various	$(9.9–10.3) \times 10^6$	$(3.7–3.9) \times 10^6$
Titanium (99.0 Ti), annealed bar	$(15–16) \times 10^6$	6.5×10^6

Source: Avallone, E. A. and Baumeister, T., Eds., *Marks' Standard Handbook for Mechanical Engineers*, 10th Ed., McGraw-Hill, New York, 1996. Reproduced with permission of the McGraw-Hill Companies.

Table 5.2 Strength of Common Structural Materials

Metal	Ultimate Tensile Strength (psi)	Yield Strength (psi)
Cast iron	18,000–60,000	8000–40,000
Wrought iron	45,000–55,000	25,000–35,000
Structural steel, ordinary	50,000–65,000	30,000–40,000
Stainless steel, 18–8	85,000–95,000	30,000–35,000
Aluminum, pure, rolled	13,000–24,000	5000–21,000
Aluminum alloy, 17ST	56,000	34,000
Titanium 6–4 alloy, annealed	130,000	120,000

Source: Avallone, E. A. and Baumeister, T., Eds., *Marks' Standard Handbook for Mechanical Engineers*, 10th Ed., McGraw-Hill, New York, 1996. Reproduced with permission of the McGraw-Hill Companies.

5.2.3 Strength of Materials

Referring again to Figure 5.2, the two most important values on the graph for most applications are the yield strength and the ultimate tensile strength. Although Figure 5.2 was representative of steel under tension, stress–strain curves for other materials and for different types of loading, such as compression and shear, are available [2,3]. With access to these values, how does the engineer use them to make sure the structure is sufficiently strong?

The first thing the engineer has to decide upon is the allowable or working stress. This is the maximum safe stress that the material should carry. For PV systems, the allowable stress should be well below the yield strength and proportional limit. A common formula for computing the allowable stress is

$$S_a = \frac{S_y}{N}, \tag{5.5}$$

where S_a is the allowable stress, S_y is the yield strength in psi, and N is a safety factor to compensate for any minor deviations of materials from the ideal.

For PV systems, N should be at least 2 to avoid accidental overloading. In areas where fatigue may be a problem due to high, gusty winds, a higher value should be used. Also, when fatigue is a concern, the structural engineer will often choose steel over aluminum for the support structure.

Table 5.2 [2] shows the ultimate tensile strength and yield strength for selected metals. Note that titanium is more than twice as strong as steel.

5.2.4 Column Buckling

Often structural members used to support a PV array have one dimension, their length, which is much larger than the other dimensions. Such members or columns, if long enough and slender enough, can fail due to buckling under compressive loading. These compressive forces may be considerably smaller than those that would cause failure due to crushing. Although there is no firm rule, generally a long, slender

MECHANICAL CONSIDERATIONS

structural member is considered to be a column if its length is more than 10 times as large as its least lateral dimension. For a column made of a material with a modulus of elasticity, E, a minimum moment of inertia (of its cross-sectional area), I, and length, L, there exists a critical compressive force, P, that will cause the column to buckle. This critical load is calculated using the following formula:

$$P = \frac{EI\pi^2}{L^2}. \qquad (5.6)$$

This equation is known as *Euler's formula* [4].

5.2.5 Thermal Expansion and Contraction

Changes in temperature cause materials to expand or contract. For example, the linear deformation of a structural member, including electrical conduit, of length L can be represented by

$$\delta_t = \alpha L \Delta T, \qquad (5.7)$$

where α is the coefficient of linear expansion and ΔT is the change in temperature with units consistent with the units of α.

If the deformation is allowed to occur freely, no stresses will be induced in the structure. But if this deformation is restricted or constrained, internal *thermal stresses* result. The general procedure for handling thermal stresses is as follows [4]:

- Assume the structure is relieved of all applied forces and constraints.
- Allow the temperature deformations to occur freely and calculate the expansion or contraction using Equation 5.7.
- Show these deformations on a sketch.
- Determine the mechanical forces necessary to restore the structure to its constrained condition by solving Equation 5.4.
- Show these forces and deformations on the same sketch used previously.
- Use the equations of static equilibrium, thermal deformation, force deformation, and the definition of stress to compute all unknowns.

The computed stresses from using this procedure are the thermal stresses.

EXAMPLE 5.1

Consider the case of a steel bolt in a bronze sleeve as shown in Figure 5.4. At 40°F, the sleeve fits snugly between the head of the bolt and the steel nut, but with no stress in any of the three parts. For the following dimensions and material properties, compute the thermal stresses in both the bolt and the sleeve when the temperature rises to 160°F:

D_1 = nominal diameter of bolt = inside diameter of sleeve = 1.00 inch
D_2 = outside diameter of sleeve = 1.50 inch
E_1 = modulus of elasticity for steel = 30×10^6 psi

E_2 = modulus of elasticity for bronze = 10×10^6 psi
α_1 = coefficient of thermal expansion for steel = 6.1×10^{-6} per °F
α_2 = coefficient of thermal expansion for bronze = 10.1×10^{-6} per °F

To solve this problem, note first that the elongation of the bronze sleeve would be greater than that for the steel bolt if both were unconstrained. For any length L, these elongations could be calculated using Equation 5.7. However, the bolt head and nut constrain the sleeve so that the elongations of both the sleeve and bolt must be equal. Hence, the sleeve is in compression and the bolt is in tension. Also, from the principle of action and reaction, the compressive force in the sleeve must be equal and opposite to the tensile force in the bolt.

From the sketch in Figure 5.4, the total elongations of the bolt and sleeve are, respectively,

$$\delta_1 = \delta_1(\text{thermal}) + \delta_1(\text{forces}) = \delta_{1t} + \delta_{1f}(\text{steel bolt})$$

and

$$\delta_2 = \delta_2(\text{thermal}) - \delta_2(\text{forces}) = \delta_{2t} - \delta_{2f}(\text{bronze sleeve}).$$

Equating the forces and elongations and applying Equations 5.4 and 5.7 yield

$$\frac{(\alpha_1)(L)(T_2 - T_1) + PL}{[(A_1)(E_1)]} = \frac{(\alpha_2)(L)(T_2 - T_1) - PL}{[(A_2)(E_2)]}$$

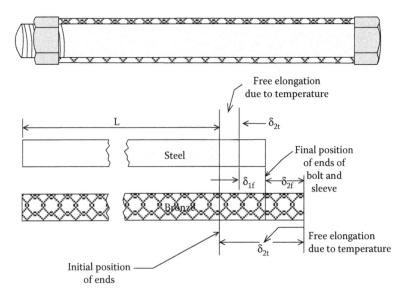

Figure 5.4 Elongations in steel and bronze due to a change in temperature.

MECHANICAL CONSIDERATIONS

The length, L, is a common term and cancels. Solving for the force, P, gives
P = 3330 lb (1510 kg).
The normal stresses in the bolt are computed using Equation 5.1 and are

$$S_1 = 4240 \text{ psi } (29{,}220 \text{ kpa}) \text{ (tension)}$$

and

$$S_2 = 3390 \text{ psi } (23{,}363 \text{ kpa}) \text{ (compression)}.$$

5.2.6 Chemical Corrosion and Ultraviolet Degradation

A PV system structure that can initially withstand all anticipated mechanical and thermal stresses with an adequate margin of safety might not be a good engineering design. The engineer must also consider the environment in which the system operates and how its component materials interact with the environment and with each other. Failure of metal components by corrosion is as common as failure due to mechanical stresses.

To design for corrosion resistance, knowledge of the various forms of corrosion is necessary.

Uniform attack is the most common form of corrosion. Chemical or electrochemical reactions proceed uniformly over the entire surface area. The metal becomes thinner and thinner and eventually fails. On a tonnage basis, uniform attack is the greatest cause of metal destruction, especially steel. However, it is not particularly difficult to control through proper material selection and appropriate use of protective coatings.

Galvanic corrosion occurs because of the potential difference that exists between two dissimilar metals in contact with a corrosive or conducting solution. Of the two dissimilar metals, the less-resistant, or anodic, metal is corroded relative to the more-resistant, or cathodic, metal. Table 5.3 [5] presents an abbreviated version of the galvanic series for commercially available metals and alloys. The relative position in the galvanic series depends on the electrolytic environment and on the metal's surface chemistry. To minimize galvanic corrosion, the design engineer should use similar metals or metals close to each other in the galvanic series.

The use of a small anodic metal in contact with a cathodic metal of larger surface area should be avoided. If two metals far apart in the galvanic series must be used in near contact with each other, they should be electrically insulated from each other. Coating the anodic material may not protect it, because coatings are susceptible to pinholes, causing the coated surface to corrode rapidly in contact with the large cathodic area. If a galvanic couple is unavoidable, a third metal that is sacrificial to both of the other metals may be used. Note that zinc is sacrificial to both aluminum and steel and, consequently, is often used to protect these two most common structural materials.

Table 5.3 Galvanic Series for Metals and Alloys

Most noble metal (cathodic) ↓	Platinum
	Gold
	Titanium
	Silver
	316 stainless steel
	304 stainless steel
	410 stainless steel
	Nickel
	Monel
	Cupronickel
	Cu–Sn bronze
	Copper
	Cast iron
	Steel
	Aluminum
	Zinc
Least noble metal (anodic)	Magnesium

Source: Dieter, G. E., *Engineering Design: A Materials and Processing Approach*, 2nd Ed., McGraw-Hill, New York, 1991. Reproduced with permission of the McGraw-Hill Companies.

Crevice corrosion is an intense form that frequently occurs at design details such as holes, gasket surfaces, lap joints, and crevices under bolts and rivet heads. Small quantities of stagnant liquid often form in these areas and allow this very destructive electrochemical process to occur. Unfortunately, stainless steel is especially susceptible to crevice corrosion.

Pitting is an extremely localized attack that produces holes in the metal. It is serious because it may lead to very premature failure of structural members. Pitting may require a relatively long initiation period but, once the process is started, it accelerates rapidly.

Intergranular corrosion is a localized attack along the grain boundaries of metals but not (or only slightly) over the grain faces. It is common in steel that has been heat treated at high temperatures or heat sensitized during welding. Intergranular corrosion due to welding is known as *weld decay*.

Selective leaching is the preferential removal of one or more of the alloying elements in a metal by the action of an electrolyte. The most common example of this phenomenon is the selective leaching of zinc from brass, leaving a spongy, weak matrix of copper—a process known as *dezincification*. Aluminum, iron, cobalt, and chromium are also susceptible to leaching. Where and whenever selective leaching occurs, the process leaves the alloy in a weakened, porous condition.

Erosion corrosion occurs when corrosive fluid flows over a metal surface and causes the gradual wearing away of the surface. Usually, the velocity of the fluid is high and mechanical wear and abrasion may be involved.

Stress corrosion cracking is caused by the combination of tensile stress and corrosion and leads to the cracking or embrittlement of a metal. The stress may result from applied forces or may be residual. Only specific combinations of alloys and the chemical environment produce stress corrosion cracking. Combinations include aluminum alloys and saltwater, some steel alloys and saltwater, and mild steel and caustic soda. Preventing stress corrosion cracking involves selecting alloys that are not susceptible to cracking under expected operating conditions. If this is not possible, the structural members should be sized for low stress levels.

Dry corrosion involves reaction of the material directly with air. Virtually every metal and alloy reacts with the oxygen in air to form an oxide. This occurs in the absence of any liquid electrolyte and is an important form of corrosion in high-temperature applications.

In addition to corrosion, ultraviolet radiation can cause degradation and deterioration of some of the exposed materials used in PV systems. These include weather-sealing materials, conduit for wiring, wire insulation, some coatings and caulking, and other miscellaneous materials.

Ultraviolet radiation is that part of the electromagnetic spectrum having wavelengths between about 100 and 400 nm. The ultraviolet portion of the spectrum can be further subdivided into three regions: UV-A (315–400 nm), UV-B (280–315 nm), and UV-C (100–280 nm). UV-B and UV-C cause sunburn (erythema) and pigmentation (tanning). UV-B produces vitamin D_3. Long-term exposure to UV radiation results in loss of skin elasticity. Also, a link has been established between exposure to UV wavelengths below 320 nm and skin cancer. Because the human body is seriously affected by UV exposure, it should come as no surprise that other materials are also affected. In fact, ultraviolet radiation degrades both the optical properties and the physical properties of many materials. Consequently, the design engineer should ensure that all materials in the PV system that are exposed to sunlight, such as wire, conduit, and connectors, are resistant to UV degradation.

5.2.7 Properties of Steel

Iron is the base element of all steels. Commercially, pure iron contains only about 0.01% carbon and is relatively soft and weak. The addition of carbon significantly strengthens iron. For example, the addition of 0.80% carbon can raise the tensile strength from about 40,000 psi (276 Mpa) to 110,000 psi (758 Mpa) [6]. Steel is an alloy of iron and carbon and usually contains small amounts of manganese and other elements. There are more than 3500 grades of steel, of which nearly 75% have been developed over the last 30 years. The many different grades of steel have many different properties: physical, chemical, and environmental. *Carbon steel* is steel that owes its distinctive properties to the carbon it contains. *Alloy steel* is steel that owes its distinctive properties to one or more elements other than carbon, or to the combination of these elements with carbon.

Low-carbon steel is often used in the support structures for PV systems. Of the many low-carbon steel products, sheet and strip steels are becoming increasingly important and presently account for approximately 60% of the total steel production in the United States [2]. The relatively low cost of these steel products gives them a major advantage over more expensive aluminum, provided they can be protected adequately from corrosion.

Many different methods are used to protect steel from corrosion. In general, these protection or reduction methods include proper selection of materials, design, coatings, inhibitors, anodic protection, and cathodic protection [2]. One method of protecting steel structural members is hot-dip galvanizing. Hot-dip galvanizing is a process in which thoroughly cleaned steel or iron is immersed in molten zinc and withdrawn to provide a smooth, even coating that typically has a crystalline appearance. It is by far the most widely used method for protecting steel against corrosion [6].

Zinc is a less noble metal than steel (see Table 5.3) and is sacrificed to protect the steel as part of the corrosion process. How long the zinc protects the steel depends on the thickness of the zinc coating and the environment to which the structural member is exposed. The average minimum weight of zinc coating is 1.5 oz/ft^2 (0.305 kg/m^2) of surface area when hot-dip galvanized according to the specifications of ASTM A123. For highly acidic industrial atmospheres, a 1.5 oz/ft^2 coating may last less than 15 years. For a rural, clean atmosphere, the same coating may last 40 years.

Lead is sometimes added to zinc to produce a surface pattern on steel called *spangle*, which is often used on unpainted surfaces. The addition of aluminum to zinc improves its corrosion protection ability. *Galfan*, which contains 5% aluminum and 95% zinc, and *Galvalume*, which contains 55% aluminum and 45% zinc, are examples of zinc–aluminum coatings that effectively protect steel from corrosion [2].

Paints, lacquers, coal-tar or asphalt enamels, waxes, and varnishes are all used as organic coatings for corrosion protection. However, all paints are permeable to water and oxygen to some degree. They are also subject to mechanical damage and eventually break down.

5.2.8 Properties of Aluminum

Aluminum is the second most commonly used structural material next to steel. Some of the properties of aluminum that make it attractive for use in PV systems are as follows:

- It is very lightweight, with a density about one-third that of steel.
- It can be fabricated into many different shapes using a variety of different methods.
- It has a wide range of properties and is available at many different tensile strengths, depending on how it is alloyed.
- It has a high strength-to-weight ratio, thus making it attractive for many applications.
- It has good corrosion resistance and can be used in a wide variety of climates and weather conditions.
- It is available in a wide variety of finishes, making it attractive as an architectural metal.

Aluminum is one of the most fascinating of all the elements. It is very chemically active, with a strong affinity for oxygen. In powdered form, it is used as a fuel for solid rocket motors. How can it be that such a chemically active material is so commonly used in an unprotected form?

What makes aluminum unique is its special relationship with oxygen, which destroys other reactive metals such as sodium and seriously attacks less reactive metals such as iron. In the laboratory, most chemistry students have witnessed the spectacular reaction between sodium and the oxygen in water, in which the sodium is left unrecognizable and even more unusable. And everybody frequently encounters rusting steel. Neither of these phenomena occurs with aluminum. Rather, as soon as aluminum is produced, its surface reacts with the oxygen in the air and forms a very thin oxide layer over the entire surface. This oxide layer is hard and tenacious and not only protects the aluminum from attack by other chemicals, but also protects it from further oxidation. Because the oxide layer is transparent, it does not detract from the metal's appearance. Aluminum oxide must be removed from aluminum wire before connections are made.

Pure aluminum is relatively weak and only about one-third as stiff as steel. However, aluminum can be strengthened significantly by alloying. The most common alloying additions to aluminum are copper, manganese, silicon, magnesium, and zinc.

Aluminum and its alloys are divided into two major classes: wrought and cast. The wrought class is broad because aluminum can be formed by virtually every known fabrication process, including sheet and plate, foil, extrusions, bar and rod, wire, forgings and impacts, drawn or extruded tubing, and others. Cast alloys are poured molten into sand (sand casting) or into high-strength steel molds (i.e., permanent molds or die casting) and allowed to solidify into the desired shape. For PV systems, wrought aluminum is more commonly used than cast aluminum. The Aluminum Association has a designation system for wrought aluminum alloys that categorizes them by major alloying additions. Table 5.4 [3] shows the various designations.

The first digit in the series classifies the alloy by alloy series, or principal alloying element. The second digit, if different from 0, denotes modification of the basic alloy. The third and fourth digits together identify the specific alloy within the series.

Table 5.4 Designation System for Wrought Aluminum Alloys

Alloy Series	Description or Major Alloying Element
1xxx	99.00% minimum aluminum
8xxx	Other element
2xxx	Copper
3xxx	Manganese
4xxx	Silicon
5xxx	Magnesium
6xxx	Magnesium and silicon
7xxx	Zinc
9xxx	Unused series

Source: Kutz, M., *Mechanical Engineers' Handbook*. 1998. Copyright Wiley-VCH Verlag GmbH & Co. KGaA. Reproduced with permission.

Alloys in the 6xxx alloy series, which use magnesium and silicon as alloying elements, possess a combination of properties including corrosion resistance, which make them well suited for structures and architectural applications. They are also commonly used for marine applications, truck frames and bodies, bridge decks, automotive structures, and furniture. From the 6xxx series, 6061 and 6063 aluminum alloys are often used as structural members for PV arrays. 6061 aluminum contains 1.0% magnesium and 0.6% silicon. For the same tempering during fabrication, 6061 aluminum alloy has higher tensile strength than 6063 aluminum, which contains 0.7% magnesium and 0.4% silicon [3]. However, 6063 aluminum has better resistance to corrosion.

5.3 ESTABLISHING MECHANICAL SYSTEM REQUIREMENTS

5.3.1 Mechanical System Design Process

Mechanical system design is the process of selecting, sizing, and configuring a variety of structural and other *components* to meet predetermined design requirements. For most PV system designs, these components are usually available "off the shelf," although in some cases completely new components may need to be developed. The challenge is to bring together components of the appropriate materials and geometric shapes and sizes to produce a *system* that best meets the design objectives.

The design requirements for the mechanical system consist of the following:

- Functional requirements
- Operational requirements
- Constraints
- Trade-offs

Developing specifications for each of the above four categories is not unique to PV system design. It is commonly used in the design of systems in general, including aircraft, satellites, and other products of modern society [7]. Determining the information to include in each of the four categories is an important responsibility assumed by the systems engineer at an early stage in the design process.

5.3.2 Functional Requirements

What are the primary functions of the mechanical parts of the PV system? First and foremost, the mechanical system must be capable of carrying all expected mechanical forces with an adequate margin of safety. In fact, the most important requirement for both the mechanical and electrical design of the PV system is the safety of individuals and the protection of structures and other property that could be damaged as the result of mechanical failure. The functional requirement related to structural considerations might take several forms, such as the specification of *factors*

MECHANICAL CONSIDERATIONS

of safety, or *maximum allowable stresses*, or *limits on deformation*. This specification will affect the materials selected, as well as the geometry and size of the structural members.

For most designs, the PV array and support structure must withstand all forces and associated stresses to which they are subjected for 20–30 years or longer. Often, they must do so in harsh weather and chemical environments. These environments cause many materials to degrade, deteriorate, and eventually fail. Consequently, the specification of the lifetimes of the structural members and other materials that make up the PV system is an important functional requirement. This specification will affect not only the materials selected, but also the selection of protective coatings that may be necessary to protect these materials from corrosion and other forms of environmental degradation over the anticipated life of the system.

Some PV systems track the sun, using either single- or double-axis tracking. The function of the tracking subsystem is to provide increased exposure to the sun's radiation. Therefore, another type of functional requirement is the specification of the required motion of the tracking system. This specification is especially important for concentrating arrays and affects the design of the drive mechanism for the tracker.

5.3.3 Operational Requirements

Operational requirements involve the more important aspects of the design that affect the people who install, own, operate, maintain, repair, and/or service the PV system. An example of operational requirements would be the specification of the maximum labor hours required to install a PV system and the specification that fuses or circuit breakers be easy to reach. These types of specifications are becoming increasingly important as industry tries to reduce the costs of PV system installations. It has a major effect on the simplicity of the design and the ease of assembly of the system.

A specification related to accessibility, especially accessibility to individual modules and other critical components for inspection, maintenance, and repair, is another type of operational requirement. It affects the size and shape of selected mechanical components and how they are configured.

For some PV system designs, the specification of security requirements may be necessary. This type of specification is used to protect against vandalism, theft, and personal injury and affects the selection and configuration of components. For example, the operational requirement to protect PV modules from flying objects has led to the frequent use of shielding on the back of small arrays for area lighting systems. Security requirements have also led the engineer to designs that include buried or tamper-proof enclosures.

Another type of operational requirement is the specification of periodic inspection and maintenance. For example, the requirement for seasonally adjusting array tilt angles affects the selection and configuration of structural support components. Anticipated rain, snow, ice, dust, and the presence of birds, predisposed to target arrays with their droppings, are all examples of conditions that have affected the operational specifications for PV system designs.

5.3.4 Constraints

Invariably, the process of engineering design is constrained or limited in some way. The most common design constraint is *cost*. In the highly competitive renewable energy marketplace, it is important to keep costs to a minimum. Cost constraints have a major effect on system design. For example, aluminum may be the preferred structural material for the mechanical system because of its resistance to corrosion. But steel may be the material of choice because it is considerably less expensive, even after it has been hot-dip galvanized. Cost constraints are omnipresent and pervasive, and they challenge the ingenuity of the design engineer. Typically, the PV system must be designed and installed within a specified budget.

Another design constraint is *time*. Good design, in all of its creative aspects, takes precious time—and sufficient time is not always available to satisfy the inner needs of the design engineer. Typically, the PV system must be designed and installed according to a specified schedule.

From time to time, the design engineer will be confronted with other types of constraints. Limited access to the sun due to existing or imminent shading (e.g., due to proposed construction) is commonplace. Restrictive building codes, covenants, and zoning may prohibit the installation of a PV system or limit the design options. The same can be said for political, regulatory, and institutional issues. And finally, the effects of weather, location, and site layout often need to be considered before getting too far into the design process.

In summary, constraints are a reflection of the real world in which the engineer operates. They usually have a major effect on design decisions and, consequently, should be included in the specification of design requirements. A clear understanding of these constraints will assist the engineer in producing a better mechanical design.

5.3.5 Trade-Offs

Specifying the functional and operational requirements establishes the desired objectives for the mechanical system design. In a sense, it is a sophisticated wish list: the structure should be strong; the materials should last a long time; they should resist corrosion and other forms of degradation; and the array should have a certain orientation or possibly track the sun. In addition, the system should be easy to assemble in a short period of time; important components should be readily accessible to work on; it should be protected from vandals and thieves; and it should be easy to maintain. On the other hand, a listing of the constraints emerges: the costs are too high; the schedule cannot be met; the array must be moved to avoid shading; restrictive covenants may be violated; the grid interconnection requirements are unreasonable; and the weather is marginal for solar. In short, because of the constraints, the desired objectives cannot be met and it is time to compromise.

When the desired objectives cannot be met, the following items are subject to compromise: quality or performance, cost, and schedule. There is an old saying: "Pick any two and trade down on the third." Assuming the cost and schedule constraints are rigid (they may not be), the engineer has to decide which trade-offs in quality and performance are acceptable. Examples of trade-offs to reduce costs and operate within budget include using steel rather than aluminum, using lower factors of safety, using lower-grade steel, using less-expensive corrosion inhibitors, eliminating tracking, using less-expensive and probably less-skilled installers, and using less-expensive security measures. Examples of trade-offs to meet schedule include substituting readily available structural members for preferred parts, eliminating certain design features, and using readily available labor.

5.4 DESIGN AND INSTALLATION GUIDELINES

5.4.1 Standards and Codes

Prior to the establishment of modern standards, structural members, nuts, bolts, screws, and so on were custom designed and manufactured. For example, one manufacturer would produce 1/2-inch bolts with 9 threads per inch, while another would use 12 threads per inch. Some fasteners had left-handed threads and had different thread profiles. In the early days of the automobile, mechanics would lay out fasteners in a row as they were disassembled to avoid mixing them during reassembly. A lack of standards leads to a lack of uniformity and precludes efficient interchangeability of parts. It is inefficient and costly.

A *standard* is a set of specifications for parts, materials, or processes intended to achieve uniformity, efficiency, and a specified quality. Another important purpose of standards is to limit the number of items in the specification and thereby provide a reasonable inventory of tooling, sizes, and varieties so custom parts will not be required.

A *code* is a set of specifications for the analysis, design, manufacture, construction, or installation of something. The purpose of a code is to achieve a specified degree of safety, efficiency, and performance or quality. It should be noted that safety codes do not imply absolute safety, which is impossible to attain. Designing a structure to withstand 120-mph (193 km/hour) winds does not mean the designer thinks 140-mph (225 km/hour) winds are impossible. It just means the designer thinks they are highly improbable.

The following organizations are involved in developing standards and codes relevant to the mechanical design of PV systems:

- Aluminum Association (AA)
- American Institute of Steel Construction (AISC)
- American Iron and Steel Institute (AISI)
- American Society of Civil Engineers (ASCE)
- American Society of Metals (ASM)

- American Society of Mechanical Engineers (ASME)
- American Society of Testing and Materials (ASTM)
- Society of Automotive Engineers (SAE)

Building codes are design, construction, and installation guidelines. They are adopted and enforced by local jurisdictions to help ensure the safety of individuals and the protection of structures and other property in and around buildings. Building and electrical codes have value only if they are followed and enforced. The process begins prior to the installation of a PV system with an application for a building permit. After the installation, the building official inspects the system to determine compliance with the relevant codes. Unfortunately, many solar systems have been installed in the past without first obtaining a building permit. Most PV systems installed prior to the mid-1990s were stand-alone systems and probably would not meet present code requirements. However, with the renewed interest in PV systems for buildings as a result of financial incentive programs and government initiatives, most new systems comply with local codes. And utility-interactive PV systems are more likely to be inspected by building code officials than are stand-alone systems because utilities will typically not allow an interconnect without a signed interconnect agreement that requires inspection of the system, usually by the authority having jurisdiction and sometimes by the utility as well.

5.4.2 Building Code Requirements

The most important standard affecting the mechanical design of PV systems is entitled *Minimum Design Loads for Buildings and Other Structures* [8]. It is a standard of the American Society of Civil Engineers and is periodically updated. It provides the minimum load-carrying requirements for buildings and other structures, and these requirements apply to PV systems. It is important to note that standards have meaning only when they are adopted by an enforcing jurisdiction and become part of their code. Many local jurisdictions reference parts of this ASCE standard in their building codes. However, local building codes are less uniform than local electrical codes, almost all of which use the *NEC*.

The basic requirements of building codes are that the PV system and any building or other structure to which it is attached shall be designed and constructed to safely support any mechanical load or combination of loads to which they are subjected. In other words, the ASCE standard provides formulas for computing the mechanical loads from wind, snow, ice, and so on. The engineer computes these loads for the conditions expected at the site of the PV system. The stresses these forces produce must not exceed the appropriately specified allowable stresses for the structural materials being used.

In addition to establishing strength requirements, building codes often limit the deformations that can occur with physical structures such as PV systems. For example, there may be restrictions on the amount of deflection or lateral drift of a structure. Such a deflection might have an adverse effect on the use of the system or attached buildings or structures. Because vibration can lead to fatigue failure, the local building code may contain provisions that address the stiffness of the structure and the associated frequencies and amplitudes of structural vibration.

5.5 FORCES ACTING ON PV ARRAYS

5.5.1 Structural Loading Considerations

Because solar energy is a relatively dilute resource, PV modules and arrays are area intensive. These large-area solar electric devices are capable of transmitting a variety of forces to themselves, their support structures, buildings, and other frames or foundations to which they are mechanically attached. These forces include the dead weight of the array, the weight of array installation and maintenance people and their equipment, aerodynamic wind loading, seismic effects, and the forces due to rain, snow, hail, and ice. Determining these forces under prescribed conditions can be an interesting and challenging mechanics problem.

Of the various types of forces mentioned above, aerodynamic wind loading presents the most concern. At most locations around the world, the force effects due to wind loading are much higher than the other forces acting on the structure. For example, both dead and live weight loads due to a PV array on a building are usually less than 5 pounds per square foot (psf) (239 PA). In contrast, the computed wind loads are typically between 24 psf (1149 PA) and 55 psf (2630 PA), but sometimes greater than 100 psf (4790 PA), depending on location and the associated design wind speed for that location.

Because of the geographic variation in environmental conditions such as rain, snow, ice, wind speed, seismic activity, and so on, local building codes tend to be much less uniform than electrical codes. Consequently, it is important for the PV systems engineer not only to become familiar with local building code requirements, but also to compute the minimum structural design loads that are relevant for a given location.

In addition to the above considerations, the design engineer must select and configure the array mounting materials so that corrosion and ultraviolet degradation are minimized. Also, if the array is mounted on a building, the structural integrity of the building must not be degraded and building penetrations must be properly sealed so that the building remains watertight over the life of the PV array.

The ASCE standard [8] provides instructions and formulas for computing the following types of loads:

- Dead loads
- Live loads
- Soil and hydrostatic pressure and flood loads
- Wind loads
- Snow loads
- Rain loads

- Earthquake loads
- Ice loads—atmospheric icing
- Combinations of the above loads

5.5.2 Dead Loads

Dead loads consist of the weight of all the materials that are supported by the structural members, including the weights of the structural members themselves. As an example, consider a PV array mounted on the roof of a building. The modules are structurally tied together in panels using steel or aluminum structural members. The weight of the modules is carried by the structural members and transmitted to the roof structure of the building. The total dead load that the roof must carry is the combined weight of the modules, structural members used to form the panels, attachment hardware, and mounting brackets. These loads can be expressed in psf or PA and assumed to act uniformly over the area covered by the array. Or the equations of static equilibrium can be used to determine the individual forces at each of the brackets that connect the panels to the roof. For PV systems, the dead loads are usually assumed to act uniformly over the supporting structure and are expressed in psf. Note that these are external loads—not internal stresses—even though similar units are used.

Dead loads for PV systems generally fall within the range of 2–5 psf (95.8–239 PA) and, even though they cannot be discounted, do not pose serious structural problems for the engineer.

5.5.3 Live Loads

Live loads associated with PV systems are those produced by individuals and their equipment and materials during installation, inspection, and maintenance. In designing structures to carry live loads, the engineer may treat them as uniformly distributed loads, that is, in psf or PA, as concentrated loads (lb or kg) or as a combination of uniformly distributed and concentrated loads. For PV arrays, live loads are usually assumed to be distributed uniformly and are small, on the order of 3 psf (144 PA) or less.

5.5.4 Wind Loads

The forces from the wind acting on PV arrays are *aerodynamic forces*. As such, their magnitudes depend partly on the properties of the atmosphere. Atmospheric properties include static pressure, temperature, viscosity, and density. Of these, density is one of the more important ones affecting the mechanical forces acting on PV systems. At any given location, density is usually assumed to be constant. However, density does vary with altitude—decreasing as altitude is increased. For example, the density in the mile-high city of Denver, CO, is about 86% of the atmospheric density for Florida cities, which are only slightly above sea level. The mass density of air for the standard atmosphere at sea level is 0.00256 lb-sec^2/ft^4.

MECHANICAL CONSIDERATIONS

The air flowing over a PV array produces two types of forces: *pressure forces* normal to the surface and *skin friction forces* along the surface. Skin friction forces are simply surface shear forces resulting from the viscosity of the air as it contacts the surface.

Most PV arrays are essentially large flat plates that are tilted at various angles to the direction of flow of the wind. Also, the buildings and structures to which they are attached and the terrain in their vicinity may significantly affect the flow around them. This flow often produces complex velocity and pressure distributions that, in turn, produce forces and stresses in the structural members. The determination of wind loads on PV structures does not lend itself to theoretically derived solutions. Rather, aeronautical and civil engineers have together developed empirical formulas for estimating the structural loading resulting from exposure to various wind conditions.

At this point, it is useful to summarize the important variables and factors that affect the aerodynamic forces acting on PV arrays and structures. They include the following:

- Wind speed
- Effects of wind gusts
- Density of the air
- Orientation with respect to the wind direction
- Shape and surface area
- Elevation above the ground
- Effects of topographical and man-made features

The list is somewhat intimidating and shows that the best wind load analysis is not as accurate as some engineers would like. Nonetheless, with "ballpark accuracy" and adequate factors of safety, good design is definitely within the engineer's grasp.

Figure 5.5 shows a view of a modular array field designed by Hughes Aircraft Company for Sandia National Laboratories [9]. Figure 5.6 shows the *net* pressure distribution on the ground-mounted array and support structure for the combined wind and dead loads. The wind speed is 100 mph (161 km/hour) and is directed toward the front of the array. The resultant force acts down and puts structural member BC in compression. Hence, the structure must be analyzed for buckling under this loading. From the pressure distribution, based on wind tunnel testing, the forces and stresses in all the structural members can be computed.

Figure 5.7 shows the same array structure, but with the wind directed toward the back of the array. Note that the resultant force is trying to lift the array from its supports and the structural support member BC is in tension.

Work done by Boeing showed that for single arrays, the wind forces were a minimum at array tilt angles of about 20° above the horizontal [10]. Surprisingly, maximum wind forces occur at array tilt angles of 10–15° and again at 90°. At 10–15° tilt, the array acts as a reasonably efficient airfoil and large aerodynamic lift forces result. For tilt angles of 20° and higher, the air separates from the array—similar to the flow over the wing of an aircraft when it is going into a stall [11].

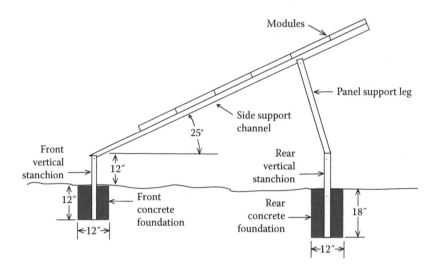

Figure 5.5 Sketch of the Hughes modular array field. This design was developed for Sandia National Laboratories. (From Marion, B. and Atmaram, G., Preliminary Design of a 15–25 kWp Thin-Film Photovoltaic System, prepared for the U.S. Department of Energy, Florida Solar Energy Center, Cape Canaveral, FL, November 13, 1987.)

Ground clearance also affects wind loading. For example, it has been found that increasing the ground clearance from 2 to 4 ft increases the normal wind forces by 10%–15% for an array with an 8-ft panel height [12].

For systems consisting of multiple rows of subarrays, wind tunnel tests have shown that the interior rows experience several times less wind loading than the

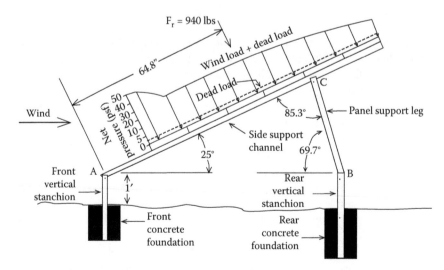

Figure 5.6 Net pressure on the array due to 100-mph winds toward the front of the array. Structural member BC is in compression.

MECHANICAL CONSIDERATIONS

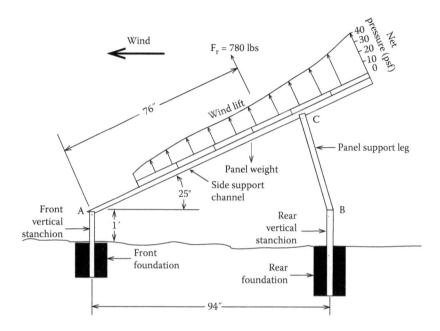

Figure 5.7 Net pressure distribution due to 100-mph winds toward the back of the array.

exterior rows [13]. Consequently, wind breaks or wind fences are sometimes used to reduce the wind loading on the exterior rows.

Because of their large exposed surface areas, their elevation off the ground and their orientation with respect to wind direction, PV arrays are often subjected to exceptionally high mechanical forces. Not only do the modules in the array have to resist these forces, but also the attachments to roofs and/or other structures must be well secured.

In designing PV arrays and structures to meet wind loads, the engineer must complete the following steps:

1. Establish the basic wind speed.
2. Determine the velocity pressure.
3. Determine the gust effect factor.
4. Determine the appropriate pressure or force coefficient.
5. Determine the wind loads on the array.
6. Determine the forces acting on critical members and attachment points.
7. Establish an appropriate factor of safety based on risk.
8. Select structural members and attachment hardware that can carry all loads with the prescribed factor of safety.

Fortunately, the American Society of Civil Engineers has developed standards and procedures for completing the steps above [8]. For example, the ASCE has developed maps indicating the basic wind speeds throughout the United States (Step 1); formulas, tables, and charts for computing the velocity pressure (Step 2); gust effect factors for various downtown, urban, rural, and flatland sites (Step 3); force coefficients for different shapes and structures (Step 4); and equations for computing the wind loads (Step 5). From these wind loads (forces per unit area) on the array, the equations of static equilibrium are used to compute the forces acting on the structural members (Step 6). Knowing these forces, the engineer selects the materials, structural shapes, and attachment hardware that meet all design requirements.

The ASCE procedures may be used to analyze the wind loads on standoff-mounted rooftop arrays, such as shown in Figure 5.8. Wind directions from both the front and rear of the array-roof structure should be considered, resulting in both uplifting and downward pressure distributions. Uplifting pressure is usually of greater concern because of the tensile and withdrawal forces it exerts on supporting hardware.

In the ASCE standard, the highest basic wind speed in the continental United States is 180 mph (290 km/hour). The velocity pressure, q, is computed using the following equation:

$$q = 0.00256 K_z K_{zt} K_d V^2 I, \qquad (5.8)$$

where K_z is velocity pressure exposure coefficient at height z, K_{zt} is topographical factor, K_d is wind directionality factor, V is basic wind speed in mph, and I is importance factor.

Using the equations, tables and charts in the standard yields $K_z = 0.85$, $K_{zt} = 1.00$, $K_d = 0.85$, $I = 1.00$, and $q = 60.48$ psf (2.9 kpa).

The design wind pressure, p, is computed using

$$p = qGC_f, \qquad (5.9)$$

where G = gust effect factor = 0.85, C_f = force coefficient = 0.70 (for a normal gable or hip roof), and p = 36 psf (1.72 kpa).

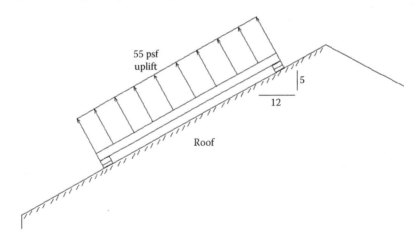

Figure 5.8 Standoff-mounted array with an uplifting pressure distribution.

MECHANICAL CONSIDERATIONS

As an example, for an array having an area of 200 ft^2 (18.6 m^2), the total uplifting (resultant) force acting on the array would be 36 psf × 200 ft^2 = 7200 lb (3266 kg). Knowing this resultant force, the design engineer can now determine the number of attachment points and the size of the mounting hardware necessary to safely carry this load.

Rather than computing the pressure distribution for each standoff array and roof combination, a more prudent approach is to assume a worst-case scenario for wind loading and design a mounting system capable of surviving the associated loading. Such a scenario might include maximum expected wind speeds (say 150 mph [241 km/hour]), array tilt angles that produce maximum uplifting pressure distributions (say 0–7°), topographical and shape effects that increase the pressure loading on the array and mounting system, and an adequate factor of safety. Using this approach and applying the methodology in the ASCE standard yields a reasonable upper limit on the design wind pressure of 55 psf (2.63 kpa) for most situations.

Many manufacturers of PV modules supply structural members to support their arrays with an adequate factor of safety. In most cases, a registered professional engineer has performed a structural analysis based on a worst-case scenario such as that described above. However, such analyses are usually not site specific. Note that the array will be physically attached to and supported by a roof or some other type of structure at a specific location. It is important that the entire site-integrated system, including the attachment hardware and the roof or support structure to which the array assembly is attached, is capable of surviving the worst-case scenario of wind loading—not just the array and mounting members that come with the PV system package.

In summary, the largest mechanical forces acting on PV arrays and structures are due to high wind speeds. Suppliers of packaged PV systems should assume a worst-case wind loading of 55 psf (or higher) in designing their array mounting systems. Many have already done so. In addition, the entire integrated PV and building system must be able to survive the worst-case scenario of wind loading and comply with all site-specific building code requirements.

5.5.5 Snow Loads

Snow loads on PV arrays and on sloped roofs decrease as the tilt angles and slopes increase. There are three reasons for this. First, as the tilt angle increases, the component of the weight force parallel to the array surface increases relative to the component normal to the array surface, thus helping to shed the snow. Second, wind forces tend to shed the snow from the array. Third, melt water reduces the friction between the array surface and helps to shed the snow. Most PV arrays have "slippery" glass surfaces that also aid the snow-shedding process. On the other hand, obstructions and rough surfaces hinder the shedding of snow and its weight. For example, some tile roofs contain built-in protrusions or rough surfaces that prevent the snow from sliding. The snow that accumulates on arrays mounted on such roofs may be obstructed from shedding [8].

An extreme case of snow loading occurs when unusually large amounts of snow accumulate on an array at a relatively small tilt angle. However, this is rare because shallow tilt angles are uncommon in colder, snow-prone parts of the world. In most cases, snow loads on PV arrays are less than 10 psf (479 PA) unless the snow is unusually dense (wet).

5.5.6 Other Loads

The PV array and support structure may experience mechanical forces due to other loads, such as rain, ice, hydrostatic pressure, and seismic activity. The latter is critical in earthquake-prone regions, and the ASCE has detailed procedures for designing for seismic loads.

Foundations, footings, or slabs that support PV arrays may experience uplift forces due to hydrostatic pressure in locations where ground water is close to the surface. If expansive soils are present, the hydrostatic pressures can be quite large. In such cases, the expansive soils are sometimes excavated to a depth of at least 2 ft (0.61 m) below ground level, followed by back filling with freely draining sands and gravel.

Regardless of the location, the PV engineer is responsible for ensuring that all of the relevant loads and load combinations have been properly accounted for in the mechanical system design.

5.6 ARRAY MOUNTING SYSTEM DESIGN

5.6.1 Introduction

Previous sections of this chapter discussed various properties of materials, the strength of materials, functional and operational requirements, standards and codes, and the mechanical forces acting on the array and support structure. Emphasis was placed on material selection and compatibility based on concerns for structural integrity, safety, and code compliance. This section focuses on those aspects of the array mounting system that affect costs, performance, maintenance, and appearance.

5.6.2 Objectives in Designing the Array Mounting System

In addition to meeting structural, code, and safety requirements, good design of the array mounting system should achieve the following:

- Minimizing installation costs
- Enhancing array performance
- Providing reasonable accessibility for installation and maintenance
- Making the system aesthetically appropriate for the site and application

5.6.2.1 Minimizing Installation Costs

Both good design and good installation practices are being employed to reduce installation costs. Studies in the 1990s by Sandia National Laboratories have shown

MECHANICAL CONSIDERATIONS   211

that only about half the cost of PV systems was for modules, and that was at a time when module costs were upwards of 20 times the current cost per watt. The other half was for nonmodule components and labor, with labor being the major contributor. For example, suppose a journeyman electrician installs a PV system. The electrical contractor may charge $75 per hour for the electrician's time. The value of reducing the number of hours for installation should be obvious. PV manufacturers and system suppliers have been reducing installation costs by doing one or more of the following:

1. *Developing packaged kits:* For about the first 2 decades of applying PV to terrestrial applications, system design was essentially customized for each installation. Since the mid to late 1990s, the trend has been to develop packaged kits that can easily be adapted to a wide variety of installations. Building applications of PV have helped to accelerate this trend. PV system suppliers now deliver standard models, which include modules, the structural support systems, attachment hardware, batteries, and inverters. Some include the module interconnect wiring, junction boxes, and other balance-of-system components. Continued development and refinement of packaged kits should help reduce the cost of PV system installations.
2. *Minimizing the total number of parts:* A second design approach to reducing installation costs is to minimize the total number of parts included in the design. As an example, some suppliers are moving toward the use of larger modules, so fewer will be needed to obtain a desired power goal. Once upon a time, modules did not come with leads, but this has not been the case since about 2005. In fact, the module leads are equipped with touch-safe male and female connectors, so it is essentially impossible to make any connection other than positive to negative. Furthermore, module power tolerances are now so precise that it is no longer necessary to sort the modules at the site by measuring V_{OC} and I_{SC} of each module to try to balance source circuits. This significantly reduces the skill level required for wiring source circuits.
3. *Minimizing part variations:* A third design approach to reducing installation costs is to minimize part variations. This concept is easily applied to the size and type of structural members, fasteners, and attachment hardware. It often eliminates the need for special tooling.
4. *Designing parts to be multifunctional:* Designing systems such that some parts are multifunctional is a fourth approach to cost reduction. For example, a structural member may be used not only to carry the mechanical loads, but also to enhance the flow of air over the back surface of the array.
5. *Designing parts for multiple uses:* Structural members, brackets, attachment hardware, and fasteners should be designed for possible use with different kits, model numbers, and product lines.
6. *Designing for ease of assembly:* This not only reduces the time required to install systems, but also often results in higher system reliability and reduced system maintenance.
7. *Avoiding separate fasteners:* One popular approach to reducing installation costs is to use slotted roof brackets that allow PV modules or panels to be simply dropped into the slots. It eliminates the need for on-site attachment of nuts and bolts to connect the panels to roof brackets. Another alternative approach is to use captive fasteners.

8. *Minimizing assembly direction:* This means that parts should be designed so that they fit or can be assembled only one way—the correct way. This is common practice in many industries.
9. *Maximizing compliance in assembly:* The installer should not have to struggle to assemble the panel framework, attach the modules to the panel, or attach the panels to the support structure. To maximize compliance in assembly, the designer should avoid tolerances that are too tight. Parts should fit together with a minimum of effort.
10. *Minimizing handling in assembly and installation:* This concept follows from a number of the other items, but also includes locating holes, slots, and other points of attachment to reduce handling.

In addition to the design practices suggested above [5], good installation practices can also be employed to reduce costs. Already mentioned is the practice of preassembling and prewiring panels off-site using less-expensive personnel. Another consideration for roof-mounted arrays is the number, type, and placement of roof attachment points. Often this decision is made by the installer at the site, rather than by the designer, and serious trade-offs must be considered. For example, minimizing the number of roof attachments saves time, but it also means that individual attachment points will be subjected to larger forces.

5.6.2.2 Building Integration Considerations

As the population grows and building density increases, space becomes more valuable and rooftops become more attractive locations for PV arrays. The question arises: How can we most effectively integrate PV systems and buildings? For existing buildings, the most prudent approach is to use designs that minimize costs by incorporating the measures discussed above.

Building-integrated PV (BIPV) products are solar electric devices that replace conventional building materials. They include solar shingles, tiles, skylights, windows, overhangs, facades, and PV laminates bonded directly to metal roofing panels. They can be used on either new or existing buildings in a manner that is aesthetically pleasing and very appealing to buyers.

For new manufactured buildings, one interesting approach to building integration involves factory-installed PV systems. Approximately 30% of new housing stock in the United States is manufactured housing. To be effective, the manufacture of buildings in a factory requires careful planning, time-motion studies, and process optimization—areas of interest to industrial engineers. Not only are the buildings manufactured quickly, but also quality control becomes an ongoing part of the process. Opportunities exist to include PV systems in the building manufacturing process. The idea is to accelerate the installation process, reduce costs, and improve quality control. The complete PV system can be installed in the factory, part of it can be installed, or the manufactured building can be specially configured to more easily accept a PV system. The latter approach has been successfully demonstrated and involves installing the following items in the factory prior to shipping the building to the site: double roof trusses (to both increase load-carrying ability and allow for the use of larger mounting brackets); wiring conduit from the roof through the attic and walls to the crawl space under the building; a rain-protected opening and flashing (i.e., vent stack flashing) to accommodate wiring from the array into the conduit; and a prepared, structurally enhanced location for mounting the inverter. Once the building has been delivered and set up on the site, installation of the prewired array panels, inverter, and other balance-of-system components is achieved at a fraction of the time and cost of a custom installation [14].

5.6.2.3 Costs and Durability of Array-Roof Configurations

Life-cycle costing is used in economic analyses of renewable energy technologies because these technologies realize their value over time. They are usually characterized by high capital costs and low operating costs. For PV systems on buildings, the combined array-roof configuration is a major contributor to the life-cycle cost. It is important that these configurations last 20–30 years. However, this is not always achieved without unexpected added cost. For example, asphalt shingles often have to be replaced about every 10 years in warm, humid climates. The life-cycle cost analysis should include the added cost of removing and reinstalling the PV array when reroofing.

The durability of some of the relatively new BIPV products is also a concern. For example, PV laminates bonded to either structural or architectural standing seam metal panels produce architecturally attractive arrays, and installation costs have been decreasing. However, the long-term durability of the bond between the laminate and the metal needs to be demonstrated [14].

5.6.3 Enhancing Array Performance

The mechanical system can affect the array performance in several ways:

- Increasing the amount of incident solar radiation
- Avoiding shading
- Allowing the array to operate at lower cell temperatures
- Protecting the array from vandalism

5.6.3.1 Irradiance Enhancement

Irradiance enhancement can be achieved by optimizing the array orientation, tracking the sun, and using concentrating collectors. The optimum array orientation depends on the type of system, application, and end user. For example, to maximize annual energy production, the optimum tilt angle is about 90% of the latitude of the site and the optimum azimuth is true south for the Northern Hemisphere and true north for the Southern Hemisphere.

However, utility companies may be more interested in peak load reduction than total energy produced. Consequently, the optimum azimuth angle might point the array more closely to the west than to the south (in the Northern Hemisphere), since utility peak loads generally occur during summer afternoons. And, as noted in Chapters 2 and 4, on an annual basis, as long as the array is tilted within 15° of latitude and within about 30° of south, only a few percent of the maximum annual array energy production is lost.

Concentrating and tracking arrays work best in areas receiving a higher-than-average percentage of direct radiation, such as the Desert Southwest United States, and may not be an attractive option in many locales. As discussed under the section on design requirements, the engineer must weigh the trade-offs.

Another consideration on tracking arrays as the cost of PV modules continues to decrease is the cost of the land on which the array is mounted. Since tracking arrays are nearly always ground mounted, the rows need to be adequately spaced to avoid shading from adjacent rows. This may result in a relatively small total percentage of the land upon which the array is located actually being covered by the array. So a desired design objective may be maximizing the annual kWh/acre, which can usually be best achieved with fixed arrays with smaller tilt angles. Simply put, the cost of land is now a more significant component of total PV system cost than it was only a few years ago, so efficient use of land is important. In some cases, it is even desirable to generate the maximum number of kWh on a given area of rooftop, even if the cost/watt of the system may be a bit higher.

5.6.3.2 Shading

Closely related to irradiance enhancement is the need to avoid shading. The performance characteristics of shaded PV arrays vary, depending upon module cell configuration and shadow specifics. Consequently, shading should be avoided to the greatest extent possible. Methods of determining shading have been discussed in other chapters.

5.6.3.3 Array Cooling

PV modules, especially those using relatively thicker, crystalline cells, work better at lower cell operating temperatures. Consequently, the array mounting should be designed to allow air circulation along the back surface of the modules. Rack-mounted arrays typically operate at lower temperatures than other configurations. Experiments at Sandia National Laboratories show that the operating temperatures of direct-mounted arrays may be 18°C higher than the cell's nominal operating cell temperature (NOCT). For standoff-mounted arrays, a standoff height of 4–6 inch (10–15 cm), with no significant obstruction to air flow, will permit adequate passive cooling of the modules. Most commercial standoff array mounts meet this criterion.

In general, the output power of thin-film materials is degraded less by high operating cell temperatures.

5.6.3.4 Protection from Vandalism

Unfortunately, PV arrays are sometimes used as targets by vandals. Therefore, it may be necessary to include protection features. Back shields are often used for arrays on area lighting systems. Some of these vandal guards do not permit adequate cooling of the array and can have a noticeable deleterious effect on performance. Back shields that allow more air circulation might be considered. For larger array fields, fencing, security lighting, motion sensors, and other protection measures are sometimes employed.

5.6.4 Roof-Mounted Arrays

PV array mounting can be categorized according to where the arrays are mounted, how they are supported, and whether they have a fixed or changing orientation. Arrays can be either roof-mounted or ground-mounted. Roof-mounted arrays typically use one of four different methods of support: rack, standoff, integrated, or direct. With the surge of interest in PV and building applications, better designs for each of these four roof-mounting approaches are being pursued.

5.6.4.1 Standoff Mounting

Standoff arrays are mounted above and parallel to the roof surface as illustrated in Figure 5.9. Standoff mounts work well for buildings with sloping roofs. When installing a standoff-mounted array, the PV modules are often attached to the roof using point connections—usually along the edges of the modules. As a minimum, the standoff height between the roof and the bottom of the module frame should be at least 3 inch. Four to six inches is preferable. Lag screws that penetrate the roof rafters 2–3 inch (5.1–7.6 cm) may be used to fasten the mounting brackets to the roof. Spanner attachments, as shown in Figure 5.10, are stronger than lag bolts or lag screws and are commonly used in areas when no truss is located where the mounting bracket needs to be placed.

To promote passive cooling of standoff arrays, the engineer should consider the following designs:

- Designs that allow both lateral and vertical airflow along the back surface of the modules

Figure 5.9 Standoff-mounted array using commercial array mount. The brackets that support the frame provide height to promote good air circulation under the array.

Figure 5.10 When trusses are not at locations where brackets are mounted, spanners can provide secure anchor points.

- Designs that induce pressure differences between air inlet and exit regions
- Arrays with larger lateral dimensions than vertical dimensions (i.e., higher aspect ratios)

5.6.4.2 Rack Mounting

Rack-mounted arrays are above and tilted with respect to the roof, as illustrated in Figure 5.11. Rack mounts work well on flat roofs and roofs with a slope of 2:12 or less. They may be mechanically attached to the building structure or may employ ballast to resist wind and other mechanical loads. Rack mounts are usually subjected to higher structural loads, incur higher costs for mounting hardware, and are often

Figure 5.11 Example of rack mount on roof in 140-mph wind zone. Mounts are anchored to trusses below. (Photo courtesy of Vergona-Bowersox Electric, Inc.)

MECHANICAL CONSIDERATIONS

less attractive than standoff mounts. However, for the same array area, the total energy output is sometimes higher because of better orientation and lower average operating temperatures, especially at northern latitudes, where the added slope helps to clear snow loads more quickly.

Ballasted rack assemblies offer the distinct advantages of simplicity and avoidance of roof penetrations, and should be considered if acceptable by local code jurisdictions. However, most rack mountings are firmly attached to the roof substructure. These are usually point connections, although several distributed attachment methods are sometimes used. Rack-mounted arrays typically run relatively cool compared with other mounts and they can reduce heat gain through roofs. Since the primary mechanism of rack array cooling is convection from the front and back surfaces of the modules, cooling can sometimes be enhanced by locating the rack in natural air channels and by reducing obstructions to airflow, such as screens, grates, and walls.

5.6.4.3 Integrated Mounting

For integrated arrays, also referred to as BIPVs or integral mounting, the array replaces conventional roofing or glazing materials as illustrated in Figure 5.12. Because integrated arrays replace conventional roofing and glazing materials, they become a significant architectural feature of a building and can be aesthetically very pleasing.

Dimensional tolerances may be tighter for some integrated arrays than for standoff or rack mounts. If commercial curtain-wall glazing techniques are used, larger modules are preferable to smaller modules.

Figure 5.12 Solar tiles replace conventional roof tiles at a South Florida residence. Note that some tiles are located in Roof Zone 2, but will still withstand a 140-mph wind.

Figure 5.13 Front and back of solar roof tile. Note the 10 multicrystalline Si cells on the front view, which produce approximately 4 V at P_{max}. It is one of a variety of new products that can replace conventional roofing materials.

5.6.4.4 Direct Mounting

In direct mounting, the array is affixed directly to the roofing material or underlayment, with little or no airspace between the module and the roof. Figure 5.13 illustrates a direct-mounted solar roofing tile.

Array operating temperatures for direct mounts are usually higher than for other mounting techniques. The use of direct mounting may become increasingly popular for new thin-film products that are not as sensitive to operating temperature. Some PV materials, such as solar shingles, fall into both the direct and BIPV mounting categories.

5.6.5 Ground-Mounted Arrays

Ground-mounted arrays are supported by racks, poles, or tracking stands. These arrays are secured to the ground to resist uplifting caused by wind loads. All ground-mounted arrays run relatively cool because good airflow is possible over both the front and back surfaces of the modules. This cooling can be enhanced by minimizing obstructions to airflow such as shrubbery and fences.

5.6.5.1 Rack Mounting

Rack mounting is commonly used for mounting arrays on the ground. Simple structural hardware such as angles, channels, and metal tubing can be used for both small and large PV arrays. Most module manufacturers and equipment suppliers offer hardware for rack-mounting PV arrays on the ground. Figure 5.14 is an example of a rack-mounted array that has been designed for ground mounting.

5.6.5.2 Pole Mounting

If the array consists of only a few modules, it can be mounted on a pole as shown in Figure 5.15. Depending on the number of modules and their height above the ground, the pole may need to be set in concrete to resist being overturned during windy conditions. Among others, PV-powered outdoor lighting is a good application for pole-mounted arrays.

MECHANICAL CONSIDERATIONS 219

Figure 5.14 Ground-mounted racking system. (Photo courtesy of Urban Solar Group, Inc.)

Figure 5.15 Pole-mounted arrays are often used for PV lighting systems.

5.6.5.3 Tracking-Stand Mounting

Because tracking arrays receive more sunlight than stationary arrays, especially in areas with high percentages of direct sunlight, each individual module produces more energy. Whether to use a tracking array depends on the trade-off between additional energy produced and added cost and complexity and space needed between trackers to avoid one row shading another.

Figure 5.16 Pole-mounted arrays with trackers. (Photo courtesy of Direct Power and Water, Corp., Albuquerque, NM.)

There are two types of trackers: active and passive. Active trackers use drive mechanisms, such as electric motors and gears, to point the array toward the sun. They may track about either one or two axes. The tracking motion may be controlled by a computer or by sun-seeking sensors.

Passive trackers normally track about only one axis and use a two-phase fluid, such as a refrigerant, which vaporizes and expands when it is heated by the sun. The expanding fluid causes the tracker to pivot toward the sun as the fluid's weight shifts from one side of the tracker to the other. An alternative design uses a hydraulic cylinder and linkage arrangement. For both tracker designs, sunshades are used to regulate the heating of the fluid and control the motion. A passive tracker is shown in Figure 5.16. Compared with a fixed-tilt array, these trackers show the greatest energy enhancement in the summer when the days are long and are less beneficial in the winter. Tracking arrays are often used for water pumping because they can produce higher starting currents early in the day and because of the higher demand for water during the summer months.

5.6.6 Aesthetics

It is important for arrays on buildings to be aesthetically pleasing. They should be designed to blend into the building lines and colors. In designing PV arrays for building applications, the architect and engineer should consider the following suggestions:

- Mounting the array parallel to the roof (if appropriate)
- Using the roof dimensions to establish the array aspect ratio

MECHANICAL CONSIDERATIONS

- Avoiding harsh contrasts and patterns
- Making the mounting hardware as inconspicuous as possible
- Avoiding shading, even if appearance suffers somewhat

As mentioned previously, many of the new BIPV products add to the architectural attractiveness of buildings. However, even in the case of standoff mounting, the additional lag screws driven into the trusses result in a more secure fastening of the roof decking to the trusses.

5.7 COMPUTING MECHANICAL LOADS AND STRESSES

5.7.1 Introduction

As indicated previously, the PV array is subjected to a wide variety of mechanical influences that affect the stresses and strains in the modules, structural support members, and attachment hardware. These include deadweight loads, live loads, wind loads, and snow and ice loads. In this section, we assume a pressure distribution on the array and discuss the methods used to calculate its effects on critical elements of the support structure.

5.7.2 Withdrawal Loads

A high percentage of standoff-mounted PV arrays are attached to the roof using lag screws or bolts. The lag screw typically passes through a hole in a mounting bracket, possibly a mounting pad, shingles, waterproof membrane, sheathing and, finally, into the truss, spanner, or other primary structural wood support member. The withdrawal strength of this and other attachment points is a function of the diameter of the lag screw, the length of thread embedded in the primary structural wood support, and the specific gravity of the wood. The allowable withdrawal load of lag screws and bolts per inch of embedded thread is given by [2]

$$p = 1800 D^{3/4} G^{3/2}, \tag{5.10}$$

where p = allowable withdrawal load (lb/inch), D = shank diameter of lag screw or bolt (inch), and G = specific gravity of oven-dry wood.

Values for p computed using Equation 5.10 include a safety factor of 4.5. This relatively high value provides added protection just in case the actual wood used may have reduced strength due to variations in specific gravity, knots, grain irregularities, holes, and other defects [15].

The typical diameters of lag bolts or screws used to attach PV arrays are 1/4 inch (6.35 mm) and 5/16 inch (7.9 mm), although they may be as large as 3/8 inch (9.5 mm) or 1/2 inch (12.7 mm) if the trusses consist of larger beams, rather than 2 inch × 4 inch (5.1 cm × 10.2 cm) or 2 inch × 6 inch (5.1 cm × 15.3 cm) lumber. The specific gravities of various types of lumber can be found in the *Wood Engineering*

Table 5.5 Allowable Withdrawal Loads for Lag Screws

Lumber	White Oak	Southern Yellow Pine	White Spruce	Douglas Fir
Specific gravity, G	0.71	0.58	0.45	0.41
Diameter, D (inch)		Allowable Withdrawal Load, p (lb/inch)		
1/4	381	281	192	167
5/16	450	332	227	198
3/8	516	381	260	226
7/16	579	428	292	254
1/2	640	473	323	281

Note: Values for p were computed using Equation 5.13 and include a safety factor of 4.5.

and Construction Handbook [16]. Table 5.5 [16] was developed using the above equation and the specific gravities for various types of lumber.

5.7.3 Tensile Stresses

In addition to withdrawal loads, uplifting pressure distributions produce tensile stresses in hardware used to secure the array to a roof or other support structure. This hardware could be lag screws, J-bolts, or threaded rods. The forces acting on the attachment hardware are computed using the equations of static equilibrium, and the tensile stresses are computed using Equation 5.1.

5.7.4 Buckling

In contrast to uplifting pressure distributions, forces distributed downward toward the plane of the array are usually not as serious a concern. However, an exception occurs when long, slender structural members are subjected to compressive forces that may result in structural instability known as buckling. In such cases, which may occur due to wind loads on rack-mounted arrays, the engineer must compute the critical buckling load using Equation 5.6, which can be expanded into the following form:

$$P = \frac{EI\pi^2}{L^2} = \frac{EAr^2\pi^2}{L^2} = \frac{EA\pi^2}{(L/r)^2}, \quad (5.11)$$

where I = moment of inertia of the cross-sectional area = Ar^2 (inch^4), A = cross-sectional area of the column (inch^2), r = radius of gyration of the cross-sectional area (inch), and L/r = slenderness ratio of the column (dimensionless).

Note from the above equation, larger slenderness ratios and smaller cross-sectional areas yield lower critical buckling loads. If the length L of the structural member cannot be significantly changed, the engineer should consider selecting members with cross sections having larger radii of gyration. Conceptually, the radius of gyration is a measure of an equivalent distance from a reference axis at which the entire

MECHANICAL CONSIDERATIONS

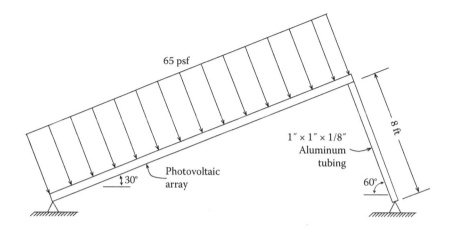

Figure 5.17 Rack-mounted array with rear support column in compression.

cross-sectional area may be assumed to be concentrated to produce the same moment of inertia. The type of cross section that yields higher radii of gyration is a hollow tube in which all the cross-sectional area is distant from the axis. Consequently, hollow metal tubing is often used for column support. Radii of gyration can be either calculated or obtained from tables for steel, aluminum, or other columns.

EXAMPLE 5.2

Figure 5.17 shows a rack-mounted array supported by square 6063 T-6 aluminum tubing having cross-sectional dimensions of 1 inch × 1 inch × 1/8 inch (2.54 cm × 2.54 cm × 0.3175 cm). The radius of gyration for this cross section is 0.361 inch² (2.33 cm²). The modulus of elasticity for the aluminum is 10×10^6 psi (68.9×10^3 Mpa) and the cross-sectional area is 0.50 inch² (3.23 cm²). If the length of the rear structural support member is 8 ft (96 inch [2.44 m]), compute the maximum force the member can withstand without buckling.

Substituting directly into Equation 5.11

$$P = \frac{10^7 (0.50) \pi^2}{(96/0.361)^2} = 698 \text{ lb } (316 \text{ kg}).$$

Thus, each rear support member should not be subjected to compressive forces that approach 698 lb.

5.8 STANDOFF, ROOF MOUNT EXAMPLES

5.8.1 Introduction to ASCE 7 Wind Load Analysis Tabular Method

In ASCE 7 versions published prior to ASCE 7-10, tabular methods of determining component and cladding wind loads were limited to buildings less than 60 ft

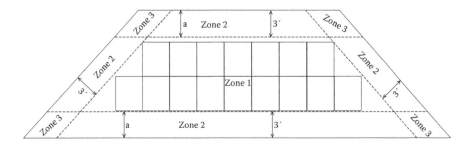

Figure 5.18 Module layout for high wind region showing roof zones as defined by ASCE 7-10.

(18.3 m) tall with a limited number of roof configurations. ASCE 7-10 significantly expanded the tabular method to include additional roof shapes, taller structures, and additional exposures.

Figure 5.18 shows the PV array layout of Section 4.5. Recall that this array is located in a hurricane-prone wind region, so the modules have been located in Roof Zone 1 as shown in the figure. Roof zones are identified in ASCE 7 [8] and are shown in Figure 5.18. ASCE 7-10 defines the parameter, a, which defines the edges of the zones as *"10 percent of least horizontal dimension or 0.4 h, whichever is smaller, but not less than either 4% of least horizontal dimension or 3 feet."* In this definition, h is the distance from ground to roof. Figure 5.19 shows how roof zones are defined in ASCE 7-10 for four different roof types. For this example, it will be assumed that the roof has a 4:12 slope, the design wind velocity is 160 mph, and the roof, which is over an enclosed building, is located in a neighborhood characterized by Exposure B, has a mean roof height of 12 ft (3.66 m) and has a = 3 ft (0.91 m).

The exposure of a roof depends upon the surface roughness of the land in the vicinity of the roof. According to ASCE 7, Roughness B is characterized by areas with numerous obstructions the size of single family dwellings or larger. Roughness C consists of flat, open country, grasslands and all water surfaces in hurricane-prone regions. Roughness D has flat, unobstructed surfaces *outside* hurricane-prone regions.

Exposure B, then, has Roughness B for a distance of 2600 ft (790 m) upwind, unless the mean roof height is less than 30 ft, in which case Roughness B prevails for at least 1500 ft (457 m) upwind.

Exposure D has Roughness D for a distance of 5000 ft (1524 m) upwind and distance of 600 ft (183 m) downwind, even if the downwind roughness is B. Note, however, that Exposure D applies to only nonhurricane-prone regions since Roughness D applies to only nonhurricane-prone regions.

Exposure C applies whenever Exposure B or D does not apply. Thus, in a hurricane-prone region, any exposure different from Exposure B will necessarily be Exposure C.

Chapter 26 of ASCE 7-10 presents two methods of determining wind load for components and cladding on all buildings and other structures. One method involves analytical procedures and applies to "regular" buildings that are defined in Parts 1

MECHANICAL CONSIDERATIONS

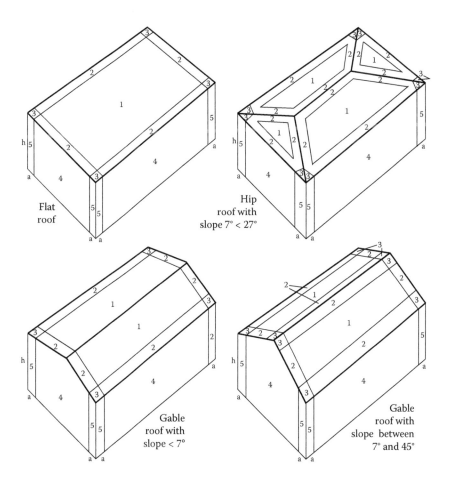

Figure 5.19 Roof zones as defined by ASCE 7-10 for flat, hip, and gable end roofs. Refer to text for definitions of zones and "a." (Adapted from Figure 30.5.1, ASCE 7-10, with permission from American Society of Civil Engineers, Reston, VA. For educational purposes only. For design and permitting, refer to the current edition of ASCE 7.)

through 6 of Chapter 30 of the standard. The second method involves wind tunnel testing and also can be applied to any structure. Since almost all enclosed buildings end up being classified as "regular," this example will assume the roof to be over a regular building. It will also use the table look-up method, since the method is very straightforward and also results in a conservative estimate of wind forces on the array. Since tables apply to a range of situations, the table entries must necessarily represent worst-case situations, so the designer can be comfortable that the table entries represent conservative estimates of wind forces on the array.

The basis of the table look-up of wind forces is Figure 30.5.1 of ASCE 7-10, part of which is shown as Table 5.6. The data of ASCE 7 apply to enclosed buildings, for components and cladding on walls and roofs in Exposure B, for roof heights less than 60 ft (18.3 m). The data are separated into three roof zones, with three ranges

Table 5.6 Wind Pressures for Exposure Category B. Excerpts from Figure 30.5-1 of ASCE 7-10

Roof Slope	Roof Zone	EWA (ft²)[a]	Basic Wind Speed (Pressures in psf with + [Downward] and − [Upward])				
			110 mph	115 mph	120 mph	130 mph	
0–7°	1	10	8.9 / −21.8	9.7 / −23.8	10.5 / −25.9	12.4 / −30.4	
	2	10	8.9 / −36.5	9.7 / −39.9	10.5 / −43.5	12.4 / −51.0	
	3	10	8.9 / −55.0	9.7 / −60.1	10.5 / −65.4	12.4 / −76.8	
7–27°	1	10	12.5 / −19.9	13.7 / −21.8	14.9 / −23.7	17.5 / −27.8	
	2	10	12.5 / −34.7	13.7 / −37.9	14.9 / −41.3	17.5 / −48.4	
	3	10	12.5 / −51.3	13.7 / −56.0	14.9 / −61.0	17.5 / −71.6	
27–45°	1	10	19.9 / −21.8	21.8 / −23.8	23.7 / −25.9	27.8 / −30.4	
	2	10	19.9 / −25.5	21.8 / −27.8	23.7 / −30.3	27.8 / −35.6	
	3	10	19.9 / −25.5	21.8 / −27.8	23.7 / −30.3	27.8 / −35.6	
			140 mph	150 mph	160 mph	180 mph	
0–7°	1	10	14.3 / −35.3	16.5 / −40.5	18.7 / −46.1	23.7 / −58.3	
	2	10	14.3 / −59.2	16.5 / −67.9	18.7 / −77.3	23.7 / −97.8	
	3	10	14.3 / −89	16.5 / −102.2	18.7 / −116.3	23.7 / −147.2	
7–27°	1	10	20.3 / −32.3	23.3 / −37.0	26.5 / −42.1	33.6 / −53.3	
	2	10	20.3 / −56.2	23.3 / −64.5	26.5 / −73.4	33.6 / −92.9	
	3	10	20.3 / −83.1	23.3 / −95.4	26.5 / −108.5	33.6 / −137.3	
27–45°	1	10	32.3 / −35.3	37.0 / −40.5	42.1 / −46.1	53.3 / −58.3	
	2	10	32.3 / −41.2	37.0 / −47.3	42.1 / −53.9	53.3 / −68.2	
	3	10	32.3 / −41.2	37.0 / −47.3	42.1 / −53.9	53.3 / −68.2	

Source: Figure 30.5-1 of ASCE 7-10, *Minimum Design Loads for Buildings and Other Structures*, ASCE Standard, SEI/ASCE 7-10, American Society of Civil Engineers, Reston, VA, 2010.

[a] Effective wind area. Note that definition of EWA in ASCE 7 requires use of 10 ft² for all modules and arrays unless interpolation is used for module areas between 10 and 20 ft². The data in this table is intended for instructional use only. For design and permit documents, refer to the current edition of ASCE 7.

of roof slopes: 0–7°, 7–27°, and 27–45° as well as vertical. Pressures are tabulated for wind speeds between 110 and 200 mph and are evaluated at mean roof heights. Correction factors are incorporated for Exposure C, depending upon effective wind area (EWA). Correction factors are also included for heights between 30 and 60 ft (9.15 and 18.3 m). Slopes and heights outside this range must be analyzed by either use of other figures in Chapter 30 or by use of the computational method or the wind tunnel method.

5.8.2 Array Mount Design, High Wind Speed Case

The array mount will be based on a commercial mounting frame that has been designed to withstand wind speeds up to 125 mph (201 km/hour) in Exposure C. The mount uses two mounting rails under the modules for wind speeds less than 125 mph. Figure 5.20 shows the use of three mounting rails for the 150-mph (241 km/hour) wind region. Note that since wind pressures are proportional to the square of the wind speed, the wind pressure at 150 mph is $(150/125)^2$ times higher than the 125-mph wind pressure, or 44% higher. But the third rail provides 50% more holding capacity, so the use of three rails on a two-rail system extends the use of the mount to 150-mph winds in Exposure C locations. If, instead, the system to be analyzed is in an Exposure B location, it should be noted that at a height of 30 ft, the Exposure C wind load, that is, the array mount capacity, is 1.4 times as high as the Exposure B wind load at a given wind speed (see Table 5.7). Thus, using the array mount in an Exposure B situation means that the Exposure B forces can be as high as the 150-mph Exposure C forces, or, in other words, this array mount will withstand Exposure B forces at wind velocities in excess of 150 mph. Specifically, the allowable Exposure B wind speed will be $150\sqrt{1.4} = 177$ mph. Since ratings are generally very conservative, this mounting system should withstand Exposure B winds up to 180 mph (290 km/hour). The specific forces will now be calculated for this array and the array mounting points.

Commercial array mounts are generally designed to have mounting brackets located 4 ft (1.22 m) apart along the rails, which is common when roof truss spacing is 24 inch (61 cm). The proposed locations of the mounting feet for the rails of the example are shown in Figure 5.20. So the exercise here is to first determine the wind

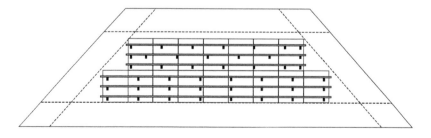

Figure 5.20 Standoff array mount for module layout of Figure 5.18 showing locations of rails and mounting feet.

Table 5.7 Adjustment Factors for Building Height and Exposures

Mean Roof Height (ft)	Exposure B	Exposure C	Exposure D
15	1.00	1.21	1.47
20	1.00	1.29	1.55
25	1.00	1.35	1.61
30	1.00	1.40	1.66
35	1.05	1.45	1.70
40	1.09	1.49	1.74
45	1.12	1.53	1.78
50	1.16	1.56	1.81
55	1.19	1.59	1.84
60	1.22	1.62	1.87

Source: ASCE 7, Figure 6.3, *Minimum Design Loads for Buildings and Other Structures*, ASCE Standard, SEI/ASCE 7-10, American Society of Civil Engineers, Reston, VA, 2010.

Note: The data in this table is intended for instructional use only. For design and permit documents, refer to the current edition of ASCE 7.

pressure on the array, then determine the total wind force on the array, and then determine the force that will be distributed among the mounting brackets.

The wind load can now be determined using the known information on roof slope, roof zone, and mean roof height. The pressures shown in Table 5.6 are valid for Exposure B for mean roof heights up to 30 ft (9.14 m). These tabulated pressures are based on an importance factor of 1.0 for the location of the roof. The 4:12 slope represents an angle of $\tan^{-1}(4/12) = 18.4°$, which falls within the 7–27° range.

The EWA is the area of the smallest component, which, in this case, will be a single module, even though the array itself is made up of more than one module. The reasoning here is that because of the way the modules are clamped to the array mount, if one should blow away, the next one will be loosened and a domino effect can take place. However, the table allows for interpolation if the EWA lies between two values in the table. In this case, the module area is 13.35 ft^2 (1.24 m^2), since the Zone 1 wind pressures for 10 ft^2 EWA at 160 mph are 26.5/–42.1 psf and the 20 ft^2 EWA wind pressures are 24.2/–41.0 psf, interpolation results in 25.7/–41.7 psf (1230/–1997 PA) for 13.35 ft^2 EWA at 160 mph.

In these expressions, the positive number represents the downward pressure on the array and the negative number represents the upward lift on the array. In other words, depending upon wind direction, the pressure may vary between these two extremes.

Since there are two rows of modules, separate calculations of wind loading will be needed for each row. For the row of seven modules, the total area is $7 \times 13.35 = 93.4$ ft^2 (8.68 m^2). The total upward force on the array is $93.5 \times 41.7 = 3899$ lb (1769 kg). The maximum downward force on the array at 160-mph wind speed is $93.5 \times 25.7 = 2403$ lb (1090 kg).

MECHANICAL CONSIDERATIONS

Normally, a roof structure is designed to and can easily withstand the downward force, so the test is to be sure that the array can withstand the upward force without being lifted off the roof. This means that the 21 mounting brackets must be able to hold down the 3899 lb of upward pull. Although there will be some nonuniformity in the forces on each mounting bracket, it is acceptable practice to assume the load to be evenly distributed among the brackets. Thus, the force per bracket is simply 3899/21 = 186 lb/bracket. So, the next check is to see how much thread must penetrate the roof truss to ensure adequate pull strength.

Most installers agree that for 2 inch × 4 inch or 2 inch × 6 inch trusses, a 5/16-inch lag screw is a good choice for attaching a mounting bracket to the trusses. Thus, depending upon the type of wood used in the trusses, Table 5.5 can be used to determine the length of lag screw needed to provide adequate pull strength to hold down the mounting bracket. In hurricane-prone regions, it is common practice to construct roof trusses of southern yellow pine, so it will be assumed that this will be the case in this example. Thus, using 332-lb/inch (59.3 kg/cm) pull strength for that portion of the thread embedded in the truss, it is found that 0.56 inch (1.42 cm) of thread must penetrate the truss. If, instead of 21 mounting brackets are used for the top row, 18 brackets are used, then the force/bracket will be 217 lb (98.4 kg) and a truss penetration of 0.65 inch (1.67 cm) will be required. To determine whether 18 brackets can be used, one must check the distance from the end of the rail to the first bracket on each end. The rail manufacturer specifies a maximum cantilever, often about 18 inch (46 cm), so if the distance is less than this, the 18-bracket solution is acceptable.

Of course, this does not mean the length of the lag screw only needs to be 0.65 inch, since the screw must also pass through the thickness of the mounting bracket, the roof underlayment, the roof decking, and, perhaps, even some foam insulation, before it enters the truss. So the installer and designer need to verify the roof composition prior to selecting lag screws. Typically, the screw length will be 3.5–4 inch (8.9–10.1 cm), with 1.5–2 inch (3.8–5.1 cm) of thread embedded into the truss.

The bottom row of nine modules in Figure 5.20 has an area of 120.2 ft^2 (11.2 m^2), is held by 24 mounting brackets, and experiences a maximum upward total force of 5013 lb (2274 kg). Thus, the maximum upward force on each bracket is 209 lb (95 kg). So for this row of modules, the lag screw threads must penetrate 0.63 inch (1.6 cm) into the trusses. Once again, if 1.5 inch (3.81 cm) of thread is in the truss, a comfortable safety factor is obtained.

5.8.3 Array Mount Design, Lower Wind Speed Case

As a final example, consider the array layout of Section 4.5 for the low wind speed case, as shown in Figure 5.21 in which 30 modules are used. It will be assumed that the roof measurements are the same, but the wind speed is now 110 mph (177 km/hour) and the fire marshal has not made an issue of the modules reaching to near the top and the bottom of the roof section. Two mounting rails will be used per row of modules, as shown in Figure 5.22.

In this case, note that the top row has six modules that have approximately two-thirds of each module in Zone 1 and one-third in Zone 2. In addition, the top row

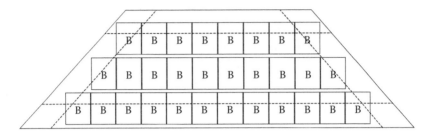

Figure 5.21 Lower wind speed region array layout allowing modules in Roof Zones 1, 2, and 3.

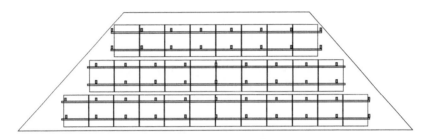

Figure 5.22 Lower wind speed region two-rail mounting system showing locations of rails and mounting feet.

has two modules that have approximately one-third of each module in each of Zones 1, 2, and 3. The middle row has eight modules in Zone 1 and two modules that are approximately half in each of Zones 1 and 2. The bottom row has 10 modules, of which approximately one-third of each module is in Zone 1 and two-thirds of each module is in Zone 2, along with the two end modules in the bottom row, each of which is approximately 15% in Zone 1 and 85% in Zone 2.

One method of estimating the forces on each row is to consider each row separately. In the top row, the module area in Zone 1 is two-thirds of the total area of the first six modules and one-third the total area of the two end modules. The module area in Zone 2 is approximately one-third the total area of the first six modules and one-third the area of each of the two end modules. The total module area in Zone 3 is approximately one-third the total area of the two end modules.

Thus, the top row has the equivalent of 4.67 modules in Zone 1, 2.67 modules in Zone 2, and 0.67 modules in Zone 3, for a rounded off total of 8 modules. Similarly, of the 10 modules in the middle row, the equivalent of 9 modules is in Zone 1 and 1 module is in Zone 2. Finally, the bottom row has an equivalent of 3.63 modules in Zone 1 and 8.37 modules in Zone 2.

Using a somewhat more conservative approach of using the EWA = 10 ft^2 table entries, the upward forces for Zones 1, 2, and 3 are, respectively, −19.9 psf (−953 PA), −34.7 psf (−1661 PA), and −51.3 psf (2456 PA). Thus, we have, for each row,

Top row total maximum uplift force = 13.35((19.9 × 4.67) + (34.7 × 2.67) + (51.3 × 0.67)) = 2936 lb (1332 kg)

MECHANICAL CONSIDERATIONS   231

Middle row total maximum uplift force = 13.35((19.9 × 9) + (34.7 × 1) + (51.3 × 0))
= 2854 lb (1295 kg)
Bottom row total maximum uplift force = 13.35((19.9 × 3.63) + (34.7 × 8.37) + (51.3 × 0)) = 4842 lb (2196 kg)

Noting that the top row has 16 mounting brackets, the average force per bracket is 184 lb (83.5 kg). The middle row has 18 brackets, so the average force per bracket is 159 lb (72.1 kg). The bottom row has 22 brackets, so the average force per bracket is 220 lb (99.8 kg).

5.8.4 Exposure C, Exposure D, and Other Correction Factors

The simplified method of the previous section can be applied to other exposures and to heights between 30 and 60 ft by applying appropriate multipliers to the pressures presented in Figure 30.5.1 of ASCE 7-10. Table 5.7 indicates the multipliers to be used for various building heights and exposure categories.

If the building is near a hill (escarpment), additional multipliers apply. If the building is partially enclosed, or if the structure is open, such as a carport or walkway, then the computational or wind tunnel approach must be used to determine the wind load. Design of array mounts for these situations is generally best left to the expertise and experience of a structural engineer.

PROBLEMS

5.1 Consider the following facts concerning a steel alloy and aluminum alloy: (a) steel is stronger, (b) steel is stiffer, and (c) aluminum stretches much more than steel. Sketch the stress–strain curves for both on the same stress–strain graph (see Figure 6.3). Show the relative positions of the yield strength, ultimate strength, and rupture strength. Also, indicate on the graphs how the moduli of elasticity compare for steel and aluminum.

5.2 For two bars having cross-sectional areas $A_1 = 0.25$ inch^2 and $A_2 = 25$ inch^2, subjected to uniform axial forces $P_1 = 100$ lb and $P_2 = 1000$ lb, compute the uniform stresses in each bar.

5.3 During a simple pull test of a bar, the deformation per unit length was measured at 0.000167 inch/inch. The corresponding stress was 5000 psi. Further in the test, the elongation per unit length and corresponding stress were 0.000667 inch/inch and 20,000 psi. If the proportional limit is 30,000 psi, what is the modulus of elasticity? Would this result be valid if the proportional limit was 18,000 psi? Explain.

5.4 A uniform bar has length = L and cross-sectional area = A. Rather than a force P acting on the end of the bar, the only force is the uniformly distributed weight of the bar. If the total weight of the bar is W, show that the total elongation is $\delta = WL/2AE$. If its weight per unit volume is w, also show that $\delta = wL^2/2E$.

5.5 A steel bar having a cross-sectional area of 0.5 inch^2 and a length L = 600 ft is suspended vertically. The force P acting on the lower end is 5000 lb. If steel weighs 490 lb/ft^3 and $E = 30 \times 10^6$ psi, find the elongation of the bar. (*Hint*: Use the results of Problem 5.4.)

5.6 A PV array is to be mounted on the roof of a nursing home, which is adjacent to a hospital. It is located in the Topeka, KS, area and is subjected to high wind loading.

List at least three possible functional requirements and three operational requirements. Also identify four constraints the designer must consider. Identify the trade-offs that might be weighed.

5.7 Why are building codes that affect the mechanical and structural design of PV systems much less uniform throughout the United States than electrical codes? Explain.

5.8 Give two examples of cases when the design loading for the PV array and structural supports is not dominated by wind loading.

5.9 The design wind load calculation involves the product of the wind speed, a force coefficient, and an importance factor. How is the design wind speed determined? If the force coefficient (i.e., fudge factor) cannot be derived analytically, how might it be determined experimentally? The importance factor is related to the consequences of failure—the greater the consequences, the higher the factor. How would the importance factor for a PV system on a rural home compare with the application in Problem 5.6?

5.10 List four objectives in designing an array mounting system.

5.11 For a fixed-orientation array to be mounted on a south-facing roof, discuss the advantages and disadvantages of each of the four roof-mounting configurations.

5.12 Using Figure 5.4, assume the sleeve is made of aluminum for which $\alpha = 12.8 \times 10^{-6}$ per °F and $E = 10 \times 10^6$ psi. The initial temperature at which there are no stresses is 59°F, and the final temperature is 200°F. All other dimensions, properties and information are the same as in the example for Figure 5.4. Compute the stresses in both the stainless steel bolt and the aluminum sleeve at 200°F.

5.13 Four lag screws of 5/16 inch diameter and 3.5 inch length are used to attach a 5 ft by 13 ft standoff array to an asphalt shingle roof. The total thread length is 2.5 inch and the combined thickness of the mounting bracket, pad, shingles, roof membrane and plywood sheathing is 1.0 inch. If the lag screw penetrates into the roof truss made of southern yellow pine, what is the allowable withdrawal load on each lag screw? What is the maximum uplift loading in psf that the array can withstand?

5.14 A four-module PV array has an area of 70 ft^2 and will be secured using eight mounting brackets. Each bracket uses two 5/16 inch lag screws to secure the standoff-mounted array. The roof trusses are made of white spruce. The design wind load is 55 psf. What is the minimum thread penetration for each fastener?

5.15 A standoff-mounted array has a surface area of 50 ft^2 and is secured to an asphalt shingle roof with four steel J-bolts of 3/8 inch diameter. If the allowable tensile stress for the steel is 40,000 psi, are the J-bolts of sufficient strength to carry a maximum uplifting wind load of 65 psf? What is the factor of safety?

5.16 An array of 10 PV modules, each with dimensions of 1 m × 1.6 m, is to be mounted in a single row above and perpendicular to a standing seam metal roof using clamps attached to the seams. The modules themselves can withstand a load of 75 psf. The spacing of the seams is 16 inch. If each clamp has an allowable withdrawal resistance of 120 lb, what is the minimum number of clamps that must be used?

5.17 Assume the ratio of direct to total irradiance is the same for two different locations, one at 30° north latitude and one at 60° north latitude. Which location would benefit the most from a two-axis tracking array (vs. a fixed-orientation array)? Explain.

5.18 For more effective cooling of standoff-mounted PV arrays, would you recommend higher or lower aspect ratios, that is, portrait or landscape? Explain.

5.19 A rack-mounted array similar to the one in Figure 5.17 is supported at the rear by two 4-ft columns of 6063 T-6 aluminum tubing. The modulus of elasticity of the aluminum is 10×10^6 psi. The cross-sectional area is 0.5 inch^2 and the radius of gyration is 0.361 inch. What is the moment of inertia of the cross-sectional area in units of inch^4? Compute the critical buckling load, P.

REFERENCES

1. Hanks, R. W., *Materials Science Engineering: An Introduction*, Harcourt, Brace & World, New York, 1970.
2. Avallone, E. A. and Baumeister, T., Eds., *Marks' Standard Handbook for Mechanical Engineers*, 10th Ed., McGraw-Hill, New York, 1996.
3. Kutz, M., *Mechanical Engineers' Handbook*, 2nd Ed., John Wiley & Sons, New York, 1998.
4. Singer, F. L., *Strength of Materials*, 2nd Ed., Harper & Brothers Publishers, New York, 1962.
5. Dieter, G. E., *Engineering Design: A Materials and Processing Approach*, 2nd Ed., McGraw-Hill, New York, 1991.
6. Clauser, H. R., editor-in-chief, *The Encyclopedia of Engineering Materials and Processes*, Reinhold Publishing, New York, 1963.
7. Larson, W. J. and Wertz, J. R., *Space Mission Analysis and Design*, 2nd Ed., Microcosm, Inc. and Kluwer Academic Publishers, Torrance, CA, 1992.
8. *Minimum Design Loads for Buildings and Other Structures*, ASCE Standard, SEI/ASCE 7-10, American Society of Civil Engineers, Reston, VA, 2010.
9. Modular Photovoltaic Array Field, prepared for Sandia National Laboratories, SAND83-7028, Hughes Aircraft Company, 1984.
10. Miller, R. D. and Zimmerman, D. K., Wind Loads on Flat Plate Photovoltaic Array Fields, Phase II Final Report, prepared for Jet Propulsion Laboratory, Contract No. NAS-7-100-954833, Boeing Engineering and Construction Company, Seattle, WA, 1979.
11. Marion, B. and Atmaram, G., Preliminary Design of a 15–25 kWp Thin-Film Photovoltaic System, prepared for the U.S. Department of Energy, Florida Solar Energy Center, Cape Canaveral, FL, November 13, 1987.
12. Wind Design of Flat Panel Photovoltaic Array Structures, prepared for Sandia National Laboratories, SAND79-7057, Bechtel National, Inc., June 1980.
13. Miller, R. D. and Zimmerman, D. K., Wind Loads on Flat Plate Photovoltaic Array Fields, Phase III and Phase IV Final Reports, prepared for Jet Propulsion Laboratory, Contract No. NAS-7-100-954833, Boeing Engineering and Construction Company, Seattle, WA, April 1981.
14. Ventre, G. G., A Program Plan for Photovoltaic Buildings in Florida, prepared for the Florida Energy Office/Department of Community Affairs and Sandia National Laboratories, FSEC-PD-25-99, Florida Solar Energy Center, Cocoa, FL, January 1999.
15. *Wood Handbook—Wood as an Engineering Material*, U. S. Department of Agriculture, Forest Service, Forest Products Laboratory, Madison, WI, 1999.
16. Faherty, K. F. and Williamson, T. G., *Wood Engineering and Construction Handbook*, 3rd Ed., McGraw-Hill, New York, 1997.

SUGGESTED READING

Building Integrated Photovoltaic Project in the 21st Century Townhouses, Advanced Housing Technology Program, National Association of Home Builders Research Center, Inc., Upper Marlboro, MD, June 1997.

Epstein, S. G., Aluminum and its alloys, Aluminum Association, Inc., Presented at the Annual Liberty Bell Corrosion Course, sponsored by the National Association of Corrosion Engineers, Washington, DC, September 1978.

Healey, H., An Investigation of the Feasibility of a National Structural Certification Process, Contract Report for the Solar Rating and Certification Corporation, Healey and Associates, Merritt Island, FL, September 25, 1998.

Hollister, S. C., *The Engineering Interpretation of Weather Bureau Records for Wind Loading on Structures*, National Bureau of Standards, Washington, DC, November 1970.

IEEE Recommended Practice for Qualification of Photovoltaic (PV) Modules, IEEE Standard 1262-1995, IEEE Standards Coordinating Committee 21 on Photovoltaics, Institute of Electrical and Electronics Engineers, Inc., New York, April 12, 1996.

Manual of Steel Construction, 8th Ed., American Institute of Steel Construction, Chicago, 1980.

McCluney, W. R., *Introduction to Radiometry and Photometry*, Artech House, Inc., Norwood, MA, 1994.

Mehta, K. C. and Marshall, R. D., *Guide to the Use of the Wind Load Provisions of ASCE 7-95*, ASCE Press, Reston, VA, 1998.

Metal roofing, *Roofer Magazine*, July 1997.

Orlowski, H., Corrosion protection for solar systems: Methods of eliminating an old problem applied to solar components, *Heat/Piping/Air Conditioning*, July 1976.

Photovoltaic System Design, Course Manual, Florida Solar Energy Center, Cocoa, FL, FSEC-GP-31-86, April 1996.

Post, H. N., Low cost structures for photovoltaic arrays, In *Proc. 14th IEEE Photovoltaic Specialists Conference*, San Diego, CA, January 7–10, 1980.

Post, H. N. and Noel, G. T., A comparative evaluation of new low-cost array field designs for grid-connected photovoltaic systems, In *Proc. 19th Intersociety Energy Conversion Engineering Conference*, San Francisco, CA, August 19–24, 1984.

Roark, R. J., *Formulas for Stress and Strain*, McGraw-Hill, New York, 1954.

Russell, M. C. and Kern, E. C., *Stand-Off Building Block Systems for Roof-Mounted Photovoltaic Arrays*, Sandia Contract 58-8796, Massachusetts Institute of Technology, December 1985.

Shigley, J. E. and Mischke, C. R., *Mechanical Engineering Design*, 5th Ed., McGraw-Hill, New York, 1989.

Sick, F. and Erge, T., Eds., *Photovoltaics in Buildings: A Design Handbook for Architects and Engineers*, International Energy Agency, Paris, 1996.

Smith, W. F., *Principles of Materials Science and Engineering*, 2nd Ed., McGraw-Hill, New York, 1990.

Stefanakos, E. K. et al., Effect of row-to-row shading on the output of flat plate south facing photovoltaic arrays, In *Solar Engineering—1986, American Society of Mechanical Engineers*, Anaheim, CA, April 13–16, 1986.

The Design of Residential Photovoltaic Systems, SAND87-1951, Photovoltaic System Design Assistance Center, Sandia National Laboratories, Albuquerque, NM, December 1988.

CHAPTER 6

Battery-Backup Grid-Connected Photovoltaic Systems

6.1 INTRODUCTION

This chapter focuses on the design of grid-connected PV systems with battery backup. All one needs to do to discover an interest in these systems is to spend a week or two after a hurricane, an ice storm, or some other natural disaster without electric power. And, believe it or not, there are places in the world where the grid is not available on a 24-hour per day basis. Those who have purchased fossil-fueled generators for backup power generally appreciate the idea of a quiet, reliable, pollution-free, nonfuel-consuming source of electricity. Rather than waiting in long lines for gasoline for their generator, they can sit by the pool drinking cold beer. If they also have a solar water heater, they will have hot showers as well. In fact, they will probably be the most popular family on the block.

Battery-backup grid-connected PV system components still must comply with UL 1741 [1]. In the event of a utility disturbance or outage, the inverter must shut down its grid connection for the allotted 5-minute interval, monitor the grid, and reconnect again only if the grid has been stable in voltage and frequency for 5 minutes. This means that the inverter must be a multipurpose device. It supplies power to the grid as a current source that continuously monitors the grid whenever the grid is energized. It also acts as an alternate means of charging the system batteries in the event that the PV system for any reason has not kept them charged. Finally, it acts as a stand-alone voltage source to power designated standby loads in the occupancy while the utility is not available, making this switchover in a few milliseconds.

Recently, ac-coupled battery-backup PV systems have generated interest. For those who have installed nonbattery-backup systems and then found they would have preferred a battery-backup system, it is possible to add a few extra components, including a battery-backup inverter, batteries, and a standby distribution panel to achieve battery backup. There is also an argument that this type of system is inherently more efficient than the dc-coupled systems that have been popular, since they immediately invert the PV array output to useful ac power, rather than have batteries as an intermediate stop. These systems are also adaptable to multiple ac sources, such as wind generators and low head hydro systems.

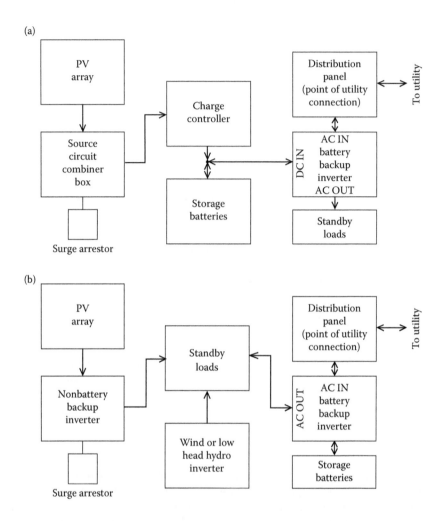

Figure 6.1 Block diagram of dc-coupled and ac-coupled battery-backup grid-connected PV systems. Grounding conductors and disconnects not shown. (a) dc-coupled. (b) ac-coupled.

Both types of systems will be discussed in this chapter. Figure 6.1 shows block diagrams of both types of system.

The final step in system design is to understand how to program charge controllers, inverters, generator starters, and battery monitors and to understand monitoring options that may be available for the system. While newer versions of software for inverters and charge controllers may involve revised programming procedures, the designer should still understand how to determine the proper settings for charging parameters, sell voltages, and the relationships among these parameters when the system has more than one inverter or charge controller that contributes to either charging or discharging of the batteries.

6.2 BATTERY-BACKUP DESIGN BASICS

6.2.1 Introduction

Just as in the case of the nonbattery-backup system, the selection of the right system for a given owner can involve a variety of selection criteria, such as the desired backup loads, the available roof space, the available space for batteries, or the budget. Often those interested in battery-backup PV systems expect the systems to pay for themselves, even though no fossil standby generator ever paid for itself with the electricity it generated. The payback from a battery-backup system must be measured in terms of food not spoiled after a utility outage, the satisfaction of having a "green" source of standby power, or the comfort level of knowing that power will be available after a storm without having to start and fuel a fossil generator.

Since most battery-backup systems are designed for use on buildings, new *NEC* 2014 safety requirements will likely apply to the system design [2]. In particular, provisions will need to be made for rapid shutdown of the system, including shutting down the array output within 10 ft of the array (*NEC* 690.12) and providing arc fault protection on any system having maximum voltages higher than 80 V (*NEC* 690.11). Techniques for satisfying these *NEC* requirements are evolving rapidly in 2016. For example, rapid shutdown might be accomplished at the module, at the rooftop junction box, or, perhaps, at the source-circuit combiner box, possibly via remote control of contactors or equivalent switching technology. The same switches used for rapid shutdown might conceivably be used for multiple purposes, such as GFDI, AFCI, and rapid shutdown. In any case, because of the many possibilities that may become available, this chapter will focus on the general design of battery-backup systems, with the understanding that suitable means of accomplishing these three safety functions will be incorporated into the design.

6.2.2 Load Determination

At some point in the design process, it will be necessary to determine the desired standby loads for the system. This is simply a tabulation of everything the system owner would like to have available on a standby basis. It is also a time of reckoning that probably the owner may need to forego the use of the 5-ton (60,000 BTU/hour) central air conditioning unit, the 4500-W electric water heater, the 12 kW electric range, and the 5600-W electric clothes dryer. On the other hand, it will usually be found as a pleasant surprise that the refrigerator, TV, washing machine, microwave oven, computer, some lights, and some fans can be used. Sometimes even a small, mini-split air conditioning unit or two can be incorporated into the standby load collection.

An important item to consider is whether there will be any 240-V loads on the standby list. While some battery-backup inverters deliver 120/240-V single-phase output, others deliver only 120 V unless they are connected in a "stacked" mode with a second inverter. Some battery-backup inverters can be configured for three-phase operation and some cannot. The important issue here is having the inverter(s) maintain the same phase sequence as existed while the grid was in operation after the grid drops out. This is critical for proper rotation of three-phase motors, for example. The designer must also consider the total power requirements and the total energy requirements, as well as how the loads might be connected to a standby electrical panel to properly balance the loads. The example system designs to follow will demonstrate how to properly tabulate standby loads.

6.2.3 Inverter Sizing

The size of the inverter(s) will be determined by the anticipated maximum instantaneous power requirements of the standby loads and whether there will be any 240-V or three-phase loads. The selection of inverters will also be influenced by subtle items, such as whether the load served is a single-phase load on a 120/240-V single-phase supply, a 120-V or a 208-V single-phase load on a three-phase supply, or a three-phase load on a three-phase supply. These possibilities will be sorted out in the design examples that follow.

In addition, whether to use ac-coupling or dc-coupling of inverters will need to be decided. This decision may be based on economics, desired system performance, desired system configuration, or the available electric service. One thing for certain—the inverter industry is growing so rapidly that additional options will no doubt be available either by the time this book goes to press or soon thereafter. So the most important part of this chapter will be to understand the design issues, so as new technology emerges, the designer will be able to understand how to evaluate the choices.

6.2.4 Battery Sizing

The sizing of the batteries will depend on the anticipated daily kWh consumption of the standby loads and on how many days of storage are desired. Once these two considerations have been established, there will likely be a trip back to the drawing board when it is discovered that either the choice of batteries is more costly than had been allowed for in the budget or there is not sufficient space for the batteries.

Another important consideration in battery selection is whether to use deep-discharge, flooded lead-acid batteries that will need maintenance and will require special venting or whether to use more-expensive maintenance-free batteries. If maintenance-free batteries are chosen, whether to use gel or absorbent glass mat (AGM) battery types will need to be determined. With the advancement of hybrid vehicle and electric vehicle technologies, it is possible that battery technologies other than lead-acid, such as lithium iron phosphate (LFP) and various flow technologies, are becoming available as lead-acid system substitutes.

The basic procedure for battery sizing involves the following:

1. Determine the standby load in kWh/day. This is the energy that must be supplied by the inverter *output*.

2. Allow for a 2% energy loss in the wiring from inverter to loads.
3. Determine the energy required by the inverter to meet the standby load requirements. This takes into account the inverter efficiency by dividing the inverter output energy by the inverter efficiency to obtain the inverter *input* energy.
4. Determine the load on the batteries in kWh/day using kWh/day(batt) = kWh/day(loads)/(inverter efficiency)/(wiring efficiency).
5. Convert the load to Ah/day at the system battery voltage, using Ah/day = kWh/day(batt) × (1000/V_{batt}), where V_{batt} is the nominal battery voltage, that is, 12, 24, 48, and so on.
6. Multiply the battery Ah load by 1.25. This factor accounts for the fact that it is desirable to ensure that the batteries do not supply more than 80% of their capacity before they are shut down for recharging.
7. If the batteries are located outside where they may be subjected to cooling, it may be necessary to use an additional multiplier to account for loss of capacity when the batteries are cold.
8. If additional days of storage are desired, multiply by the number of days desired. Fractional days are acceptable as multipliers.
9. Determine suitable batteries to connect in series and parallel as necessary to achieve the calculated storage capacity. Normally it is desirable to have no more than four parallel groups of batteries, unless an advanced battery management system is used, in order to ensure closely balanced currents in each series group of batteries.

There are a few other factors that are also important in designing grid-connected PV systems with batteries, such as rates of charging and discharging, since charging too fast or discharging too fast can result in more losses due to the internal battery resistance. This can lead to the need for additional PV array size to overcome these losses. It may also lead to the need to increase the size of the battery bank to reduce losses from discharging too quickly. These considerations will be accounted for in the design examples that follow.

6.2.5 Sizing the Array

Once the standby loads have been determined, and the batteries have been selected, the array must be designed to provide sufficient daily energy to operate the standby loads when grid power has been lost. In other words, the array should be sized so that the system can be operated as a stand-alone system. In this part of the exercise, it may be desirable to establish the month or months when it is most likely that the grid may fail, since these will be the times when the array will need to produce the needed standby energy. So if NREL SAM PVWatts option [3] is used to estimate system performance, the estimated monthly system production can be checked against the anticipated monthly standby energy requirements.

Proper sizing of the array must account for the loss mechanisms described in Chapter 4 plus additional losses encountered in charging and discharging the batteries. It is common to estimate all loss mechanisms and then calculate a default overall loss percentage for the system for use in array sizing. For example, loss mechanisms will typically include the following:

1. Array temperature and soiling losses, typically about 15% worst case
2. System wiring losses, typically overall about 4% worst case
3. Inverter losses, typically about 6% worst case, depending on whether ac- or dc-coupling is used
4. Battery charging and discharging losses, typically assumed to be overall 10%
5. Charge controller losses, typically in the 2%–4% range

Using these loss factors, the worst-case dc:ac conversion efficiency can be estimated to be $0.85 \times 0.96 \times 0.94 \times 0.9 \times 0.98 = 0.68$, or 68%. On a cool, breezy, clear day, the conversion efficiency will probably be closer to 72% and when the grid is operational, the conversion efficiency should approach 77% or higher. It should be noted that these performance figures are based on the use of a maximum power point tracker (MPPT) charge controller, so the array size can be computed in watts, as opposed to amperes. Also, noting the difference between worst-case and better-case performance, a reasonable choice for average system operation under standby conditions is 72%. This figure will be used in all the examples that follow as a default value.

So to determine the array size, first convert the computed daily kWh needs of the batteries to monthly kWh needs for the months most likely to require standby power. Then note that the NREL SAM PVWatts version already accounts for all loss mechanisms between the array and the inverter when it performs its simulation. Thus, even though the simulation result represents inverter output, since the array and wiring up to the inverter is all dc, if the inverter is replaced by an MPPT charge controller of comparable efficiency, the kWh/day output of the charge controller will be nearly identical to the kWh/day output of the inverter. So, once the monthly kWh to the batteries requirement has been established, one only needs to vary the array size in NREL SAM until the simulation matches the desired monthly kWh to the batteries.

The designer can then vary the array size in NREL SAM PVWatts, using the 1.39 dc/ac ratio, until the array size yields the desired monthly kWh for the standby loads. Alternatively, if the appropriate loss factor is used, the array size can be calculated on the basis of the charge needed by the batteries.

There is one fine point, however. The array must be broken down into source circuits that have output voltages that fall within the input voltage and current limits of the system charge controller, preferably at a point close to the maximum operating efficiency of the charge controller.

At this point, the system will have been designed, except for BOS components. All that is left is to determine whether the cost of the system is within budget, whether there is room for the batteries and power conditioning equipment, and whether there is room for the PV array. If any of these requirements cannot be met, then it is back to the drawing board. If they can be met, then the design can be completed by selecting all the BOS components in conjunction with *NEC* and ASCE-7 requirements, and any other local, regional, or national code requirements that may apply to the installation.

6.3 A SINGLE INVERTER 120-V BATTERY-BACKUP SYSTEM BASED ON STANDBY LOADS

6.3.1 Determination of Standby Loads

For this example, it will be assumed that the owner would like to operate a shiny new Energy Star refrigerator, a microwave oven, a washing machine, a desktop computer, two lighting circuits, and two ceiling fans in the event of a power outage. The system will be located in Tampa, FL, where the greatest likelihood of a hurricane-induced power outage is during the months of August through October.

Since the desired standby loads are known, the first step in the design is to construct a table that lists the loads, as shown in Table 6.1. The refrigerator is rated at 420 kWh/year, which equates to 1.15 kWh/day. To determine the power consumption of the refrigerator, the best method is to find the model number and check the specifications. To estimate the power consumption of the refrigerator, it is possible to assume the compressor will be running 33% of the time and will be off 66% of the time. If this is the case, then the 1.15 kWh is consumed over an 8-hour period. So dividing the 1.15 kWh by the 8 hours results in 0.1438 kWh/hour = 0.1438 kW = 144 W. Since the unit operates at 120 V, and as it has a motor, it can be assumed to operate at an 80% power factor. Thus $P = IV(pf)$ can be solved to find a current of 1.5 A.

For the remainder of the loads, either the nameplate power and current, or estimates of power and current based on requirements of *NEC* Article 220, along with the estimated operating hours per day, are used to obtain the required table entries. For example, for the lighting circuits, local code administrators will generally require that the full possible load for the circuit be used for load determination purposes, since most of the time the system will be operating under utility power as opposed to standby power.

6.3.2 Inverter Selection

The inverter must be capable of *passing through* adequate power to meet the load demands when the utility is operational and be capable of meeting a reduced demand from standby loads when the utility is down. An important section in *NEC*

Table 6.1 Summary of Standby Loads for 120-V DC-Connected Battery-Backup System

Load Description	Load Power (W)	Load Current (A)	Load (kWh/Day)
Refrigerator	144	1.5	1.15
Microwave oven	600	5.0	0.6
Washing machine	1500	12.5	0.4
Computer	250	2.1	1.0
Lighting ckt 1	1200	10	1.2
Lighting ckt 2	1200	10	1.2
Ceiling fans	150	1.3	1.8
Total	5044	42.4	7.35

is Section 690.10(A), in which it states that in a stand-alone PV system, the inverter does not need to be sized to meet the total load of the occupancy, as would be done in accordance with *NEC* Article 220 for utility power.

Battery-backup inverters have one power rating for when they are powering up the standby loads when the utility is down and another output current rating when the utility is operational. It is important to find each of these power ratings on the inverter specification sheet when the inverter is selected. For this example, one popular battery-backup inverter is rated at 3600-W backup power at 120 V, but will pass through 60 A on a continuous basis when the utility is available. Comparing the 60-A pass-through current with the total load current requirements of Table 6.1 shows that the load current requirements are easily met with this inverter. In fact, if a 60-A circuit breaker is used as the point of utility connection (PUC) for the inverter output, for the purpose of determining load current, the load current must be considered to be continuous, which means that the circuit breaker must be rated at 125% of the load current. As a check on this requirement, 125% of 42.4 A = 53.0 A < 60 A, so the selected standby loads are acceptable. Furthermore, since it is very unlikely that all loads will be simultaneously operational, an additional safety factor is introduced.

If the utility is down, then the provisions of *NEC* 690.10(A) apply, meaning that although the maximum standby load on the inverter is 5044 W, the inverter will only supply 3600 W. So in this case, not all loads can be operated simultaneously while the grid is down. If the load on the inverter exceeds 3600 W, the inverter will shut down to protect itself and display an error message that it has been overloaded. When the overload is corrected, the inverter will restart. Normally, the system owner will have installed compact fluorescent or LED lighting and will have incorporated other energy efficiency measures, so the loads may not be as large as allowed in Table 6.1. Furthermore, the owner will know to use standby loads sparingly in order to be sure the system batteries do not discharge too far before the sun gets another chance to recharge them.

Now that the standby loads have been defined and the inverter has been selected, the batteries can be selected. Note that the selected inverter is designed for operation with a 48-V battery bank.

6.3.3 Battery Selection

The information in Table 6.1 is also needed for battery selection, following the procedure outlined in Section 6.2.4. Since the daily load requirements are 7.35 kWh, the energy that must be supplied *to* the inverter input is thus given by (7.35/0.98)/0.94 = 7.98 kWh, where 0.98 represents wiring losses between inverter and loads and 0.94 represents the inverter efficiency. Converting this figure to Ah that must be supplied by the batteries to meet the standby load requirements results in Ah = 7.98 × (1000/48) = 166.2. Finally, accounting for 80% depth of discharge, divide again by 0.8 (equivalent to multiplying by 1.25), to obtain Ah(batt) = 207.8.

There is a subtle point that should be noted here that the reader has probably already observed. That is, the rating of a battery depends on how fast it is discharged, as was explained in Chapter 3. Fortunately, battery manufacturers rate their batteries at specified discharge rates. If 1 day of storage is desired, then the 24-hour discharge rate should be used in determining the battery rating. If 2 days of storage are desired, then the 48-hour discharge rate is used, and so on. So before a battery can be selected, the system owner must decide on how many days of storage are needed. Admittedly, there are even finer points that might be argued in the battery selection process, such as how much battery discharging occurs during the day versus how much discharging occurs during the night. Obviously, all the nighttime loads must be supplied by the batteries, but during the day, all of the loads must be supplied by some of the energy from the PV array, while the remaining fraction of the PV array output will be used to make up the battery charge that was used the previous night. But, then, there are also those periods soon after sunrise and just before sunset where the standby loads are being satisfied partly by the PV array and partly by the batteries. The ambitious designer may feel the need to document all the energy use habits of the PV system owner and then construct a minute-by-minute estimate of the power consumption of the loads, the power production of the PV array, and the amount of battery charging taking place. And, by the way, be sure to take into account the randomness of cloud cover. In other words, if 1 day of storage is desired, use the 24-hour discharge rate unless it is known for sure that all of the standby loads will only operate after sundown.

The decision of storage days may be influenced by space, since more storage days need more batteries, or by budget, as the cost of batteries is not trivial. In any case, unlike a stand-alone PV system, where the batteries are needed every day, assuming the loads are used every day, the batteries in a grid-connected system are hopefully rarely used for long periods of time. Thus, 1 day of storage is a common choice for battery-backup grid-connected systems, especially in hurricane-prone regions where the sun generally shines brightly after the storm has passed, with no evidence of the storm except for debris all over the place. The idea is that if the sun shines every day following whatever caused the utility to fail, then the batteries will be recharged each day. If the sun does not shine enough for a full charge, then the owner needs to use the loads more sparingly, as would be done if a fossil-fuel standby generator were in use for backup, but its fuel were running low. For this example, 1 day of storage will be used, so the 24-hour discharge rate should be used for the batteries selected.

The final consideration before battery selection is whether the batteries will likely be cold when they are needed. If so, larger batteries or additional batteries will be needed for the system. In this example, since the system is in Florida and the most likely time the batteries will be needed is the fall, no temperature correction will be needed.

So it is now time to look through the battery specification sheets. When doing so, the designer will note that nearly every manufacturer of PV batteries has a 6-V battery rated in the range of 200–250 Ah. So all that is needed is to decide whether to use flooded batteries or sealed batteries if lead-acid batteries are chosen. If sealed batteries are chosen, it generally is preferable to select the AGM type, since these batteries are designed to discharge somewhat faster than gel batteries. Gel batteries are generally preferred. For this example, it will be assumed that the owner has shown a preference for maintenance-free batteries.

Thus, any maintenance-free deep-discharge battery with a 24-hour discharge rating of 207.8 Ah or more is satisfactory. Coincidentally, one manufacturer makes a 208-Ah, 6-V, 24-hour discharge rate sealed AGM unit, which comes close enough to the 207.8 Ah that is needed. If the battery is discharged in 8 hours, instead of 24 hours, according to the manufacturer's specifications, the capacity is reduced to 186 Ah. If this number feels uncomfortable to the designer, a somewhat larger battery can be selected.

Of course, since the batteries are for a 48-V system, eight of the batteries will be needed for connection in series to obtain the necessary 48 V. Just as the current is the same for components in series, the Ah is also constant as batteries are added in series. The Wh, of course, increases with the addition of each battery, since the Wh is obtained by multiplying the Ah by the voltage.

6.3.4 Array Sizing

The array must be sized so that it will provide sufficient energy each day to meet the energy requirements of the standby loads. If the array is sized to fully charge the batteries, then it must supply enough energy to make up for the energy supplied by the batteries to the standby loads, inverter losses and ac wiring losses, or 166.2 Ah. So the question is: How much energy must be supplied by the array so that 166.2 Ah can be removed from the batteries? This is where the battery charging and dc wiring efficiency enter the process. If it is assumed that the round-trip efficiency of the batteries is 90%, that is, 90% of what goes into the batteries comes out of the batteries, then the array must supply 166.2/0.9 = 185 Ah per day to the batteries. Since the 185 Ah must be supplied at 48 V, it is convenient to convert this charge into energy by multiplying by the battery voltage. This gives 8880 Wh, or 8.88 kWh, delivered by the array to the batteries. So the next question is: What size array is needed to supply 8880 Wh/day?

Using NREL SAM PVWatts version with default loss values, accounts for all loss mechanisms between array and charge controller output. The task, then, is to determine the array size needed to provide 8880 Wh/day (8.88 kWh/day) to the batteries, either for the months assumed to be at greatest risk of power loss, or for all 12 months of the year. This is easily accomplished via a 3-step process that begins with running a simulation using a 1.0 kW array using the tilt and orientation proposed for the Tampa, FL site. The array size is then adjusted to achieve 8.88 kWh/day on the month of the hurricane season with the lowest 1.0 kW array performance. Finally, the array size is adjusted to produce 8.88 kWh/day on the worst month of the year.

For Tampa, FL, using a south-facing roof with a slope of 18.43°, and starting the NREL SAM simulation with an assumed array size of 1 kW, one obtains monthly kWh production values of 134 for August, 119 for September, and 121 for October. Keep in mind that these numbers do not represent ac output from the system, but, rather, dc energy supplied to the batteries. Since the goal of this design exercise is to size the array to meet the standby load requirements during the most likely hurricane months, the array needs to be increased in size from 1000 W to 308.88 1000/119 = 2238.655462 W to meet the September load requirements. A closer look at the NREL SAM PVWatts output data indicates that an array of this size will provide only 212/31 = 6.84 kWh/day for December. If the array size is increased to 8.88 2239/6.84 = 2907 W, it will meet minimum monthly needs all year,

BATTERY-BACKUP GRID-CONNECTED PHOTOVOLTAIC SYSTEMS

with annual kWh production of 3,303 kWh. Since inverter and charge controller can handle 2907 W, this array size will be used.

6.3.5 Charge Controller and Module Selection

Before selecting modules and designing the array, the designer needs to know about the charge controller. Typical MPPT charge controllers are rated at either 60-A output or 80-A output. The 60-A output charge controllers normally specify a maximum array size of 3200 W and limit the maximum array open-circuit voltage to 150 V. For a 48-V battery system, these controllers generally operate at maximum efficiency when the array voltage is in the 70–80 V range. So at this point, an exercise similar to those used in sizing the arrays in Chapter 4 can be used.

As a check on charge controller output, when the array is operating at 2907 W, the charge controller *output* current will be approximately 2907 W × (0.98/53 V) = 53.8 A, which is within the limits of the controller. Since the controllers are rated for continuous output current, no multipliers are needed for the 53.8 A, other than for sizing the wire that carries the current, which will be done in the next section. So a 60-A MPPT charge controller can be used.

The careful reader might be wondering about the origin of the 53 V figure in the last calculation. This voltage was chosen because it is approximately what would normally be selected for the "sell" voltage of the inverter. This means that as soon as the battery voltage reaches 53 V when the grid is up, the inverter begins to use array current for producing electricity to sell to the utility. So the battery voltage is limited to 53 V. Other possible battery voltages will be discussed when the charge controller programming is discussed in Section 6.3.7.

Some charge controllers provide GFDI and/or AFCI and some do not. Some accommodate positive grounded, negative grounded, or ungrounded arrays and some do not. So, before specifying a charge controller, it will be necessary to determine how or if the array will be grounded and then whether the proposed charge controller will interface properly with the array.

Table 6.2 shows performance characteristics for four different modules. The design temperature range for the modules will be from −5°C (23°F) to +65°C (149°F), so a table can be generated that summarizes the number of each module that can be connected in series to keep V_{OC} of the array below 150 V. This information is shown in Table 6.3.

In Table 6.3, the maximum values of V_{OC} per source circuit are calculated using Equation 4.8 and the minimum values of V_{mp} per module are calculated using Equation 4.9. Examining this table indicates that none of the four modules when connected for maximum charge controller efficiency can be configured into an array that will be rated at exactly 2907 W. A reasonable design decision would be to avoid array power ratings less than this value, but also to keep the array power below 3200 W so an additional charge controller will not be needed. So it appears that reasonable choices for the array would be four source circuits of module A, five source circuits of module B, or four source circuits of module C. If the price per watt of all modules is the same and the warranties are the same, then any of the modules

Table 6.2 Performance Characteristics for Four Different PV Modules

Module	Rated Power (W)	V_{OC} (V)	I_{SC} (A)	V_{mp} (V)	I_{mp} (A)	$\Delta V_{OC}/\Delta T$ (%/°C)	$\Delta V_{mp}/\Delta T$ (%/°C)
A	235	37	8.48	29.42	7.99	−0.344	−0.52
B	320	64.8	6.24	54.7	5.86	−0.316	−0.46
C	265	38	9.2	32	8.31	−0.316	−0.46
D	300	40.1	10.23	31.6	9.57	−0.3	−0.45

Table 6.3 Determination of Maximum Number of Series Modules per Source Circuit

Module	Rated V_{OC} (V)	V_{OC}(max) (V)	Rated V_{mp} (V)	V_{mp}(min) (V)	Max # Series Modules	Source Ckt Max V_{OC} (V)	Rated Source Ckt Power (W)
A	37	40.8	29.4	23.3	3.7	122.5	705
B	64.8	70.9	54.7	44.6	2.1	141.9	640
C	38	41.6	32.0	26.1	3.6	124.8	795
D	40.1	43.7	31.6	25.9	3.4	131.1	900

would represent good choices. Since module A requires only four source circuits, this means fewer wires and series circuit breakers, meaning a lower installation cost, so module A will be selected. The array will thus consist of 12 modules.

Once again, there is a subtle point. It is possible that one of the modules will be more efficient than the others and will thus have a smaller module area. This could result in the need for less mounting hardware, fewer roof penetrations, and lower array mounting cost, so this is also a consideration in module selection. Since the module dimensions are not given, the designer is reminded to check module dimensions and include the cost of the mount in the overall system cost for all modules and then select the module that results in the lowest overall installation cost, including wiring to the combiner box and mounting the modules to the roof.

6.3.6 BOS Selection and Completion of the Design

The remaining system components include the array mounting equipment, extension cables for array wiring, a rooftop junction box, a source-circuit combiner box, a GFDI device, input and output disconnects for the inverter and charge controller, battery cables, battery enclosure, inverter input and output disconnects, inverter bypass assembly, circuit breakers and possibly a separate distribution panel for the standby loads, the equipment for connecting to the utility, and all conduit and wire for the system, including grounding conductors.

6.3.6.1 Array Mounting Equipment

Noting that the occupancy is in an urban setting, it is subject to wind load exposure B. Using the roof slope of 18.43°, a mean roof height of 15 ft, and the 140 mph design wind speed for Category II buildings in Tampa, FL [4]. Table 5.7 shows a

BATTERY-BACKUP GRID-CONNECTED PHOTOVOLTAIC SYSTEMS 247

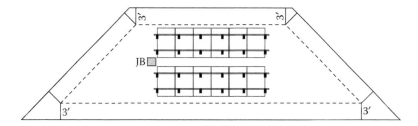

Figure 6.2 Module layout on roof showing mounting rails, mounting brackets, and junction box.

wind load of 20.3/–32.3 psf in Roof Zone 1 for EWA = 10 ft^2. Thus, the modules can be mounted with two rails per row, as noted in Figure 6.2. The roof trusses are constructed of southern yellow pine, and 5/16-inch (0.7938 cm) stainless steel lag screws will be used to attach the array mounting brackets to the roof.

The area of each module is 1.63 m^2 (17.5 ft^2), so the total area of each row of six modules is 9.78 m^2 (105 ft^2). Since both rows are completely in Roof Zone 1, the total upward force on each row is thus 105 × 32.3 = 3392 lb (1539 kg).

With the mounting brackets spaced at 4 ft (1.22 m) intervals, Figure 6.2 shows a total of 12 mounting brackets for each row. The force per bracket is thus 3392/12 = 283 lb (128.4 kg). Thus, from Table 5.5, the pull strength of a 5/16-inch lag screw is 332 lb/inch (59.3 kg/cm), so for each row of modules, the lag screws will need to penetrate 0.85 inch (2.16 cm) into the trusses. So, depending upon the roof composition, lag screws in the 3.5-inch (8.9 cm) to 4-inch (10.16 cm) length range will probably be more than adequate. At this point, the designer need only to check the manufacturer's installation instructions for the mounting rails to be sure that the mounting brackets are not mounted too far from the ends of the mounting rails. And, of course, it will be necessary to verify that either trusses are present close to the indicated locations or that the attic is accessible for any additional bracing that may be needed between trusses.

6.3.6.2 Rooftop Junction Box

The sizes of junction boxes are determined by the number and size of wires that enter the box as well as the volume of the devices that are installed in the box. Table 314.16(B) in the *NEC* specifies the volume required in a junction box for conductors smaller than #6. For #10 (5.261 m^2) wire, a volume of 2.5 inch^3 (41 cm^3) is required for each conductor that terminates in the box. In this example, since there are four source circuits and a grounding conductor entering the box from the array, there will be a total of nine wires entering the junction box and nine wires leaving the junction box in conduit that will be run to the source-circuit combiner box. Allowing 9 inch^3 (148 cm^3) for an approved dc splicing mechanism, such as terminal blocks rated at a minimum of 90°C, plus 2.5 inch^3 for each of the 18 wires entering the junction box, results in a minimum required volume of 54 inch^3 (885 cm^3) for the box. A popular size junction box measures 6 inch × 6 inch × 4 inch (15.2 cm × 15.2 cm × 10.2 cm), providing adequate space for a neat wiring job in the box, assuming the array will be wired with #10 wire. In fact, even if #8 (8.367 mm^2) wire is used between junction box and source-circuit combiner box, the junction box will provide adequate room since #8 requires 3 inch^3 (49 cm^3) per conductor. For 18 conductors, this is 63 in^3 (1032 cm^3) for #8 wire versus 54 in^3 (885 cm^3) for #10 wire.

6.3.6.3 Source-Circuit Combiner Box and Surge Arrestor

A number of manufacturers can supply source-circuit combiner boxes. In particular, the manufacturers of the inverters generally also provide source-circuit combiner boxes. In this case, a box capable of combining four source circuits will be needed. Since the array voltage will not exceed 150 V, 15-A dc circuit breakers can be used, per module manufacturer's specifications, in the combiner box as shown in Figure 6.3. Not shown in Figure 6.3 is the surge arrestor, which fits into a 1/2-inch electrical knock-out hole in the combiner box, with leads that attach to the positive array output busbar, the negative array output busbar, and the ground busbar.

6.3.6.4 Wire and Circuit Breaker Sizing—DC Side

At this point, it is convenient to specify all the wire sizes for the system so the conduit sizes and disconnect sizes can be established. The most convenient way to do this is to construct a table to be sure no wires are missed. Table 6.4 shows one possible method of presenting the data and calculations.

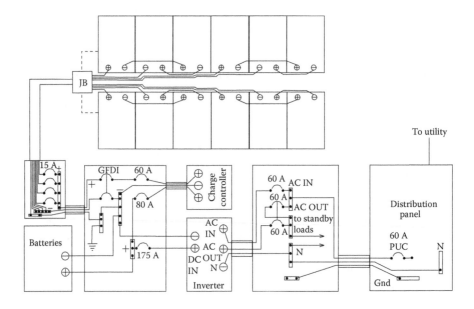

Figure 6.3 Schematic diagram of 2820-W battery-backup grid-connected PV system. Not all conduit and grounding conductors are shown.

BATTERY-BACKUP GRID-CONNECTED PHOTOVOLTAIC SYSTEMS 249

Table 6.4 Determining Wire Sizes for the 2820-W Battery-Backup Grid-Connected PV System

Circuit	I_{sc} (A)	$1.56I_{sc}$ (A)	Wire Size	Wire 30°C Ampacity (A)	Temp Derate	Cond fill Derate	Derated Ampacity (A)
PV source circuits	8.48	13.2	10	40	0.71	0.7	21.3
PV output circuit	33.92	52.9	8	55	1	1	55

Circuit	I	1.25I	Wire Size	Wire 30°C Ampacity (A)	Temp Derate	Cond fill Derate	Derated Ampacity (A)
Ch cont output ckt	51.1	63.9	6	75	1	1	75
Inverter dc input I_{max}	85	106	2/0	195	1	1	195
Inverter ac input (A)	60	75	6	75	1	1	75
Inverter ac output (A)	60	75	6	75	1	1	75

Voltage Drop Calculations

Circuit	I_{op}	V_{op}	Wire Size	Max % VD	Wire Ω/kft	Max Wire L (ft)	Actual Wire L (ft)
PV source circuits	7.99	88.2	10	2	1.21	75.3	60
PV output circuit	31.96	88.2	6	0.1	0.51	2.7	2
Ch cont output ckt	50.8	53	6	0.2	0.51	2.0	2
Inverter ac input	60	120	6	2	0.51	39.2	25
Inverter ac output	60	120	6	2	0.51	39.2	6

[a] Derated ampacity is the larger of $1.56I_{sc}$ and $1.25I_{sc}$ derated for temperature and conduit fill.

The first calculations deal with the PV source circuits and PV output circuits. For these circuits, the ampacity of the wiring must be the larger of $1.56I_{SC}$ or $1.25I_{SC}$ derated for ambient temperature and conduit fill per *NEC* 690.8. Thus, for an ambient temperature of 35°C, plus an additional 22°C for conduit run 2 inch above the roof, the ambient temperature derating factor for 57°C is 0.71, and the derating factor for eight source-circuit conductors in a conduit run down to the source-circuit combiner box is 0.7. Thus, for $I_{SC} = 8.48$ A, $1.56I_{SC} = 13.2$ A, and $1.25I_{SC}/0.71/0.7 = 21.3$ A. This means the 30°C ampacity of the source-circuit extension wire needs to be at least 21.3 A, which is easily met by the choice of #10 (5.261 mm^2) wire, in anticipation of needing this size to meet voltage drop requirements.

For voltage drop calculations in source circuits, first the %VD in the module leads should be calculated. With 2 m (6.56 ft) total lead length of #12 (3.31 mm^2) PV wire on each of the three modules in each source circuit, the %VD of the module wiring is found to be 0.35%. This leaves 1.65% for the rest of the source circuits and output circuit. Assuming the longest source-circuit extension equivalent length is 10 ft (3.05 m) and assuming a distance of 50 ft (15.24 m) from rooftop junction box to source-circuit combiner box, the %VD in the #10 wire over a distance of 60 ft (18.3 m) is 1.29%. Finally, with the source-circuit combiner box located next to the charge controller, the voltage drop between these two points will be about 0.1%. Also, since the charge controller will be very close to the inverter, the charge controller to inverter voltage drop will be approximately 0.2%. So the overall voltage drop from array to inverter is found to be 1.94%, which is within the allowable limits. Ampacity and voltage drop calculations for all circuits on the dc and ac side of the inverter are summarized in Table 6.4.

With the source-circuit combiner box located near the power conditioning equipment to make it easier for start-up and troubleshooting the system, the PV output circuit will consist of a short run of wire inside the occupancy from the source-circuit combiner box to a circuit breaker and then to the charge controller, so no derating is necessary for these conductors. For circuit breakers, which also serve as disconnects, the PV source circuits will have the 15-A dc circuit breakers in the source-circuit combiner box and the PV output circuit will have a 60-A circuit breaker that also serves as the charge controller input disconnect and the array disconnect. This circuit breaker will be located in a dc component enclosure along with the charge controller output disconnect and the inverter input disconnect, as shown in Figure 6.3.

The next calculations deal with the charge controller output circuit and the inverter input and output circuits. First of all, it needs to be noted that the charge controller will begin to limit current if its output current exceeds 60 A, but the charge controller is rated at a continuous 60 A at its output. Thus, the wiring of the output circuit must be sized at 125% of this value, or 75 A. Furthermore, unless a continuous-duty circuit breaker is used as the charge controller output disconnect and overcurrent protection, the circuit breaker must also be rated at 125% of the charge controller output current, which is shown in the table to be 71.6 A. Since #6 (13.3 mm^2) wiring will be used, with a rating of 75 A, but since 75 A is not a standard circuit breaker rating, an 80 A circuit breaker can be used, according to *NEC* 240.4(B).

BATTERY-BACKUP GRID-CONNECTED PHOTOVOLTAIC SYSTEMS 251

The inverter input is rated at 85 A dc, but on occasion, the inverter may need to supply a short surge current to start a motor or other load. It is thus commonplace to use 2/0 (67.43 mm^2) battery cables to the inverter that are protected with a 175 A circuit breaker. It is also advisable to keep the battery cables at lengths less than 10 ft (3 m) to keep the voltage drop at an absolute minimum between batteries and inverter.

An important requirement of the 2014 *NEC* (690.71(H)) is that if the batteries are more than 5 ft (1.5 m) from the inverter, disconnects are required at both the battery end and the inverter end of the battery cables.

6.3.6.5 Wire and Circuit Breaker Sizing—AC Side

The inverter AC IN and AC OUT ports are both rated at 60 A at 120 V. Note that for a dc-coupled system, ac power will only flow out of the AC OUT port, while ac power can flow either into the AC IN or out of the AC IN port. When the utility is down, the AC OUT port will supply a maximum of 30 A to the standby loads, when the utility is up, the AC IN port will pass through 60 A to the AC OUT port. So wiring of both ports must be sized for 60 A, which means #6, that is normally protected with a 60-A circuit breaker. Note that if the 60-A circuit breakers used are not rated for continuous duty, then the load at the AC OUT port must be limited to 80% of the 60 A, or 48 A. Alternatively, since #6 (13.3 mm^2) wire is rated at 75 A, an 80-A circuit breaker could be used at the PUC, as it is the next available size circuit breaker above 75 A.

The other component on the ac side that is mounted along with the circuit breakers that provide overcurrent protection and a disconnecting means for the inverter AC IN and AC OUT ports is the inverter bypass assembly, which is also shown in Figure 6.3. This combination of circuit breakers is installed such that it is not possible to have both on simultaneously, but it is possible to have one of them on at a time or both of them off. It is left as an exercise for the reader (Problem 6.1) to determine the positions of the circuit breakers that route power through the inverter and that bypass power from the AC IN to the AC OUT busbars in the ac component enclosure.

6.3.6.6 Wiring of Standby Loads

The standby loads indicated in Table 6.1 and implied by the two arrows in Figure 6.3 deserve some discussion, especially if the installation is to be a retrofit in an existing occupancy.

There are essentially two types of 120-V circuits in an occupancy. The simplest is when a hot wire and a neutral wire, generally run as a piece of "romex" (Type NM) cable, run from an existing electric panel to a load or, perhaps to as many as ten 120-V outlets connected in parallel. Normally, these circuits are connected to single-pole 15- or 20-A circuit breakers, or, sometimes to fuses. In these cases, the circuits can be transferred to either a separate distribution panel that has been wired for 120 V, or to a circuit breaker in the PV system ac component enclosure. That is, except for one detail. In the 2014 *NEC*, nearly every circuit in a dwelling unit must be protected by either an arc fault circuit interrupter or a ground fault circuit interrupter. The arc fault circuit interrupter is capable of detecting arcing at current levels below the circuit breaker trip rating and interrupting the circuit. The ground fault circuit interrupter is capable of detecting a difference of as little as 5 mA between the hot and neutral conductors of a circuit and interrupting the circuit if such a fault should occur.

The possibility is that a local inspector may interpret the installation of the PV system and subsequent transfer of circuits from the main panel to a standby panel to be an upgrade. In this case, even though the old circuit may have been protected by a standard circuit breaker, the moved circuit may be required to be fed by either an arc fault circuit breaker or by a ground fault circuit breaker. Before specifying the wiring or installing the wiring, it is advisable to discuss the situation with the local electrical inspector to determine whether the circuits will need to be upgraded.

The circuits that must be carefully checked, however, are the multiwire branch circuits. A multiwire branch circuit uses a common neutral as the return for wiring from opposite sides of a 120/240-V single-phase supply or for two or three lines that are connected to two or three phases of a three-phase system. These situations are illustrated in Figure 6.4. In the single-phase case, the neutral carries the *difference* between the two line currents, so the current on the neutral never exceeds the current on either line. For the three-phase case, the neutral carries the *sum* of the currents, since the voltages in a three-phase system are 120° out of phase, if the currents on the three lines are balanced, the neutral current is zero. If only two lines are used, and if the line currents are equal, then the neutral current will be equal in magnitude to the line current. If the line currents are unbalanced, the neutral current will be less than the larger of the two line currents (see Problem 6.2). Again, under worst-case unbalanced load situations, the current in the neutral does not exceed the current in any of the lines unless a significant third harmonic component is present (see Problem 6.3).

The problem with multiwire branch circuits is that they have been commonly used to feed bedroom circuits. A single three-conductor cable is run to a convenient location in or near two bedrooms and then one line is fed to one bedroom and the

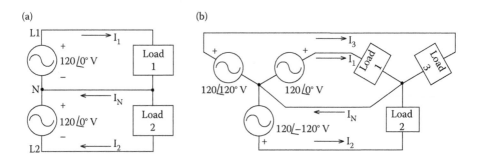

Figure 6.4 Illustration of multiwire branch circuits in single-phase and three-phase circuits. (a) $I_N = I_1 - I_2$. (b) $I_N = I_1 + I_2 + I_3$.

other line is fed to the other bedroom. It is probably not too difficult to imagine one of the circuits going to the daughter's bedroom and the other going to the son's bedroom. So when the time comes to determine who gets power if the utility goes out, the answer is neither. The reason is that if both are fed via the three-wire system from a single 120-V source, the neutral will return the *sum* of the two currents rather than the difference between the two, since both hot leads will be connected to the same phase. So the only way that multiwire branch circuits can be connected to standby power is if the standby power is either 120/240-V single-phase or 120/208-V three-phase. Otherwise, with a single 120-V standby source, multiwire branch circuits cannot be used. Furthermore, the local inspector will need to agree that these circuits will not need to be upgraded. If any new or existing circuits need to be upgraded as they are transferred over to the standby power source, it may be necessary to protect them with either arc fault protection or ground fault protection, depending upon the specific circuit use.

In new construction, it is smart design to run standby circuits so, for example, a lighting circuit with five duplex outlets on it will have one outlet in each of five rooms so each room will have one standby power outlet. The bottom line is that owner, builder, architect, and PV designer need to carefully plan the locations of standby power outlets.

6.3.6.7 Equipment Grounding Conductor and Grounding Electrode Conductor Sizing

Equipment grounding conductors are sized in accordance with *NEC* Article 250.122. For this system, since most dc circuit breakers are 60 A or less, #10 (5.261 mm^2) equipment grounding conductors are acceptable. If the batteries are in a metal enclosure, then the grounding conductor for the battery enclosure will need to be sized according to the 175-A inverter disconnect, which means sizing to a circuit breaker between 100 and 200 A, which requires a #6 (13.3 mm^2) equipment grounding conductor.

Finally, if the PV power conditioning equipment is located more than 50 ft (15.2 m) from the PUC, it may be advisable to install a separate grounding electrode for the PV equipment. For example, if the PV equipment is in an outbuilding, then a separate grounding electrode is required. The grounding electrode conductor is then sized according to *NEC* Table 250.66, noting exceptions for when the grounding electrode is a rod or a plate.

This completes the design of the system. The schematic diagram, for those who have not looked at it yet, is shown in Figure 6.3. All that remains now is to program the inverter and charge controller once the system has been installed, and the installer may be depending on the designer to provide programming instructions for the system.

6.3.7 Programming the Inverter and the Charge Controller

In battery-backup systems, both the inverter and the charge controller need to be programmed to account for the types of batteries used in the system. While inverter and charge controllers normally come with default programs, and while the default programs of the charge controllers are generally close to the correct values, it is important for the designer to understand the programming capabilities of the inverter and charge controller(s). Most importantly, however, the designer should be aware that the programming instructions that go into every small detail of programming possibilities for inverter and charge controller may be many pages long in the instruction manuals. The serious designer should make it a point to read and understand the entire installation and maintenance manuals for inverters and charge controllers in a system. The goal of this relatively short section is to stress the importance of understanding the purposes of the various programmable settings of the inverter and charge controller. The authors have already encountered a sufficient number of battery-backup system owners who read fervently the recommendations on various blogs and proceed to reset their inverter and charge controller programs without understanding what they are really doing. At some point, someone who understands the meaning of the settings will need to explain to them the reasons for the various programmed set points.

Inverters that are used with battery-backup grid-connected PV systems can also be used as stand-alone inverters or as inverters that are grid connected only for charging batteries, but not for selling to the grid. So the first item in programming the inverter is to establish the use of the inverter. If there is no PV system, then there is no need to sell to the grid. If there is no grid, then the system design will follow the procedure of the next chapter. But when there is a PV system to charge the batteries, then the inverter needs to be programmed so that the grid will not be trying to charge the batteries when the PV system is trying to do so.

The battery charging process was discussed in Chapter 3. Recall that initially, the batteries are charged with the controller acting as a current source. If the current comes from the PV array, then presumably all the array current and, for that matter, maximum array power will be used during the bulk charging stage. When the battery terminal voltage reaches the bulk voltage limit, which is programmable, then the charging process enters the absorption stage for a time that is also programmed. During the absorption stage, the charger acts as a constant voltage source at the bulk (or absorption, depending upon whose instruction book is being followed) voltage level. After the absorption time has elapsed, then the charger drops its voltage to the float voltage and continues to act as a constant voltage source at the float voltage as long as current is available from the charger.

The charging process can be achieved with either the inverter, when grid power is available, or the charge controller. So the idea is to program the inverter at a relatively low bulk charging limit, which will depend on whether flooded or maintenance-free batteries are being used. Since it is preferred to have the inverter sell electricity back to the grid, normally the "sell" voltage for the inverter, assuming a 48-V nominal system, will be set at approximately 52–53 V. The inverter bulk voltage is then set just below the sell voltage so the inverter will only charge the batteries if their voltage drops below the sell level. The inverter float voltage will be set slightly lower than the inverter bulk voltage, and, if the inverter has a refloat voltage setting, this setting will be set somewhat lower than the float setting. The purpose of the refloat setting on the inverter is to have the inverter begin a float charge cycle if the battery voltage should drop to the refloat level for any reason. Of course, if the grid is down, the inverter is not able to charge batteries.

In normal operation, with the inverter in the "sell" mode, as the PV system wakes up in the morning and begins to supply current to the charge controller, the charge controller will try to charge the batteries to the bulk charge level. But as soon as the batteries reach the sell voltage, the array current will be shifted from the batteries to the inverter for inversion to ac to power ac loads. So, under normal operation, the battery voltage will not reach the bulk voltage. Again, remember that the battery *terminal* voltage is not the same as the battery *cell* voltage. Under conditions of high array current, if any of the array current flows into the battery, then the battery terminal voltage will exceed the battery cell voltage. Thus, initially, the switch from charging batteries to selling through the inverter is not necessarily an instantaneous event. As charging current is drawn away from the batteries, the battery terminal voltage will tend to drop below the sell voltage, thus shutting off the sell function. But then the array current flows to the batteries and raises the terminal voltage above the sell level again. So initially, the array current is divided between the inverter and the batteries until the battery *cell* voltage reaches the battery *sell* voltage.

The system designer should check with battery specification sheets for recommended set points for bulk and float voltages as well as absorption time, even if these settings are determined by the inverter and/or the charge controller(s) once they "know" the battery configuration. If the batteries are flooded batteries, then an additional equalize cycle should be run about once a month. During equalization, the batteries are purposely overcharged in order to cause the electrolyte to bubble as hydrogen and oxygen (and probably some acid) are given off. When the electrolyte bubbles, it tends to clean the electrodes and also to mix the electrolyte, which may have stratified with a higher acidity at the bottom and a lower acidity level at the top. But equalization is only used for flooded batteries. Maintenance-free batteries are not designed for gassing, and if pressure builds up from equalizing, the batteries can be damaged.

Recommended set points seem to change from manufacturer to manufacturer and from year to year. It may even be the case that one set of charge controller settings may be recommended for one country and a different set of settings may be recommended for another country [5].

Table 6.5 shows typical set points for inverter and charge controller for a 48-V battery-backup system that uses lead-acid batteries. Two inverter set points not yet mentioned are the low battery cut out (LBCO) voltage and the low battery cut in (LBCI) voltage. The purpose of the LBCO is to disconnect the inverter from the batteries when the batteries reach approximately 20% of full charge, since further

Table 6.5 Typical Default Battery Charging Set Points for a 48-V Battery Backup PV System

	Bulk V	Float V	Refloat V	Sell V	LBCO V	LBCI V
Charge controller	57.6	54.4	–	–	–	–
Inverter	52	51.6	50	52.4	44	50

discharge could damage the batteries. Of course, as soon as the batteries disconnect from the inverter, the battery terminal voltage will rise above the LBCO voltage, since no further discharge current is flowing to lower the battery terminal voltage below the LBCO level. This would, of course, cause the inverter to think the battery voltage is not really as low as it thought it was. But inverters are smarter than that. They know perfectly well that if they turn on again, the battery voltage will drop below the LBCO again. So when the battery voltage reaches the LBCO and stays there for a programmed time, usually 15 minutes, the inverter shuts down and does not invert again until the battery voltage has reached the LBCI level. The inverter LBCI level is set sufficiently high that the batteries must be nearly fully charged before the inverter will begin to supply the standby loads again. The LBCO condition, of course, will only be realized if the grid is down and the system is operating as a stand-alone system. If the grid is up, then as soon as the battery voltage drops to the refloat voltage, the inverter will recharge the batteries using grid power.

6.3.8 Fossil-Fuel Generator Connection Options

Battery-backup grid-connected inverters can usually use a standby fossil-fuel generator for backing up the sun on a cloudy day when the grid is down. The generator may be connected to a separate generator input on some inverters, or it may be connected in place of the grid by using a transfer switch on other inverters. The separate generator input is particularly convenient, since there is no need either for the owner to manually flip the transfer switch or to install a more-expensive automatic transfer switch. But there are also subtle differences. For example, many generators do not have very stable frequency, so if the inverter is programmed to sell, it will be looking for a power source that meets the requirements of UL 1741 with tight frequency limits. Another subtle possibility is for the PV system to try to sell back to the utility, which, in this case, is just a small generator. This means the PV system will try to operate the fossil generator as a motor, which the prime mover may not appreciate. So if a fossil generator is used as a replacement for the utility at the AC IN port of the inverter, the inverter should be programmed out of the sell mode so that it will not try to run the fossil generator backwards and it will not be sensitive to the frequency of the fossil generator. Of course, use of an inverter-based fossil generator with stable frequency should solve any frequency doubts the PV inverter may have.

Once the generator hookup has been established, then the inverter can be programmed to start the generator at a predetermined battery voltage level and shut down the generator at a higher battery voltage level, allowing for the PV system to complete the charging process. The inverter can also be programmed to establish an operating time window for the generator so it will not keep the neighbors awake at night (or during the day if they work nights). More generator options, such as automatic generator start (AGS), will be discussed in Chapter 7 where hybrid stand-alone PV systems will be discussed.

As a final note, it is also possible to incorporate battery state of charge (SOC) monitoring and control into the battery system. The battery management systems used with newer technologies, such as LFP, monitor and manage battery charge at the cell level and display the battery system SOC as well as the battery system voltage. This is particularly useful when a fossil-fuel generator is used as a backup for maintaining battery charge in the absence of the grid, since battery voltages are dependent on the charge rate or discharge rate of the batteries and may not be reliable indicators of actual battery SOC, especially in the case of LFP battery systems.

6.4 A 120/240-V BATTERY-BACKUP SYSTEM BASED ON AVAILABLE ROOF SPACE

6.4.1 Introduction

Since battery-backup grid-connected systems essentially use the utility grid to store/use any excess PV power generation, there is no reason to limit the size of an array to a size that will just keep a set of batteries charged or operate a particular set of standby loads. It may be the case that a larger array will be desired so the system can run certain loads during daytime hours if the grid is down but run a minimum of loads at night, thus reducing the need for a large battery bank. For this example, it will be assumed that a new house is being planned to make the greatest possible use of PV along with efficient overall design. It will be assumed that the house is designed to have enough useable south-facing roof space for mounting a PV system and a 4 ft (1.22 m) by 10 ft (3.05 m) solar domestic water heater. The house will be located in Burlington, VT, where the owner wants to have secure electrical power in the event of a winter power outage. The roof will have a 45° slope to discourage snow buildup and will face directly south. The distance between rooftop junction box and power conditioning equipment is 40 ft (12.2 m). The power conditioning equipment is located next to the PUC, so all ac wiring distances are less than 10 ft (3.05 m). Since roof mounting details have now been covered in several examples, this example will assume that the roof has been specifically designed to accommodate an array of twenty-four 300-W modules. Details of source circuit calculations will follow. Problem 6.4 offers the opportunity to design the mounting system for the array.

6.4.2 Module Selection and Source Circuit Design

The electrical characteristics of the selected PV module appear in Table 6.6. The reader might recognize that these are the characteristics of Module D of the previous example. Since Burlington is not exactly a tropical winter paradise, the

Table 6.6 PV Module Electrical Characteristics

Rated Power (W)	V_{oc} (V)	I_{sc} (A)	V_{mp} (V)	I_{mp} (A)	$\Delta V_{oc}/\Delta T$ (%/°C)	$\Delta V_{mp}/\Delta T$ (%/°C)
300	40.1	10.23	31.6	9.57	−0.30	−0.45

design must incorporate a minimum array temperature of –30°F (–34.4°C) and a high temperature of 100°F (37.8°C) [6]. Using Equation 4.8 yields, for one module,

$$V_{OC}(max) = 40.1[1-(-0.003)(25-(-34.4))] = 47.25 \text{ V}.$$

Since a source-circuit $V_{OC}(max) < 150$ V, this means no more than $150/47.25 = 3.17$ modules can be connected in series. Thus, rounding down, the maximum number of modules in a source circuit is 3. For 3 modules, $V_{OC}(max) = 142$ V. Next, the minimum source circuit V_m on a hot day must be determined from Equation 4.9, yielding

$$V_m(min) = 3 \times 31.6[1-(-0.0045)(25-(70.3))] = 75.5 \text{ V},$$

where the maximum cell temperature, 70.3°C (158.5°F), is determined from the NOCT for the module, which is 46°C (114.8°F). Thus, at full sun (G = 1.0 kW/m²), the cell temperature is found from

$$T_C = T_A + \frac{\text{NOCT}-20}{0.8}G = 37.8 + 32.5 = 70.3°C.$$

Keeping in mind that this calculation is based on the hottest day on record, and that a more typical ambient temperature is 75°F in the summer, a more likely cell operating temperature will be $24 + 32.5 = 56.5°C$ (133.5°F), which equates to $V_m = 3 \times 31.6[1 - (-0.0045)(25 - 56.5)] = 81.36$ V. These values of V_m fall within acceptable limits for operation with an MPPT charge controller operating with a nominal 48-V output.

Thus, with 3 modules per source circuit, if 24 modules are used, then 8 source circuits will be required. Note that since the original plan was to use 24 modules, 24 is divisible by 2, 3, 4, 6, and 12, so the choice of 24 was not a very great risk. On the other hand, if it turned out that 5-module source circuits would have been preferable, then it would have been necessary to either add 1 module and have a 25-module array or to subtract 4 modules and settle for an array of 20 modules.

6.4.3 Source-Circuit Combiner Box and Charge Controller Selection

Since the array size is 7200 W, with eight source circuits, if an MPPT charge controller is used, the available current will be approximately $7200/48 = 150$ A. This suggests that two 80 A MPPT charge controllers will be needed, since these charge controllers are rated for continuous duty. Since there are eight source circuits, the system can be nicely balanced by feeding each charge controller with four source circuits, as shown in Figure 6.5a.

This also means that a suitable source-circuit combiner box or boxes must be found to feed the charge controller inputs. Since the maximum array output voltage is less than 150 V, it may be possible to use a combiner box that uses circuit breakers rated at 150 V dc. The maximum circuit breaker current rating is determined by the PV module specifications, which, in this case, is 20 A. The minimum size circuit

BATTERY-BACKUP GRID-CONNECTED PHOTOVOLTAIC SYSTEMS

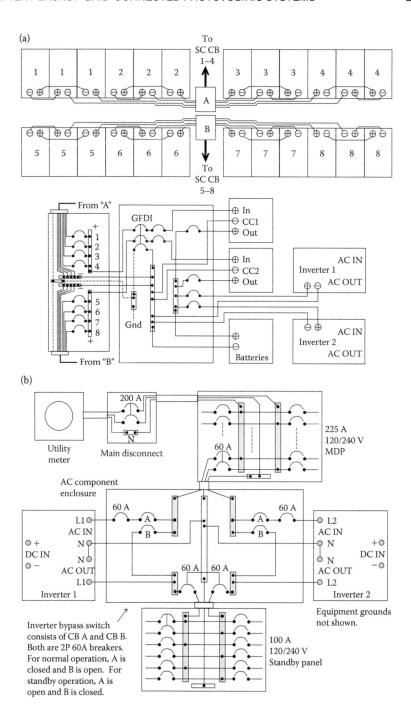

Figure 6.5 (a) Wiring of dc portion of 7200-W battery-backup system. (b) Wiring of ac portion of 7200-W battery-backup system.

breaker would be 156% of I_{SC}, which is 15.96 A. The output busbars of the combiner box will need to be rated at a minimum of 156% of the combined short-circuit current of four source circuits, or 63.8 A.

It turns out that at least one such source-circuit combiner box is manufactured that will accommodate all eight source circuits and conveniently divide them into two separate output circuits, one for each charge controller. It is left as an exercise for the reader to find such a combiner box on the Internet (Problem 6.5).

Finally, the designer should note that a maximum battery charging current of 150 A may occur. Normally, the charging current would be less, since the array will not normally be operating at full rating, and the charge controllers will likely have rated efficiencies of approximately 97%. Nevertheless, a substantial charging current is available from the array, and this should be noted when batteries are selected, since, if the battery system is lead-acid, the battery charging rate should generally not exceed C/5.

6.4.4 Inverter Selection

It makes sense to choose an inverter or inverters that will fully utilize the array output once the system batteries are charged. Several inverters are currently available that provide for grid connection and battery backup and will meet the design requirements of the system. It turns out that one manufacturer markets a 3600-W, 120-V inverter that has all the features required by the design. Furthermore, this inverter can be combined in a "classic stacked" configuration with another identical inverter and programmed to provide an overall single-phase ac output of 120/240 V. So when the system is operating independently of the utility grid, it will provide 7200 W of 120/240-V power to standby loads. The units are designed to pass through 60 A each when the grid is energized, so under these circumstances, a total of 14,400 W can be passed through to the standby loads as long as they are not continuous. However, if the standby loads were to run for 3 hours continuously, then the pass-through load must be reduced to 80% of 14,400 or 11,520 W. Since the load must be derated, this means that a two-pole, 60-A circuit breaker will be the correct size for connecting to the standby load distribution panel.

6.4.5 Determination of Standby Loads and Battery Selection

Once again, the standby loads are generally determined by discussing the options with the owner and generating a table that lists the kW and kWh associated with all the desired loads. In this case, since a 120/240-V source is available, the table is expanded to list which loads are connected to which side of the line, as shown in Table 6.7. Note also that in Table 6.7, loads are expressed in VA rather than in W, to account for the presence of motor loads with lagging power factors in the refrigerator and the pool pump.

Next, notice that the total VA of all connected loads exceeds the total inverting capacity of the two inverters, but is well within the pass-through capacity of the two inverters. Thus, the requirements of *NEC* 690.10(A) are satisfied since the inverters

BATTERY-BACKUP GRID-CONNECTED PHOTOVOLTAIC SYSTEMS

Table 6.7 Summary of Standby Loads for 120/240-V DC-Connected Battery-Backup System

Load Description	Load (VA)	Line 1 (A)	Line 2 (A)	Load (kWh/Day)
Refrigerator	144	1.5		1.2
Microwave oven	600		5.0	0.6
Washing machine	1500	12.5		0.4
Computer	250		2.1	1.0
Lighting ckt 1	1200	10		1.2
Lighting ckt 2	1200		10	1.2
Ceiling fans	150	1.3		2.4
Appliance ckt	1500		12.5	1.8
Mini-split AC	500	4.6		4.0
TV	120		1.0	1.0
Furnace fan (0.75 HP)	700	3.6	3.6	1.4
Total	7864	33.5	34.2	16.2

can supply the largest load connected and, when the utility is down, the inverters are not required to supply all the standby loads simultaneously. Also, notice that the load currents are balanced reasonably well, if all loads are operating and that one load, the pool pump, is connected to operate on 240 V. Finally, the total anticipated daily kWh requirements of all the standby loads are 16.2 kWh. It is thus important to verify that the batteries and array are capable of meeting the daily standby loads or that it is possible to reduce the standby loads on days of low sunshine levels.

To calculate the battery size, noting that the inverter efficiency is approximately 94% and the wiring efficiency is approximately 98%, the batteries must thus supply (16.2/0.94)/0.98 = 17.6 kWh/day to the inverters. If 1 day of storage is allowed, assuming an 80% depth of discharge, the batteries must be capable of storing 22.0 kWh, which, at 48 V, equates to 458 Ah.

The student is encouraged to search for batteries that will meet this storage requirement. One such battery will supply 256 Ah at 12 V at a C/24 discharge rate. Thus, eight of these batteries connected in two parallel sets of four in series will deliver 512 Ah at 48 V, which exceeds the system storage requirements. It might also be argued that since the most likely time the batteries will be needed will be winter, and that in winter the batteries might be colder than in summer, so the additional 54 Ah may provide a good safety factor in the event that utility power should be lost on a cold day.

As a final note, when the array delivers full power, the charging current will be 150 A, as determined in Section 6.4.3. This corresponds to a 512 Ah/150 A = C/3.4 charging rate when maximum array current is flowing. If this rate is felt to be too high, it may be desirable to limit the battery charging current to C/5, which is 102 A. This will likely not be a problem anyway, since with the number of standby loads in this system, it will be likely that when the array is generating maximum power, at least 50 A of the charge controller output current will be used to deliver power to standby loads.

6.4.6 BOS Selection and Completion of Design

Once again, the remaining system components include the array mounting equipment, extension cables for array wiring, a rooftop junction box, provisions for rapid disconnect, a source-circuit combiner box, a GFDI device, arc fault protection, input and output disconnects for the inverter and charge controller, battery cables, battery enclosure, inverter input and output disconnects, inverter bypass assembly, circuit breakers and possibly a separate distribution panel for the standby loads, the equipment for connecting to the utility, and all conduit and wire for the system, including grounding conductors. It is also likely that the owner will choose a system controller that is capable of establishing communication/control of all the electronic system components and provide remote monitoring capability. The design of the array mount is the subject of Problem 6.4, so this section will focus on the remaining electrical design.

6.4.6.1 Rooftop Junction Box

The previous example used four source circuits. This example has eight source circuits, so at least 17 wires (16 circuit conductors and a grounding conductor) must be run to the source-circuit combiner box. While it is possible to pull 17 wires in a single conduit, or, alternatively, to install the source-circuit combiner box on the roof, the sizes of the conductors will first be determined and then suitable routing of conductors will be selected.

Since the distance between rooftop junction box and source-circuit combiner box will be 40 ft (12.2 m) if the source-circuit combiner box is mounted near the charge controllers and inverters, the ampacity and voltage drop will be calculated for this configuration.

Noting first that 125% of I_{SC} = 12.8 A and 156% of I_{SC} = 16 A, the 30°C ampacity of the source-circuit conductors will need to be the larger of 16 or 12.8 A derated for ambient temperature and conduit fill. For wire temperature up to 60°C (0.71 derating factor) and a conduit fill of 16 (0.50 derating factor), this would be (12.8/0.71)/0.5 = 36.02 A. Thus, source-circuit wiring must have an ampacity of at least 36.02 A at 30°C (86°F), which means #10 (5.261 mm^2) will be the smallest allowable conductor.

Before settling on the use of #10, it is useful to check voltage drops with #10 source-circuit wiring. Using Equation 4.1, it is found that #10 wire can be run a distance of 82 ft with a voltage drop of 2%. Thus, over the 40-ft distance between rooftop junction box and charge controllers, the voltage drop will be about 1%, allowing an additional 1% voltage drop in the open wiring between modules and rooftop junction box as well as the short wire runs between source-circuit combiner boxes and inverter input.

In this case, it makes sense to use two rooftop junction boxes and a conduit from each junction box to one of the two combiner boxes. If this is done, then the number of current-carrying wires is reduced to 8 for each junction box and each conduit. This allows for the use of a derating factor of 0.70 rather than 0.50 for

BATTERY-BACKUP GRID-CONNECTED PHOTOVOLTAIC SYSTEMS   263

conduit fill and makes it easier for the installer to keep track of all the wires as well as to use smaller, easier to work with, conduit runs, one to each source-circuit combiner box.

The junction boxes can thus be sized based on 2.5 inch^3 (41 cm^3) per wire plus the volume of terminal blocks. With 9 wires entering each box, and 9 terminal blocks in each box, the box volume needed will be a minimum of $(9 \times 2.5) + 9 = 31.5$ inch^3 (516 cm^3).

6.4.6.2 Source-Circuit Combiner Box and Surge Arrestors

In this case, a box capable of combining two groups of four source circuits will be needed. Since the array voltage will not exceed 150 V, 20-A dc circuit breakers can be used, per module manufacturer's specifications, in the combiner box as shown in Figure 6.5a. In this case, since there are two sets of source circuits and two charge controllers, each will need a separate surge arrestor. If the reader has tried Problem 6.5 and found a single-source-circuit combiner box, then that would be the box of choice for this example. If the reader has not found such a box, it will be necessary for the reader to believe that at least one such box exists.

6.4.6.3 Wire and Circuit Breaker Sizing—DC Side

The wire sizing for the PV output circuits from the combiner box to the system charge controllers is based upon four circuits being combined at each section of the combiner box with no derating factors for ambient temperature or conduit fill. Thus, the wire size must be adequate for four times 156% of I_{SC}, which is 64 A for each set of combined circuits. Each section of the combiner box can thus be wired with #6 copper and 70-A circuit breakers at charge controller inputs and 80-A circuit breakers at charge controller outputs.

The remaining wire and circuit breaker sizing on the dc side of the system is shown in Table 6.8 and Figure 6.5a. The only differences are that each inverter is wired with 2/0 (67.43 mm^2) cable at the input and each inverter requires a separate 175-A circuit breaker as its input disconnect. The 175-A circuit breakers are supplied with a predrilled busbar that is attached to the circuit breaker battery connection sides and attached to a 4/0 (107.2 mm^2) cable that is routed to the batteries, as shown in Figure 6.5a. An alternative to the 4/0 cables is to run parallel 2/0 cables from the circuit breaker input busbar back to the batteries. The outputs of the charge controllers are also connected, via 100-A circuit breakers, to the input positive busbar.

The inverter input is rated at 85 A dc, but on occasion, the inverter may need to supply a short surge current to start a motor or other load. It is thus commonplace to use 2/0 battery cables to the inverter that are protected with a 175-A circuit breaker. It is also advisable to keep the battery cables at lengths less than 10 ft (3 m) to keep the voltage drop at an absolute minimum between batteries and inverter. If large surge currents are anticipated, parallel 2/0 cables will experience lower voltage drop during the surges. Otherwise, 4/0 cables are rated in excess of 125% of the combined inverter maximum ratings of 170 A.

Table 6.8 Determining Wire Sizes for the 7200-W Battery-Backup Grid-Connected PV System

Circuit	1.25I$_{sc}$ (A)	1.56I$_{sc}$ (A)	Wire Size	Wire 30°C Ampacity (A)	Temp Derate	Cond Fill Derate	Required Ampacity (A)
PV source circuits	12.79	15.96	10	40	0.71	0.7	25.7[b]
PV output circuit	51.2	63.8	6	75	1.0	1.0	63.8

Charge Controller and Inverter Calculations

Circuit	I	1.25I	Wire Size	Wire 30°C Ampacity (A)	Temp Derate	Cond Fill Derate	Derated Ampacity (A)
Charge controller output circuit	75	93.75[a]	3	110	1.0	1.0	110
Inverter dc input I$_{max}$	85	106.3	2/0	195	1.0	1.0	195
Inverter ac input (A)	60	75	6	75	1.0	1.0	75
Inverter ac output (A)	60	75	6	75	1.0	1.0	75

Voltage Drop Calculations

Circuit	I$_{op}$	V$_{op}$	Max % VD	Wire Size	Wire (Ω/kft)	Max Wire L (ft)	Actual Wire L (ft)
PV source circuits	9.57	94.8	2	10	1.290	77	40
PV output circuit	38.28	94.8	2	6	0.5100	48	4
Charge controller output circuit	67.9	53	2	3	0.2540	31	3
Inverter ac input	60	120	2	6	0.5100	39	6
Inverter ac output	60	120	2	6	0.5100	39	5

[a] Charge controller output limited to 80 A continuous.
[b] Derated Ampacity applied to 1.25I$_{sc}$ per NEC 690.8(B)(2).

6.4.6.4 Wire and Circuit Breaker Sizing—AC Side

The AC IN and AC OUT ports of each inverter are both rated at 60 A at 120 V. In this case, the inverters will be programmed so that if the utility goes down, they will operate 180° out of phase. So one inverter will act as Line 1 (L1) of a 120/240 V single-phase supply system and the other inverter will follow instructions, via the system controller, to operate as Line 2 (L2). Since the inverters are connected in tandem, as opposed to parallel, line currents are still limited to 60 A, but now the total available voltage is 240 V.

When the utility is operational, the AC IN port of one of the inverters will be connected to utility L1 and the AC IN port of the other inverter will be connected to utility L2. Since these phases are passed through each inverter to the AC OUT port, the AC OUT ports of each inverter supply power in the form of a 120/240 V single-phase supply to the AC OUT L1 and L2 busbars, since this is the utility configuration. The two outputs have a common neutral at the AC IN and AC OUT ports. In any case, since the AC IN and the AC OUT ports each must carry 60 A, #6 (13.3 mm^2) wire and a 60-A circuit breaker will be needed for each of the ports for each inverter.

An inverter bypass assembly is also included to bypass utility power directly to the AC OUT L1 and L2 busbars of the AC component enclosure. The inverter bypass assembly operates identically to the simpler assembly of the 120 V system, but requires four circuit breaker poles to achieve the bypass function, as shown in Figure 6.5b.

Figure 6.5b also shows the PUC in the main distribution panel. In this case, the main distribution panel has busbars rated at 225 A and is fed by a 200-A main circuit breaker. As a check on the largest allowable circuit breaker at the PUC, note that 120% of 225 is 270 A. Since the sum of all circuit breaker ratings of circuit breakers feeding power to the panel plus 125% of the inverter output current must not exceed 120% of the busbar rating and since the sum of 125% of the combined inverter output current and the main circuit breaker ratings is 237.5 A, the PUC is compliant with *NEC* 705.12(D)(2). Furthermore, note that the PUC circuit breaker has been located on the busbar at the opposite end of the utility feed as required by *NEC* 705.12(D)(2)(3)(b). Wire sizing on the ac side of the system is also shown in Table 6.8.

6.4.6.5 Wiring of Standby Loads

With a 120/240-V standby system, the only consideration with regard to multiwire branch circuits is that if one phase of a multiwire branch circuit is selected for standby operation, then the other phase must also be selected for standby operation. The reason is because the *NEC* requires that for multiwire branch circuits, the circuit breakers that supply these circuits must emanate from the same distribution panel and must be supplied by either a double-pole circuit breaker or by two adjacent circuit breakers that are supplied with a handle tie.

The standby loads should be wired in the standby panel as shown in Table 6.7 in order to properly balance the loads on the two phases of the panel. Again, be sure to check with the local code authority to determine whether any of the standby circuits will need to be protected with either ground fault circuit interrupters or with arc fault circuit interrupters.

6.4.6.6 Equipment Grounding Conductor and Grounding Electrode Conductor Sizing

Once again, equipment grounding conductors are sized in accordance with *NEC* 250.122. For this system, since most circuit breakers are 60 A or less, #10 (5.261 mm^2) equipment grounding conductors are acceptable. If the batteries are in a metal enclosure, then the grounding conductor for the battery enclosure will need to be sized according to the 175-A inverter disconnect, which means sizing to a circuit breaker between 100 and 200 A, which requires a #6 (13.3 mm^2) equipment grounding conductor. The same criteria of the previous example apply to whether an additional grounding electrode conductor and grounding electrode should be required.

When the wiring design is complete, then the designer will need to provide recommendations for programming set points for the installer to check. The same considerations used in the previous example also apply to this example, with the additional requirement that the two inverters be programmed in the "stacked" configuration to produce 120/240 V when operating in the backup mode.

6.4.6.7 Rapid Shutdown

As mentioned in Section 6.2.1, methods used for rapid shutdown are evolving rapidly and will be assumed to be incorporated somehow into the designs when required by a local code authority or by the design engineer. For battery-backup systems, it needs to be remembered that the rapid shutdown sequence is primarily intended to shut down the conductors from array to combiner box, as close to the array as possible, but not to exceed 10 ft (3 m) from the array, or 5 ft (1.5 m), if the array output passes through the building at a distance less than 10 ft from the array. Normally, this will shut down the system and eliminate shock or arcing hazard if the building main ac disconnect is shut down, *unless* the PV system has battery backup. In this case, shutting down the array and shutting down the building main disconnect will not shut down the standby loads. To shut down the standby loads, it will also be necessary to shut down the inverter or the battery connection to the inverter. The important thing to remember when creating a suitable design is to follow the *NEC* marking and labeling requirements as stated in *NEC* 690.51-55 so emergency responders will be able to find all switches necessary to shut down the system.

An example of a rapid shutdown system available at the time of this writing is shown in Figure 6.6.

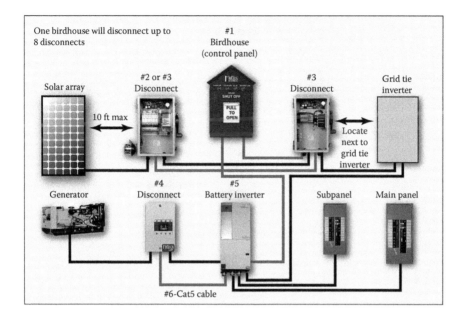

Figure 6.6 Block diagram of a commercial rapid shutdown system. (Courtesy of Midnight Solar, Arlington, WA.)

6.5 AN 18-KW BATTERY-BACKUP SYSTEM USING INVERTERS IN PARALLEL

6.5.1 Introduction

As commercial enterprises experience prolonged power outages, the interest in larger battery-backup systems is becoming more common. Only a few years ago, most battery-backup systems were less than 11,000 W. It is now possible to find fully UL 1741-compliant battery-backup inverters in the 80-kW range [7]. This example will show the design of an 18-kW battery-backup system that supplies 120/240-V backup power.

Fitting 18 kW on a residential installation would require quite a substantial roof. But it is not at all unusual for a commercial installation to have a roof that will accommodate an 18-kW system. This design will be for a commercial building with a flat roof in Lexington, KY, located at 38°N latitude. The ambient temperature range for the design, based on historic Lexington weather data [8], has been chosen to be from −10°F (−23.3°C) to +100°F (37.8°C).

The roof is a membrane roof applied over a layer of insulation that is attached to a corrugated steel structure supported by metal trusses that will support an average of 20 psf (958 pa). In fact, the maximum daily snowfall rarely exceeds 9 inch (22.9 cm) [9]. Assuming a "wet" snowfall would imply a density of approximately 20% [10], or 9.4 psf (450 pa) for a 9-inch snowfall. The ASCE7 occupancy category II building

Table 6.9 Summary of Standby Loads for 18-kW, 120/240-V DC-Connected Small Commercial Battery-Backup System

Load Description	Load (VA)	Line 1 (A)	Line 2 (A)	Load (kWh/Day)
Refrigerator #1	144	1.5		1.2
Refrigerator #2	144		1.5	1.2
Furnace blower #1	700	3.6	3.6	6.0
Furnace blower #2	700	3.6	3.6	6.0
Furnace blower #3	700	3.6	3.6	6.0
Lighting ckt 1	1200	10		2.5
Lighting ckt 2	1200		10	2.5
Lighting ckt 3	1200	10		2.5
Lighting ckt 4	1200		10	2.5
Computer ckt 1	300	2.5		2.4
Computer ckt 2	300		2.5	2.4
Computer ckt 3	120	1.0		1.0
Computer ckt 4	120		1.0	1.0
Coffee maker	1000	8.3		1.0
Microwave oven	900		7.5	1.0
Total	**9928**	**44.1**	**43.3**	**39.2**

is located in exposure category B with a design wind load of 115 mph (185 km/hour). The roof height is 15 ft (4.57 m), and the roof has a 3-ft (0.91 m) parapet around the roof perimeter.

After a serious ice storm that left them without utility electrical power for 10 days, the owners have decided to invest in a battery-backup PV system to provide standby power for their furnace blower motors and ignitors as well as some lighting and commercial refrigeration equipment, as shown in Table 6.9. The owner already has a 15-kW fossil-fuel backup generator, and the equipment of Table 6.9 is already connected to the backup generator via a manual transfer switch. The building is served by a 400-A, 120/240-V single-phase system. The plan is to configure the standby system so that the generator will operate only if the PV system does not meet the standby load requirements. Figure 6.7 shows the roof, including air conditioning units and vent fans.

6.5.2 Inverter and Charge Controller Selection

6.5.2.1 Inverter Selection

Since it has already been determined that it is desirable to install an 18-kW PV system, this time the inverters will be selected first, along with accompanying charge controllers. The inverting for this system will be accomplished by a parallel combination of three 6.8-kW inverters that operate with a 48-V nominal dc input from batteries. Each inverter has a 120/240-V single-phase output. The three inverters can be connected in parallel such that a system controller will ensure that the inverters all provide outputs that are in phase in the event that utility power is lost. When the

BATTERY-BACKUP GRID-CONNECTED PHOTOVOLTAIC SYSTEMS

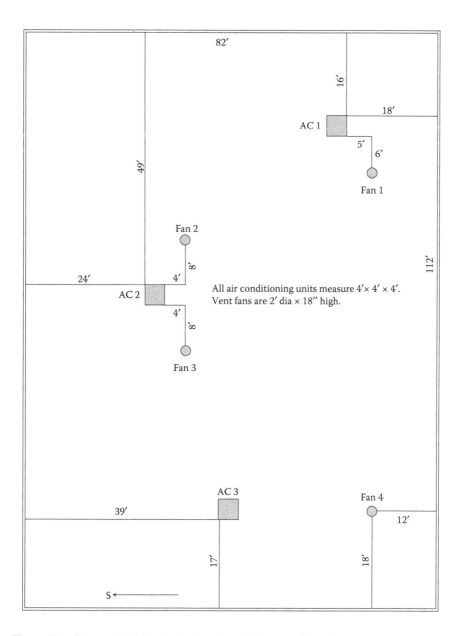

Figure 6.7 Commercial flat roof showing air conditioners and vent fans.

utility is operational, the utility synchronizes the inverter outputs. It should be noted that the installation manual for this particular inverter is over 100 pages long. This illustrates the importance of developing an understanding of the capabilities of this type of system with the knowledge that when designing a system, the designer will need to be familiar with the contents of the installation manual.

Each inverter has the following basic specifications. Additional specifications will be mentioned as needed:

Output power	6800 W continuous or 12,000 W for 60 seconds
Maximum surge output current	102 A at 120 V or 52 A at 240 V
AC input connections	Grid and generator
AC input breaker	2P 60 A
CEC weighted efficiency	92.5%
AC output voltage	120/240 V
DC input current at rated power	142 A
Nominal DC input voltage	48 V
DC input voltage range	42–60 V

The inverters come with optional conduit enclosures and an ac/dc power distribution panel that houses all the system disconnects. The overall integrated power conditioning system also includes MPPT charge controllers, a system control panel (SCP), and automatic generator start (AGS). The system control panel and AGS will be discussed as the system design progresses.

6.5.2.2 Charge Controllers

In order to determine the number of charge controllers that will be needed in this system, it will first be necessary to look at the specifications of the charge controllers that are available. It makes sense to select charge controllers manufactured by the inverter manufacturer, since they can then be installed to communicate with the inverters.

In this case, at the time of this writing, only a 60-A charge controller is available from this manufacturer for use with a 150-V or lower PV array. Since 60-A MPPT charge controllers should normally be limited to 3200 W of array power for a 48-V system, and since this system will have a nominal 18-kW array, the system will require six charge controllers as long as the array size does not exceed 19,200 W and as long as the array output power can be distributed evenly among the charge controllers.

The inverter manufacturer also offers an 80-A charge controller for use with an array up to 600 V. This charge controller can be used with array power up to 4800 W, which would require four units to handle a 19,200-W array. Since the last two examples have been designed with 150-V arrays, the 600-V charge controller will be used in this example. If the arrays are designed carefully, it will be possible to eliminate source-circuit combiner boxes and implementation of rapid shutdown may be less complicated.

As a final note, both charge controllers have internal ground fault detection and interruption, so no additional GFDI devices will be needed.

6.5.3 Module Selection and Array Layout

The roof of the building is typical of many flat-roofed buildings. It has an abundance of other equipment on it, all of which needs to be checked for shading of the PV array. This book could not possibly be complete without presenting such an example. So the first step in the design is to determine how to locate the PV modules, which means doing a shading analysis on the existing rooftop mounted equipment. As mentioned in a previous chapter, there are a number of good shading analysis computer programs, but in this case, use will be made of the equations of Chapter 2 that describe the position of the sun in order to prove that the equations can be useful.

Usually, it is adequate to look at the position of the sun on the first days of each of the four seasons over approximately an 8-hour period centered on solar noon. More than likely, the exposure period will need to be reduced to 6 hours on December 21, since otherwise shadows will be quite long. To optimize system performance, it is desirable to be sure that the array will receive at least 6 hours of unshaded sunlight every day of the year, neglecting cloud cover. This analysis can be done quite readily with the help of a spreadsheet that is equipped with the formulas for solar altitude and azimuth as functions of latitude, solar declination, and time of day. In other words, it will be necessary to tabulate values of ϕ, δ, and ω in order to compute values of α and ψ using Equations 2.9 and 2.10. Once α and ψ are known, the length and direction of shadows can be determined in order to determine both the spacing between module rows and the clearances between modules and rooftop equipment. But before the spacing between module rows can be determined, a module must be selected.

6.5.3.1 Module Selection

In this case, the building owner's cousin is dating the sister of a distributor who can supply a particular multicrystalline silicon module at an attractive cost. The modules are cosmetically blemished, but carry the full UL listing and full manufacturer's warranty. The local building official has approved their use. The modules are 240 W units that have $V_{OC} = 37.5$ V, $V_m = 30.6$ V, $I_{SC} = 8.38$ A, and $I_m = 7.84$ A. The temperature coefficients are $\Delta V_{OC}/\Delta T = -0.35\%/°C$ and $\Delta V_m/\Delta T = -0.51\%/°C$ and the module NOCT = 46°C. The modules measure 1 m × 1.62 m (39.3 inch × 63.7 inch).

The maximum number of modules, based upon a 19,200-W array will be 80, suggesting 20 modules per charge controller if charge controller operating voltages can be met.

At the lowest ambient temperature, V_{OC} will increase to $37.5[1 - (25 - (-23.3))(-0.0035)] = 43.8$ V. Noting that the array V_{OC} must not exceed 600 V, so the charge controllers will not be damaged, the maximum number of modules in a source circuit is limited to $600/43.8 = 13.7$, which rounds down to 13. Thus, the maximum array V_{OC} will be 569 V.

At the highest temperature and highest irradiance level, the module will operate at a cell temperature that is 31°C above ambient, or 70.3°C (see Problem 6.7). This will reduce V_m to $30.6[1 - (25 - 70.3)(-0.0051)] = 23.53$ V.

The charge controllers have a 195-V minimum MPPT voltage. Thus, the minimum number of modules in a source circuit will be $195/23.53 = 8.28$, which must be rounded up to 9. Thus, source circuits may use between 9 and 13 modules. With 20 modules connected to each of four charge controllers, this suggests using 10-module source circuits, with 2 source circuits feeding each charge controller.

6.5.3.2 Array Layout

In this case, the owner has expressed great concern over roof penetrations and has shown a strong preference for somehow mounting the array without penetrations. Since this owner is not the first to express such concerns, racking manufacturers have developed a collection of ballasted array mounts that use concrete blocks in pans to hold down the array rather than bolts through the roof. So it is simply a matter of selecting a suitable array mount and determining the row spacing.

Since the latitude of Lexington is 38°, it would be nice to have a tilt within 15° of latitude to minimize annual losses. So the search will be for an array mount that will allow a tilt of approximately 25° without roof penetrations, and will have a dead load low enough so that the roof will not collapse under the combined weight of the array mount, ballast, array, and a good snowfall. Since the building is in a 115-mph (185 km/hour) design wind speed zone, which is a typical wind speed for most of the United States, it can be expected that if a commercial array mount can be found that will allow a nonpenetrating, 25° tilt, that it would be engineered to withstand a 115-mph wind speed. It will also be assumed that if the wind were to blow at a speed of 115 mph, it would blow most of the snow off the roof, so maximum wind load and maximum snow load will not occur simultaneously. In fact, the likelihood of a 115-mph wind in the winter is extremely small, but cannot be ruled out.

Figure 6.8 shows how the row spacing can be determined. The procedure is to solve Equations 2.9 and 2.10 on an hour-by-hour basis over the period during which it is desired to have the array unshaded. This is probably most easily done by developing a spreadsheet to do the repetitive calculations. By solving for altitude and azimuth angles during the period over which the modules are desired to be unshaded, it is possible to determine the length and direction of shadows during this period. Then, either

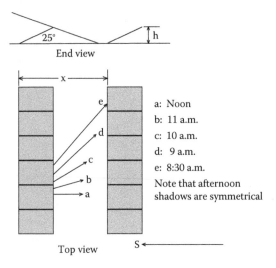

Figure 6.8 December 21 shadow analysis to determine row spacing.

a little trigonometry or a little playing with a scaled drawing can produce the desired shadows, and, subsequently, the necessary spacing of modules from each other and from objects on the roof, such as fans, air conditioners, and parapets. Normally, the spacing will be determined by the shadows on the first day of winter, but since the sun follows a limited arc from southeast to southwest in the winter, there may be obstructions to the east, northeast, west, or northwest that might cause shading during either the swing season or during summer months. So these possibilities should be checked. It is left as an exercise for the reader to calculate the necessary distance between rows if it is desired to have 7 hours of unshaded operation on December 21 (Problem 6.8).

Figure 6.9 shows the analysis of shading of the three air conditioners and the parapet walls. Each air conditioner is a cube, 4 ft (1.22 m) on a side. The parapet walls are 3 ft high. The dotted line in the figure shows the extent of shadows caused by the parapet wall on December 21 at 8:30 a.m. and 3:30 p.m. sun time. On all other days, the shadows are shorter and during the 7-hour interval between 8:30 and 3:30, the shadows are also shorter.

The air conditioner and fan shade patterns show the extent of shading on the first day of winter, the first day of fall, and the first day of summer. So the modules must be placed outside these shadow zones. Finally, Figure 6.9 shows one possible configuration of the modules to avoid shading as much as possible. Note that the array layout has been designed to be as installer-friendly as possible. Since source circuits consist of groups of 10 modules, the modules are grouped in multiples of 5 or 10. The nine small dark rectangles indicate reasonable locations for junction boxes to transition from open wiring to wiring in conduit. The idea is to bring two source circuits down in each of four conduits to each of four charge controllers. To avoid roof penetrations, the conduit to the source-circuit combiner boxes will pass through the parapet, down along the outside of the building, and into the equipment room.

For ballasted array mounts, the manufacturer will normally specify the number of ballast blocks needed for a given module orientation and a given module tilt. In this case, the modules are mounted in portrait at a tilt of 25°. The average additional dead load imposed by the modules and array mounts, including ballast, will typically not exceed 4 psf (191.5 pa), especially in this case where the array covers only a small percentage of the roof.

Finally, one can note that the per module weight of a ballasted system is normally calculated to be 150% of any uplift from wind forces. Furthermore, the downward wind forces will not exceed the upward forces, so any additional downward wind force that must be added to the dead load will not exceed the dead load itself, so it is safe to add another 4 psf as a wind load to obtain a total load on the roof of 8 psf as a worst-case load on the roof by the PV system. Thus, the maximum load from wind, snow, and the PV system will not exceed 20 psf (958 pa), so the roof is adequately designed to withstand all the forces simultaneously.

6.5.3.3 Array Performance

Now that the array has been defined, NREL SAM PVWatts [3] can be used to estimate the monthly and annual system energy production. Using an array size

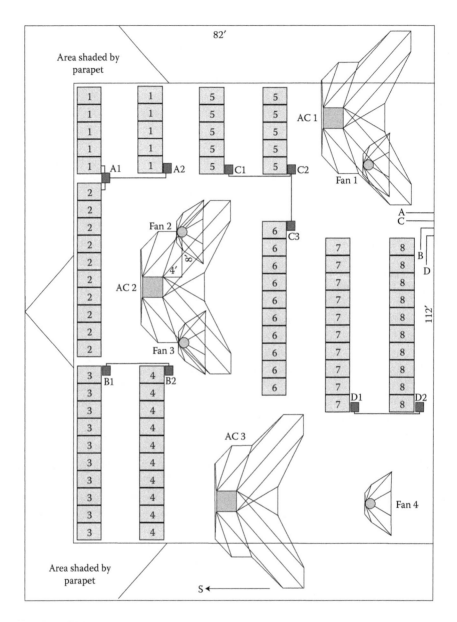

Figure 6.9 Commercial flat roof 19.2-kW array layout showing equipment and parapet shading patterns.

of 19,200 W, with a 25° tilt in Lexington, KY, and using 81.6% of array energy to be delivered to the batteries, as in Section 6.3, results in a dc/ac conversion ratio of 1.23. Thus, the array can be expected to deliver approximately 25,360 kWh/year to the batteries, with a December–February monthly average of 1470 kWh, a March–May average of 2440 kWh, a June–August average of 2630 kWh, and a

September–November average of 1910 kWh. The lowest monthly performance is in December, with 1200 kWh from array to batteries.

On a daily basis, the average daily kWh available to the batteries for winter is 49, with a December worst-case average of 38.7 kWh.

It should be noted that kWh to the batteries implies that if the batteries are charged, then any kWh not used by the batteries will go directly to the inverter for delivery to building loads, which, if satisfied, will result in any excess PV energy being sold to the utility. So the next step is to determine the energy requirements of the batteries in order to meet standby load requirements.

6.5.4 Battery and BOS Selection

The battery size is determined by examining the energy requirements of the standby loads as shown in Table 6.9. Assuming a wiring efficiency between inverter and loads of 97%, an inverter efficiency of 92.5%, and a 90% battery charging/discharging efficiency, this means that the amount of energy that must be delivered to the batteries to meet the load requirements is ((39.2/0.97)/0.925)/0.90 = 48.5 kWh/day. Thus, based upon the simulation done in Section 6.5.3.3, December is the only month during the year for which the PV array may not meet 100% of the standby load requirements.

Since the owner is serious about keeping the business open, but also since the owner has a fossil generator for additional backup power, only 1 day of battery storage will be included in the design. The batteries are sized according to the energy that must be delivered by the batteries, which is 0.9 × 48.5 = 43.7 kWh. Allowing an 80% depth of discharge, the batteries must be capable of storing 54.6 kWh at 48 V at a 24-hour discharge rate. This means the Ah rating of the batteries must be 54,600/48 = 1138 Ah.

Since good design practice is to design the battery bank with fewer than four parallel sets of batteries, a search can be made for batteries that are either rated near 1140 or 570 or 380 or 285 Ah. A decision also needs to be made as to whether the batteries should be flooded or sealed.

Flooded or sealed, it turns out that a typical L-16 size battery will be rated at approximately 400 Ah at 6 V at 24-hour discharge rate. So 24 of these batteries, connected in 3 parallel sets of 8 series-connected batteries each, will store the necessary energy. It will be up to the owner to check the prices and then decide whether to purchase flooded or sealed units.

6.5.5 Wire Sizing

As in previous examples, Table 6.10 summarizes the calculation of wire ampacities. Note that for this system, it will be assumed that the power conditioning equipment, including the batteries, will be located in a room in the center of the north side of the building. The cabling between batteries and equipment will be less than 10 ft (3 m). Source-circuit combiner boxes will be located near the power conditioning equipment. Lengths of source-circuit conductors include the equivalent two-way

Table 6.10 Determining Wire Sizes for the 18-kW Battery-Backup Grid-Connected System

Circuit	I_{sc} (A)	1.56I_{sc} (A)	Wire Size	Wire 30°C Ampacity (A)	Temp Derate	Cond fill Derate	Required Ampacity (A)
PV source circuits	8.38	13.07	#12	30	0.71	0.80	18.4
PV output circuit	16.8	26.2	#10	40	1.0	1.0	26.2

Charge Controller and Inverter Calculations

Circuit	I	1.25I	Wire Size	Wire 30°C Ampacity (A)	Temp Derate	Cond fill Derate	Derated Ampacity (A)
Ch cont output ckt[a]	80	100[a]	#4	95	1	1	95
Inverter dc input I_{max}	142	177.5	#4/0	260	1	1	260
Inverter ac input (A)	60	75	#6	75	1	1	75
Inverter ac output (A)	60	75	#6	75	1	1	75

[a] Charge controller output limited to 80 A continuous.

length of open wiring plus the length of conductors from junction box to source-circuit combiner box. Four separate conduit runs will carry the source-circuit outputs to four source-circuit combiner boxes, each of which will feed one of the four charge controllers. Thus, each conduit will have four current-carrying conductors and a ground wire. In order to determine voltage drop, it will be necessary to determine the length of the wires.

Starting with the modules, each module has #12 (3.31 mm²) positive and negative leads, each of which is 42-inch (1.07 m) long. Thus, for a 10-module source circuit, the total length of module leads will be 840 inch, or 70 ft (21.33 m). Note that this is equivalent to a one-way distance of 35 ft (10.7 m) for two side-by-side conductors. This will be the distance, d, used in the voltage drop formula.

Next, note that most of the junction boxes shown in Figure 6.9 are located at the ends of source circuits. Thus, one of the source-circuit leads will connect directly into the junction box and the other will need a short extension to reach the junction box. Conduit connections between junction boxes are also shown in Figure 6.9. Note that in the case of Source Circuits 3, 4, 6, 7, and 8, each source circuit is a continuous row of 10 modules. For Circuits 3 and 4, two wires plus a ground for Circuit 3 are run in conduit from Junction Box B1 to Junction Box B2. Then, from Junction Box B2 to the equipment room, four wires plus a ground wire are run in a single conduit spaced approximately 2 inch above the roof.

In the special cases of Source Circuits 1 and 5, the circuits are split in the center such that the first set of Circuit 1 modules are connected in series with the second set via the conduit between Junction Boxes A1 and A2. Source Circuit 2 extension wires are connected through Junction Boxes A1 and A2 and then to the equipment room along with the extension conductors from Source Circuit 1. It should now be evident from Figure 6.9 how the remaining source-circuit conductors are routed to the equipment room.

Since no conduit will contain more than four current-carrying conductors, the conduit fill derating for these conduit sections will be 0.8. For ambient temperature derating, 22°C must be added to the 37.8°C highest expected ambient temperature, resulting in a conduit temperature of 59.8°C (139.6°F), which requires a derating factor of 0.71. Since I_{SC} = 8.38 A, 125% of I_{SC}, derated for temperature and conduit fill, is 18.44 A, and 156% of I_{SC} is 13.07 A. Thus, the minimum ampacity of source-circuit extension conductors is 18.44 A. This means #14 wire is adequate for ampacity, but voltage drop must also be considered. Since the source circuits are relatively high voltage and since the module leads are #12, it is reasonable to try #12 for the source-circuit extension conductors as well for an initial voltage drop check.

Assuming a 2% overall voltage drop in each source circuit, the distance each circuit can be run can be calculated. The result, using the familiar voltage drop formula with source circuit I_m = 7.84 A and V_m = 306 V, is 204 ft (62 m). Since the equivalent length of the module leads in each source circuit is 35 ft (10.7 m), this means the extension conductor distance can be up to 169 ft (51.5 m). Noting the building dimensions of 82 ft (25 m) × 112 ft (34.1 m), it is clear that all conduit runs will be less than 169 ft. Therefore, #12 extension conductors will be adequate between source circuit and charge controllers in the equipment room.

Technically speaking, it is not necessary to use fused combiners when only two source circuits are combined. The reason is that if one circuit is faulted, the maximum current that would backfeed this circuit is I_{SC} of the other circuit, which is significantly less than the rated fuse current. Thus, even if there were a problem, no fuse would blow. On the other hand, if fused combiners are used, then it is easier to separate the two source circuits for troubleshooting purposes by removing one of the fuses. Thus, small, fused combiner boxes will be incorporated into the design as a convenience to the maintenance technician.

The inverter manufacturer recommends using 4/0 (107.2 mm^2) battery cables with 250 A circuit breakers for each inverter in order to handle surge currents. Under normal operation, the battery cable current will not exceed 142 A.

Table 6.10 also shows the wire sizes for the ac side of the inverters. Up to the busbar side of the inverter terminals, #6 (13.3 mm^2) is used, protected by 60-A circuit breakers. The outputs of all the inverter inputs are combined at the busbars, shown in Figure 6.9b, so that the wires to the utility interconnect point, the generator, and the standby loads will be larger than #6. These wire sizes will be determined in the next section.

6.5.6 Final Design

Figure 6.10 shows the final system design. Note that the dc schematic of Figure 6.10a shows the array to be broken down into four subarrays of 20 modules each, such that 4800 W of array is connected to each charge controller array input. Source circuits are protected from backfeed by the 15-A fuses in the source-circuit combiner boxes, each of which combines two source circuits for one charge controller. The disconnects at the inputs of the charge controllers serve as array disconnects. Note that these disconnects can be implemented as two 2-pole switches in order to

278   PHOTOVOLTAIC SYSTEMS ENGINEERING

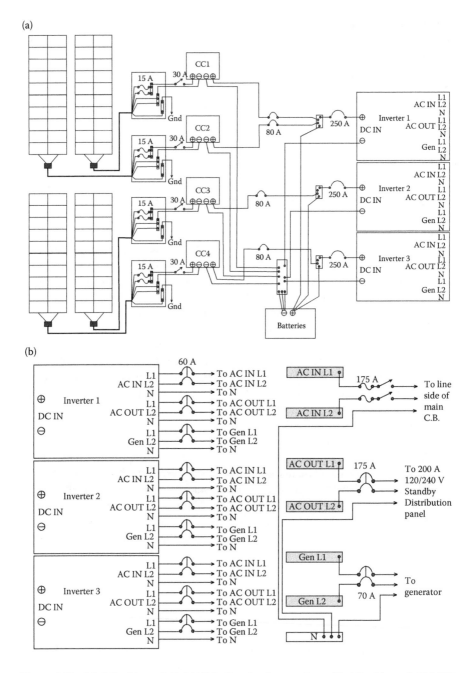

Figure 6.10 (a) DC wiring of 19.2-kW battery-backup system. (b) AC wiring of 19.2-kW battery-backup system. Grounding conductors and control wiring not shown.

reduce the number of switches that need to be opened to disconnect the array. The 80-A circuit breakers at the outputs of the charge controllers serve as disconnects for the charge controller outputs. The inverter inputs can be connected as shown with three busbars to provide connections to the battery for an inverter and one or two charge controllers each, or by a single large busbar with connections to all inverter inputs, three parallel 4/0 (107.2 mm^2) or equivalent to the batteries, and #4 (21.15 mm^2) wiring from the charge controller outputs. Note also the three parallel 4/0 conductors from battery negative to the negative busbar. The busbars must be rated to carry 125% of the full load inverter input currents, depending upon the number of inverters connected to each busbar.

It should also be noted that the combined outputs of the four charge controllers, at 80 A each continuous, is 320 A. This a pretty high charging current for the batteries. Since the batteries have a total capacity of 1200 Ah at 48 V, a charging current of 320 A represents a C/3.75 charging rate. Of course, this is when the charge controllers are operating at full power, which will likely not be the case, at least on a continuous basis. The charge controller manuals should be consulted for appropriate settings of absorption time for this charging rate.

The ac sides of the inverters are wired as shown in Figure 6.10b. In this figure, the authors had to decide whether to present a confusing mess of wires crossing wires, or to simply indicate the appropriate busbar to which each inverter ac-side wire should be attached. The figure shows which choice was the winner. The bottom line is that all of the ac ports of the three inverters are paralleled, so the busbars simply provide the points for achieving the parallel connections. Thus, the input/output currents can be triple the ratings of single inverters.

The generator circuit breaker is only 70 A, since the generator is only rated at 15 kW. Note that 15 kW/240 V = 62.5 A. It is good practice to avoid overloading mechanical generators, since overloading can cause the prime mover to stall. Hence, the circuit breaker is sized at just under 115% of this value, or 70 A. The inverters can be programmed to limit the generator current to 90% of its power rating so that it will operate at maximum efficiency when it is running. When the generator runs is another programming task. The inverters can be programmed to start the generator when the battery voltage reaches a predetermined value, such as 48 V, and to stop the generator when the batteries reach 52 V. The system controller also incorporates a real-time clock so that the generator can be prevented from starting after a certain hour in the evening, or before a certain hour in the afternoon. Note that the generator is only expected to be used on cloudy days or at night when there is a utility outage. This means that monthly maintenance/starting of the generator will be necessary to ensure that the generator is available when it is needed.

Since the AC IN and AC OUT busbar outputs are protected by 175-A circuit breakers, this means that 2/0 (67.43 mm^2) copper will be needed, since the distance between inverter output and either the utility connection or the standby distribution panel is relatively short.

For this system, the manufacturer provides a system control panel and an AGS accessory [11]. The system control panel enables programming of all the programmable system components, including inverters, charge controllers, and the AGS accessory. The AGS has three built-in relays to support 13 preset generator configurations. It monitors the battery voltage as well as the inverter charge mode. It also monitors the system output power and can start the generator if the system output power needs are higher than the inverters can supply. It can also be set up to automatically exercise the generator so that the generator field will not become demagnetized or the generator gasoline will not turn into gelatin.

Note that the system output must be connected to the line side of the 400-A building main disconnect, because 175 A exceeds 120% of the busbar rating of a 400-A electrical panel. The only way this 175 A of PV current could be fed to a 400-A electrical panel would be if the panel main circuit breaker were no larger than 300 A, since 120% of 400 A is 480 A and (300 + 175) < 480. If the system AC IN is connected to the line side of the main breaker, then the AIC (asymmetrical interrupting capacity) rating of the 175-A overcurrent devices at the PUC must be no smaller than the AIC rating of the main breaker. The AIC rating of a circuit breaker indicates the maximum short-circuit current that can flow through the circuit breaker and still cause the circuit breaker to trip.

The critical reader may also have noticed the absence of any inverter bypass circuit breakers in this system. There are several reasons for this. One is that the component enclosure only has room for nine 2-pole ac circuit breakers. The other reason is that the likelihood of all three inverters failing simultaneously is remote. Thus, if one inverter should fail, it can be isolated and repaired while the other two inverters continue to operate.

Finally, note that the schematic diagrams do not show all the grounding conductors and do not show any surge arrestors. Good design practice would involve incorporating surge arrestors at the source-circuit combiner boxes at the outputs to each charge controller. The grounding conductors between various system components are selected in accordance with *NEC* Table 250.122. For example, since the wiring from busbars to the PUC is protected by a 175-A circuit breaker, the grounding conductor must be no smaller than #6 (13.3 mm^2) copper. On the array side of the charge controllers, where the conductors are protected by 60-A circuit breakers, #10 (5.261 mm^2) grounding conductors will suffice. If the batteries are in a metallic case, then the case will need to be grounded with a #1/0 (53.49 mm^2), since the total circuit breaker rating for battery cables is 750 A.

6.6 AC-COUPLED BATTERY-BACKUP SYSTEMS

6.6.1 Introduction

While the dc-coupled battery-backup systems of the previous examples have been popular, there are situations where ac-coupled battery-backup systems may be preferred. One reason might be that a system owner has installed a straight grid-connected system without batteries and has then experienced a lengthy power outage. Regrettably, the system owner then had an idle PV system until the utility came back on. Another possibility is that the designer or owner has noticed that dc-coupled battery-backup inverters tend to have efficiencies in the 92%–94% range, while straight grid-connected inverters have efficiencies in the 96%–98% range. Furthermore, the dc-coupled systems encounter additional losses to the batteries, even though they remain essentially fully charged. If the charge controller outputs are pulsed, to some extent the pulse enters the battery and then discharges to the inverter while the pulse is off, causing charge/discharge battery losses to the battery internal resistance. If the array output is converted to ac first, then feeds the standby loads, and then passes through to the grid, the process can be more efficient.

Another possibility is having more than one renewable source in the system, such as wind or low head hydro. Then it will likely be desirable to incorporate all of these as sources of standby power if the grid should fail. Since all of these sources must comply with UL 1741 for grid connection, the easiest way to incorporate them all is via ac coupling. Another reason for ac-coupled systems is to take advantage of high-voltage source circuits.

These systems are not without disadvantages. One disadvantage is higher first cost that will presumably be justified by any or all of the preceding reasons. Another disadvantage is the possible power limitations of the battery-backup inverter, as well as the energy storage limitations of the batteries. These systems tend to be best suited when the standby loads are coincident with the power available from the PV array and the nighttime standby loads are minimal.

The trick is to keep the straight grid-connected inverters operating when the utility grid is down by introducing a battery-backup source that will look like the utility to the other inverters and will pass excess power production back to the utility when the standby loads are satisfied. This section will not attempt a complete system design, since all the component systems have already been designed. Rather, the focus will be on the basic ac-coupling concept. The array, load, and battery calculation procedures are the same for these systems as they were for previous systems. It should be emphasized that before attempting to design ac-coupled systems, the designer should be very familiar with the installation manuals and any associated application notes available from the inverter manufacturers [12].

6.6.2 A 120/240-V Battery-Backup Inverter with 240-V Straight Grid-Connected Inverter

The ac coupling problem can be straightforward if a 120/240-V battery-backup inverter is used in conjunction with a 240-V straight grid-connected inverter. The system takes advantage of the well-kept secret that the AC OUT port of most battery-backup inverters can allow power flow in either direction. Normally, these ports are considered as supply-only ports. Ideally, the battery-backup inverter will have a contactor or electronic switch that will connect directly from AC OUT to AC IN ports when the grid is up. This way, when the grid is up, the grid functions as the synchronization signal for the straight grid-connected inverters. If the grid goes down, then the contactor opens and the AC OUT port functions as an ac voltage source that has utility voltage characteristics, namely, stable amplitude and stable frequency of a good sine wave, while the AC IN port is disconnected from the utility except for monitoring capability. There are limits to this system, since the grid can only supply as much power as the battery-backup inverter will pass through, which is typically limited to 60 A for each battery-backup inverter used. The other limits are the battery capacity and the inverter output power rating when the grid is down, which may be less than the pass-through capacity. During hours of sunshine, depending upon the power ratings of the array and the straight grid-connected inverters, the maximum loads that can be served are essentially dictated by the array power and the dc/ac conversion efficiency. But at night, the loads must be restricted to whatever the batteries can supply unless another source is present, such as wind or hydro.

This system is shown in Figure 6.11, which shows the internal switching arrangement of the battery-backup inverter along with a necessary control link between the two inverters. The diagram shows that it is also possible to incorporate an external fossil-fuel generator as well. No grounding conductors are shown in the diagram.

The first thing to notice is that the PV array is connected in the usual fashion to Inverter 2, as though Inverter 2 were operating in a straight grid-connected configuration. No PV is connected to the input of the battery-backup inverter. If it were to be connected, then the system would involve dc coupling of the array to the inverter via batteries and the ac coupling advantage would be lost.

Next, note the internal switching of Inverter 1. The AC OUT port is always connected directly to the ac output of the invert/charge electronics of the inverter. The AC IN port will be connected to the AC OUT port if the utility stability requirements of UL 1741 are met. If the utility power is lost, then the switch to the AC IN port is opened automatically by the inverter and power to the AC OUT port is supplied by the inverter, using battery backup if no power is available from the PV array. The generator input port switch is left open unless the inverter is programmed to accept power from a generator.

The system has several possible operating modes. Normal daytime operation with the grid available involves the grid being directly connected to the ac port of Inverter 2

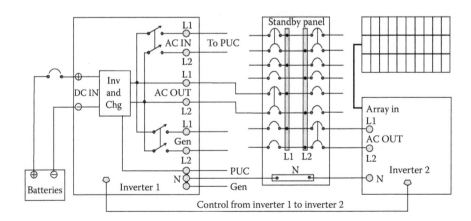

Figure 6.11 Basic ac-coupled system with 120/240-V battery-backup inverter and 240-V straight grid-connected inverter.

via Inverter 1 and the standby panel. This enables the operation of Inverter 2. If the PV array produces sufficient power to operate the standby loads with leftover power, then the leftover power flows back to the grid through Inverter 1 as long as the batteries are fully charged. Note that this means the batteries are charged by rectifying power from the ac side of Inverter 1, the origin of which may be either the grid or the output of Inverter 2. It also means that Inverter 1 will have the charger settings that would normally be assigned to a charge controller if a charge controller were present. But since the array is on the ac side of this inverter, no external charge controller is needed.

Another possible operating mode is when the grid is present, but the sun is not. Thus, the straight grid-connected inverter is not capable of supplying any power to the standby loads. This power must be supplied by pass-through of the AC IN grid connection to the AC OUT port. This means that the pass-through current rating of Inverter 1 must be adequate to supply the nighttime standby loads. If the load on the AC OUT port of Inverter 1 is too large, the inverter will shut down unless the system has a generator that is programmed to assist the battery-backup inverter in such cases.

Another possible operating mode is when the grid is not present, but the sun is present. As long as the battery voltage is high enough to operate Inverter 1, Inverter 2 will interpret Inverter 1 as the grid and will deliver power as soon as power is available from the PV array. If the sun produces adequate power via Inverter 2 to operate the standby loads, any excess power now cannot be sold back to the utility, but can only be used for charging the batteries. But unless a mechanism is present to prevent overcharge of the batteries, it is possible to overcharge and possibly damage the batteries. This is why the control cable is present between Inverter 1 and Inverter 2. If the batteries are fully charged and power is still flowing to Inverter 1 from Inverter 2, then Inverter 1 tells Inverter 2 to shut down to prevent overcharging the batteries. But, then, what happens to the standby power if Inverter 2 is shut down? It is supplied by the batteries via Inverter 1 until the battery voltage has dropped to a level such that Inverter 1 will allow Inverter 2 to start up again.

Finally, it is possible that both the grid and the sun may be down. In this case, the batteries need to supply the standby loads via Inverter 1. If the standby loads are too large, it is possible that the batteries will discharge to the low voltage cut-out point and the standby loads will go without power. This also presents a control dilemma for the inverters, since if Inverter 1 is shut down, then Inverter 2 will think the utility is down and will remain off. The result is no power available to recharge the batteries. So this challenge must also be met by communication between the two inverters. If an attempt is made to construct such a system, then it is important for the designer to determine whether the inverters are capable of overcoming this situation. If not, it means back to the drawing board.

6.6.3 A 120-V Battery-Backup Inverter with a 240-V Straight Grid-Connected Inverter

Figure 6.12 shows an ac-coupled system that uses a 240-V straight grid-connected inverter and a 120-V output battery-backup inverter. An important design challenge

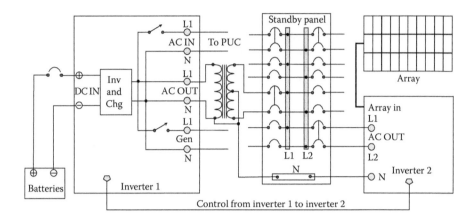

Figure 6.12 Basic ac-coupled system with 120-V battery-backup inverter and 240-V straight grid-connected inverter.

for this system is to meet *NEC* requirements for line-to-line and line-to-neutral voltages. When an inverter with 120-V output is used to simulate the utility for 240-V output inverters, a transformer must be used. Electrically speaking, the transformer can be either a double-winding transformer with isolated primary and secondary windings, as shown in Figure 6.12, or an autotransformer that has two common terminals between primary and secondary, as shown in Figure 6.13.

When a transformer with isolated windings is used, usually the center tap of the secondary winding is used as a common neutral for a 120/240-V output. Since, in the ideal case, the output power of the transformer will be equal to the input power, if the output supplies a 240-V load, the output current will be half the input current and the rated power of the isolated winding transformer will need to be equal to the power rating of the battery-backup inverter.

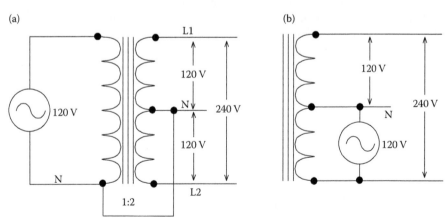

Figure 6.13 Comparison of isolated winding and autotransformer used to achieve 120/240 V from 120 V. (a) Isolated primary and secondary. (b) Autotransformer.

If an autotransformer is used, care must be taken to use the center-tapped winding of the autotransformer as the neutral connection as shown in Figure 6.13. This means that the input and output circuits of the autotransformer will have a common neutral and a common hot leg, with the second leg of the output connection carrying the other half of the 120/240 V secondary. Once again, half the available output current of the 120-V inverter is available to each side of the output of the transformer if the load is balanced, and the power rating of the autotransformer must equal the power rating of the 120-V inverter.

Just as in the previous example, the battery-backup inverter must be programmable to sell power to the utility when power is being generated by the PV array, the standby loads have been satisfied and the batteries are fully charged. It must also be programmable to shut down battery operation if the battery voltage is too low and to limit charging if the battery voltage becomes too high if the grid is disconnected. The inverter would be expected to sell to the grid when the grid is available and the batteries are fully charged. It is possible that auxiliary relay contacts on the battery-backup inverter will need to be used to achieve some of these features if they are not directly incorporated into the inverter. The possible operating conditions of this system are the same as the possible operating conditions of the previous example.

6.6.4 A 120/208-V Three-Phase AC-Coupled System

In larger buildings, it is common for the service to be at 277/480 V three-phase. This voltage is then transformed down to 120/208 V to serve lower voltage loads, while the medium voltage is used for 277-V lighting or 480-V motors. The 120/208 V supply is sometimes delivered as a single-phase, three-wire supply to a single-phase distribution panel and sometimes is fed as a three-phase, four-wire supply to a three-phase distribution panel. For the single-phase supply, the phases remain 120° out of phase, rather than 180° out of phase, which accounts for the 120 V line-to-neutral voltage and the 208 V line-to-line voltage. Although essentially all straight grid-connected inverters will operate at either 208 or 240 V (and sometimes 277 and even 480 V), only a few battery-backup inverters can be controlled to operate 120° out of phase if the utility fails. Thus, the primary requirement for a three-phase system is to have battery-backup inverters that can produce a three-phase supply that has the same phase sequence as the utility line so that any three-phase motors on the system will not change rotation direction if the utility is down. Thus, the battery-backup inverters need to remember the phase sequence in the event of utility power loss. Figure 6.14 shows a balanced system that uses three single-phase 208-V straight grid-connected inverters and three 120-V battery-backup inverters. Once again, the design constraints and possible operation modes are the same as those of previous examples.

With three-phase supplies, however, several other interesting options exist. The reader may be familiar with the open delta utility transformer configuration in which a proportionately larger, 120/240-V center-tapped, transformer is used to supply a predominance of 120/240-V single-phase loads. If this transformer has Utility Phases A and B connected to its primary, then a second, smaller transformer with a 240-V secondary winding is connected between either Phases B and C or between C and A,

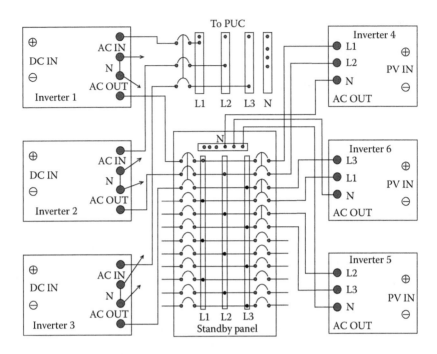

Figure 6.14 Three-phase battery-backup ac-connected system. Grounding conductors and control wiring not shown.

such that its output will be 120° out of phase with the larger transformer. This transformer configuration is shown in Figure 6.15a. Note that if the voltage drop from A to B is 240∠0°, then the voltage drop from B to C will be 240∠–120° and the voltage drop from C to A will be 240∠120°, for a positive phase sequence. The phasor diagram of these voltages is shown in Figure 6.15b. Thus, a three-phase set of voltages can be produced with two transformers. This suggests that two inverters with the proper phase relationships between their outputs could also produce a three-phase output. This should be true for either the battery-backup inverters or the straight grid-connected inverters as long as they are properly synchronized. Once again, before attempting to design any such system, it is important to check installation manuals and application notes for any idiosyncrasies in the operation of the inverters. It is also important to verify whether the three-phase supply is a balanced 120/208-V system or a 120/240-V delta system with a hi leg (Phase C in Figure 6.15b).

6.7 BATTERY CONNECTIONS

6.7.1 Lead-Acid Connections

An important, but often overlooked, component of good PV system design and installation is the proper connection of batteries to ensure a balanced current flow

BATTERY-BACKUP GRID-CONNECTED PHOTOVOLTAIC SYSTEMS 287

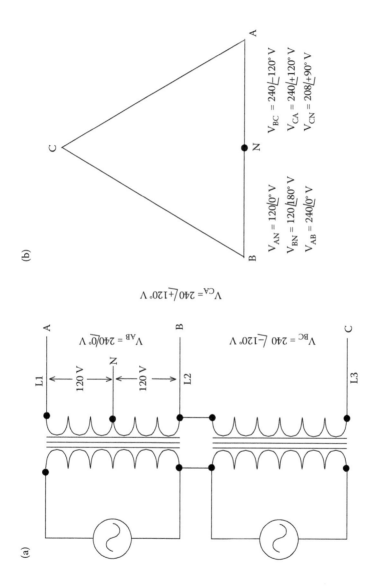

Figure 6.15 Open delta three-phase system. (a) Transformer configuration. (b) Phasor voltage relationships.

in all batteries in the system. If connecting wires did not have resistance, the manner in which batteries are connected would be relatively unimportant. But wire does have resistance, and therefore one needs to consider this resistance when hooking up batteries. In fact, even the terminal lugs have resistance, but this resistance is more difficult to characterize, since it will depend upon the specific lug type and how tightly the lug is connected to the wire and to the battery. Connecting lug resistance will also increase over time if any corrosion should occur at the lug. In the examples to follow, the connecting lug resistance will be assumed to be incorporated into the Thevenin equivalent (internal) resistance of the batteries and it will be assumed that the battery system consists of either flooded or sealed lead-acid batteries.

Figure 6.16 shows three possible ways to connect eight batteries in a series-parallel configuration. If the batteries are 12-V batteries, the system will produce 24 V. Note that Options 1 and 2 show battery-to-battery parallel connecting wires to be of equal length, and thus of equal resistance. The series-connecting wires are also of equal length. If 2/0 (67.43 mm^2) copper cables are used, and if $\ell_1 = \ell_2 = 1$ ft (30.5 cm), then the cable resistance will be 0.0000967 Ω for each of these 1-ft lengths of cable. Typically, the distance from battery to inverter input is about 6 ft. Thus, if $\ell_3 = 6$ ft (1.83 m), the resistance will be 0.00058 Ω.

It is interesting to calculate the currents that will flow under charging and discharging conditions in the series battery strings for Connection Options 1 and 2 if all batteries in the system are identical. For example, consider lead-acid batteries for which the open-circuit battery voltages are all 12.60 V and the Thevenin equivalent resistances of the batteries are 0.01 Ω each. According to Table 3.1, this would mean the batteries are about 25% discharged. Note that these parameters suggest a possible short-circuit current of 1260 A.

Consider first the charging situation. Figure 6.17 shows the equivalent circuit for Option 1. Note that if the current source negative lead is connected to Point A rather than to Point D, then the circuit will be equivalent to Option 2. Setting I = 60 A as a nominal value for either charge or discharge and enlisting the assistance of a convenient network analysis program yields the results for charging and discharging for Options 1 and 2 as shown in Table 6.11.

It is thus evident that Option 2 should be the preferred option for several reasons. First of all, the currents are more closely balanced for all series strings of batteries. Furthermore, the charging currents are equal to the discharging currents, so even though the A and the D batteries are cycled somewhat deeper than the B and the C batteries, the starting and ending points of a full cycle of charge and discharge are the same for all the batteries. Furthermore, when the batteries are in a state of higher discharge, the cell voltage decreases and the Thevenin equivalent resistance increases such that the rate of discharge of the batteries tends to be self-regulated. In other words, as the batteries become more discharged than a parallel set, the batteries at higher charge supply more current to the load than the batteries at lower charge. Under charging conditions, the batteries at lower charge levels should tend to charge faster, depending on their internal resistance.

The problems with the Option 1 connection are also somewhat mitigated by batteries at a lower SOC delivering less current. But there is still a greater difference

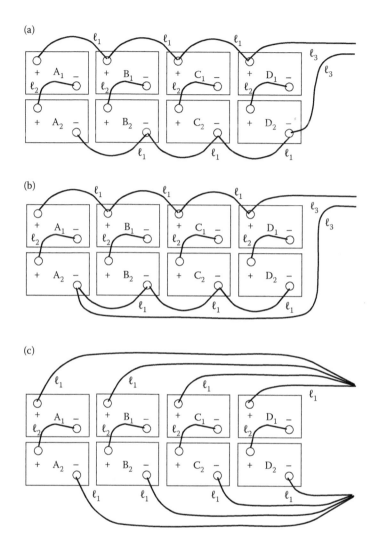

Figure 6.16 Three possible battery hookup configurations. (a) Option 1. (b) Option 2. (c) Option 3.

in charging and discharging currents for Option 1 than for Option 2. The Option 1 connection causes the "D" batteries to cycle between a higher level of charge and a lower level of discharge than the "A," "B," or "C" batteries. Over many cycles, this may cause the "D" batteries to degrade sooner than the other batteries. The battery life versus average daily discharge curve of Figure 3.11 illustrates this point.

Option 3 presents a somewhat different approach to equalizing battery currents. For this option, smaller wire, such as #6, is used to connect to each series battery string. The idea is to terminate the #6 ends at terminal blocks near the inverter and then use very short lengths of larger wire between the terminal blocks and the

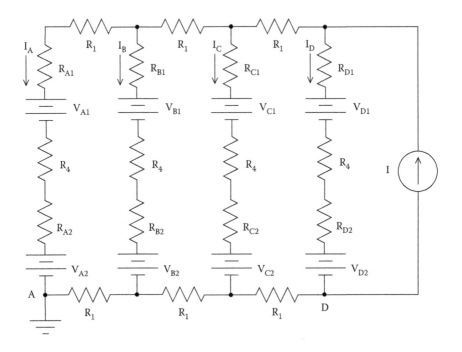

Figure 6.17 Equivalent circuit for battery connection Option 1.

Table 6.11 Comparison of Charging and Discharging Currents for Options 1 and 2

Situation	I_A (A)	I_B (A)	I_C (A)	I_D (A)
Option 1 charge	14.65	14.79	15.07	15.50
Option 1 discharge	14.65	14.79	15.07	15.50
Option 2 charge	15.07	14.93	14.93	15.07
Option 2 discharge	15.07	14.93	14.93	15.07

inverter. If all the lengths of wire are the same on a round-trip basis, then each string of batteries will experience the same voltage drop in the battery cabling, and currents will be exactly balanced under charge or discharge conditions. The higher resistance of the #6 wire tends to produce a current limiting effect. So if for some reason one series set tends to discharge or charge at a higher rate than another series set, the discharge or charge current will be limited by the resistance of the connecting wires, thus creating a balancing effect.

The disadvantage of Option 3 is that each individual battery string will require a separate fuse or circuit breaker, similar to the source-circuit fuses in a PV array, but much larger. So in this case, there would be four battery disconnects instead of the single disconnect needed for Options 1 and 2. However, it is possible that the higher resistance of the connecting wires of Option 3 will limit the battery short-circuit current sufficiently that fuses or circuit breakers of a lower interrupting capacity (AIC), and thus lower cost, can be used. Recall that the interrupting capacity of an overcurrent device is a measure of the ability of the device to interrupt the circuit under short-circuit conditions, where there is a possibility that arcing may occur across open switch contacts. As a result, the overall cost of Option 3 may still be attractive. However, *NEC* 230.71(A) limits the number of switches allowed to disconnect a circuit to no more than six. An interesting approach to this problem is to use two- or three-pole circuit breakers, since each device has a single lever for operating the device and hence will count as a single disconnect even though it will disconnect two or three strings of batteries with a single flip of the switch handle.

In summary, the key to optimizing battery performance is proper cabling. The use of equal lengths of cables is essential. Problems 6.13 through 6.15 offer the reader an opportunity to explore the effect of unequal cable lengths and battery parameters on battery system charging and discharging. Option 2 is clearly better than Option 1 for battery wiring, and Option 3 is also potentially attractive.

Furthermore, new batteries tend to have lower internal resistance than older ones of the same type, resulting in greater unbalance of currents in parallel strings of new batteries. This generally means that if one battery is replaced, all should be replaced, probably with the exception of an early failure due to a battery defect. In any case, if not all batteries are replaced, the condition of all batteries should be carefully checked to ensure that currents are well balanced and that individual battery terminal voltages are equal. If this is not the case, future premature failures may continue to occur. A clamp-on dc ammeter is most useful for this sort of analysis.

6.7.2 Other Battery Systems

In lead-acid systems, generally balancing the system involves attempting to keep individual series strings of batteries at the same SOC as defined by string voltage. More modern systems, such as LFP systems, are generally shipped with accompanying battery management systems. The battery management systems are wired and monitored at the cell level and full charge is measured in terms of accumulated Ah of charge per cell as measured by integrating the current into or out of the cells. In a lead-acid battery system, if one cell goes bad, it affects the entire series battery combination, just as if one PV cell is shaded or defective, it can affect the entire source circuit if no bypass diodes are present in the module. In an LFP system, if one cell goes bad, only that cell is affected. The other cells have alternate charging and discharging paths via the battery management system.

Some battery-backup inverters now advertise compatibility with LFP battery systems. The question to ask in these situations is whether the compatibility involves monitoring SOC or monitoring voltage for use in controlling the charging and discharging process, starting or stopping backup fossil generators, or any other intended function of the monitoring information.

PROBLEMS

6.1 Explain the operation of the inverter bypass switch in the ac component enclosure of the example in Figure 6.3.

6.2 Assume that a three-wire branch circuit is derived from a 120/208-V distribution panel where the voltage on one phase is $120\angle 0°$ and the voltage on the other phase is $120\angle -120°$.
  a. Prove that, for resistive loads on either side of the circuit, the current in the neutral will not exceed the current in either phase.
  b. Prove that when the loads on each phase are equal, that the magnitude of the neutral current will equal the magnitude of the current in either phase.

6.3 Show that in the case of a three-phase, four-wire set of branch circuits,
  a. The current in the neutral will not exceed the current in either of the phases if only the fundamental component of the current is present on each branch circuit.
  b. If a balanced, nonlinear load across all three phases causes a third harmonic current component to flow in each phase conductor, that the neutral current can exceed the individual line current.

6.4 Design a stand-off mounting system for the array of Section 6.4 if the 24 modules are arranged in rows of 6, 8, and 10 modules. The dimensions of the modules are 39.3 inch × 66 inch (1 m × 1.68 m). You may assume a wind speed of 115 mph and exposure category B with all modules located in either Roof Zone 1 or Roof Zone 2 of the Category II building. The mean roof height is 20 ft (6.1 m) and the roof slope is 23°.
  a. Specify the rail lengths and the numbers and locations of mounting brackets.
  b. Calculate the total lifting force on each section of array and the force on each mounting bracket.
  c. Determine the necessary penetration depth of the threads of a 5/16-inch lag screw if the trusses are made of Douglas fir.

6.5 Find a suitable source-circuit combiner box that meets the requirements of Section 6.4.3. List the specifications of the unit that prove that the unit is suitable for the application.

6.6 Determine the size of schedule 40 PVC conduit that will be needed to accommodate nine #10 THWN-2 conductors between rooftop junction box and source-circuit combiner box. Then, assuming that the source-circuit wiring will be routed through an attic to the power conditioning equipment and array disconnect inside the structure, determine the size of electro-metallic tubing (EMT) that will be required to accommodate the nine wires. Finally, if a metallic pull box is a part of the raceway between rooftop junction box and source-circuit combiner box, recommend a suitable box size if the wires entering the box pass through the box without splicing. You may want to refer to *NEC* 314.16(B)(1).

6.7 Verify that the modules of the array of Section 6.5 will have maximum operating temperatures of 70.3°C when ambient temperatures reach their maximum design values and the irradiance level is 1000 W/m^2.

6.8 For the 19.2-kW array of Section 6.5, calculate the necessary distance between rows if it is desired to have 7 hours of unshaded operation on December 21.

6.9 Verify the lengths of the open-wire extension cables that will be needed for each of the source circuits shown in Figure 6.9.

6.10 Calculate the overall voltage drops for each of the eight source circuits of the 19.2-kW array of Section 6.5. You might want to construct a table to summarize all source-circuit lengths and wire sizes similar to Table 6.4.

6.11 Search the Web for at least one battery-backup inverter that can be configured to produce a three-phase output. If necessary, multiple units may be used.

BATTERY-BACKUP GRID-CONNECTED PHOTOVOLTAIC SYSTEMS   293

6.12 Prove that the voltage between Phase C and neutral as shown in Figure 6.15b will be 208∠90°.

6.13 Use a network analysis program to explore the following effects of unequal battery cable lengths for Battery Connection Options 1, 2, and 3 of Figure 6.17. Solve using charging/discharging currents of 60 A.
   a. In Option 1, change l_2 as follows: from A_1 to A_2, 1 ft; from B_1 to B_2, 2 ft; from C_1 to C_2, 3 ft; and from D_1 to D_2, 4 ft.
   b. In Option 2, use the replacement values for l_2 that were specified in Part a.
   c. In Option 3, assume the l_1 cables are #6 Cu and the distance from the charging/discharging point to the A_1 and A_2 batteries is 12 ft; the distance from the charging/discharging point to the B_1 and B_2 batteries is 10 ft; the distance from the charging/discharging point to the C_1 and C_2 batteries is 8 ft; and the distance from the charging/discharging point to the D_1 and D_2 batteries is 6 ft.

6.14 Use a network analysis program to explore the effect of unequal open-circuit battery voltages that might be expected for charge levels of 50% and 75%. Assume, for example, that the A batteries of Figure 6.16 are charged to 50% and the rest of the batteries are charged to 75%. Then solve for the battery string currents in Options 1, 2, and 3 under charge and discharge conditions. Use charge/discharge currents of 60 A. Cable lengths remain unchanged.

DESIGN PROBLEMS

6.15 Design a nominal 12.0 kW, customer-owned, commercial rooftop utility-interactive PV system with a 120/240-V output and 30 kWh of battery backup. Specify all components and connections.

6.16 Using PV roofing tiles that are characterized by $V_{OC} = 3.7$ V, $I_{SC} = 8.0$ A, $V_{mp} = 2.8$ V, $I_{mp} = 7.8$ A, $\Delta V_{OC}/\Delta T = -0.4\%/°C$, and $\Delta V_m/\Delta T = -0.5\%/°C$,
   a. Determine the maximum number of tiles that can be installed in series with $V_{OC} < 150$ V, for the series combination if the lowest annual temperature will not fall below −20°C.
   b. Determine the minimum number of tiles that can be installed in series and still have $V_{mp} > 65$ V for the string if the highest anticipated module (cell) temperature will be 70°C.
   c. Determine the range of power available from a single-source circuit at STC, and then determine how many tiles (modules) will be needed to achieve an array power of approximately 3 kW at STC. Choose source circuits to achieve a power as close to 3 kW as possible without letting the values of V_{OC} or V_{mp} fall out of range for the system charge controller(s).
   d. Select a utility-interactive battery-backup inverter, charge controller(s), and batteries to provide a minimum of 12 kWh of storage with an 80% depth of discharge of the batteries at a C/24 discharge rate. The inverter must supply a maximum load of 3 kW.
   e. Complete the system electrical design and draw a schematic showing all wire sizes, disconnects, and overcurrent protection.
   f. If the system is located in Birmingham, AL, estimate the monthly and annual system performance if the roof has a tilt of 30°, facing directly south.

6.17 Design a 5-kW STC, ac-coupled, grid-connected, battery-backup PV system that will supply power to a maximum load of 4 kW at 120/240 V ac. The system should have 16 kWh of battery backup at 80% depth of discharge. Use a battery-backup inverter that has a 120/240-V AC OUT port. Be sure to specify all the control functions that need to occur to avoid overcharging or overdischarging of the batteries.

6.18 Design a battery-backup, grid-connected PV system for use in Nashville, TN, that will provide adequate power to supply a collection of 120- and 240-V ac loads that are expected to consume 15 kWh/day during summer months. The maximum power to be consumed at any one time will be 4 kW. The system can be ground mounted at whatever angle the designer specifies, on a mounting system to be specified by the designer. Show the system electrical schematic diagram and specify all system components, including the array mount. Upon completion of the design, run NREL SAM and show the expected monthly and annual system performance.

6.19 a. Find an acceptable, listed method of providing rapid shutdown for one of the previous design problems or one of the examples in this chapter.

b. Show how and where arc fault protection can be incorporated into one of the previous design problems or one of the examples in this chapter.

REFERENCES

1. UL 1741, *Inverters, Converters and Controllers for Use in Independent Power Systems*, Underwriters Laboratories, Inc., Northbrook, IL, 2005.
2. *NFPA 70 National Electrical Code*, 2014 Ed., National Fire Protection Association, Quincy, MA, 2013.
3. System Advisor 2016 (SAM 2016.3.14), National Renewable Energy Laboratory, Golden, CO, 2016.
4. ASCE Standard 7-10, *Minimum Design Loads for Buildings and Other Structures*, American Society of Civil Engineers, Reston, VA, 2010.
5. FLEXmax™60, *FLEXmax™80 Maximum Power Point Tracking Charge Controller User's Manual*, Installation and Programming, Outback Power Systems, Inc., Arlington, WA, 2008, www.outbackpower.com/pdf/manuals/flexmax.pdf.
6. NOAA, National Weather Service Forecast Office, Burlington, VT, http://www.erh.noaa.gov/btv/climo/BTV/extremes/extremetemps.shtml.
7. AEE solar article on battery-backup PV systems with references to several different manufacturers, http://www.aeesolar.com/grid-tied-solar-systems-with-backup-power.
8. NOAA, Lexington, KY climate information, http://www.crh.noaa.gov/images/lmk/lexington_cli_pdf.
9. NOAA, Information on snowfall extremes for Lexington, KY, http://www.crh.noaa.gov/lmk/?n=lexsnowextremes.
10. CSAC Snow and Avalanche Glossary, http://www.avalanche-center.org/Education/glossary/density.php.
11. Manufacturer information on automatic generator start and battery state of charge monitors, http://solar.schneider-electric.com/.
12. OutBack Power Radian Series Inverter/Charger GS8048A Installation Manual, http://www.outbackpower.com/downloads/documents/radian_8048a_4048a/gs_8048a_4048a_install.pdf.

CHAPTER 7

Stand-Alone Photovoltaic Systems

7.1 INTRODUCTION

There are many similarities between stand-alone PV systems and grid-connected, battery-backup systems. As previously discussed, when a grid-connected, battery-backup system is disconnected from the grid, it becomes a stand-alone system until grid power is restored.

To some extent, stand-alone systems have more flexibility, since if an inverter is used, it does not necessarily need to comply with UL 1741. Stand-alone systems may have the option of being designed for dc only, for a mix of dc and ac loads, or for ac loads only. Some stand-alone systems will have battery storage and some will not. Grid-connected, battery-backup systems are normally limited to ac loads only.

On the other, in the previous chapter design examples have shown that the tilt of the array for a grid-connected system is generally not critical for annual performance, as long as it is within about 15° of latitude. But stand-alone systems may well need to be designed for seasonal performance, meaning more critical consideration of array tilt angle since no grid is available for backup. In fact, in the far north or the far south, it may be necessary to supplement the energy production of the PV system with another form of generation, such as fossil or wind.

In the grid-connected, battery-backup examples, only 1 day of battery storage was normally included in the design. In stand-alone systems, it is common to have several days of backup, again, since the PV system is the only source of electrical power year-around.

Furthermore, monthly electrical loads may differ. This means a more critical consideration of the loads that must be served by the PV system, often on a monthly basis. For example, for solar street lighting, there is generally abundant sunlight in the summer to charge batteries, but the light is on less because the sun is up longer. The opposite is true for winter. Thus, the array will probably need to be tilted for optimal winter performance with the hope that the winter tilt will provide adequate battery charging for summer use.

This chapter will introduce the design process for several complete self-contained PV systems. The first system will be a simple dc fan powered by a PV module, followed by a dc water pumping system that incorporates a linear current booster (LCB). A parking lot lighting system, a cathodic protection system, and a highway sign will also be designed as dc systems. Finally, a collection of ac systems will be designed, including a critical need ac refrigeration system, a remote cabin, and a hybrid permanent residence that will operate off the grid. The hybrid residence will be designed with an emphasis on the process of selecting an appropriate mix of PV and propane electrical generation.

In any stand-alone design, the first task is normally to determine the load. Often there are several choices. For example, a choice might be made between an inexpensive incandescent lamp and a more-expensive fluorescent or LED lamp. In almost all such cases, the choice of the more efficient load is more cost-effective. In fact, the most-efficient load may not necessarily be the most expensive one.

The load voltage, current, and power are needed for proper sizing of fuses, wires, batteries, and other system components. In some cases, not all of these quantities are given. For example, the specifications for a refrigerator may only give the voltage and kWh per month, or Ah per day. In these cases, it may be necessary to either calculate or estimate the other parameters.

Once the load has been determined, then the amount of battery backup needs to be determined. Some systems will not need batteries, some will have minimal storage, and some will require sufficient battery storage to meet critical performance requirements in which the system must operate more than 99% of the time. It will also be the responsibility of the engineer to determine whether to use flooded lead-acid batteries or to use maintenance-free batteries at nearly twice the initial cost, or, for that matter, to use new technology batteries such as lithium iron phosphate (LFP) to achieve a greater energy density per unit weight or per unit volume.

After battery selection, the size of the PV array must be determined. Then the electronic components of the system, such as charge controllers and inverters, are selected. Finally, the BOS components are selected, including the array mounts, the wiring, switches, fuses, battery compartments, lightning protection, and, perhaps, monitoring instrumentation.

While computer programs may be available for assistance in system design, as in previous chapters, emphasis in this chapter will be placed on sound, common-sense reasoning and the use of good engineering judgment. In order to keep the design examples as realistic as possible, many of the system components described in the examples of this chapter have been found on an assortment of websites. It is important for the reader to note that the use of a particular component or website in an example does not constitute an endorsement of the component or the company. Vendors and component availability change on a daily basis, so when a real system is designed, the designer should carry out an extensive search for the components that best meet the design goals. Hopefully the procedures for component selection outlined in this chapter will prove to be useful when the opportunity to design a real system presents itself.

7.2 THE SIMPLEST CONFIGURATION: MODULE AND FAN

Figure 7.1 shows the simplest of PV systems, a fan motor connected to a PV module. The figure also shows the superimposed performance (I–V) characteristics of the fan and the module. The operation is simple: as the sun shines brighter, the fan turns faster. If the designer has no concern for the exact quantity of air moved, the design becomes nearly trivial. However, if the amount of air moved must meet a code requirement or other constraint, then it will be necessary to consider the design in more detail. Obviously, this system is useful only during daytime hours. Nighttime operation will require adding battery storage to the system.

The system operation point is determined by the intersection of the performance characteristics. Note that as the sun shines brighter, making more PV current and voltage available, the fan consumes more power. It is reasonable to assume that as the fan consumes more power it will move more air.

Perhaps the second observation the reader will make regarding the performance characteristics intersections in Figure 7.1 is that the module is not operating anywhere near maximum power at low light levels. As irradiance approaches 1 kW/m^2, the fan operating point is close to the module maximum power point. If a module having higher short circuit current is used, the fan will remain at a more constant speed as the fan current exceeds 3.5 A, but at high illumination levels, the module will be acting more as a voltage source and again will deliver only a fraction of its maximum power capability. Hence, the designer must decide how much air movement is needed at various irradiance levels, and choose the module and fan accordingly. So, even in this relatively simple design example, the designer must use discretion. Larger modules will cost more, but will deliver more air at lower irradiance levels.

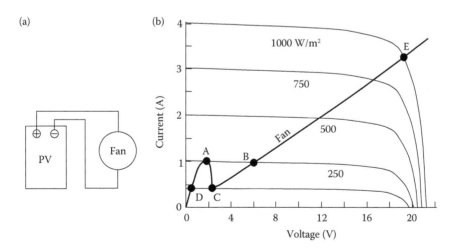

Figure 7.1 A simple, PV-powered fan, showing performance characteristics for fan and PV module. (a) Schematic diagram. (b) Fan and module performance characteristics.

Figure 7.1b also shows the hysteresis effect encountered in starting the fan. Under stalled rotor conditions, the fan motor does not produce a back EMF and thus the fan will draw stalled rotor current until sufficient armature current is present to overcome the starting torque. The irradiance level at Point A on the curve is just adequate to provide this current, and the operating point then jumps to Point B. As the irradiance level continues to increase, the operating point moves toward Point E. When irradiance levels decrease, fan performance follows the fan characteristic to Point C, after which the fan stalls and the operating point jumps to Point D and eventually approaches the origin as darkness falls.

Another question for the designer to ask is whether it would be better to use a different fan to meet the design requirements. The obvious answer is "maybe." And that is what makes the design of PV systems so much fun. It should be clear from Figure 7.1b that regardless of the choice of fan or module, there will be a significant power mismatch over a relatively wide range of irradiance. Thus, no matter what the choice, there will be some portion of the fan or PV characteristic where maximum power will not be transferred to the fan. If it is desired to optimize fan power for all illumination levels, a maximum power point tracker (MPPT) will need to be incorporated into the design.

The MPPT can be particularly useful at irradiance levels between the start and stop irradiance levels of Figure 7.1b, where it will enable the fan to start at a lower irradiance level and to stop at a lower irradiance level, with greater air flow at irradiance levels between these points. Here the interesting part of the trade-off is whether including an MPPT with a smaller module will cost less than using a larger module to obtain comparable system performance.

For example, the PV curve upon which Points A and B lie has an approximate power available at the maximum power point of 18.13762 W. But at the intersection of the fan curve and the PV curve, the fan is using only 5.78623 W, which means that an additional 12.35139 W is available, but not used. So the fan is only using 31.9% of the available power. In the ideal case, $P \propto v^3$, where v is the air velocity moved by the fan and P is the power consumed by the fan. So as the power triples, the air velocity increases by a factor of 1.44. Thus, if the simple fan system does not move enough air at low sunlight levels, then an MPPT can make a difference in fan performance.

7.3 A PV-POWERED WATER PUMPING SYSTEM

7.3.1 Introduction

Another common stand-alone PV application is water pumping, especially when the water to be pumped is a long distance from a utility grid. Water pumping applications do not necessarily require battery backup unless the water source will not produce an adequate supply of water to meet the pumping needs during daylight hours. Another reason for using batteries is that the water can be pumped over a longer time at a slower rate with a smaller, less costly pump. Under these circumstances, it may be necessary to charge a battery so the pump can run for an extended period. When the water supply can meet the pumping capacity of the system, then it may be more cost-effective to pump all the water the pump is capable of delivering and store any excess in a storage tank. In effect, the storage of water replaces the storage of electricity in batteries. It still represents conversion of kinetic to potential energy.

When designing a water pumping system, it is necessary to determine a number of parameters in order to properly size the system components. First of all, the daily water needs must be determined. Secondly, the source must be characterized in terms of available water and vertical distance over which the water must be pumped. Once these factors are known, along with the number of hours per day available for pumping, the pumping rate can be determined. The pumping rate along with the pumping height equates to the pumping power, once again the product of a pressure quantity with a flow quantity.

Once the size of the pump motor is known, the ampere-hour requirements of the motor can be determined, and, finally, the size of the PV array needed to provide the ampere-hours can be determined. As in the fan example, inclusion of an LCB or an MPPT extends the useful pumping time of the pump motor and enables the use of a smaller motor and a smaller array that is utilized more efficiently.

7.3.2 Selection of System Components

To quantify the pumping problem, it is useful to note that a gallon (3.785 L) of water weighs 8.35 lb (3.73 kg) and that one horsepower = 550 ft-lb/second = 746 W, assuming 100% conversion efficiency. This means that pumping a gallon of water to a height of 1 ft involves 8.35 ft-lb of work. In the MKS system, converting gallons to liters and feet to meters gives the result that pumping a liter of water to a height of 1 m requires 7.23 ft-lb = 9.83 J.

Although this information can be used to determine the horsepower of the pump, it is more common to determine pump size from manufacturers' specifications when daily pumping requirements and pumping height are known. For the design of a pumping system, it is also useful to understand the pressure versus flow characteristics of piping as well as to understand that pumps can be designed with trade-offs between pressure and flow.

Piping friction loss is determined by the type and diameter pipe used, just as voltage drop is determined by the size and material of the wire used, although the relationship between pressure and flow for a water pipe tends to be somewhat more nonlinear than the I–V relationship for a wire. However, at relatively low flow rates, the flow versus pressure curve for a piping system can be approximated by a linear relationship. Figure 7.2 shows pressure versus flow curves for several sizes of piping [1].

Although it is reasonably straightforward to select the horsepower for a pump, it is somewhat more involved to select the pumping system that will perform at maximum efficiency. The reason is that some pumps are designed to deliver higher pressure than others. The pumps that can deliver higher pressure are needed for lifting

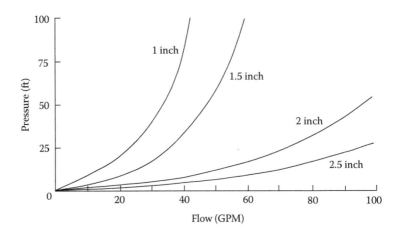

Figure 7.2 Pressure versus flow curves for equal lengths of piping of different diameters.

water to greater heights. Figure 7.3 shows performance curves for two pumps of equal horsepower, one of which is a high head (pressure) pump and the other is a medium head pump [2].

Note that the medium head pump will deliver more volume than the high head pump at low pressure, but the high head pump will overcome a greater pumping height. Note also that the performance of a pump depends on the speed at which the pump is operated. If the pump speed decreases, both pressure and flow capacity decrease. It is thus important to select a pump that will be able to overcome the lift requirement under low sun conditions, unless energy storage is included in the design.

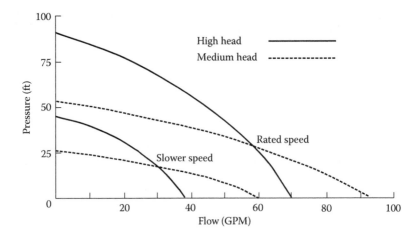

Figure 7.3 High head and medium head pump performance characteristics at two different speeds.

STAND-ALONE PHOTOVOLTAIC SYSTEMS

7.3.3 Design Approach for Simple Pumping System

7.3.3.1 Pump Selection

If pumping data are available from a pump manufacturer, then all the designer needs to know is the daily amount of water needed and the overall pumping height. The daily water needs can be converted to gallons or liters per minute over the time the pump will operate and a suitable pump can be selected from manufacturers' tables.

Ideally, the desired system will have the lowest cost and the highest reliability. Component choice to achieve this involves deciding upon smaller versus larger pump, dc versus ac pump, batteries versus no batteries, LCB versus inverter, and maybe even tracker versus fixed mount.

The minimum pump size is determined by assuming continuous pumping, which would require battery storage if the pump is to run during the night. Using the 2000-gallon/day (GPD) water requirement means pumping an average of 83.33 gallons/hour (GPH) over a 24-hour period. So the pump must be able to deliver 83.33 GPH with a lift of 210 ft. Table 7.1 tabulates the GPH delivered at specific pumping heights when a particular dc submersible pump is operated at 24 V. As expected, the pump current increases as the pumping height increases.

For this pump, the question is whether it will deliver 2000 GPD at a pumping height of 210 ft. Linear interpolation between the 200- and 230-ft data results in GPD(210 ft) = 2112, which is close to the desired pumping rate, since the desired 2000 gallons will be delivered in 22.7 hours. Thus, this pump will meet the design criteria, so in order to evaluate the overall system, the rest of the system can now be designed.

Since the pump will be operating continuously, it will be necessary to provide battery storage. Battery storage generally requires the use of a charge controller to optimize the battery charging process and to prevent overcharging of the batteries.

Table 7.1 Flow versus Lift Showing Operating Conditions for a Specific 24-V DC Submersible Pump

Lift (ft)	Flow (gph)	Current (A)	Power (W)	Gallon/Day	Ah/Day	kWh/Day
20	117	1.5	36	2808	36	0.864
40	114	1.7	40.8	2736	41	0.979
60	109	2.1	50.4	2616	50	1.210
80	106	2.4	57.6	2544	58	1.382
100	103	2.6	62.4	2472	62	1.498
120	101	2.8	67.2	2424	67	1.613
140	99	3.1	74.4	2376	74	1.786
160	98	3.3	79.2	2352	79	1.901
180	93	3.6	86.4	2232	86	2.074
200	91	3.8	91.2	2184	91	2.189
230	82	4.1	98.4	1968	98	2.362

If the pump were an ac pump, it would require an inverter and the inverter would likely have built-in protection for overdischarging the batteries. The dc pump does not have an inverter, so some sort of switching will be needed to protect the batteries from overdischarge. The system will also likely incorporate a water storage tank and may also use a float switch on the tank to prevent the tank from overflowing.

Table 7.1 also shows the daily energy consumption of the pump for various pumping heights. Again, linear interpolation between the 200-ft (61 m) and 230-ft (70.1 m) energy requirements results in a daily energy requirement of 2.25 kWh/day supplied to the pump. Thus, the PV array must produce sufficient energy, such that losses in wiring, charge controller, and batteries are taken into account; there will still be 2.25 kWh/day left over for the pump.

7.3.3.2 Battery Selection

Since batteries are sized in Ah, the first step is to convert the 2.25 kWh/day to Ah/day. This is accomplished by simply converting kWh to Wh and then dividing the result by the battery voltage, yielding a requirement of 93.75 Ah/day at 24 V. Allowing for 80% depth of discharge of the batteries, a single day of storage will require 93.75/0.8 = 117 Ah of battery capacity. Since most deep-discharge solar batteries are either 6 or 12 V units, the next step is to look for either four 6 V units or two 12 V units having 117 Ah ratings. The question is to what discharge rate does the 117 Ah rating apply? The answer is determined by the number of storage days (days of autonomy) that are desired. One 12-V AGM battery, for example, is rated at 108 Ah for 24-hour discharge, 118 Ah for 48-hour discharge, and 122 Ah for 72-hour discharge. So, if 2 or 3 days of storage are desired, then this battery will be acceptable, with two units connected in series and additional sets of two in parallel with the first set for the extra days.

Another choice of batteries, however, could be a 12-V AGM battery rated at 235 Ah at a 48-hour discharge rate. With this battery, a total of two in series will provide 2 days of autonomy. So it boils down to overall cost of batteries, battery cables, and a battery compartment.

7.3.3.3 Module Selection

Module selection will depend upon the location of the system and the tilt of the array. If the system is located near Lubbock, TX, then an analysis using the PVWatts model in NREL SAM yields the data of Table 7.2 for an array tilted at 40° [3]. Note that this data is a result of assuming a grid-connected system, which calculates monthly performance data in terms of kWh delivered to the grid, that is, at the inverter output. Assuming a 96% inverter efficiency, the inverter output is based upon a 91% overall conversion efficiency, which is close to what would be expected as the output of an MPPT charge controller. Thus, if a battery round-trip efficiency loss of 10% is assumed, the overall efficiency from array output to battery output will be $0.91 \times 0.9 = 0.82$, or 82%. Column 3 of Table 7.2 shows this figure on a monthly basis. Column 4 shows the monthly kWh consumption of the pump and

STAND-ALONE PHOTOVOLTAIC SYSTEMS

Table 7.2 Lubbock, TX Monthly kWh Production for PV Array at 40° Tilt

Month	Inv Out (kWh/kW)	Batt Out (kWh/kW)	Load (kWh/Month)	Min Array (kW)
Jan	128	115.2	69.75	0.61
Feb	129	116.1	63.00	0.54
Mar	157	141.3	69.75	0.49
Apr	144	129.6	67.50	0.52
May	146	131.4	69.75	0.53
Jun	136	122.4	67.50	0.55
Jul	132	118.8	69.75	0.59
Aug	136	122.4	69.75	0.57
Sep	132	118.8	67.50	0.57
Oct	152	136.8	69.75	0.51
Nov	135	121.5	67.50	0.56
Dec	127	114.3	69.75	0.61

Column 5 shows the array size needed to provide the monthly kWh consumption. The table shows that for December and January, the array size needs to be a minimum of 0.61 kW to deliver 69.75 kWh to the pump. For all other months, the array can be smaller, which provides oversizing of the array during all other months if it is sized for December and January. If it is felt that this is too close for the summer months, additional PV power can be added to the array or the tilt can be lowered, which will also result in additional PV power required to meet the December and January pumping requirements.

Depending upon selection of charge controller, the array voltage can likely be somewhere between 40 and 150 V, just so the array power rating is at least 610 W.

A 610-W array is easily achieved with two 305-W modules in series. One such module is rated at 305 W with $V_{OC} = 46.3$ V, $V_m = 37.5$ V, $I_{SC} = 8.7$ A, and $I_m = 8.1$ A. So two in series will have $V_{OC} = 92.6$ V, $V_m = 75$ V, $I_{SC} = 8.7$ A, and $I_m = 8.1$ A.

7.3.3.4 Charge Controller Selection

If an MPPT charge controller with an input voltage range up to 150 V is used, then, for a 610-W array, the maximum charge controller output current will be approximately 30.8 A, allowing for a 97% efficiency of the charge controller and a 125% cloud focusing factor for array power. A number of 40-A MPPT charge controllers are available on the market, so at this point it is a matter of how many features are desired on the unit.

7.3.3.5 BOS and Completion of the System

Figure 7.4 shows a schematic of this system. Now that all the voltages and currents are known, it is left as an exercise for the reader to determine the sizes of wires and switches. Note that the intent of this design is to show one way to design a water pumping system. Just as the case of the simple PV fan, water pumping can also be

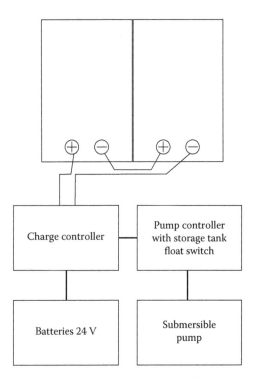

Figure 7.4 Electrical schematic diagram of water pumping system.

accomplished with an array and a pump and only the wire and piping to deliver the water. When it comes to a serious design, it is recommended that the designer explores more than one option for the system in order to achieve a system with maximum cost-effectiveness.

7.4 A PV-POWERED PARKING LOT LIGHTING SYSTEM

7.4.1 Determination of the Lighting Load

It should be evident that if a PV system is to power a parking lot lighting system, a battery storage system will be needed, since if the sun is shining, the lights are not needed, and vice versa. The first step is to determine the lighting load, followed with battery selection and, finally, the number and type of PV modules to use. In order to determine the lighting load in watts, it is first necessary to determine the amount of light needed and the area of the space to be illuminated. Hence, the design begins with the determination of the necessary illumination level.

While illumination levels could be measured in W/m^2, in the United States, illumination levels are most commonly measured in foot-candles. A foot-candle is the amount of light received at a distance of 1 ft from a standard candle. A standard

STAND-ALONE PHOTOVOLTAIC SYSTEMS

candle is a candle that emits a total amount of light equal to 4π lumens. The lumen is thus the basic quantity of light in the foot-candle system of measurement of light intensity. It compares to the coulomb in the electrostatic realm.

If a closed surface surrounds the standard candle, then all of the 4π lumens of light must ultimately pass through the surface. If the light from the source is emitted uniformly in all directions, and if a sphere of radius 1 ft is centered on the light source, then the light will be uniformly distributed over the surface of the sphere with a density of (4π lumens)/(4π ft^2). This light intensity of 1 lumen/ft^2 defines the foot-candle (f-c).

The Illumination Engineering Society publishes guidelines for illumination levels for various spaces [4]. For example, parking lot lighting should normally be lighted to an average illumination level of approximately 1 f-c, depending on the degree of security desired. A desk for normal work is generally adequately lighted with 50 f-c. Direct sunlight provides about 10,000 f-c [5].

The *luminous efficacy* of a source is a measure of the efficiency with which the source transforms electrical energy to light energy. It is measured in lumens per watt. Table 7.3 [6–8] shows the luminous efficacies for several light sources.

Determination of the wattage of light needed to accomplish a specific lighting task, then, will depend on the required illumination level and the area to be lighted. It also depends on the luminous efficacy of the source. Other factors include whether the available light can be directed only to the area where the light is needed and whether some of the light will be absorbed by walls or other absorbers, such as the light fixture itself, before it reaches the surface to be illuminated. Dust on the fixture and lamp also absorbs useful light output.

In addition to the intensity of light, the color temperature of light may also a factor to be considered. The color temperature of light refers to the equivalent spectral content of radiation from a blackbody at a particular temperature. The color temperatures with which the reader is most familiar are the 5800 K (daylight) temperature of the sun, which produces its characteristic white color, and the 2700 K (warm white) temperature of a tungsten incandescent light filament, which is more

Table 7.3 Approximate Luminous Efficacy for Several Light Sources

Source	Luminous Efficacy (L/W)	Lamp Lifetime (h)
25-W incandescent	8.6	2500
100-W incandescent	17.1	750
100-W long-life incandescent	16.0	1125
50-W quartz incandescent	19.0	2000
T-8 fluorescent	75–100	12,000–24,000+
Compact fluorescent	27–80	6000–10,000
Metal halide	80–115	10,000–20,000
High-pressure sodium	90–140	10,000–24,000+
3.6-W LED array	~130	100,000+

Source: *9200 Lamp Catalog*, 22nd Ed., GE Lighting, General Electric Co., 1995; *Lamp Specification and Application Guide*, Philips Lighting Co., Somerset, NJ, 1999; www.cetsolar.com.

toward orange. Not all light sources can be characterized by a color temperature, since the concept is based on blackbody radiators. Sources with discrete spectral components, such as lasers or gas discharge lamps, can be assigned equivalent color temperatures to indicate the temperature to which the spectrum of the source is most closely matched, but the color temperature is not a precise measure of the color of the source. For example, xenon produces a very white flash, which, on photographic film, appears to be close to the color of daylight. Although the output spectrum of a xenon lamp differs from the AM 1.5 solar spectrum, xenon lamps are commonly used in solar simulators with appropriate correction factors.

Consideration of color temperature can be an important factor in the choice of light sources. For example, low-pressure sodium has a very high luminous efficacy, but the light is composed primarily of the sodium d_2 lines. When low-pressure sodium sources are used, anything not yellow in color will not be accurately perceived, since, if a source does not contain a particular color, then that color cannot be reflected back to the eye to be perceived as such. For PV applications, generally the most popular and efficient sources are fluorescent, metal halide, high-pressure sodium, and LED. Occasionally incandescent sources are used for special purpose applications, while LED sources are now becoming increasingly popular as replacements for fluorescent and incandescent sources.

In any case, once a light source and a suitable fixture for the source are chosen, for an outdoor lighting system, the available light from the fixture, expressed in lumens, can be obtained from the formula

$$\text{Lumens} = (FC \times \text{area}) / (CU) / MF / RCR, \qquad (7.1)$$

where FC is the desired illumination level in foot-candles, area is the area to be illuminated, measured in ft^2, CU is the *coefficient of utilization* of the fixture, MF is the *maintenance factor* of the fixture, and RCR is the *room cavity ratio*. The coefficient of utilization of a fixture is a measure of the fraction of light available from the lamp that is directed to the surface to be illuminated. The maintenance factor of the fixture provides a means of estimating the amount of light from a fixture that can be lost as the fixture and lamp get dirty and the room cavity ratio is a measure of the amount of light from the fixture that will be absorbed by the room and its contents. A good fixture will have a CU of 0.8 or more and a MF of 0.9 or more. Of course, the MF is dependent upon the environment in which the fixture operates and the maintenance interval for the fixture. The RCR depends upon room size, wall color, floor color, and room contents. A room full of dark colors will appear darker because it *is* darker. Outdoors, the RCR is generally assumed to be 1.0. Indoors, the RCR is generally <1.

7.4.2 Parking Lot Lighting Design

7.4.2.1 Introduction

Figure 7.5 shows a parking lot for which it is desired to provide an average illumination of 2 FC. The figure also shows suitable locations for fixtures and an

STAND-ALONE PHOTOVOLTAIC SYSTEMS 307

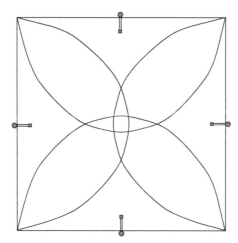

Figure 7.5 Parking lot showing locations of fixtures and approximate coverage patterns for each fixture.

approximate lighting pattern that would be obtained with a well-designed sharp cut-off fixture. The fixture has a CU of 0.8 and a MF of 0.9. The sharp cut-off feature of the fixture helps to prevent spillage of light onto adjacent property, so it will be assumed that any loss of light to adjacent property will be accounted for by the CU. The lot measures 160 ft (48.8 m) × 160 ft, so the total area of the lot is 25,600 ft^2 (2378 m^2). With the spacing of the fixtures as shown, the distance between fixtures is 113 ft (34.4 m). For parking lots, the spacing of lighting poles is typically set at 4 pole heights. Thus, in this case, the pole height should be approximately 28 ft (8.5 m).

7.4.2.2 Determination of Lamp Wattage and Daily Load Presented by the Fixture

First, the total lumens needed to illuminate the parking lot must be determined, then a suitable lamp can be selected. The total lumens can easily be found from Equation 7.1 to be ((2 × 160 × 160)/0.8)/0.9 = 71,111. If four fixtures are to be used, as shown in Figure 7.5, then each lamp will need to supply 17,778 lumens. Next, a search can be made for LED parking lot fixtures that are rated at a minimum of 17,778 lumens. One such fixture is a 126-W, 17,724-lumen fixture designed to replace 300- to 525-W HID fixtures. For about $40 additional per fixture, the lumen output can be increased to 22,155 with a 158-W fixture. The LEDs are rated to last more than 100,000 hours.

If this parking lot is in Miami, FL, and if the lights are to be on from sunset to sunrise, then the maximum amount of time the fixture will need to operate will be 13 hours during the month of December if the lights come on 15 minutes after sunset and shut off 15 minutes before sunrise. During the month of June, the fixture will only need to operate for 9.8 hours [9]. Thus, the winter worst-case daily Wh = 126 × 13 = 1638 Wh, and the summer daily energy consumption is

Table 7.4 Miami NREL SAM PVWatts Simulation for Parking Lot Lighting System with 41° Array Tilt

Month	PV to Batt (kWh/kW)	Lights Max Daily on Time	Load (kWh/Month)	Min Array (kW)
Jan	120	12.9	53.0	0.442
Feb	115	12.5	46.4	0.404
Mar	134	11.9	48.9	0.365
Apr	126	11.1	44.2	0.351
May	113	10.4	42.8	0.378
Jun	98	9.8	39.0	0.398
Jul	113	10.2	41.9	0.371
Aug	113	10.8	44.4	0.393
Sep	112	11.6	46.2	0.412
Oct	120	12.3	50.6	0.421
Nov	109	12.8	50.9	0.467
Dec	112	13.0	53.5	**0.477**

$126 \times 9.8 = 1235$ Wh. If the fixtures are 120-V fixtures, then a pure sine inverter will be needed to supply the fixtures. Assuming 95% inverter efficiency, this would increase the energy required from the batteries to 1724-Wh winter and 1300-Wh summer loads. Maximum loads for each month are shown in Table 7.4.

7.4.2.3 Determination of Battery Storage Requirements

The value of stand-alone outdoor lighting systems is enhanced if each fixture can operate independently of others. This saves the effort of running wires from one fixture to another. So all that is needed is to design one system and then build three more identical to it. Since Miami does not normally have many consecutive totally cloudy days, especially in the winter, 2 days of storage will be used. For a 24-V system, this means the winter daily *connected* Ah for the load will be $1724/24 = 71.8$ Ah, which needs to be supplied by the batteries. Allowing for 2% wiring losses and an 80% depth of discharge, the batteries will need to have a 24-hour discharge rate capacity of $((71.8 \times 2)/0.98)/0.8 = 183$ Ah at 24 V. Note that the 24-hour discharge rate is used because over the 2-day period, the batteries will be discharging for approximately 12 hours each day. Total discharge will thus take approximately 24 hours. If an inverter with 24 V dc is used, one way to achieve this is to use four 12 V at 89 Ah (24-hour rate) batteries in two parallel sets of two in series. If an inverter with 48-V dc input is used, the four batteries are then connected in series and the charge controller is set for a 48-V dc input rather than a 24-V dc input.

7.4.2.4 Determination of Array Size

The array size can be determined by first calculating the energy that must be supplied to the batteries by the array. Since the connected load on the batteries is 1724 Wh, and since losses of 10% can be expected for battery charging/discharging

STAND-ALONE PHOTOVOLTAIC SYSTEMS 309

and an additional 2% for wiring losses, a total of (1724/0.98)/0.9 = 1955 Wh must be supplied to the batteries by the array each winter day. For summer days, (1300/0.98)/0.9 = 1470 Wh/day will be needed.

Again, the PVWatts version of NREL SAM can be used to determine the array size [3]. Assuming inverter efficiency of the SAM simulation to be the same as the charge controller efficiency of the lighting system, then the resulting ac kWh of the SAM simulation will be essentially the same as the lighting system dc kWh input to the batteries. So an array is needed that will supply 1.955 kWh/day in winter and 1.470 kWh/day in summer. (Note the change from Wh to kWh.)

If the array is tilted at latitude +15° to optimize winter performance, Table 7.4 shows the resulting monthly kWh/kW delivered to the batteries along with the maximum on time each month, the monthly load kWh, and the corresponding array size needed to meet the energy needs of each light fixture. Note that the maximum array size needed is 477 W for December, the month with the least sun and the most nighttime hours, which should come as no surprise. If a typical 60-cell module rated at 240 W, with $V_{OC} \cong 37.3$ V, $V_m \cong 30$ V, $I_{SC} \cong 8.65$ A, and $I_m \cong 8.1$ A, is selected, then two in series will be adequate. Higher wattage 60-cell modules of similar physical size are also suitable choices. Note, however, that if a 48-V system is desired, these modules may not have adequate V_m for charging the batteries at elevated array temperatures.

7.4.2.5 Charge Controller and Inverter Selection

This system will use an MPPT charge controller and a pure sine inverter, so the inverter input voltage needs to be consistent with the charge controller output voltage. A wide variety of both devices is available, so either 24- or 48-V inverter input could be used, depending upon the battery configuration. For a 24-V system, the charge controller will need to supply 477 W × 1.25/24 V = 24.8 A, when taking cloud focusing into account. The MPPT charge controllers with current outputs in the 25–40 A range are readily available. Pure sine inverters that will supply the necessary fixture power are also readily available, but if the inverter will also be required to shut off the light if the batteries discharge too far, then an inverter with this feature will need to be selected.

7.4.2.6 Final System Schematic

At this point, all major components have been selected, so the next step is selecting all wire sizes and any switches or controllers that may be needed for proper system operation.

The method here is the same as it has been in all previous design examples: calculate all deratings due to temperature and then check voltage drops.

If the two modules are operated at STC, then $V_m = 60$ V and $I_m = 8.1$ A for the suggested module choice. For a distance of 25 ft (7.62 m) between array and charge controller, with a 2% limit on voltage drop, the wire must have Ω/kft < 2.96. The ampacity of the wire must be at least 156% of I_{SC} of the modules, or 1.56 × 8.1 = 12.6 A.

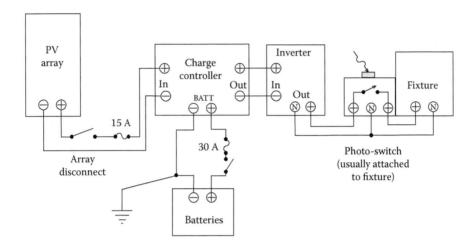

Figure 7.6 Schematic diagram of dc parking lot lighting system.

A temperature derating will not be needed for this wire run since it will not be run over a roof space. Thus, in this case, #12 will be required, since #14 has $\Omega/kft = 3.07$.

Next, appropriately sized fuses and switches need to be selected for the input and output of the charge controller and also for the inverter. Figure 7.6 shows a schematic diagram of the system. Note that it includes a photo-control to turn on the light when it is dark outside. Finally, it needs to be noted that the inverter or the charge controller must also be capable of shutting off the light if the battery voltage drops too low. In this circuit, the 15-A fuse is provided at the charge controller input, since the 30-A battery fuse will not protect the PV array from excessive backfeed, the only possible origin of which might be the batteries. As a practical matter, both the 15-A fuse and accompanying switch and the 30-A fuse and accompanying switch can be replaced by 15- and 30-A circuit breakers, since dc circuit breakers with voltage ratings less than 150 V are relatively inexpensive. The circuit breakers then serve dual roles as overcurrent protection and disconnect.

Since the array voltage is in excess of 50 V, it is required to be grounded, which is also shown in the schematic. All metal parts, including the mounting pole if it is metal, will also be connected to this same ground, making sure that the grounded conductor is connected to ground at only one point.

7.4.2.7 Structural Comment

Although the optical and electrical design requirements have been met for this system, it is important to note that it will also be necessary to design a sufficiently robust 28-ft (8.53 m) pole to hold all the system components. Since the design wind speed for Miami is 170 mph (274 km/hour), this may be a nontrivial exercise. The actual design of such a structure is beyond the scope of this text, and, as such, should be done by a structural engineer.

STAND-ALONE PHOTOVOLTAIC SYSTEMS

In the event that the structure should prove to be too expensive, the electrical engineer will be confronted with a redesign of the system. This process may involve exploring the use of more fixtures on shorter poles, shorter lamp operating times, or a compromise on the light level. The point here is that the engineer should not be discouraged if the first try does not work. Often the development of an acceptable solution to a problem will involve a number of redesigns.

7.5 A CATHODIC PROTECTION SYSTEM

7.5.1 Introduction

Material can be electroplated onto another material by immersing the two materials in a suitable electrolyte and applying a voltage between an anode composed of the desired plating substance and a cathode consisting of the material to which the material is to be plated. The result is transfer of material from the anode to the cathode.

When a metal is buried in the ground, it is highly likely that it will become a part of an electroplating system resulting from galvanic action between two dissimilar metals. If the metal assumes a higher potential than its surroundings, that is, becomes an anode, then metal will be removed as a result of ion loss from the metal. However, if the metal is deliberately connected as the cathode of a system, then electrons will flow from the voltage source negative terminal to the metal. The positive terminal of the voltage source is connected to a buried anode material so electrons flow from the anode material to the positive terminal of the voltage source. Removal of electrons from the anode material creates positive ions that can enter the electrolyte (i.e., the ground) and flow toward the cathode. The process is shown in Figure 7.7.

The U.S. government requires that any underground storage of toxic materials or petrochemicals must have cathodic protection. Cathodic protection involves using the material to be protected, usually steel, as a cathode, while an anode (or anodes),

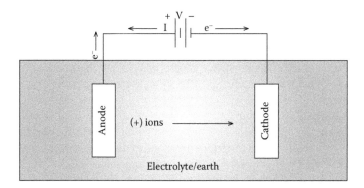

Figure 7.7 Schematic diagram of flow of current and charge carriers in an electrolyte system.

typically of graphite, is buried nearby. In addition to the protection of toxic waste containers, billions of dollars of infrastructure are built with steel reinforced concrete, much of which is under water. Cathodic protection of the steel in the concrete can prolong the life of buildings, bridges, and other important infrastructure components.

To prevent ion loss from the cathode, different current densities are required for different materials, ranging generally from a fraction of a mA/ft^2 to several mA/ft^2. The total current needed to protect a cathode is thus the product of the necessary current density and the surface area of the cathode. The voltage needed to supply this current is thus determined by the product of the current and the resistance from anode to cathode. In this example, the methods for determining current density, resistance, and voltage are explored, along with determining necessary battery storage and array size.

If all soils were identical, cathodic protection system design would be nearly trivial. Fortunately for the corrosion engineer, the earth has been blessed with a wide range of soil types, often with diversity in a small area. This relatively wide range of soil conditions, sometimes varying with time in the same location, makes the job of protection of critical systems sufficiently challenging to warrant the high fees of the corrosion engineer. Considering the fines and/or lost revenues that can result from a spill of petrochemicals or other pollutants, design of systems to protect these systems cannot be left to amateurs. While this example will not convert the reader into a professional cathodic protection designer, it will at least convey some background needed for cathodic protection design.

7.5.2 System Design

The first step in the design of a cathodic protection system is to determine the system current needs. Suppose the item to be protected is an uncoated steel tank in sandy soil. Suppose the tank has an exposed surface area of 100 ft^2 (9.3 m^2). The current density required for use in this environment for exposed steel is 1 mA/ft^2, so the total current needed will be 100 mA. If more current is generated, the cathode will remain protected, but the anode will be sacrificed at a higher rate.

The next step is to choose an anode. A typical anode will carry a maximum current of 2 A, so the choice of anodes will presumably not be significantly affected by current. The other consideration in choosing an anode is the resistance between the anode and the cathode. This resistance depends on the soil resistivity and on the size of the anode. Since the anode is generally cylindrical, the current from the anode travels more or less radially outward. The reader may recall from an electromagnetics course that a cylindrical geometry with an infinitely long charged cylinder produces an electric field that varies inversely with the distance from the cylinder. This results in a logarithmic variation in voltage and a nice, nonlinear relationship between the length and diameter of the anode and the resistance from anode to ground. Fortunately, the resistance to ground for anodes of different diameter and length in a uniform soil with resistivity, $\rho = 1000$ Ω-cm, is tabulated [10]. The resistance to ground for different soil resistivity is then in proportion to the resistance at

Table 7.5 Anode Resistance to Ground in Standard 1000 Ω-cm Soil

Anode Diameter (in.)	Anode Length (ft)				
	4	5	6	7	8
3	5.0 Ω	4.3 Ω	3.7 Ω	3.3 Ω	3.0 Ω
4	4.7 Ω	4.0 Ω	3.5 Ω	3.1 Ω	2.8 Ω
6	4.1 Ω	3.5 Ω	3.1 Ω	2.8 Ω	2.5 Ω
8	3.7 Ω	3.2 Ω	2.9 Ω	2.6 Ω	2.3 Ω
10	3.5 Ω	3.0 Ω	2.7 Ω	2.4 Ω	2.2 Ω

Source: Stand-Alone Photovoltaic Systems: A Handbook of Recommended Design Practices, Sandia National Laboratories, Albuquerque, NM, 1995.

standard conditions. The resistance to ground for the anode is used as the resistance between anode and cathode.

In this case, suppose a 3-inch-diameter, 5-ft-long anode is chosen. Table 7.5 shows the resistance to ground for such an anode in 1000 Ω-cm soil to be 4.3 Ω. But the resistivity of sandy soil is closer to 25,000 Ω-cm, so the resistance to ground of this anode at the tank location will be approximately 25 times higher, or 25 × 4.3 = 107.5 Ω.

The required voltage is thus the product of the current and resistance, which, in this case, is 0.1 × 107.5 = 10.75 V, which can readily be supplied by a standard nominal 12-V module (i.e., $V_{OC} \approx 20$ V). Assuming the current is needed on a 24-hour/day basis, a 12-V storage battery will be needed.

Using a battery charging efficiency factor of 0.9 and a wire efficiency factor of 0.98, the daily corrected system load is determined to be (0.1 A × 24 h)/(0.9 × 0.98) = 2.72 Ah.

The next step is to determine the size of the battery needed, along with the current rating of the PV array. Assuming deep-discharge batteries with 5 days of storage time and an allowable discharge of 80%, the battery needs will be (2.72 Ah × 5 days)/0.8 = 17 Ah. This is about the size of the battery in a small uninterruptible power supply for computer backup power.

Next, for a 12-V system, the array/module will need to provide 32.64 Wh/day during the worst sun month of the year. The PVWatts version of NREL SAM can be used again here, assuming that a charge controller is used instead of the inverter of the simulation. As an example, if the system will be installed in Laramie, WY, then, if the module tilt is 50°, then the minimum module size will be 9.1 W, while if the tilt is 60°, then the minimum module size decreases to 8.9 W.

Finally, the charge controller needs to be selected. If the module selected will provide 8.9 W at about 15 V, then the charge controller output current will be about 0.6 A. If an MPPT unit is used, then it is important that the module/array selected will have sufficient voltage to meet the specified charge controller input voltage range. For such a small system, it may not make economic sense to use an MPPT charge controller. If a conventional (PWM) charge controller is used, the module power may need to be increased by 25%, but PWM controllers that will charge 12-V batteries are readily available. Increasing the module size by 25% still results in a charge controller output less than 1 A, and a number of small PWM chargers are available to handle this requirement.

For the design of small systems, one might wonder whether it may be possible to avoid using a charge controller and simply charge the battery directly from the module. The test for this possibility is to recognize that *NEC* 690.72(A) requires a charge controller whenever the PV array provides more than 3% of the battery rating in 1 hour at maximum array current.

Suppose the combination of protection current and anode resistance to ground had resulted in the need for more than 12 V. Several means of solving this problem are available. One is simply to use modules with higher V_m. Another is to use a larger anode and another is to use anodes in parallel. One might expect an anode with twice the surface area to have half the resistance to ground, but due to the nonlinearities of the system, this is not the case, as shown in Table 7.5.

The parallel anode solution results in a lower resistance to ground, but, perhaps not surprisingly, the resistance to ground of two identical anodes is not simply half the resistance of a single anode. The cylindrical geometry again adds a nonlinear twist to the problem, resulting in the resistance to ground of two anodes depending on the separation of the anodes. Table 7.6 [10] shows multiple anode adjustment factors, assuming all anodes are identical and that the soil is uniform in composition. The factors in the table are multipliers for the single-anode resistance to ground. Thus, for example, if the resistance to ground for a single anode is 100 Ω, then the resistance to ground for three anodes spaced 15 ft (4.57 m) apart will be 100 × 0.418 = 41.8 Ω. This reduces the voltage required to 41.8% of that required for a single anode. It is thus a matter of calculating the life-cycle cost (coming up

Table 7.6 Multiple Anode Adjusting Factors

Number of Anodes	Separation of Anodes (ft)				
	5	10	15	20	25
1	1.000	1.000	1.000	1.000	1.000
2	0.652	0.576	0.551	0.538	0.530
3	0.586	0.460	0.418	0.397	0.384
4	0.520	0.385	0.340	0.318	0.304
5	0.466	0.333	0.289	0.267	0.253
6	0.423	0.295	0.252	0.231	0.218
7	0.387	0.265	0.224	0.204	0.192
8	0.361	0.243	0.204	0.184	0.172
9	0.332	0.222	0.185	0.166	0.155
10	0.311	0.205	0.170	0.153	0.142

Source: *Stand-Alone Photovoltaic Systems: A Handbook of Recommended Design Practices*, Sandia National Laboratories, Albuquerque, NM, 1995.

STAND-ALONE PHOTOVOLTAIC SYSTEMS

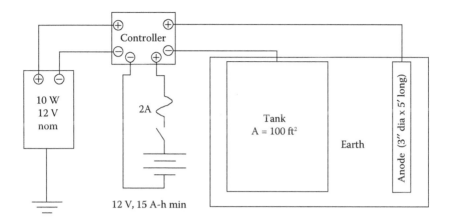

Figure 7.8 Cathodic protection system schematic diagram.

in Chapter 8) of the three-anode system versus the single-anode system. Keeping in mind that each anode now will only carry one-third of the system current, the anodes should last three times longer.

A somewhat more elegant solution to the problem would involve an electronic constant current source that would provide the required current regardless of soil conditions or PV output. However, wet soil requires more current than dry soil, so the current source would need to be compensated for soil resistivity, rendering the current source design somewhat more challenging.

Whenever possible, it is advisable to make soil resistance measurements so that empirical data can be used to size and locate system components properly. This eliminates many of the assumptions made and provides for greater confidence in the performance of the system. Figure 7.8 shows the final system.

7.6 A PORTABLE HIGHWAY ADVISORY SIGN

7.6.1 Introduction

Once upon a time, illuminated highway signs either had to be connected to the power grid or else had to be self-contained with their own portable fossil-fueled generator. Furthermore, their messages were often hard-wired, so the sign could only convey a single message. In this example, rather than determining the number of PV modules and number of batteries for a given load, the energy available for a load and the corresponding average load power will be determined. This approach is taken for several reasons, one of which is that it has not been used yet and the other being that sometimes a PV system may be limited by size or cost.

Anything to be transported on a roadway should normally be 8 ft (2.44 m) or less in width. Assuming a sign of 8 ft width and assuming that PV modules will be mounted horizontally on top of the sign since the orientation of the sign will be random suggests that two modules, each 40 inch (1.02 m) × 72 inch (1.83 m), can be used conveniently. The sign will be mounted on a trailer of standard size along with the battery pack and BOS. While it may be possible to mount additional modules elsewhere on the trailer, it is assumed that to avoid module shading and minimize damage from vandalism, only the modules on top of the sign are practical.

Since each month has different peak sun hours (psh), each month will have different available average power for the sign. Thus, during some months, it may be possible that the sign will not be used 24 hours/day or it may be possible to convey longer or brighter messages during some months. For this example, it will be assumed that the sign will be used in the vicinity of Atlanta, GA, where freeway construction has been in progress for the past 60 years and may continue into the foreseeable future.

7.6.2 Determination of Available Average Power

If modules with 18% efficiency are used, then the modules will be able to generate approximately 16.7 W/ft^2 (180 W/m^2) under full sun irradiance. The two modules will thus have a maximum power output of (2 × 40 × 72 × 16.7)/144 = 668 W, provided that they are kept clean and provided modules meeting this specification can be found.

Using the NREL SAM PVWatts simulation again yields average available monthly kWh to the batteries from the 668-W horizontal array as shown in Table 7.7. Note that the default array parameters have been adjusted in the model to allow for 6% soiling, since the signs will typically be used at construction sites, but module mismatch has been reduced to 0%, since only two modules will be used and exit testing of the modules upon manufacture will guarantee the modules will have a minimum power rating of 334 W each.

Table 7.7 also shows the average daily Wh available to the load, assuming wiring losses of 2% and battery in–out losses of 10%. Assuming the sign will remain

Table 7.7 Highway Warning Sign Monthly Maximum Daily Load Power

Month	PV Array to Batt (kWh/month)	Max Load (Wh/day)	Max Cont P$_{load}$ for 24-Hour Period (W)
Jan	42	1195	49.8
Feb	51	1451	60.5
Mar	72	2048	85.4
Apr	92	2617	109.1
May	98	2788	116.2
Jun	96	2731	113.8
Jul	95	2702	112.6
Aug	88	2503	104.3
Sep	71	2020	84.2
Oct	66	1877	78.2
Nov	47	1337	55.7
Dec	39	1109	46.2

STAND-ALONE PHOTOVOLTAIC SYSTEMS 317

illuminated continuously, the average power available over a 24-hour period for each month can also be determined. Not surprisingly, the winter available power is less than half the summer available power because of the horizontal array orientation.

7.6.3 Determination of Battery Requirements

Battery requirements will normally be based upon the worst-case daily array performance, which, in this case, is for December, in which the array, via an MPPT charge controller, will deliver 39 kWh/month to the batteries. Since this averages 1.26 kWh/day, battery sizing can be based on this average daily charging rate. Assuming a 12-V battery system, this amounts to 1260/12 = 105 Ah. Thus, allowing for 5 days of autonomy and 80% depth of discharge, the battery requirements are found to be (105 × 5)/0.8 = 656 Ah. Battery selection is thus based upon a 120-hour discharge rate. This can almost be achieved with two-size 8D, 305 Ah at 12 V AGM batteries in parallel, if 4.65 days of autonomy are considered to be acceptable. If 656 Ah is considered to be an absolute minimum, then the next size batteries from this manufacturer will result in either 759 Ah for three-size 4D batteries in parallel or in 716 Ah for four-size 30H batteries in parallel.

7.6.4 Additional Observations and Considerations

The wide range of available average power opens up some interesting options for system operation. Clearly, when the sun produces more energy, more energy can be used by the display. However, if this choice is made, say, to illuminate the sign with 100 W of power in August, then the sign will use 2.4 kWh/day, which equates to 200 Ah/day. Since the batteries have been sized for about 600 Ah, this reduces the battery storage time to about 3 days. A decision will need to be made whether 3 days is adequate. Also, for 3 days of storage, a 72-hour discharge rate should be used to reevaluate the battery capacity under this faster discharge. For the 8D battery, the capacity is reduced to 295 Ah at the C/72 rate.

With a microcontroller in the system, it is straightforward to program the unit to inform the user of the average power that will be used to implement any particular program and the resulting storage time. For that matter, the system can even be programmed to give a warning to the programmer if the average daily power is exceeded by the proposed announcement. If the sign is not programmed to use maximum available power, then the controller needs to have the capability to disconnect the PV array from the batteries, which will typically be the case.

7.7 A CRITICAL NEED REFRIGERATION SYSTEM

7.7.1 Introduction

It is desired to provide electric power for 10 ft^3 of refrigeration for storage of medication. The medication is costly and difficult to obtain, so the system must have

99% availability. The system will be located in a village in Kenya where villagers will be able to keep a careful watch on system performance so any problems will become immediately evident. This design will be based on an ac refrigerator that has one of the highest energy efficiency ratings. The refrigerator is 10.12 ft^3 (0.29 m^3) unit that is rated at 171 kWh/year, which equates to 0.47 kWh/day [11]. It will be powered by an inverter that has a 94% efficiency rating that produces a pure sine wave output that is preferred as a source of power for the refrigerator. Supplying anything other than a pure sine wave to the refrigerator will result in harmonics in the power to the compressor motor, which, due to heat from increased hysteresis and eddy current losses, may shorten the life of the compressor motor. The inverter will have a 12-V input and will have low-voltage cut-out and cut-in capability.

7.7.2 Load Determination

To convert 0.47 kWh/day to Ah/day, multiply by 1000 to get 470 Wh/day. Then assume the refrigerator is well-insulated such that the compressor only needs to run for 6 hours/day. This means the compressor will require 470/6 = 78 W for operation. Note that this estimate is for the purpose of determining the output power rating of the inverter. So a 300–500 W range for inverter power should easily cover any starting surge of the compressor. In fact, if the inverter size is much larger, the low power efficiency of the inverter may drop below the 94% level.

The next step is to determine the daily load on the batteries. Starting with the refrigerator load and then compensating for inverter and wiring losses results in the load on the batteries of 470/(0.94 × 0.98) = 510 Wh/day. This will be defined as the *connected load*, meaning the actual load connected to the batteries.

7.7.3 Battery Sizing

In the previous examples in this chapter, a relatively simple sizing formula has been used. For more critical loads or for high- or low-temperature battery use, additional factors need to be incorporated into the battery sizing procedure. The batteries must be sized to power the connected load over the duration of time determined as the critical load storage period. For a single day, the amount of energy that must be transferred to the batteries will be the *connected* load adjusted for battery charge/discharge losses, wiring losses, and possibly losses due to lower operating temperature or rapid charging rates. In equation form, this is

$$\text{Ah} = \left(\frac{\text{Ah}}{\text{day}}\right)\left(\frac{\text{days}}{D_T D_{ch}(\text{disch})}\right), \tag{7.2}$$

where Ah/day represents the *corrected load* on the batteries, days represents the number of days of autonomy, D_T is the temperature derating factor, D_{ch} is the charge/discharge derating factor, and disch is the depth of discharge expressed as a fraction. In Section 3.5.2, it was noted that lead-acid battery capacity decreases at lower temperatures and higher discharge rates. Higher charge rates also can result in greater losses.

The capacity of a battery decreases as temperature decreases. For lead-acid batteries, the capacity reduction can be approximated by the following empirical relationship, for battery temperatures between 20°F and 80°F (−6.67°C and 26.7°C).

$$D_T = \frac{C}{C_o} = 0.00575T + 0.54, \qquad (7.3)$$

where C is the battery capacity at temperature T (in °F) and C_o is the rated battery capacity at 80°F [11].

If the anticipated load will exceed the specified discharge rate for more than 10 minutes, an additional correction factor should be applied to the total corrected Ah. For example, if the battery is rated at a discharge current of 20 A, and the actual discharge rate will be 30 A, then D_{ch} = 20/30 = 0.67. This factor reflects the fact that there are greater I^2R losses to the internal battery resistance as the current increases, thus decreasing the energy available from the batteries to power the load.

If the battery charge rate exceeds the rated charge rate, an alternate calculation of battery capacity should be performed and compared with Equation 7.2. Deep-discharge lead-acid batteries are typically designed for charging at a C/10 rate of charging. Hence, if 10 hours times the PV charging rate in amperes exceeds the battery capacity in Ah calculated in Equation 7.2, it means the PV array is capable of fully charging the batteries faster than the C/10 rate. If this condition should occur, the battery capacity should be determined by the product of the rated charging time and the available charging current, rather than by Equation 7.2. The capacity calculated in this manner should result in a value lower than the C/10 rated capacity since greater battery internal I^2R losses occur at the higher charging rate.

For the present design, assuming a combined loss of 12% for battery charge/discharge and wiring losses and neglecting other factors means that the *corrected* load will be 510/0.88 = 580 Wh/day. Next, allowing for 80% depth of discharge, the amount of battery storage for 1 day of operation must be 580/0.8 = 725 Wh. Dividing by the nominal battery voltage (12 V) converts the 725 Wh to 60.41 Ah.

At this point, it will be necessary to determine the number of storage days needed for critical system availability. Table 7.8 shows the results for monthly average daily psh values for Makindu, Kenya, obtained by determining the monthly average kWh/kW delivered to the batteries from the PV array, then dividing by inverter and wiring efficiencies to obtain the monthly kWh/kW from the array before any system losses are accounted for and then dividing by the number of days in each month to convert the monthly array kWh/kW to daily kWh/kW. Note that the units of psh are hours. Using the minimum daily average psh for Makindu from Table 7.8, as obtained from NREL SAM PVWatts version, in Equation 3.14 yields D_{crit} = (0.2976 × 3.77²) − (4.7262 × 3.77) + 24 = 10.4 days when the array is tilted at latitude +2°. Thus, a total of 10.4 × 60.41 = 628 Ah at C/120 will be needed.

Since most battery manufacturers only specify discharge rates up to C/120, this is the discharge rate to use when selecting the batteries.

Table 7.8 Calculation of Peak Sun Hours from NREL SAM Data for Makindu, Kenya

Month	PV Array to Batt (kWh/month)	Array Daily (psh)	1 kW Array Daily Wh to Batt	Batt Daily Wh Needed	Minimum Array Size (W)
Jan	134	4.75	4323	580	134
Feb	123	4.83	4393	580	132
Mar	135	4.79	4355	580	133
Apr	120	4.40	4000	580	145
May	117	4.15	3774	580	154
Jun	103	3.77	3433	580	169
Jul	108	3.83	3484	580	166
Aug	117	4.15	3774	580	154
Sep	131	4.80	4367	580	133
Oct	135	4.79	4355	580	133
Nov	116	4.25	3867	580	150
Dec	123	4.36	3968	580	146

Additional factors to consider when selecting batteries include the additional connections required when using larger numbers of batteries and the corresponding additional time required for installation and maintenance. Furthermore, the more connections, the greater the possibility of failure of a connection. And the cost of battery cables is not negligible. On the other hand, with a smaller number of batteries, if one should fail, the system is impacted more significantly than in the case of a larger number of batteries. The weight of the battery and the difficulty of obtaining the battery may also be considered. A local supplier with the ability to provide timely delivery and warranty service is generally more desirable.

Once again, the 12 V, 305 Ah at C/120 8D battery from the highway advisory sign example nicely meets these requirements. So two of these in parallel will provide 610 Ah of storage, as long as they are available at an acceptable cost.

7.7.4 Array Sizing

The array must be sized to meet the daily *corrected* load requirements (580 Wh) during the month with the lowest average daily solar harvest.

If an MPPT charge controller is used, then the array size can be calculated in watts. Thus, after array losses, losses in wiring between array and charge controller, and charge controller losses, 580 Wh must still be left for the batteries. The second column in Table 7.8 shows the monthly kWh delivered to the batteries for a 1-kW array with a latitude tilt. Thus, to deliver 580 Wh/day in June, the array size must be 169 W, which can be achieved with a single module. But this could present a problem for the design. Since the module has been sized based upon the use of an MPPT charge controller, the array must have sufficient voltage to operate the charge controller at a 12-V output level. This presents a marginal situation for a standard, 36-cell (12 V) module operating at elevated temperatures. Fortunately, with the hundreds of different modules now available, there is at least one module that is rated at

175 W that has V_m = 36.63 V. This module will work conveniently with a number of different MPPT charge controllers.

7.7.5 Charge Controller and Inverter Selection

The two remaining electronic components are the MPPT charge controller and the pure sine inverter. To match these components, the charge controller output voltage must match the inverter input voltage, both of which will be 12 V. There is an interesting consideration for this system. Normally, it would be desirable for the inverter to shut off in the event of battery state of charge (SOC) dropping below 20%. Depending upon the value of the contents of the refrigerator versus the value of a new set of batteries, it may be desirable to risk damage to the batteries to protect the content of the refrigerator. However, given this concern, it can be addressed alternatively by using an oversized module, say, 220 W rather than 175 W. This provides the capability of producing an additional 190 Wh/day on sunny days in the event of a prolonged cloudy period. The charge controller will still work with the new module, even if the module has $V_m \cong 30$, which would be the case with a common 60-cell module. For a 175-W module, the charge controller output current rating needs to be at least (175 × 1.25)/12 = 18 A, and for a 220-W module, the charge controller output current rating needs to be at least 23 A. Either charge controller is readily available.

The inverter needs to be able to handle the running current as well as the starting current of the refrigerator compressor, which may be as high as 250% of the running current. Thus, using the running power estimated earlier (78 W), 250% of this power is 195 W, which means that at 12 V, the inverter input current rating must be at least 16.25 A and the inverter output power rating needs to be at least 195 W. Thus, a 120-V, 300-W pure sine inverter with a 12-V input will be more than adequate for the purpose.

7.7.6 BOS Component Selection

At low voltages, careful attention must be paid to voltage drop in the wiring. It is generally useful to lay out the system and then tabulate the lengths of the various wire runs along with the allowable voltage drops. This enables a simple calculation of the required Ω/kft of wire to keep the wire voltage drop within allowable limits. Once the Ω/kft is known, the proper wire size can be selected. As in previous examples, the correct wire size is the larger of either the size needed to carry the rated current or the size needed to meet system voltage drop constraints.

Table 7.9 summarizes the wire sizing for a system with fixed array tilt and an MPPT charge controller. A word of explanation may be useful for some of the table entries. In the first column, current used to calculate voltage drop, the array to controller current shown, is I_{mp}, which is generally used for voltage drop calculations. It is recognized that $1.56I_{SC}$ is required for sizing based on ampacity, so this check will be made after the voltage drop calculation for wire size is done. Note that all but the

Table 7.9 Summary of Wire Sizes and Fuse Sizes for the Refrigerator System

Wire Location	VD A	One-Way Length (ft)	System Volts	Max Allow R (Ω)	Wire (Ω/kft)	Wire Size	Fuse Size (A)
Array to controller	4.78	40	36.6	0.0766	0.9575	#8	10
Controller to batteries	13.9	6	12	0.00863	0.719	#6	25
Controller to inverter	6.9	8	12	0.0174	1.0875	#8	10
Inverter to refrigerator	0.65	15	120	1.846	61.5[a]	#14	15

[a] Wire size must be based on ampacity rather than voltage drop.

inverter to refrigerator wire sizing ends up being determined on the basis of voltage drop as opposed to ampacity.

The controller to battery current is 95% of the quotient of the 175-W array power and the battery voltage. The 95% figure compensates for array to controller wiring loss plus controller loss. The controller to inverter current is the inverter input power divided by the inverter input voltage. The inverter input power is the power to the refrigerator divided by the wiring efficiency (close to 100%) and the inverter efficiency.

The next two columns are self-explanatory. The allowable resistance is the wire resistance that would produce a 1% voltage drop, and the Ω/kft column is the allowable resistance divided by the round-trip kft length of the wire run. Wire sizes are based upon the allowable Ω/kft value except for the wire to the refrigerator, which is set at #14 since #14 is the smallest wire size that is allowed to be concealed in a 120-V system.

7.7.7 Overall System Design

At this point, all system components have been selected for the system. It is now possible to develop a system schematic diagram. Figure 7.9 shows the schematic diagram of the system that uses a fixed array and an MPPT charge controller. Since the maximum system dc voltage is below 50 V, grounding of the wiring is optional for systems wired in accordance with the *NEC*. However, if the system is left ungrounded, then disconnect switches must open all ungrounded conductors. It is thus generally better to establish a single point where the *grounded* conductor is connected to the *grounding* conductor, since all metallic enclosures must be grounded anyway.

The grounded conductor carries current and is normally chosen as the negative conductor of the PV system. The grounding conductors do not carry current and are used for the purpose of connecting all exposed metal system parts to ground to protect against shock hazards. As long as the grounded and grounding systems are connected at only one point, current will not flow in the grounding conductors under conditions of normal operation. Grounding conductors are shown along with the

STAND-ALONE PHOTOVOLTAIC SYSTEMS

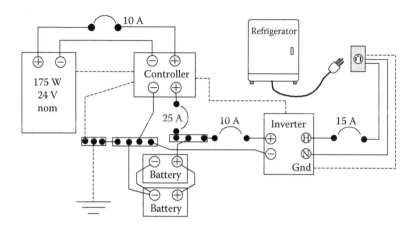

Figure 7.9 Refrigerator system schematic diagram.

single point of connection between grounded and grounding conductors. Sizing of grounding conductors will be considered in later design examples.

The single grounding point is usually a terminal block with multiple-set screw terminations for connecting grounded and grounding conductors. In systems that require a GFDI device, the grounded conductors will generally terminate at one terminal block and the grounding conductors will terminate at a separate terminal block. The two blocks are then connected to each other through the trip mechanism of the GFDI device. Any current flowing through the GFDI device would indicate that more than one point of the grounded system is connected to the grounding system.

7.8 A PV-POWERED MOUNTAIN CABIN

7.8.1 Introduction

Figure 7.10 shows the floor plan of the cabin, located west of Boulder, CO, to be outfitted with PV-powered electrical loads. The cabin will have a kitchen light, a dining room light, a light in each bedroom, a bathroom light, a living room light, a motion sensor light outside each door, a ceiling fan in each bedroom, an entertainment center, a deep-well-water pump, and an energy star 16 ft^3 (0.4531 m^3) refrigerator/freezer. The PV array will be located approximately 10 ft (3 m) from the back of the cabin as shown. The 200-ft-deep (61 m) well will be located 30 ft (9.14 m) from the cabin in the same relative direction. The pump will be a submersible unit, so the total distance to the pump will be 230 ft (70.1 m). All loads will operate at 120 V ac, so an inverter will be needed to power the loads. The inverter input voltage will be determined after the inverter power requirements have been determined.

The system loads, with the exception of the refrigerator, will operate 3 days/week to accommodate weekend use over long weekends. The refrigerator will be left on

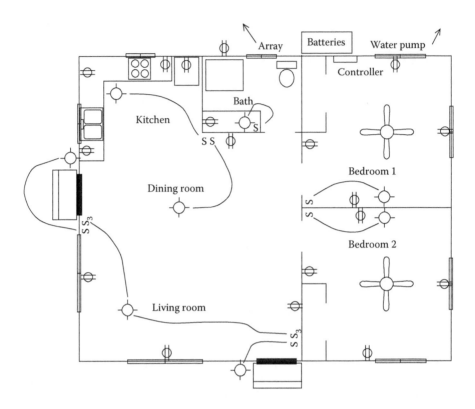

Figure 7.10 Floor plan of mountain cabin.

continuously in order to prevent inadvertent odor build-up in the event it should get too warm inside the unit. At least this is the stated reason. It is highly suspected that the real reason is to have cold beer ready as soon as someone arrives at the cabin.

Since the *NEC* requires that a building be wired as though it were to be connected to the grid, the cabin will be wired for all 120 V loads. The PV array will be rack mounted at ground level, so the feasibility of seasonal adjustment of array tilt will be explored as a means of optimizing system performance. Also, maintenance-free, valve-regulated AGM lead-acid batteries will be used as a part of keeping system maintenance to an absolute minimum.

7.8.2 Load Determination

At this point, it should be abundantly clear that the only reasonable choice for the refrigerator and other loads will be high-efficiency units. The refrigerator that will be used has separate freezer and refrigerator compartments and is rated at 333 kWh/year, which equates to 0.912 kWh/day. Assuming 6 hours of compressor operation per day equates to a power consumption of 152 W.

For the water pump, it is first necessary to determine the water needs for the cabin. This determination can be quite a wild guess unless the uses of the water are

STAND-ALONE PHOTOVOLTAIC SYSTEMS

reasonably well defined. Assuming the only means of heating water is a propane stove, it is reasonable to assume that showers will not be too lengthy. However, if a solar water heater is added later, shower length may increase. A reasonable estimate for water usage might be 40 gallons (151 L) per day per person, with anticipated occupancy of four persons. This amounts to a total water requirement of 160 gallons (606 L) per day. If more people use the cabin, it may be necessary to do some water rationing. In the worst case, where drinking is the only use made of the water, 160 people would still have a gallon (3.785 L) each per day for drinking over the 3-day weekend. It is assumed that less water is used in winter, spring, and fall than the 160 gallons/day of summer use.

The next step is to go to the manufacturers' catalogs or to the Internet. For example, Reference 12 lists a number of different water pumps. For deep wells, normally a submersible pump is the best choice. The first check to make on a pump to be used in a system with battery storage is whether it can pump the needed water in either a day or a week. If the needed water can be pumped in a day, then only a 160-gallon (606 L) storage tank will be needed. If it takes all week to pump the water for the weekend, then a 480-gallon (1817 L) storage tank will be needed. The pumping power will depend on the flow rate and the lift distance. If the top of the storage tank is 10 ft (3.05 m) above ground to provide pressure for the system, and, allowing for 5% piping losses, the equivalent lift for the pump will be $210 \times 1.05 = 220.5$ ft (67.21 m).

In Section 7.2, it was observed that it is generally less costly to use a small pump that pumps over a longer period of time than a larger pump that pumps the needed water in a short time. Of the many pumps on the market, the ac pumps tend to be higher horsepower than the dc pumps. One pump, however, that will run on either dc or ac will pump 1.15 gpm at a 225-ft head, using 115 W. This is not exactly an inexpensive pump, but has a reputation for performance and efficiency. The 1.15-gpm pumping rate amounts to about 31.5 ft-lb/second, or 0.057 HP. This means the wire-to-water pumping efficiency is approximately 37%, which is, indeed, quite efficient for this size pump. At 1.15 gpm, the time to pump 160 gallons is 139 minutes, or 2 hours 19 minutes. So a 160-gallon storage tank should be adequate. The energy required by the pump for pumping this amount of water is $(125 \text{ W}) \times (2.317 \text{ h/day}) = 290$ Wh/day.

The remaining loads are listed in Table 7.10 as corrected for inverter and wiring losses. Thus, the actual loads are divided by 0.92 to determine the load on the batteries. Notice that in this example, the monthly variations in loads are accounted for. For example, the ceiling fans will be used more in the summer and the lights will be used more in the winter. The monthly variations in loads are important to use when determining the array tilt.

Note that the 120-V branch circuit loads are computed at the values required by *NEC* 690.10, even though the total power and energy used will be limited by the array size, the inverter size, and the battery size. The idea is that one day when the grid is everywhere, some future owner may elect to connect to the grid. For the moment, however, *NEC* 690.10(A), (B), and (C) will apply while the system is a stand-alone system. In other words, the inverter will not be required to simultaneously power

Table 7.10 Summary of Monthly Variation in Weekly Ah @ 48 V Loads for Mountain Cabin

Load Description	P (W)	Days/Week	Nov–Feb Hour/Week	Nov–Feb Ah/Week	Mar Hour/Week	Mar Ah/Week
Kit light	20	3	12	5.0	10.5	4.4
BR 1 light	10	3	6	1.3	4.5	0.9
BR 2 light	10	3	3	0.6	3	0.6
LR light	10	3	15	3.1	12	2.5
Outdoor lights	10	3	1.5	0.3	1.5	0.3
DR light	15	3	12	3.8	9	2.8
Bath light	10	3	6	1.3	6	1.3
Refrigerator	152	7	35	110.8	38.5	121.9
Water pump	125	7	5.5	14.3	6	15.6
BR 1 fan	32	3	0	0.0	0	0.0
BR 2 fan	32	3	0	0.0	0	0.0
Receptacles	2500	3	2	104.2	1.75	91.1
Total	2926			245		242
Corrected loads				277		274

Load Description	Apr, Oct Hour/Week	Apr, Oct Ah/Week	May, Sep Hour/Week	May, Sep Ah/Week	Jun, Jul, Aug Hour/Week	Jun, Jul, Aug Ah/Week
Kit light	10.5	4.4	9	3.8	7.5	3.1
BR 1 light	4.5	0.9	3	0.6	3	0.6
BR 2 light	3	0.6	3	0.6	3	0.6
LR light	9	1.9	6	1.3	6	1.3
Fr Dr light	1.5	0.3	1.5	0.3	1.5	0.3
DR light	7.5	2.3	6	1.9	4.5	1.4
Bath light	5	1.0	4	0.8	3	0.6
Refrigerator	38.5	121.9	42	133.0	45.5	144.1
Water pump	6.5	16.9	7	18.2	8	20.8
BR 1 fan	3	2.0	8	5.3	24	16.0
BR 2 fan	3	2.0	8	5.3	24	16.0
Receptacles	1.5	78.1	1.25	65.1	1	52.1
Total		232		236		257
Corrected loads		263		268		291

all loads and the power distribution panel will need to be labeled that the panel is a single 120 V supply.

Per *NEC*, each duplex outlet, except in kitchen or bathroom, must be rated at 180 VA, so the total load assigned to the duplex outlets in the two bedrooms and the living/dining room area is 2160 W. The load assigned to the kitchen circuit is 1500 W as is the load assigned to the bathroom (hair dryers). Even though all the

STAND-ALONE PHOTOVOLTAIC SYSTEMS

lighting outlets will have no more than a 20-W LED lamp, they still must be rated at 90 W each. But to calculate the load on the inverter and the batteries, the expected off-grid loads on these circuits are used in Table 7.10, with the understanding that the circuits are wired adequately to handle the loads as required by *NEC* 210 and 220.

For each load, the average Ah/week if the load operates for n hours per day and d days per week is determined by

$$\frac{Ah}{week} = \frac{Pnd}{\eta V}, \qquad (7.4)$$

where P is the load power, η is the conversion efficiency of the inverter, and V is the 48-V system dc voltage. The 48-V system voltage at this point is a guess, since a specific inverter has not yet been selected. In fact, the total power in Column 1 suggests that an inverter rated at approximately 3000 W will be suitable for the cabin, and it is likely that a 3000-W inverter will operate on 48 V. Note that the power assigned to the receptacles is unusually high, but the hourly use of the receptacles is estimated to be comparatively low to correspond to a few minutes of microwave or hair dryer time and longer times for other, relatively low power plug-in devices, such as a TV or a computer or a small LED reading lamp.

The entries in Table 7.10 are based on 3 days/week of usage for all loads except the refrigerator, which remains on for 7 days/week, and the water pump, which will pump as needed, to keep the storage tank full. The pump is controlled by a float switch in the storage tank.

It is important to note that the average Ah/week represent the weekly average, taking into account that the water pump and the refrigerator may run 7 days/week, but the other loads are only operational for 3 days/week while the cabin is occupied. The array and batteries will thus be selected to meet the average weekly system corrected loads.

The next step is to compute the system corrected loads. Using a 98% wire efficiency factor and a 90% battery efficiency factor, the corrected Ah loads for each month are obtained by dividing the load Ah/week by 0.98 × 0.9. The corrected loads are shown in Table 7.10.

7.8.3 Battery Selection

Since the batteries need to hold sufficient charge to power the cabin loads for a week, the batteries will not discharge in less than 10 hours. In addition, since the batteries will take a week to charge fully, the charge rate will also not be excessive. Thus, the charge and discharge derating factors will both be unity.

The factors that affect battery selection in this case are the monthly variation in average daily energy use and the monthly temperature variation of the batteries. In winter months, the batteries will be colder, even though they will be in an insulated enclosure, so temperature correction factors are needed. Table 7.11 shows the calculation of required battery capacity on a monthly basis. It is assumed that a deep-discharge unit will be chosen with a maximum depth of discharge of 80%. Once

Table 7.11 Determination of System Battery Capacity Requirements

	Jan	Feb	Mar	Apr	May	Jun
Ah/week	277	277	274	263	268	291
Temp derate	0.8	0.8	0.85	0.9	0.95	1
Total cap req	433	433	402	366	352	364
% cap used	89%	89%	82%	75%	72%	75%

	Jul	Aug	Sep	Oct	Nov	Dec
Ah/day	291	291	268	263	277	277
Temp derate	1	1	0.95	0.9	0.85	0.8
Total cap req	364	364	352	366	408	433
% cap used	75%	75%	72%	75%	84%	89%

the total battery capacity is determined, a battery can be selected that has a capacity close to an integral divisor of the total capacity. Since the batteries are allowed 3 days to discharge, the rating at C/72 can be used for sizing. In this case, a 244-Ah, 12-V, 8-year lifetime sealed lead-acid battery is a reasonable choice, as indicated by the ratio of total capacity to battery capacity. The system will require 8 of these units configured in two parallel sets of 4 units in series. The last row in Table 7.11 indicates the percentage of the battery capacity used each month.

7.8.4 Array Sizing and Tilt

Array sizing involves determining the optimum system design current by computing the design current for each month of the year at each of three tilt angles, then choosing the tilt angle that yields the lowest design current. Since the corrected weekly loads shown in Table 7.10 represent the energy that needs to be supplied to the batteries, once again it is possible to use the NREL SAM PVWatts model to determine the array sizes needed to meet the monthly load requirements. To begin, Table 7.12 shows the weekly corrected load in Ah converted to monthly corrected load in kWh, where the number of days in each month is taken into account and a 48-V dc system is assumed.

Simulations are then performed using a nominal 1-kW array to determine the monthly kWh/kW for each listed array tilt. These results are shown for the latitude −15°, latitude, and latitude +15° array tilts. Then, dividing the actual corrected kWh/month by the kWh/kW gives the necessary array size to produce the monthly energy needs of the cabin.

The next step is to note the maximum array sizes needed for each array tilt. In this case, the latitude −15° tilt will require a 620-W array to meet the December cabin load requirements. For a tilt at latitude, the array size can be reduced to 531 W to meet the December requirements, and for a tilt of latitude +15°, the array can be reduced to 511 W to meet the July requirements. Thus, if a 511-W array is used, with a latitude +15° tilt, it will exceed all the monthly needs except for July, which it will just meet.

Table 7.12 Determination of Optimum Design Current and Array Tilt Angle

Month	Corr Load (kWh/month)	Latitude −15° kWh/kW	kW	Latitude kWh/kW	kW	Latitude +15° kWh/kW	kW
Jan	58.9	100	0.589	115	**0.512**	122	0.483
Feb	53.2	102	0.522	112	0.475	116	0.459
Mar	58.2	145	0.401	152	0.383	148	0.393
Apr	54.2	147	0.369	143	0.379	133	0.407
May	56.9	154	0.369	144	0.395	126	0.452
Jun	59.9	151	0.397	138	0.434	118	0.507
Jul	61.9	149	0.415	139	0.445	121	**0.511**
Aug	61.9	148	0.418	143	0.433	130	0.476
Sep	55.1	138	0.399	141	0.390	136	0.405
Oct	56.0	126	0.444	137	0.409	140	0.400
Nov	57.0	100	0.570	113	0.504	120	0.475
Dec	58.9	95	**0.620**	111	**0.531**	120	0.491

For a 48-V system that uses an MPPT charge controller, two 260-W, 60-cell modules in series might be a possibility. This will result in $V_{OC} \cong 75.4$ V, $V_m \cong 63.2$ V, $I_{SC} \cong 8.73$ A, and $I_m \cong 8.24$ A for the pair in series. Before specifying this module, however, since V_m at STC is relatively low compared to a charging voltage of approximately 57 V for a 48-V battery system, the decrease in V_m during summer months should be checked.

Assuming the highest summer module temperature to be close to 70°C and a temperature coefficient of −0.44%/°C, V_m can be expected to decrease to $63.2[1 + (70 − 25)(−0.0044)] = 50.7$ V, which is too low for charging the batteries. So, regrettably, this module needs to be rejected. So, what are the options?

One option might be to consider a 24-V dc system. Essentially all the system calculations have already been made for a 24-V system, since system requirements have been converted to Wh. This would still use the same batteries, but connected as a 24-V system. Of course, a suitable 3-kW pure sine inverter will need to be found that will operate with a 24-V input voltage, but, in fact, such inverters are relatively common for stand-alone systems.

Another option is to look for 72-cell modules that add up to the design power required. So, rather than looking for 260-W, 60-cell modules, a search for 260-W, 72-cell modules could be made. While an abundance of 72-cell modules exist with power ratings in excess of 300 W, finding a 260-W module may require a bit more searching, and if one is found, it may not be available for purchase. A good place for a search for modules is the module database in NREL SAM, which shows a 265-W, 72-cell module with $V_m = 36$ V. As long as the cell temperature of this module does not exceed 62°C, the series pair of modules will have $V_m > 60$ V. So, even this module is not a comfortable solution, although, if necessary, a higher wattage 72-cell module will have somewhat higher V_m and will generate additional Wh when module temperatures are below about 65°C to make up for losses at higher temperatures. With module prices often below $1/W, this may be a cost-effective solution.

Yet another option is to look for three smaller modules that might be hooked up in series to obtain adequate V_m at high temperature. This leads to a search for a 175-W module with high enough V_m. And, indeed, the SAM module database lists at least two 72-cell modules rated at 175 W that have $V_m = 35.8$ V, $I_m = 4.91$ A, $V_{OC} = 43.8$ V, and $I_{SC} = 5.21$ A. Three of these modules in series operating at 70°C will have $V_m = 78.4$ V, so if either of these or a similar module is available, the problem is solved, provided that on a cold, winter day the V_{OC} of the three modules in series will not exceed the input limits of the charge controller. At STC, the combined V_{OC} of the three modules will be 131.4 V. At −30°C (−22°F), with a temperature coefficient of −0.386%/°C, V_{OC} of the combination increases to 131.4[1 + (−30 − 25)(−0.00386)] = 159.3 V, which may or may not be compatible with a charge controller.

7.8.5 Charge Controller Selection

The charge controller needs to be selected on the basis of its maximum continuous output current as well as its input voltage limits. To make this estimate, no array degradation is assumed and the wire and controller efficiencies are assumed to be 100%, to be sure that a conservative estimate is made. Then, another correction is made for cloud focusing by multiplying the array rating by 125%. Finally, the batteries are assumed to be charging at their nominal voltage. So the calculation is simply $I_{out} = 1.25 P_{array}/48 = 13.7$ A for the 525-W array. So any MPPT charge controller with an output current rated at 13.7 A or more will be more than adequate for the system as long as it can handle the array size and has a 48-V output. Keep in mind that if a controller is rated at 30-A output, it can control any current up to this value. So the challenge is to find an MPPT charge controller that can handle the winter worst case V_{OC} (161 V).

A Web search reveals that most 30-A MPPT charge controllers have a maximum allowable V_{OC} of 150 V. However, it turns out that at least one manufacturer incorporates HyperVoc design, which allows the input V_{OC} to go as high as 162 V. Wow! What luck! Persistence often pays off! Admittedly, this is a pretty close match, but by the time this is read, it is anyone's guess as to what additional products may serve the purpose. The important part is to know how to determine what is needed.

Note that this controller will not be used to control the discharge of the batteries. It will do two things: (1) prevent the batteries from overcharging and (2) maximize the energy delivered by the array.

7.8.6 Inverter Selection

Since the load on the inverter may go as high as 2990 W and since there will be at least two ac motor loads, an inverter that produces a utility-grade sine wave will be selected. The selected inverter will need to (a) meet the power requirements of the loads, (b) monitor battery voltage, and provide for low battery cut-out and low battery cut-in operation. Additional bells and whistles may also come in the inverter package, but (a) and (b) along with the highest conversion efficiency possible are the

STAND-ALONE PHOTOVOLTAIC SYSTEMS

key selection criteria. For example, some units may be available in 120/240-V output models. Many will have battery charging capability, so that an external ac generator can be connected to charge batteries if the sun does not shine enough. If the inverter has battery charging capability, then it may have current limiting, automatic generator start, multiple-stage charging, and so on. Once again, there are sufficiently many inverters presently available that will meet the design specifications that the reader is encouraged to do a bit of surfing on the Web to become familiar with some of the options. For the BOS calculations, it will be assumed that the inverter selected will be rated at 3000 W, with a nominal 48-V dc input and an efficiency of 94%.

7.8.7 Excess Electrical Production

It is interesting to look at the amount of excess kWh produced by the system on a month-by-month basis, since the array has been selected to produce adequate energy during the worst month. By multiplying the monthly kWh/kW by the array size in kW (using the 525-W array), the available load kWh/month can be obtained. Then simply subtract the corrected monthly kWh from the available monthly kWh to obtain the excess monthly kWh. The results are shown in Table 7.13 as excess system weekly Ah.

The next logical challenge for the designer is to figure out what to do with this extra electricity. Comparison with the Ah requirements of the cabin loads shows that the spring and fall excess is approximately equal to the weekly receptacle Ah. So maybe a vacuum cleaner might be used in spring and fall and a broom might be used in winter and summer. Then there is the possibility that more hot coffee, tea, or other beverage might be needed during cold winter months via an electric teapot for heating the water.

Since it takes 1 BTU to raise the temperature of 1 lb of water by 1°F, and since 1 kWh is equivalent to 3413 BTU, one need simply convert the excess kWh to BTU. For example, in April, the excess 15.6 kWh will generate 53,243 BTU. With a cold water temperature of 55°F, heating a gallon of water to a temperature of 180°F will require $8.35 \times 125 = 1044$ BTU, since a gallon of water weighs 8.35 lb. Hence, the leftover 53,243 BTU is enough to heat 51 gallons of water to 180°F. Of course, if any heat is lost from the pot, then not all the available BTU will go into heating the coffee water, but even at 80% efficiency, 40.8 gallons of coffee should be enough to keep everyone happy, since this would be almost a gallon per day per person if four people stay 3 days/week.

Table 7.13 Average Weekly Excess Ah Produced by the Selected Array for the Cabin

	Jan	Feb	Mar	Apr	May	Jun
Ah	24.2	40.1	92.0	76.1	43.5	10.1
	Jul	Aug	Sep	Oct	Nov	Dec
Ah	7.7	30.0	79.4	82.4	29.1	19.2

Another possibility would be to pump extra water. It is left as an exercise for the reader to determine the additional water that could be pumped with the leftover electricity (Problem 7.22).

It is up to the engineer to decide which use, if any, will make economic sense for this extra electricity. The determination will ultimately be dependent upon specific preferences of the cabin owner. Who knows, since it is turning out to be such a nice cabin, maybe the owner will want to spend 4 days/week at the cabin during the months with the greatest excess production.

7.8.8 BOS Component Selection

7.8.8.1 Wire, Circuit Breaker, and Switch Selection

The nice thing about an ac cabin is that no extraordinary effort is required for sizing the wiring to most of the loads, as long as they are no more than 100 ft (30.5 m) from the distribution panel. In this case, everything is close with the exception of the well pump, for which a voltage drop calculation will be needed. Table 7.14 summarizes maximum system component currents and Table 7.15 summarizes one possible selection of circuits along with wire sizing and sizing of fuses. For the dc circuits, ampacity and voltage drop calculations will be needed, but the procedure is the same as in all previous examples. For example, array wiring will need to be sized at a minimum of 156% of I_{SC}. The inverter input current can be determined by dividing the inverter output power rating by the minimum inverter input voltage and the inverter efficiency. This gives the battery-to-inverter current of Table 7.14. The reason the battery-to-inverter wiring is nearly twice the size based on the inverter input current is because most inverters can produce surge currents that are double the normal maximum operating current. Thus, the larger wire and larger circuit breaker ensures that the inverter will be able to generate the surge current that may be needed to start motors or other loads that encounter large surges when starting. The inverter output current is simply the quotient of the inverter rated power and the inverter ac output voltage (120 V). The rated power of the inverter is used, since the wiring between inverter and distribution panel must be capable of carrying 125% of the rated inverter output current.

In Table 7.15, the *NEC* currents incorporate the appropriate multipliers, such as 156% for array and charge controller currents and 125% for the remaining currents. It turns out that all the calculations in this table lead to wire sizing based on necessary ampacity. Note that the cabin lights and outlets are wired with #14 (2.08 mm²) wire as would be done in a grid-connected residence. The kitchen circuits are 20-A circuits that are wired with #12 (3.31 mm²). The 2014 *NEC* requires arc fault circuit

Table 7.14 Summary of Component Maximum Currents for Cabin

Array to Controller	Controller to Battery	Battery to Inverter	Inverter to Panel
5.21 A (I_{SC})	10.5 A	69.4 A	24.4 A

STAND-ALONE PHOTOVOLTAIC SYSTEMS

Table 7.15 Summary of Circuits, Wiring, and Overcurrent Protection for the Cabin

Wire Location	Max (A)	NEC (A)[a]	Length (ft)	Max (Ω/kft)	Wire Size	C.B. Size (A)
Array to controller	5.2	8.13	30	4.75	#14	10.0
Controller to batteries	10.5	16.4	6	7.62	#10	30
Batteries to inverter	69.4	86.7	6	1.15	#2/0	175
Inverter to panel	24.4	30.5	5	3.94	#8	40
Refrigerator	16	20	<100	1.93	#12	20
Water pump	1.0	1.25	230	5.00	#14	15
Kitchen receptacles	16	20	<100	1.93	#12	15
LR receptacles	12	15	<100	3.07	#14	15
BR 1 receptacles	12	15	<100	3.07	#14	15
BR 2 receptacles	12	15	<100	3.07	#14	15
Bath and outside receptacles	12	15	<100	3.07	#14	15
Lights and fans	12	15	<100	3.07	#14	15

[a] Required for ampacity calculations. All wire sizes based on ampacity requirements.

breakers on all the 15-A circuits except the well pump and ground fault circuit interrupters on kitchen, bathroom, and outside receptacles.

Switches should have the same current rating as the fuses or circuit breakers along with adequate voltage rating for the purpose. For lighting, the switches sell for less than $1 at your favorite home improvement store. Fans come with fan controllers rated for the purpose. The pump will need to be controlled with a level switch in the storage tank and possibly with a time switch to limit pumping to certain hours. These switches need to be adequately rated to carry the pump motor current, which is a comparatively small amount.

7.8.8.2 Other Items

Other items that need to be included are array mount, distribution panel(s), lightning arrestors, ground rod, battery container, battery cables, and miscellaneous connectors and junction boxes. Most of these components, except perhaps those related to the batteries, will be used in the electrical system regardless of the source of electricity.

Once wire length and size are known, wire cost can be calculated. For the cabin, a distribution panel with ac circuit breakers rather than fuses is used to supply individual circuits to the various loads, with some doubling up of loads on a single circuit as indicated in Table 7.15. Lightning arrestors are important for protection of the total system and are normally located at the controller.

The array mount may take many shapes and forms, from commercially available units to homemade wooden frames. For the cabin, a commercial pole mount has been selected.

Figure 7.11 shows the cabin system electrical schematic diagram. Equipment grounding conductors are shown as dashed lines. On the dc side, #10 (5.261 mm^2) conductors are used because the array grounding conductors will be exposed to the

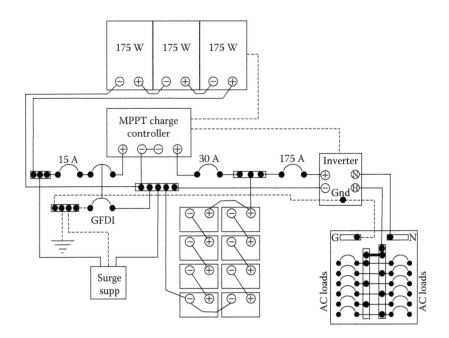

Figure 7.11 Cabin system based on MPPT charge controller and array tilted at latitude + 15°.

weather and thus will be bare wire. Also, the 30-A circuit breaker at the charge controller output requires a #10 grounding conductor for the charge controller. The remaining equipment grounding conductors will also be #10. If the batteries are in a metal container, the metal container will need to be grounded with #6 (13.3 mm^2) because of the 175-A circuit breaker. If the battery enclosure is nonconductive, then no ground is needed for the enclosure.

7.9 A HYBRID-POWERED, OFF-GRID RESIDENCE

7.9.1 Introduction

This design example investigates the method of choosing alternate generation capacity to supplement the output of the PV array when there is a large discrepancy between month-to-month system needs versus month-to-month PV generation capacity. If installation of a PV array to meet minimum sun availability results in significant excess generation for a number of months, then much of the PV output is wasted. In such cases, it may make better economic sense to use a generator to supplement the PV output during the months of low PV output and size the PV to meet most or all of the needs during months of higher peak sun.

It was observed in Chapter 3 that there is a significant cost increase between sizing a PV system to provide 95% of system electrical needs versus providing 99% of system needs. Hence, the use of a generator for increasing system availability from

STAND-ALONE PHOTOVOLTAIC SYSTEMS

general to critical may also be cost-effective. In this chapter, the design will be completed with several array and generator options. In the next chapter, a life-cycle cost analysis will be performed on the options.

In the present example, it is desired to power a residence near Bismarck, ND (lat 46.8°N), but away from the utility grid, with a combination of PV and generator. The residence will be occupied 7 days/week all year long. All loads will be either 120 or 240 V ac as in typical residences, except that energy efficiency has been taken into account in the selection of the loads and the design of the dwelling. For example, a high-efficiency refrigerator has been selected and all lighting will be LED. The dwelling is designed with fans and is highly insulated with triple-pane glass in the windows. Much of the heating will be by passive solar, but propane will be used for supplemental heating, solar water heating backup, and cooking. Since propane will be on site, it will also be used for the generator. The goal of the design is to arrive at a combination of PV and propane generator energy production that will result in the lowest life-cycle cost. The reader may wish to note that this design example is relatively unchanged from the hybrid residence example in previous editions of this book in terms of the loads and the choice of inverter and generator. One significant change will be the substitution of LFP batteries for the AGM lead-acid units that have been used in previous designs. At the time of this writing, lithium batteries are now becoming the norm for cell phones, laptop, notebook and tablet computers, hybrid and electric vehicles, and large airplanes. It is anticipated that they will also become a preferred technology for backup storage in renewable energy systems as well, at least for the near future. Use of these batteries will require a different charge control algorithm to ensure safe use of the batteries. Although initial cost of LFP batteries is higher than that of sealed lead-acid batteries, the life-cycle cost, as will be discussed in Chapter 8, is becoming more and more attractive. To reduce initial cost, storage day design will be reduced from the 3 days of previous editions to 2 days for the present design.

What is especially interesting is to observe the increase of the ratio of PV generation to fossil generation to achieve maximum system cost-effectiveness, as the cost of PV generation has dropped significantly over the past 16 years.

Since PV and generator output capability come in steps, this design process will explore several combinations of PV and generator. Up to this point, little discussion of generator properties has taken place, so a part of the design discussion will deal with the selection of generators. The design process is somewhat modified from the previous examples, since the propane generator also needs to be sized and incorporated into the system. This means a more complicated system block diagram and presents an additional challenge to the battery charge control system.

Since the generator is available for PV system backup, it may seem unnecessary to incorporate batteries into the system at all. However, for generators to operate efficiently, they need to run at close to 90% of their output capacity. Operation of a generator at a small fraction of capacity will result in significant decrease in efficiency, as was noted in Chapter 3. Hence, batteries are used so that the generator can charge them at a rate close to its capacity. Furthermore, at the time of this writing, the authors are unaware of any stand-alone inverters that will operate without batteries.

Sometimes a generator can be used to fool a straight grid-connected inverter into believing that the generator is the utility, but if the inverter then tries to "sell" electricity to the grid, it ends up trying to sell it to the generator, thus turning the generator into a motor. When this is done, the "motor" tries to run the prime mover, rather than vice versa. Prime movers are not the least bit happy with such a role reversal. They will generally change speed, and, hence, frequency, resulting in shutting down the inverter per UL 1741 requirements.

Since charging batteries too quickly tends to result in an inefficient charging process, the generator/battery system should be sized so the generator will take at least 5 hours to charge the batteries [10]. The battery selection criteria previously discussed appear to present a more conservative requirement of at least 10 hours to fully charge the batteries. Actually, these criteria are not necessarily inconsistent, since the batteries will not normally be charged from full discharge to full charge. Normally, the generator will charge the batteries from about 20% to about 70%. Charging the batteries from 20% to 70% in 5 hours requires a charging rate of C/10.

The bottom line, then, for batteries, is to provide a few days of storage so the charging rate will not be excessive. More storage will normally result in somewhat lower use of the generator, since the generator will not necessarily need to back up the PV array in the event of cloudy weather for a few days. Longer storage times may be desirable in areas where summers have periods of sunny days followed by periods of cloudy days. In general, fewer batteries will be used in a hybrid system since the propane generator will supplement the sun. Choice of the number of days of autonomy for the system, however, becomes more dependent on other factors, such as how long it may take to implement emergency repairs on the generator.

After loads and batteries are selected, the mix of PV and propane generation is determined. Then the array size and the generator size are calculated, followed by calculation of generator fuel use and maintenance costs. Then controllers, inverters, battery chargers, fuses, wires, and other BOS components are selected.

7.9.2 Summary of Loads

Table 7.16 summarizes the estimated loads for the residence by season. All loads are 120 V ac loads and the input of the inverter will be 48 V dc. The loads are tabulated in Ah at 48 V dc. A conversion efficiency of 92% is assumed for the inverter and wiring from inverter to load, so the connected loads include this figure and thus represent the energy that must be supplied by the batteries to the inverter to operate the loads. The corrected load figures in Table 7.16 represent the connected loads divided by the battery charge/discharge efficiency and are therefore the Ah needed to be supplied to the batteries by the PV array, the generator, or both.

It is assumed that the household will have common electric small appliances such as toasters, blenders, vacuum cleaners, hair dryers, and so on and that no more than 2400 W of small appliances will be used at any one time. It is also assumed that the fan motor of the propane furnace and the ceiling fans will not run simultaneously. Since the lighting has now been converted to LED lamps, additional energy savings for lighting have enabled adding a few more receptacle loads, such as

STAND-ALONE PHOTOVOLTAIC SYSTEMS 337

Table 7.16 Estimated Seasonal Loads for Hybrid Residence

Load Description	P (W)	Dec, Jan, Feb Hour/Day	Ah/Day[a]	Mar, Apr, May, Sep, Oct, Nov Hour/Day	Ah/Day[a]	Jun, Jul, Aug Hour/Day	Ah/Day[a]
Kitchen lights	20	4	1.81	3.5	1.43	2	0.91
Dining R lights	15	2	0.68	1.5	0.46	1	0.34
Living R lights	15	1	0.34	1	0.31	1	0.34
Family R lights	15	4	1.36	4	1.23	3	1.02
BR 1 lights	10	1	0.23	1	0.20	1	0.23
BR 2 lights	10	1	0.23	1	0.20	1	0.23
BR 3 lights	10	1	0.23	1	0.20	1	0.23
Bath 1 lights	10	1	0.23	1	0.20	1	0.23
Bath 2 lights	10	1	0.23	1	0.20	1	0.23
Refrigerator	150	7	23.78	7	21.46	7.5	25.48
Microwave oven	600	0.5	6.79	0.5	6.13	0.5	6.79
TV	120	4	10.87	4	9.81	4	10.87
Stereo	80	2	3.62	2	3.27	2	3.62
Water pump	200	1	4.53	1	4.09	1	4.53
BR 1 fan	50	0	0.00	0	0.00	8	9.06
BR 2 fan	50	0	0.00	0	0.00	8	9.06
BR 3 fan	50	0	0.00	0	0.00	8	9.06
Family R fan	50	0	0.00	0	0.00	8	9.06
Furnace fan	500	8	90.58	6	61.32	0	0.00
Washer	600	0.5	6.79	0.5	6.13	0.5	6.79
Vacuum and Sm App	2400	0.5	27.17	0.5	24.53	0.5	27.17
Receptacles, Misc	1000	0.75	16.98	0.75	15.33	0.75	16.98
Total	5965		196		157		142
Corrected total			218		174		158

[a] At 48 V dc.

charging computer batteries or operating a few other computer peripherals, as long as the overall power consumption of the residence does not exceed 5965 W. These assumptions may affect sizing of the generator and the wiring, depending on whether any of the load will need to be supplied by the generator while it is running to charge batteries. In this particular design example, the assumption is that the generator is used only for battery charging. Wiring, fusing, and switching between batteries and inverter and between inverter and distribution panel, however, is based on the rating of the inverter as required by the *NEC*.

7.9.3 Battery Selection

The batteries will be reasonably well protected from the cold North Dakota winter, but will experience somewhat lower winter temperatures. The design goal for the system will be to maximize PV energy production without experiencing unreasonably high system life-cycle cost or unreasonable waste of PV-generated electricity. Noncritical battery backup is considered satisfactory for this system, with a choice of 2 days of storage to allow the sun 2 days to reappear before calling in the generator as a replacement for the sun. This period also provides time for emergency repair of the generator or other system components. If a generator is warranted (and hopefully this will be the case since this design example calls for one to be used), it will be selected to charge the batteries in 10 hours or more so there will not need to be any compensation for quick charging of the batteries. Also, with 2 days of storage, it is highly unlikely that quick discharge of the batteries will occur. Finally, in this case, Equation 3.15 was not used as the means of determining the number of days of storage, even though the problem stated noncritical storage. The presence of the generator adds another source of energy that was not used in determining Equation 3.15.

On this basis, battery capacity is shown in Table 7.17 with temperature compensation taken into account, with an assumed depth of discharge of 75%, based on the corrected loads of Table 7.16. Table 7.17 also shows the number of batteries required if a 48-V LFP battery pack with a capacity of 200 Ah at C/72 is selected for the system. Note that three of these battery packs in parallel are needed to meet the required total Ah capacity. Note also that the selection of this battery will provide more than 2 days of storage for much of the year. However, at 90% capacity rating during the winter, they will provide 95% of the capacity required for 2 days of storage. Thus, during winter months, the batteries will provide 1.9 days of storage, while in the summer, they will provide nearly 3 days of storage.

Since each LFP battery pack will have its own battery management system (BMS), the final system design will need to incorporate a special control system such that the BMS, generator controller, and system charge controller(s) will interact accordingly to achieve the desired battery charging control algorithm that will minimize generator run time and provide necessary protection for the batteries. In addition, when excess PV is available, the charge control algorithm must ensure that the batteries are never disconnected from the system, since the batteries supply the necessary operating voltage for the inverter, charge controller, and other system electronic components. In simple terms, it means that rather than having battery voltage as the parameter that controls whether the array or generator is on or off, the battery SOC should be the controlling variable.

Table 7.17 Battery Storage Requirements for Hybrid Residence @ 48 V

	Dec–Feb	Mar, Nov	Apr, Oct	May, Sep	Jun–Aug
Corrected Ah/day	218	174	174	174	158
Temp derate	0.9	0.93	0.96	0.98	1
2-day cap req	647	499	483	473	421
# Batt	3	3	3	3	3
Avail capacity	600	600	600	600	600
# Days storage	1.9	2.4	2.5	2.5	2.8

7.9.4 Array Design

Sizing of the array for a hybrid system is generally an iterative process. The first step is to size the array for a system with no generator and then to gradually reduce the number of modules in the array while simultaneously computing the percentage of the annual energy needs provided by the PV array. Once the collection of designs is complete, then the economic advantages of each system can be calculated. The designs will be completed in this chapter and an economic analysis will be made in Chapter 8.

A somewhat different approach to array sizing will be taken in this example. Since December has the lowest sunlight levels for the year, regardless of array tilt, and since December also has the highest load, the NREL SAM PVWatts model [3] will be used to determine an array size and tilt that will meet the December load requirements. The output of this array will then be compared with system load requirements for the remaining months of the year. Assuming MPPT charge control, the array daily energy output can be estimated from the battery input energy requirement. Since the December corrected daily Ah requirement is 218, if the battery charging voltage is 54 V, then the charging energy is $54 \times 218 = 11{,}772$ Wh, or 11.8 kWh. The array can thus be configured to produce 11.8 kWh/day = 366 kWh/month in December with a winter tilt. Tilt can then be adjusted to minimize overproduction during other months. Table 7.18 shows the results for an array tilt of 62°. Again, this is the PVWatts model result for ac kWh from a string inverter, which we have been assuming to be equivalent to the dc output of an MPPT charge controller.

Table 7.18 shows the annual results for a tilt of 62° (latitude +15°) and array size (3.6 kW) that will meet December requirements. The table also shows the excess

Table 7.18 Determination of Array Size and Tilt for Hybrid Residence

Month	Corr Load @ 48 V (Ah)	Corr Load (kWh/month)	Array Rated (kW)	Monthly Avg Array (kWh)[a]	Excess Daily (kWh)
Jan	218	325	3.6	386	1.98
Feb	218	293	3.6	386	3.32
Mar	174	259	3.6	505	7.96
Apr	174	250	3.6	456	6.86
May	174	259	3.6	453	6.28
Jun	158	228	3.6	428	6.67
Jul	158	235	3.6	454	7.07
Aug	158	235	3.6	461	7.29
Sep	174	250	3.6	433	6.07
Oct	174	259	3.6	404	4.70
Nov	174	250	3.6	307	1.90
Dec	218	325	3.6	335	0.34

[a] Based upon 62° array tilt.

energy production of the system for each month of the year. Note that the excess energy production corresponds to *corrected* load energy requirements, as opposed to actual array energy production.

Now that the array size has been determined, a module needs to be selected for which the source circuit $V_{OC} < 150$ V when the temperature is −40°C (−40°F) and a sufficient number of source circuits can be incorporated to reach the required 3.6-kW array power.

One might imagine that it might be a good idea to look at available 300-W modules, since this would require 12 modules. If the modules can be connected in two-module source circuits, then the system will require six source circuits. Alternatively, one could look for 200-W modules, such that it would take 18 modules to achieve 3.6 kW. Then the source circuits might consist of three modules, depending upon the source circuit V_{OC} limits.

At this point, it is useful to remember that a propane generator will also be incorporated into the system. Thus, it would be helpful to have a larger number of source circuits rather than a smaller number, if it is desired to explore the effect on life-cycle system cost as PV generation is replaced with propane generation. This way, as each source circuit is removed, an additional increment of propane generation will need to be added.

Checking the possible options again suggests that the 300-W module option may meet this requirement better than the other module sizes, especially if two-module source circuits will be possible.

One module rated at 300 W has $V_{OC} = 45.4$ V, $I_{SC} = 8.7$ A, $V_m = 37.8$ V, and $I_m = 8.2$ A, with NOCT = 48°C and $\Delta V_{OC}/\Delta T = -0.34\%/K$. Thus, at −40°C (−40°F), for two modules in series, $V_{OC} = 90.8[1 + (65 \times 0.0034)] = 111$ V, which is acceptable. Also, in summer, if module temperatures reach 65°C (149°F), for two modules in series, V_m will drop to $V_m = 75.2[1 - (40 \times 0.005)] = 60.2$ V, which is also just adequate input for an MPPT charge controller charging 48-V batteries.

The next step is to tabulate the available energy for battery charging from 1 to 7 source circuits for each month of the year, as shown in Table 7.19. In this table, the excess PV is calculated on a monthly basis, but the annual percentage of the total load requirements met by the PV array counts the months with positive excess as just meeting the needs for the month, since the rest of the energy produced is either wasted or used for unanticipated purposes. But the months with negative excess are months where the generator will need to make up the difference, so the amount the generator will need to supply is subtracted from the total annual requirement, then divided by the total annual requirement, then converted to a percentage. This gives the percentage of annual energy needs supplied by the PV system. The remaining percentage must be supplied by the generator. Table 7.19 indicates that the 10-module array, which supplies 5/6 as much energy as the 12-module array, needs a boost from the generator during December and January, but supplies all the needed energy during the rest of the year. Each time a source circuit is removed, another 1/6 of the 12-module array output is removed. By the time the array is reduced to three source circuits (1800 W), it needs supplementary energy every month of the year, but the PV system still provides nearly 80% of the annual energy needs.

STAND-ALONE PHOTOVOLTAIC SYSTEMS

Table 7.19 Monthly Excess kWh Capability of PV Array for Seven Array Sizes

Month	Corr Load (kWh/ Month)	14 Mod 4200 W	12 Mod 3600 W	10 Mod 3000 W	8 Mod 2400 W	6 Mod 1800 W	4 Mod 1200 W	2 Mod 600 W
Jan	325	126	61	−3	−67	−132	−196	−260
Feb	293	157	93	28	−36	−100	−165	−229
Mar	259	331	247	162	78	−6	−90	−175
Apr	250	282	206	130	54	−22	−98	−174
May	259	270	195	119	43	−32	−108	−183
Jun	228	272	200	129	58	−14	−85	−156
Jul	235	295	219	144	68	−8	−84	−159
Aug	235	303	226	149	72	−5	−81	−158
Sep	250	254	182	110	38	−34	−106	−178
Oct	259	213	146	78	11	−57	−124	−191
Nov	250	108	57	6	−46	−97	−148	−199
Dec	325	66	10	−45	−101	−157	−213	−269
Ann PV %		100	100	98	92	79	53	26

The next step is to tabulate the amount of energy that must be supplied by the generator on a monthly basis under the seven different PV array scenarios. This listing is shown in Table 7.20 and is simply the PV deficits from Table 7.19.

At this point, it is possible to determine a size for the generator and to then determine the monthly operating hours for the generator.

Table 7.20 Required Monthly kWh to Batteries from Generator for Seven PV Array Sizes

Month	kWh Needed	14 Mod	12 Mod	10 Mod	8 Mod	6 Mod	4 Mod	2 Mod
Jan	325	0	0	3	67	132	196	260
Feb	293	0	0	0	36	100	165	229
Mar	259	0	0	0	0	6	90	175
Apr	250	0	0	0	0	22	98	174
May	259	0	0	0	0	32	108	183
Jun	228	0	0	0	0	14	85	156
Jul	235	0	0	0	0	8	84	159
Aug	235	0	0	0	0	5	81	158
Sep	250	0	0	0	0	34	106	178
Oct	259	0	0	0	0	57	124	191
Nov	250	0	0	0	46	97	148	199
Dec	325	0	0	45	101	157	213	269
Total	3168	0	0	48	250	663	1498	2333
Ann gen %		0.0	0.0	1.5	7.9	20.9	47.3	73.6

7.9.5 Generator Selection

The generator should be sized to charge the batteries at approximately C/10 as indicated previously. At this rate of charging, the generator should be operating at approximately 80–90% of its rated output in order to ensure maximum generator efficiency. Hence, it is first necessary to determine what C/10 means for this system.

The first question when analyzing Table 7.17 is which capacity should be used in determining C/10? For LFP batteries, battery capacity varies minimally with temperature over the temperature range at the installation, so in this case, it is acceptable to use 600 Ah as the total battery charge to be delivered to the batteries in 10 hours. Thus, 60 A for 10 hours will deliver 600 Ah. In fact, in order to be sure the PV array supplies maximum energy to the batteries, the generator will be programmed to start only if the battery SOC drops to 25% and will shut off when the battery capacity reaches 75%. So the generator will only operate for a maximum of 5 hours, unless it is simultaneously supplying power to inverter loads.

The generator will be rated in watts. Thus, it is necessary to determine the charging power, rather than the Ah requirements of the batteries. This is obtained by incorporating the battery charging voltage. For a C/10 charging rate, the battery charger output power will thus be the product of the charging voltage and the charging current, or $48 \times 1.2 \times 60 = 3456$ W, since the charging voltage is generally about 120% of the nominal battery voltage. If the electrical conversion efficiency of the battery charger is 90%, this would require an input power of $3546/0.9 = 3840$ W to provide 60 A of charging current. Since the generator is to be used only for battery charging, a 5000-W generator would be a reasonable choice, since it will run at 77% of its rated output when charging at a C/10 rate. If the generator is set to deliver at 90% of its rating, then the charge rate drops to C/8.5, which is still well within the acceptable rate for an LFP battery system.

If it is desired to have the generator run part of the load while charging the batteries, or to charge at a faster rate, it could be reasonable to select the next larger size generator. However, the 5000 W unit operating at 3840 W will charge the batteries from 25% to 75% in 5 hours. This is normally an acceptable charge rate, especially since it will take 1.5 days to use this amount of charge during winter months. Charging the batteries to only 75% allows for the PV array to top off the charge if it can. If not, and the batteries discharge to 25% again, the generator comes on again. For the following calculations, the C/8.5 (4500 W) rate will be used.

7.9.6 Generator Operating Hours and Operating Cost

The more the generator runs, the more it will cost to operate. Generator operating cost consists of fuel cost, oil changes, tune-ups, and rebuilding. To determine the operating cost, then, it is necessary to determine the annual operating hours of the generator.

STAND-ALONE PHOTOVOLTAIC SYSTEMS

Table 7.20 lists the monthly kWh requirements of the batteries from the generator. The generator kWh output (kWh$_{Gen}$) needed to supply the battery kWh (kWh$_{Batt}$) needs is found from

$$kWh_{Gen} = \frac{kWh_{Batt} \times 1.2}{\eta} \qquad (7.5)$$

where 1.2 is the ratio of charging voltage to nominal battery voltage and η is the efficiency of conversion of ac to dc in the battery charger.

The monthly hours of generator operation can then be determined from the monthly kWh$_{Gen}$ by the output power at which the generator is set to operate. Since most generators operate at maximum efficiency at approximately 90% of their rated output power, it will be assumed that the generator will be set at the C/8.5 rate of 4500 W. It is also reasonable to assume a battery charger conversion efficiency of 90%, as was done during the process of calculating the needed generator power.

The monthly generator fuel consumption can be determined from knowledge of either the hourly fuel consumption or the fuel consumed per kWh of generation. A utility-grade generator will generate approximately 12 kWh/gallon of fuel [13], but a small generator will generate considerably less, since the combustion efficiency of a small gasoline, propane, or diesel engine is much less than the efficiency of a large utility steam turbine. Table 7.21 estimates fuel consumption on the basis of a generator output of 5 kWh/gallon of fuel.

Table 7.21 Monthly Generator Operating Hours and Annual Fuel Use for Seven PV Array Sizes

Month	Battery kWh for Loads	14 Mod	12 Mod	10 Mod	8 Mod	6 Mod	4 Mod	2 Mod
Jan	325	0	0	0.9	20.0	39.0	58.1	77.2
Feb	293	0	0	0	10.6	29.7	48.8	67.8
Mar	259	0	0	0	0	1.8	26.8	51.7
Apr	250	0	0	0	0	6.6	29.2	51.7
May	259	0	0	0	0	9.5	31.9	54.3
Jun	228	0	0	0	0	4.0	25.2	46.3
Jul	235	0	0	0	0	2.3	24.8	47.2
Aug	235	0	0	0	0	1.4	24.1	46.9
Sep	250	0	0	0	0	10.1	31.5	52.8
Oct	259	0	0	0	0	16.8	36.7	56.7
Nov	250	0	0	0	13.5	28.7	43.8	59.0
Dec	325	0	0	13.4	30.0	46.6	63.1	79.7
Ann gen op h		0.0	0.0	14.3	74.1	197	444	691
Ann gen gal fuel		0.0	0.0	12.9	66.7	177	400	622
Ann gen fuel cost[a]		0	0	$30	$153	$407	$919	$1431

[a] At $2.30/gallon.

Table 7.22 Annual Generator Maintenance Frequency for Seven PV Arrays

Item	14 Mod	12 Mod	10 Mod	8 Mod	6 Mod	4 Mod	2 Mod
Op h/year	0.0	0.0	14.3	74.1	196.5	444.0	691.4
Oil changes/year	0	0	1	5	13	30	46
Tune-ups/year	0	0	0	0	1	1	2
Year/rebuild			209	40	15	7	4
Ann cost ($)	40	40	50	90	220	440	725

With a generator output of 4500 W, it will take 1.11 hours to generate 5 kWh$_{Gen}$, resulting in a fuel consumption rate of 0.90 gallon/hour. Assuming an ac/dc conversion efficiency for the battery charger of 0.9, use of Equation 7.5 results in 5 kWh$_{Gen}$ providing 3.75 kWh$_{Batt}$. Thus, 1.0 gallon of fuel will produce 3.75 kWh$_{Batt}$ in 1.11 hours. This information can now be used to complete Table 7.21 to determine the generator fuel consumption and generator hours of operation for the seven proposed array sizes.

Generator annual operating costs can now be estimated from the maintenance information of Table 3.7, along with known fuel cost. Table 7.22 summarizes the required generator maintenance for the seven PV arrays previously considered. The table entries are determined by the quotient of the annual operating hours and the maintenance intervals of 25 hours per oil change, 300 hours per tune-up, and 3000 hours per rebuild. Once the cost of each of these items is known, the annual maintenance cost can be determined. As will be seen in Chapter 8, annual maintenance costs can be an important component of the system life-cycle cost, depending upon the frequency of maintenance operations.

7.9.7 Charge Controller and Inverter Selection

The control for a hybrid system is somewhat more complicated than the control for a conventional PV system. It must control battery charge and discharge by both the PV array and the generator. It must provide a starting signal/voltage for the generator when the batteries have discharged to a preset level and must shut down the generator when the batteries reach a preset level of charge. Proper setting of these levels is essential. Too low a setting on discharge may render the batteries unable to provide starting current for the generator, unless a separate battery is used for generator starting. Too high a setting may result in the generator's unnecessarily replacing energy that might be available from the PV array, with the PV array then being shut off with energy to spare. Fortunately, these control functions can be accomplished with an MPPT charge controller and an off-grid version of the inverter used previously in a grid-connected, battery-backup configuration, provided that the generator operation can be controlled by the BMS controller. The MPPT charge controller provides efficient battery charging and maximizes PV array output, while the inverter provides the interface between generator and batteries by using either the AC IN port or the GEN (AC2) IN port (if available) as the generator connection port, with all the programmable charging features of the grid-connected version except for the "sell" option and its corresponding tight controls on voltage and frequency at the AC IN

STAND-ALONE PHOTOVOLTAIC SYSTEMS

port. The BMS controller must be capable of monitoring battery SOC and providing a generator start command when battery SOC drops too low and a generator stop command when the battery SOC is adequate.

Obviously, in addition to the battery charging function, regardless of the array size, the inverter must supply all the loads of the house, which have been previously tabulated at 5965 W in Table 7.16. So an inverter rated at a minimum of 5965 W is needed. Fortunately, an inverter that supplies 6800 W at 120/240 V is now available that has a separate generator input and operates on a 48-V battery system. Depending upon the array size, either one or two charge controllers will be needed, if it is desired to have the inverter and charge controller(s) communicate with each other. Perhaps by the time this text reaches publication, a larger charge controller will be available that will handle an array power of 3600 W. The controller presently available works best with array power ratings below 3200 W.

So if it is decided to use the 3000-W array rather than the 3600-W array, the cost of one charge controller as well as the cost of two modules will be avoided. The trade-off, however, is either using a kWh or two less energy during November and December, or including a generator in the system. For smaller arrays, the generator will be needed in order to prevent a significant reduction in daily kWh use.

Although the peak efficiency of the inverter is very close to 95% for 1000 W < P_{out} < 3000 W, the efficiency drops off at low power levels and at high power levels as shown in Figure 7.12. When P_{out} < 300 W, the efficiency drops below 90%.

Since the average power consumption of the loads is 414 W or less, depending upon the month of the year, this means there may be times when the inverter is

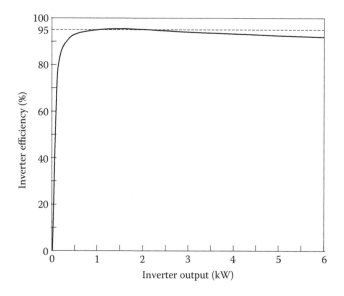

Figure 7.12 Inverter efficiency versus output power.

operating at efficiencies less than 90%. This means higher percentage losses in the inverter when the load on the inverter is small, which means the overall system load, including inverter losses, will be somewhat higher than calculated. For example, if the inverter delivers 100 W at 70% efficiency, this means the inverter input power must be 143 W, rather than the 105 W that would result if the inverter efficiency were 95% as assumed in the connected load calculation. Over a 10-hour period, this additional 38 W loss amounts to 0.38 kWh, which must be added to the daily load on the batteries. With a daily kWh consumption of approximately 10 kWh, this amounts to an additional 3.8% load on the system. The bottom line is that, depending upon the number of modules chosen for the system, there will be less excess kWh/month for months where the PV delivers excess kWh, and the generator will run slightly longer on the months when the PV does not provide excess kWh. To provide this additional 0.38 kWh/day, the generator will need to run an additional 6.7 minutes/day. Fortunately, the inverter incorporates an adjustable search mode control, so the inverter will "sleep" if the connected loads are below the sleep threshold.

Since the inverter runs on a real-time clock, it is possible to program the hours when generator operation will be permitted. It is also possible to program battery charging current so the generator will run at the design output of 4500 W.

Note, however, that this system has several components that are either programmable or that supply important information in digital form—the PV MPPT charge controller(s), the inverter, the automatic generator start, and the BMS. If these components can be successfully interfaced to communicate with each other, it will be possible for the system to operate automatically with no user intervention. However, if the BMS operates independently of the inverter and charge controller(s), then manual operation of the generator may be necessary to avoid unnecessary generator run time. The reason for this is the relative independence of LFP voltage and LFP SOC, compared with comparable lead-acid units. The voltage of the LFP battery pack will be more dependent on the load on the battery pack than on the SOC of the battery pack, which is why the LFP BMS needs to control the generator start/stop sequencing. As LFP systems become more popular, it is likely that the manufacturers of these systems will coordinate control functions with the inverter and PV charge controller manufacturers.

7.9.8 Wire, Circuit Breaker, and Disconnect Selection

All wiring on the load side of the inverter can be done in a manner consistent with conventional 120/240 V residential wiring. The *NEC* requires a distribution panel no smaller than 100 A to be installed, and the panel will need to have sufficient circuit breaker pole spaces to accommodate all the loads plus a few spare spaces for possible future loads.

Table 7.23 shows the wire sizes required for array to source-circuit combiner, combiner to charge controller, charge controller to inverter, battery to inverter, generator to inverter, and inverter to distribution panel, assuming distances of 40 ft (12.2 m), 6 ft (1.83 m), 4 ft (1.22 m), 10 ft (3.05 m), 30 ft (9.14 m), and 10 ft, respectively. As usual, the source circuit to combiner ampacity and the combiner to charge

STAND-ALONE PHOTOVOLTAIC SYSTEMS 347

Table 7.23 Summary of Wiring and Fusing for the Hybrid Residence Based on Use of 3600-W PV Array

Wire Location	VD (A)[a]	NEC (A)[b]	1-Way Length (ft)	System (V)[a]	Wire Max (Ω/kft)	NEC Wire (A)	Wire Size	C.B. Size (A)
Src ckts to Combiner	8.2	10.88	40	65	1.9817	21.6	#10	15
Combiner to controller[c]	32.8	43.5	6	65	3.3028	54.3	#6	60
Controller to inverter[c]	38.8	75	6	48	2.0619	75	#6	60
Batteries to inverter	130	163	10	48	0.369	163	#4/0	250
Generator to inverter	20.8	23.9	30	240	3.8462	23.9	#10	25
Inverter to panel	37.5	46.9	10	120	3.2000	46.9	#6	60

[a] Used for voltage drop calculations.
[b] Used for ampacity calculations.
[c] Based on use of two charge controllers, so half of array to each controller.

controller ampacity must be at least 156% of the rated module short-circuit current. The current used for voltage drop calculations is I_{mp}, since this is the expected operating current of a source circuit under full sun. One might argue that the array will never be exposed to full sun, since, in winter, the altitude of the sun is low enough so the air mass increase causes a decrease in intensity, and in spring, summer and fall, the sun is never perpendicular to the array. Thus, using I_{mp} gives a conservative estimate for the voltage drop calculation. The voltage used for voltage drop calculations is V_{mp} of a source circuit.

The minimum source-circuit wire ampacity is determined by the larger of Equations 7.6 and 7.7 per *NEC* 690.8(5)(B),

$$I(30C) = 1.56 I_{SC} \qquad (7.6)$$

and

$$I(30C) = \frac{1.25 I_{SC}}{D_T D_C}, \qquad (7.7)$$

where I(30C) is the ampacity at a temperature of 30°C of the wire to be used, that is, the ampacity as tabulated in Table 4.2; I_{SC} is the module I_{SC}; D_T is the *NEC* temperature derating factor; and D_C is the *NEC* conduit fill derating factor. Using a maximum ambient temperature of 55°C (131°F) on the roof, along with a conduit fill of 14 conductors, results, for wire with 90°C insulation (THWN-2), values of $D_T = 0.76$ and $D_C = 0.5$. The necessary ampacity of the source-circuit wiring in conduit must therefore be at least 21.6 A. Since the 30°C (86°F) ampacity of #10 THWN-2 (5.261 mm²) is 40 A, this is the recommended wire size, simply because using #12 (3.31 mm²) with dc Ω/kFT = 1.98 for stranded wire is too close to the maximum allowed wire resistivity per Table 7.23.

The charge controller output currents are determined by the portion of the rated array power that is connected to each controller. In this case, using the 14-module array, four source circuits would be connected to one charge controller and three circuits would be connected to another charge controller. The four-source-circuit figure is used as a worst-case figure. It can be assumed that 97% of this power will appear at the charge controller output at a battery charging voltage of approximately 55 V.

The battery-to-inverter current is the rated inverter output power divided by the system dc voltage *at its lowest expected value (44 V), which is normally considered to be the lowest rated inverter input voltage,* and then divided by the inverter efficiency. Alternatively, it is acceptable to use the maximum input current as listed on the manufacturer's specification sheet when this number is available.

It might seem that the current from inverter to distribution panel should be determined from the rated inverter output power, divided by 240 V. This would be the case if the inverter output were balanced, but the inverter allows a maximum unbalance of 75% between Line 1 and Line 2 currents, with a maximum line-to-neutral output current of 37.5 A. All ac wiring must be sized to carry 125% of these rated currents, except for the wiring from generator to inverter/charger, where the wire must be sized to carry 115% of the rated generator output current, per *NEC*. The rated generator output current will be 5000/240 = 20.8 A and 115% of 20.8 is 23.95 A. Thus, wiring from the generator to inverter will need to be #10 (5.261 mm^2), protected by a 25-A circuit breaker, unless the generator manufacturer will allow a 30-A circuit breaker to be used.

In a few previous examples, the wire size was determined on the basis of voltage drop for longer runs of wire. Table 7.23 shows that in this system, only the source-circuit wire size is based upon voltage drop considerations.

7.9.9 BOS Component Selection

The BOS components will include a battery storage container, an array mount, surge protection, and provisions for proper grounding of the system. Of course, the wiring from the distribution panel is also a part of the balance of the system, but its cost will not change as the mix of PV versus propane generation is varied. The cost of the array mount and the cost of wire from array to combiner box are the only items in the BOS that will vary with the number of modules.

7.9.10 Total System Design

Figure 7.13 shows the schematic diagram of the hybrid dwelling 3600-W PV system. It is assumed that the batteries will be placed in a reasonably well-insulated location so that they will remain reasonably warm in the winter when they are needed the most. The array is located as close as practical to the batteries, but free of any objects that may shade the array. Although calculations show that the generator will not be needed for this system, it is included to show how it will be connected into the system. For the smaller-sized systems, all major components remain the same with the exception of fewer modules and only one charge controller for 3000 W and less. The 6.8-kW inverter is still needed for supplying adequate power to the loads, even though the array size may drop to 600 W.

STAND-ALONE PHOTOVOLTAIC SYSTEMS 349

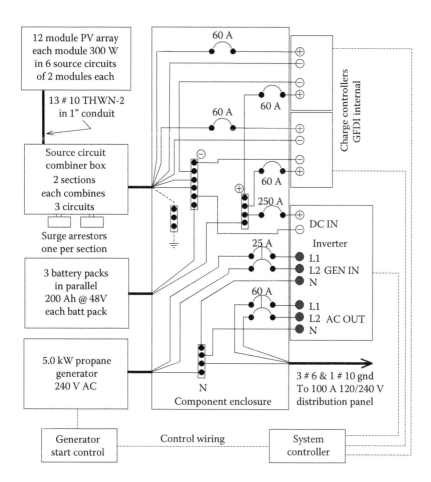

Figure 7.13 Hybrid residence electrical system.

The other components that will be included in the system will be an automatic generator start module and a system controller. The generator start module needs to be controlled by the BMS, as previously discussed, but it also monitors generator output to verify that the generator is operating properly. The inverter is programmable via the system controller so battery charging current from the generator can be varied as needed. This is a case where reading the instruction book is essential to proper operation of the generator and the rest of the system.

7.10 SUMMARY OF DESIGN PROCEDURES

At this point, a relatively wide range of stand-alone PV systems have been designed in this chapter. An effort has been made to present alternate, common-sense design procedures to illustrate that no single methodology is necessarily the best one for any given problem. However, there are many commonalities among the designs, including the procedure for determining loads, the procedure for sizing batteries and the procedure for sizing arrays. These procedures are summarized in the following listings. The designer should remember, however, that these procedures are intended to be used as guidelines. Specific design requirements may result in the need to modify these procedures.

DETERMINATION OF AVERAGE DAILY PV SYSTEM LOAD

1. Identify all loads to be connected to the PV system.
2. For each load, determine its voltage, current, power, and daily operating hours. For some loads, the operation may vary on a daily, monthly, or seasonal basis. If so, this must be accounted for in calculating daily averages.
3. Separate ac loads from dc loads.
4. Determine average daily Ah for each load from current and operating hours data. If operating hours differ from day to day during the week, the daily average over the week should be calculated. If average daily operating hours vary from month to month, then the load calculation may need to be determined for each month.
5. Add up the Ah for the dc loads, being sure all are at the same voltage.
6. If some dc loads are at a different voltage, which will require a dc-to-dc converter, then the converter input Ah for these loads needs to account for the conversion efficiency of the converter.
7. For ac loads, the dc input current to the inverter must be determined and the dc Ah are then determined from the dc input current. The dc input current is determined by equating the ac load power to the dc input power and then dividing by the efficiency of the inverter.
8. Add the Ah for the dc loads to the Ah for the ac loads, and then divide by the wire efficiency factor and the battery efficiency factor to obtain the corrected average daily Ah for the total load.
9. The total ac load power will determine the required size of the inverter. Individual load powers will be needed to determine wire sizing to the loads. Total load current will be compared with total array current when sizing wire from battery to controller. The power lost in the wiring is determined by the product of the current and the voltage drop. Depending on the nature of the load, a drop in voltage may result in either a drop or a rise in current. Normally, the current will drop slightly, but, in some motors with constant loads, the current will rise. If the load is an MPPT, the current will also rise if the voltage drops. However, for a voltage drop of only 2%, it is reasonable to assume the load current will remain essentially constant. With this assumption, the power loss can be approximated by the product of load current at nominal load voltage and the voltage drop. Hence, a 2% voltage drop implies a 2% power loss, or, in terms of Ah, a 2% Ah loss. This loss must be added to the PV array power output requirements. Thus, a 2% voltage drop implies a wiring efficiency of 98%, and the load Ah per day must be increased by dividing by 0.98 to account for the wiring loss.

STAND-ALONE PHOTOVOLTAIC SYSTEMS

BATTERY SELECTION PROCEDURE

1. Determine the number of days storage required, depending on whether the load will be noncritical or critical. The designer may use discretion for special cases such as weekend occupancy or seasonal variations. It is possible that different seasons may have different numbers of storage days.
2. Determine the amount of storage required in Ah. This is the product of the corrected Ah per day and the number of days of storage required. This amount may vary with season. If so, list all values.
3. Determine the allowable level of discharge. Divide the required Ah by the level of allowed discharge, expressed as a fraction. For example, using 80% of total charge requires dividing by 0.8. This result is the total corrected Ah required for storage.
4. Check to see whether an additional correction for discharge rate will be needed. If so, apply this correction to the result obtained in Step 3.
5. Check to see whether a temperature correction factor is required. If so, apply this to the result of Step 3 or 4.
6. Check to see whether the rate of charge exceeds the rate specified by the battery manufacturer. If so, multiply the charging current by the rated number of hours for charging. If this number is larger than the result of Equation 7.2, use this as the required battery capacity.
7. Divide the final corrected battery capacity by the capacity of the chosen battery. The result may be rounded up or down, depending on the judgment of the system designer. Often the result can be rounded down, provided that the system receives at least some diffuse sunlight over prolonged overcast periods.
8. If more than four batteries are required in parallel, it is generally better to consider higher capacity batteries to reduce the number of parallel batteries to provide for better balance of battery currents.

ARRAY SIZING AND TILT PROCEDURE

1. For each tilt angle of interest, simulate the array performance for the system geographic location using the PVWatts version of the NREL SAM simulation program and a 1-kW array. Determine the required design array power in kW for each month of the year by dividing the corrected kWh load for the system each month by the monthly average kWh supplied by the 1-kW array at each array tilt angle.
2. Determine the worst-case (highest monthly) design array power for each tilt angle.
3. For a fixed mount, select the tilt angle that results in the lowest worst-case design array power.
4. If tracking mounts are considered, then determine the design array size in kW for one- and two-axis trackers. Note that the one-axis tracker tracks from

east to west and needs to have its design power checked for the three angles of fixed arrays.
5. Select a module that has rated output current and voltage at maximum power consistent with system needs.
6. Determine the number of modules in the array by dividing the required array power by the rated power of the selected module. Alternatively, pick a reasonable integer and divide the required array power by this integer to determine the necessary module power rating. Keep in mind that the number of modules in series will be limited by the maximum value of V_{OC} of the series combination as evaluated at the lowest expected array operating temperature. In addition, the minimum number of modules in series will be limited by the value of V_{mp} of the series combination as evaluated at the highest expected array operating temperature.
7. When choosing the number of modules in the array, if there will be more than one source circuit, it is good design practice to make sure number of modules in each source circuit will be the same. Thus, the total number of modules will be the product of the number in series and the number in parallel.

PROBLEMS

7.1 For the fan example, indicate applications for which it would be desirable to have an oversized PV module and indicate when it would be satisfactory to have a smaller module that would produce a significantly lower fan speed at lower light levels. Can you envision an application for which use of an MPPT/LCB would be advantageous?

7.2 How much power could be recovered at low light levels by using an MPPT on the fan of the first example in the text, assuming the MPPT to be 95% efficient? Estimate the additional maximum power output that would be required of a PV array that would produce the same low-light-level power as the system with the MPPT. If the additional PV costs $1.50/W (installed), how much could you spend on the MPPT to produce the same effect?

7.3 It is desired to pump 50 gpm from a depth of 100 ft, using a pump that is claimed to have a wire-to-water efficiency of 35%.
   a. Determine the horsepower the pump will need to develop to achieve this pumping rate.
   b. Determine the daily kWh needed to pump 3000 gallons.

7.4 Search the Web for a suitable set of batteries that will provide 275 Ah at 24 V when the discharge rate is C/10. Sketch the wiring diagram.

7.5 Find a suitable module and charge controller to charge the batteries used in the parking lot lighting example in Section 7.4.

7.6 Using the module and charge controller of Problem 7.5, determine the wire size between array and charge controller if the distance between the two is 25 ft and the voltage drop is to be limited to 1%.

7.7 Recalculate the battery size and the array size for the parking lot fixture example if the 158-W fixture is chosen.

7.8 Determine the lamp wattage required to obtain an illumination level of 50 f-c over a 100 ft² area if a fixture is used with a CU of 0.75 and 80% of the available light

STAND-ALONE PHOTOVOLTAIC SYSTEMS

353

reaches the work surface, the rest being absorbed by walls and other items in the space. Assume a luminous efficacy of 70 lumens/W.

7.9 For a soil with a resistivity of 12,000 Ω-cm, and a cathode that requires 200 mA of current for protection, configure an anode system that will allow the use of a 12-volt system. Then select a battery to provide 5 days of storage and a module(s) to supply the system energy needs if the minimum irradiation is 3 hours/day.

7.10 For a soil with resistivity of 20,000 Ω-cm, and a cathode that requires 500 mA of current for protection, configure an anode system that will allow the use of a 12-V system. Then select a battery to provide 5 days of storage and a module(s) to supply the system energy needs for minimum peak sun of 2.5 hours/day.

7.11 To test the soil resistivity after a large steel tank has been buried, a 6-ft-long, 4-inch-diameter anode is buried about 40 ft from the tank. When a 12-V battery is connected between the anode and the tank, a current of 0.5 A flows. What is the resistivity of the soil?

7.12 The highway information sign is to be designed for use in Miami, FL, using the same PV modules. The sign still requires 5 days of autonomy under worst-case sun conditions, so the battery requirements will need to be recalculated. Then tabulate the average daily power available to the sign for each of the 12 months of the year.

7.13 Redo the highway information sign example for use in Phoenix, AZ, again calculating the battery requirements and then showing the monthly average daily power availability.

7.14 For the highway sign of Section 7.6, determine the Ah rating of batteries that will provide 5 days of backup power for the month with the highest average daily kWh generation by the array.

7.15 How many days of storage will the batteries of the highway sign example of Section 7.6 provide at summer average daily power levels?

7.16 Assume 10 days of autonomy are desired for a battery system, but the battery size chosen only allows for 9.2 days of storage, with a maximum depth of discharge of 80%. Then assume 12 consecutive days occur during which peak sun averages only 10% of the predicted worst-case average. What will be the SOC of the battery system after the end of the 12th day?

7.17 Show that the wire-to-water efficiency of the well pump used for the cabin is approximately 42%.

7.18 Determine the additional water that can be pumped with excess summer electricity produced by the cabin array.

7.19 Show why the hybrid residence will take 1.5 winter days to use up 5 hours of charging at 3840 W, as claimed in the text.

7.20 Show that the average power consumption in the hybrid residence is 436 W or less.

7.21
   a. At what temperature will lead-acid batteries hold 80% of their rated capacity?
   b. At what temperature will lead-acid batteries hold 90% of their rated capacity?
   c. What percentage of their rated capacity will lead-acid batteries hold if their temperature is 0°C?

DESIGN PROJECTS

7.22 Write a program or develop a spreadsheet that will size a pump and wire to the pump for a PV-powered pumping system if the pumping height and daily volume are known along with the distance from the array to the pump.

7.23 Design your own little off-grid hide away. Specify your own loads, occupancy, peak sun, and storage requirements and determine the number of batteries and the number of modules you would need to implement the system. Then specify BOS components.

7.24 Design your own slightly larger, off-grid hide away for a location at a latitude higher than 50°, where winter psh are significantly less than summer psh. Use winter loads such that your system will end up as a hybrid system.

7.25 Complete the design of the hybrid residence electrical system of Section 7.8 using an array of 10 modules. Be sure to show the changes in wiring that will be made in the source-circuit combiner box and in the component enclosure.

REFERENCES

1. For a table of head loss versus flow, see http://www.engineeringtoolbox.com/pressure-loss-plastic-pipes-d_404.html (accessed June 29, 2016).
2. For comparison curves for high head and medium head water pumps, see http://www.inyopools.com/blog/which-is-best-high-head-or-medium-head-pumps/ (accessed June 29, 2016).
3. System Advisor 2016 (SAM 2016.3.14), National Renewable Energy Laboratory, Golden, CO, 2016.
4. *Lighting Handbook: Reference and Application,* 8th Ed., Illuminating Engineering Society of North America, New York, 1993.
5. *IES Lighting Handbook,* Illuminating Engineering Society of North America, New York, 1984.
6. *9200 Lamp Catalog,* 22nd Ed., GE Lighting, General Electric Co., 1995.
7. *Lamp Specification and Application Guide,* Philips Lighting Co., Somerset, NJ, 1999.
8. For information on LED arrays, see www.cetsolar.com.
9. U.S. Naval Observatory, Sun or Moon Rise/Set Table for One Year, http://aa.usno.navy.mil/data/docs/RS_OneYear.php.
10. *Stand-Alone Photovoltaic Systems: A Handbook of Recommended Design Practices,* Sandia National Laboratories, Albuquerque, NM, 1995.
11. U.S. Environmental Protection Agency and U.S. Department of Energy, Energy Star Products, Refrigerators and Freezers, http://www.energystar.gov/index.cfm?c=refrig.pr_refrigerators.
12. Solar Water Pumps, submersible and surface pumps, www.sun-pumps.com.
13. Danley, D. R., Orion Energy Corporation, Ijamsville, MD, Personal communication regarding fuel efficiency of fossil fueled generators, August, 1999.

SUGGESTED READING

For additional information on array mounts, see www.power-fab.com.
For battery information, see www.batteries4everything.com.
For information on array mounts, see www.unirac.com.
For information on tracking array mounts, see www.zomeworks.com.
NFPA 70 National Electrical Code, 2014 Ed., National Fire Protection Association, Quincy, MA, 2013.

CHAPTER 8

Economic Considerations

8.1 INTRODUCTION

The costs of a PV system include acquisition costs, operating costs, maintenance costs, and replacement costs. At the end of the life of a system, the system may have a salvage value or it may have a decommissioning cost. This chapter introduces the method of life-cycle costing, which accounts for all costs associated with a system over its lifetime, taking into account the time value of money. Life-cycle costing is used in the design of the PV system that will cost the least amount over its lifetime. Life-cycle costing, in general, constitutes a sensible means for evaluating any purchase options when a decision has been made to purchase. Sometimes a potential PV system owner wants to know how long it will take for the system to pay for itself, and sometimes a potential PV system owner simply wants to know what can be expected from the system, independent of cost. For larger systems, in order to be able to compare the costs of electricity from different sources, the levelized cost of electricity (LCOE) method has been introduced. This chapter explores some of the economic considerations that enter the decision process when someone expresses an interest in owning a PV system.

If it is necessary to borrow money to purchase an item, the cost of the loan may also need to be incorporated into the total cost of a system. If incentive programs are available in any of a number of forms, they may significantly affect the life-cycle cost (LCC), or payback of a system. Often the only question in the mind of the potential PV system owner is how long it will take to pay for itself. So some attention will also be paid to the payback question as well as various forms of incentive programs currently in place in certain areas.

Another economic option is third-party ownership, in which a for-profit company offers to install and own a PV system on a property and then sell electricity from the system to the property owner at a preferred rate. As the cost of PV systems continues to fall and as tax incentives continue to be adopted, this option is becoming more attractive, since it involves minimal risk for property owner or for system owner, as long as the sun rises and as long as the equipment functions properly with high reliability.

This chapter also introduces the concept of externalities. Externalities are costs that are not normally directly associated with an item. For example, it is generally agreed that acid rain can be caused by sulfur emissions from smokestacks. It is also generally agreed that acid rain can cause damage to buildings and lakes. Yet, the cost of these damages is generally not paid directly by the entity that generates the emissions. Externalities will be considered in more detail in Chapter 9.

8.2 LIFE-CYCLE COSTING

8.2.1 The Time Value of Money

The LCC of an item consists of the total cost of owning and operating an item over its lifetime. Some costs involved in the owning and operating of an item are incurred at the time of acquisition, and other costs are incurred at later times. In order to compare two similar items, which may have different costs at different times, it is convenient to refer all costs to the time of acquisition. For example, one refrigerator may be initially less expensive than another, but it may require more electrical energy and more repairs over its lifetime. The additional costs of electrical energy and repairs may more than offset the lower acquisition cost.

Two phenomena affect the value of money over time. The *inflation rate*, i, is a measure of the decline in value of money. For example, if the inflation rate is 3% per year, then an item will cost 3% more next year. Since it takes more money to purchase the same thing, the value of the unit of currency, in effect, is decreased. Note that the inflation rate for any item need not necessarily follow the general inflation rate. Recently, healthcare costs have exceeded the general inflation rate in the United States, while the cost of most electronic goods has fallen far below the general inflation rate.

The *discount rate*, d, relates to the amount of interest that can be earned on principal that is saved. If money is invested in an account that has a positive interest rate, the principal will increase from year to year. The real challenge, then, in investing money, is to invest at a discount rate that is greater than the inflation rate.

As an example, assume an initial amount of money is invested at a rate of 100d% per year, where d is the percentage rate expressed as a fraction. After n years, the value of the investment will be

$$N(n) = N_o(1+d)^n. \tag{8.1}$$

However, in terms of the purchasing power of this investment, N(n) dollars may not purchase the same amount as this amount of money would have purchased at the time the investment was made. In order to account for inflation (or deflation), note that if the cost of an item at the time the investment was made is C_o, then the cost of the item after n years if the inflation rate is 100i% per year will be

$$C(n) = C_o(1+i)^n. \tag{8.2}$$

One might argue that if the cost of an item increases at a rate that exceeds the rate at which the value of saved money increases, that the item should be purchased right

ECONOMIC CONSIDERATIONS

away. Similarly, if the cost of the item increases more slowly, or, perhaps, actually decreases over time, then one should wait before making the purchase, since the cost will be less at a later time. The disadvantage of this purchasing algorithm, of course, is that the item to be purchased will not be available for use until it is purchased. Hence, the new computer that becomes less and less expensive while the invested money continues to increase in value is not available for computing when it should be purchased. In other words, economics may not be the only consideration in making a purchase. Sometimes people buy things simply because they want them.

It is important to remember that choosing values for d and i is tantamount to predicting the future, since d and i fluctuate over time. Depending upon the saving mechanism, the rate of return may be fixed or may be variable. Inflation is, at best, unpredictable. The worldwide recession of 2009 surprised many observers and supports the observation that the future will be different from the past. The important point is to be able to distinguish between predictable events, such as death, taxes, sunrise, and loan repayment amounts, and unpredictable events, such as the value of the stock market, the inflation rate, and available interest rates on savings plans.

Figure 8.1 [1–3] shows how the consumer price index, as a measure of inflation; the Dow Jones industrial average, as one possible measure of d; and the U.S. effective federal funds rate, as a measure of the absolute lowest minimum borrowing interest rate; have varied over the period between 1985 and 2016. It is interesting to note how the effective federal funds rate is adjusted with the intent of either controlling inflation by discouraging borrowing or stimulating the economy by encouraging borrowing. The high effective federal funds rate in the early 1980s reflects the attempt to control high inflation in the late 1970s resulting from significant increases in energy prices during this period. The low rates in 2002 and 2008 reflect the attempt to stimulate the economy and bring about a recovery of the stock market. The corresponding changes in the consumer price index and the Dow Jones industrial average suggest there just may be a correlation here.

8.2.2 Present Worth Factors and Present Worth

If $C_o = N_o$, the ratio of C(n) to N(n) becomes a dimensionless quantity, Pr, which represents the *present worth factor* of an item that will be purchased n years later, and is given by

$$\mathrm{Pr} = \left(\frac{1+i}{1+d}\right)^n. \tag{8.3}$$

The present worth of an item is defined as the amount of money that would need to be invested at the present time with a return of 100d% in order to be able to purchase the item at a future time, assuming an inflation rate of 100i%. Hence, for the item to be purchased n years later, the present worth is given by

$$\mathrm{PW} = (\mathrm{Pr})C_o. \tag{8.4}$$

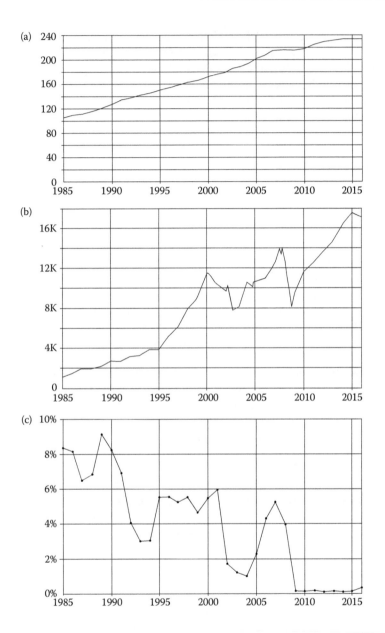

Figure 8.1 Comparison of (a) consumer price index, (b) Dow Jones industrial average, and (c) U.S. Effective funds rate, 1985–2015. (From Consumer Price Index Data from 1913 to 2016 US Inflation Calculator, http://www.usinflationcalculator.com/inflation/consumer-price-index-and-annual-percent-changes-from-1913-to-2008; Dow Jones Indexes, http://www.djaverages.com; Historical Changes of the Target Federal Funds and Discount Rates, http://newyorkfed.org/markets/statistics/dlyrates/fedrate.html.)

ECONOMIC CONSIDERATIONS

Sometimes it is necessary to determine the present worth of a recurring expense, such as fuel cost. Since recurring expenses can be broken down into a series of individual expenses at later times, it is possible to determine the present worth of a recurring expense by simply summing up the present worth of each of the series. For example, suppose a commodity such as diesel fuel is to be used over the lifetime of a diesel generator. It is desired to determine how much money must be invested at present at an annual interest rate of 100d%, under conditions of 100i% annual inflation in order to purchase fuel for n years. If the first year's supply of fuel is purchased at the time the system is put into operation, and each successive year's fuel supply is purchased at the beginning of the year, the present worth of the fuel acquisitions will be

$$PW = C_o + C_o\left(\frac{1+i}{1+d}\right) + C_o\left(\frac{1+i}{1+d}\right)^2 + C_o\left(\frac{1+i}{1+d}\right)^3 + \cdots + C_o\left(\frac{1+i}{1+d}\right)^{n-1}. \quad (8.5)$$

Letting $x = (1+i)/(1+d)$, Equation 8.5 becomes

$$PW = C_o(1 + x + x^2 + \cdots + x^{n-1}). \quad (8.6)$$

This expression can be simplified by observing that

$$\frac{1}{1-x} = 1 + x + x^2 + x^3 + \cdots = \sum_{i=0}^{\infty} x^i. \quad (8.7)$$

Now, the *cumulative present worth factor* can be defined as

$$Pa = \frac{PW}{C_o} = \frac{1}{1-x} - \sum_{i=n}^{\infty} x^i = \frac{1}{1-x} - x^n \sum_{i=0}^{\infty} x^i,$$

or, finally,

$$Pa = \frac{1 - x^n}{1 - x}. \quad (8.8)$$

It is important to recognize that Equation 8.8 is based on the assumption that the first year's supply is purchased at the beginning of the year at a time when the fuel is at its present value. The fuel is then purchased annually with the last purchase occurring 1 year before the system lifetime has expired. In other words, there are n purchases of fuel, each at the beginning of the nth year.

If the recurring purchase does not begin until the end of the first year, and if the last purchase occurs at the end of the useful life of the system, there will still be n purchases, but, using x again, the cumulative present worth factor becomes

$$Pa_1 = x + x^2 + x^3 + \cdots + x^n$$
$$= x(1 + x + x^2 + \cdots + x^{n-1}) = xPa. \qquad (8.9)$$

Since x will typically be in the range $0.95 < x < 1.05$, and since determination of i and d are at best, good guesses, and since it would be unusual to purchase an entire year's supply of many things all at once at the beginning of the year or at the end of the year, either Equation 8.8 or 8.9 will generally provide a good estimate of the present worth of a cumulative expenditure. Often the values for Pr and for Pa are tabulated. Since most engineers have programmable calculators and computers, there is little point in repeating tabulated values for these expressions. Once the values for the variables have been decided upon, the present worth of quantities can be calculated. For that matter, if an engineer wanted to assume different values for d and i for different years, the same methodology could be used to determine the present worth of a quantity.

8.2.3 Life-Cycle Cost

Once the PW is known for all cost categories relating to the purchase, maintenance, and operation of an item, the LCC is defined as the sum of the PWs of all the components. The LCC may contain elements pertaining to original purchase price, replacement prices of components, maintenance costs, fuel and/or operation costs, and salvage costs or salvage revenues. Calculating the LCC of an item provides important information for use in the process of deciding which choice is the most economical. The following example demonstrates the use of LCC.

EXAMPLE 8.1

Refrigerator A costs $600 and uses 150 kWh of electricity per month. It is designed to last 10 years with no repairs. Refrigerator B costs $800 and uses 100 kWh of electricity per month. It is also designed to last 10 years with no repairs. Assuming all the other features of the two refrigerators are the same, which is the better buy if the cost of electricity is $0.07/kWh? What if the cost of electricity is $0.15/kWh? Assume a discount rate of 3% and assume an inflation rate of 5% for the electrical costs. Also, note that this example is being written out in early 2016, a time where estimating either i or d is highly speculative.

Solution

The solution, of course, is to perform LCC analyses on each refrigerator. First note that x = 1.05/1.03 = 1.019. For a 10-year period, since electricity purchase begins at the time of purchase of the refrigerator, using Equation 8.8 gives Pa = 11.16.

ECONOMIC CONSIDERATIONS

Table 8.1 LCC Analysis for Two Refrigerators at $0.07/kWh and $0.15/kWh

	Refrigerator A			Refrigerator B		
	First Year	PW1	PW2	First Year	PW1	PW2
Purchase price	$600	$600	$600	$800	$800	$800
Electrical cost at $.07/kWh	$126	$1406		$84	$937	
Electrical cost at $.15/kWh	$270		$3013	$180		$2008
LCC		$2006	$3613		$1737	$2808

Then note that for Refrigerator A, the electrical cost for the first year will be (12 months) × (150 kWh/month) × ($0.07/kWh) = $126 and for Refrigerator B, the electrical cost for the first year will be $84. Multiplying the first-year cost by Pa yields the PW of the electrical cost. A simple table may be constructed to compare the two refrigerators. Note that separate columns are used for the PW associated with electricity at $0.07/kWh and $0.15/kWh.

From Table 8.1 it is evident that the $800 refrigerator has a lower LCC than the $600 refrigerator. It is also evident that as the price of electricity is increased, the LCC of the more-expensive first-cost refrigerator becomes more and more attractive. Federal law requires that certain appliances, including refrigerators, have labels that disclose the energy consumption. *It is not necessarily the case that the more-expensive units use less electricity.* One should check the labels carefully when contemplating a purchase.

EXAMPLE 8.2

Compare the LCC of a highway construction warning sign that is PV powered versus using a gasoline generator to power the same sign. The system is to be capable of 24-hour/day operation with minimal down time. Assume the load to be 2 kWh/day with a 20-year lifetime.

To power this load with a PV system, it will take a 500-W array of PV modules, plus array mount, at a cost of $1.25/W, $1000 worth of storage batteries, which need to be replaced every 6 years, and a $250 charge controller. Assume a system maintenance cost of $150 per year.

Although the average power requirement is only 83 W, it is unlikely that an 83-W generator will be used. For the purposes of this example, assume that a 500-W gasoline generator can be purchased for $250. Since it is running well under rated load, a generous efficiency estimate is 2 kWh/gallon, and will thus use about 365 gallons of gasoline per year at an initial cost of $2/gallon, and will require frequent maintenance with an annual cost of about $1500 for oil changes, tune-ups, and engine rebuilds. Because of the heavy use, after 5 years the generator must be replaced. Assume an inflation rate of 3% and a discount rate of 5%.

Solution

For the PV system, Pr is needed for 6 years, 12 years, and 18 years, using Equation 8.3. For the generator, Pr is needed for the generator replacement after 5 years, 10 years, and 15 years. For the PV system, Pa_1, using Equation 8.9 is needed for maintenance costs and Pa using Equation 8.8 is needed for

Table 8.2 Comparison of LCCs for PV System and Generator System for Highway Sign

PV System			Generator System			
Component	Initial Cost	PW	Component	Initial Cost	Ann Cost	PW
Array + mount	$625	$625	Generator	$250		$250
Controller	$250	$250				
Batteries	$1000	$1000	Fuel		$730	$12,271
Batt 6 years	$1000	$891	Gen 5 years	$250		$227
Batt 12 years	$1000	$794	Gen 10 years	$250		$206
Batt 18 years	$1000	$708	Gen 15 years	$250		$188
Annual maintenance	$150	$2474	Annual maintenance		$1500	$24,735
LCC		$6742	LCC			$37,876

generator fuel and maintenance costs. For the given inflation and discount figures, $x = 0.981$, $Pa = 16.81$, and $Pa_1 = 16.49$. So Table 8.2 can now be completed.

Hence, even though the initial cost of the PV system is significantly higher, its LCC is significantly lower. Could this possibly explain the rapid deployment of these signs?

EXAMPLE 8.3

Compare the LCCss of the hybrid residence electrical system of Section 7.8 for the seven different PV array options. Assume $x = 0.981$. This comparison has been done in the previous three editions of this text. In the first edition, the more PV that was used, the higher was the LCC. By the time the second edition was written, prices of PV had decreased and fuel costs had increased such that the LCC was optimized when PV supplied approximately 95% of the annual energy mix. So, naturally, the tradition must be carried on using 2016 pricing. The system size remains the same for this edition, but higher-first-cost LFP batteries are used, making it even more interesting to compare the overall LCC costs of various system options. The second edition maximum PV system consisted of 24 modules, each rated at 110 W, for a maximum PV array rated power of 2640 W, compared with the 3600 W maximum array size of the third edition using 200 W modules and the current 3600-W array, using 300-W modules.

Solution

Much of the information needed for the solution of this example is available in Tables 7.17 through 7.23. In the ideal case, a price should be affixed to the relative environmental costs of each type of generation as well as to the hardware associated with each type. While this will not be done for this example, environmental costs will be discussed in Chapter 9. In this chapter, it will simply be assumed that the less the generator runs, the less noise and air pollution will be created.

Table 8.3 presents the elements that contribute to the LCC of the various systems. Table entries are only shown for systems of 12 modules and 6 modules. It should also be noted that the component prices shown in this table are somewhere between wholesale and retail cost. The reader is encouraged to obtain the best price quotes for the individual items as a check on the prices in the table.

ECONOMIC CONSIDERATIONS

Table 8.3 Comparison of LCC for 12-Module Hybrid System and 6-Module Hybrid System

Item	12-Module System Cost	Present Worth	% Total LCC	6-Module System Cost	Present Worth	% Total LCC
		Capital Costs				
Array	2880	2880	4.9	1440	1440	2.2
Batteries	18,000	18,000	30.7	18,000	18,000	27.9
Array mount	350	350	0.6	200	200	0.3
Charge controller	675	675	1.1	675	675	1.0
Inverter/charger	4500	4500	7.7	4500	4500	7.0
Source ckt combiner	255	255	0.4	195	195	0.3
Installation	4100	4100	7.0	2300	2300	3.6
Generator	2500	2500	4.3	2500	2500	3.9
BOS	4365	4365	7.4	3465	3465	5.4
		Recurring Costs				
Annual insp	100	1677	2.9	100	1677	2.6
Generator fuel	0	0	0	407	6825	10.6
Generator maint	40	671	1.1	220	3689	5.7
		Replacement				
Batteries 10 years	18,000	14,858	25.3	18,000	14,858	23.0
Charge cont 15 years	675	506	0.9	675	506	0.8
Inverter 15 years	4500	3375	5.7	4500	3375	5.2
Gen rebuild 15 years				500	375	0.6
Total		58,712	100		64,581	100

Reasonable assumptions are made for fuel cost and maintenance cost for the generator. It is assumed that propane costs $2.25/gallon, an oil change will cost $10, a tune-up will cost $50, and a rebuild will cost $500. Array mount costs are based on the cost of typical commercial array mounts. It is assumed that the inverter for the hybrid system will be the same as the inverter for the nonhybrid system to allow for addition of a generator at a later date if desired. The installation cost is assumed to be $1.25/W for the PV system plus 20% of the cost of the generator.

Figure 8.2a is a plot of LCC versus the number of modules in the system, and Figure 8.2b is a plot of the LCC versus the percentage of annual kWh provided by the PV array. It is now up to the system owner to decide whether to spend the additional $779 for the convenience of the 100% PV system over the 98.5% PV system, or, perhaps to spend an additional $1450 for a 14-module system that allows for more usage. In any case, it makes no economic sense to install a system with fewer than 10 modules. For the LCC of the generator-only system, it must be remembered that the system also includes an inverter and batteries so the generator need not run continuously and can run at its most efficient output power level. If the batteries and inverter are eliminated, the generator must run continuously and will run well below its maximum efficiency most of the time,

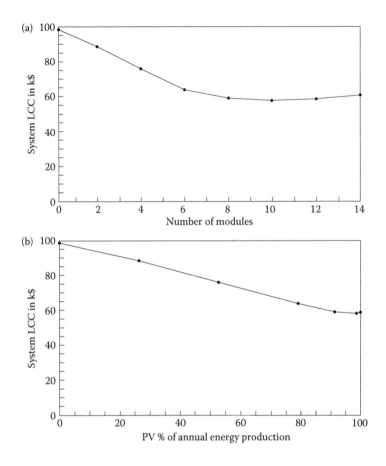

Figure 8.2 Optimizing LCC by varying mix of PV and propane generator. (a) LCC versus number of modules. (b) LCC versus % of power produced by PV.

thus significantly increasing annual fuel and maintenance costs and requiring additional cost analysis.

8.2.4 Annualized LCC

It is sometimes useful to compare the LCC of a system on an annualized basis. Dividing the system LCC by the expected lifetime of the system may appear to be the way to arrive at an annual cost. This, of course, would assume the cost per year to be the same for every year of operation of the system, which is assumed not to be the case in the original set of assumptions. Hence, to find the annualized LCC (ALCC) in present day dollars, it is necessary to divide the LCC by the value of Pa or Pa_1 used in the PW analyses for the system components.

For the refrigerators in Example 8.1, this means dividing the LCC by 11.16 for each LCC evaluated. For the PV system of Example 8.2, the

ALCC = $7085/16.81 = $421, and for the gasoline generator system of Example 8.2, the ALCC = $38,357/16.81 = $2282. The ALCC for the hybrid residence system depends upon which mix of PV versus generator is chosen.

8.2.5 Unit Electrical Cost

One especially valuable use of the ALCC is to determine the unit cost of electricity produced by an electrical generating system. Obviously, if the electricity is to be sold, it is necessary to know the price for which it should be sold to either earn a profit or at least know how much will be lost in the process. Once the ALCC is known, the unit electrical cost is simply the ALCC divided by the annual electrical production. If the annual electrical production is measured in kWh, then the unit electrical cost will be measured in $/kWh.

For the PV system of Example 8.2, the unit electrical cost is ALCC/kWh = $421/730 = $0.577/kWh. For the gasoline generator system, the unit electrical cost is $2282/730 = $3.13/kWh. Clearly, both of these costs far exceed the cost of utility-generated electricity, but since this is a portable application, where the sign is moved around from day to day, the cost of hooking up and disconnecting the system from utility power is impractical, at best.

8.2.6 LCOE Analysis

The concept of a method to evaluate the relative cost of all energy sources was introduced by Walter Short and his colleagues at NREL in March 1995 [4]. The simplicity of the method is in the determination of the cost of a unit of energy by dividing the net present value of the all costs and expenses of running a power generator through its expected life, by the total energy generated over the same period.

The method has been adopted by most planners around the world as a standard for energy unit costs whether the generation is fossil or nuclear based or renewable. It has also been modified and adapted to account for depreciation, tax credits, degradation and efficiency losses, and governmental or utility policies. The basic LCOE formula can be represented by

$$\text{LCOE} = \frac{(\text{project cost}) + \sum_{n=1}^{N} \frac{\text{OM}}{(1+\text{DR})^n} - \frac{\text{RES}}{(1+\text{DR})^n}}{\sum_{n=1}^{N} \frac{(\text{initial kWh}) \times (1 - \text{system degradation rate})^n}{(1+\text{DR})^n}} \quad (8.10)$$

where OM is the operations and maintenance cost for each year n, RES is the residual value of the generating plant in year n, and DR is the assumed discount rate.

The Department of Energy Sunshot goal for LCOE of PV-generated electricity is $0.06 US per kWh by 2020 [5]. It should be noted that if decommissioning of a generating plant involves environmental mitigation, disposal fees, or other costs, the minus (−) in the formula may become a plus (+), thus increasing the LCOE. LCOE can also be modified to include the contribution of carbon credits which would of course lower the unit cost. A simple spreadsheet exercise to determine the LCOE of solar PV plant has been left as a homework problem in this chapter.

EXAMPLE 8.4

The following LCOE Excel exercise is designed to show the utility of the method and the ease with which it can be applied to virtually any electricity generation scenario. The model is built around a 1-MW plant that has several possible implementations, beginning with a fixed crystalline Si array, followed by a single-axis tracking crystalline Si array with 20% additional annual kWh production. Other options considered include dual-axis tracking crystalline Si with 33% more annual kWh production than the fixed array, a concentrating PV (CPV) system with 40% higher annual generation, a diesel generator operating 8 hours/day that uses 120 gallon of diesel fuel per day with a first-year fuel cost of $1.50/gallon and a 1-MW gas turbine generator operating 8 hours/day that uses 102,960 ft^3/day (2894.4 m^3/day) at a cost of $5.00/ft^3.

The same basic assumptions for discount rate and inflation are used for each system. To give the model more flexibility, the inflation rate for cost of electricity is decoupled from the general inflation rate. This is because in recent years, the volatility of oil and gas prices have had more of an impact on cost of electricity than the general inflation rate both on the upside and the downside. The tracking solar PV and CPV plants are also assumed to be operating in regions where tracking or concentrating solar is competitive. These conditions have been discussed in other chapters of this book and as a reminder, they are in areas with high direct beam normal incidence (DNI).

The costs and the generation characteristics shown in this example are based on the general information currently available to the authors, but they need to be researched before any definitive analysis is conducted. For example, the assumption that 1-MW PV system generates 1.5 MWh after all the losses and derating factors in the first year of operation is based on numerous factors and is purely for the purpose of this exercise. A serious LCOE analysis should consider all the information provided in other chapters to develop a reliable kWh generation profile based on the solar conditions, regional data, module inclination, the soiling and washing cycles, and so on. The assumption that a single tracking system generates 20% more electricity than nontracking may be true in a few regions but is most likely not realistic for Florida, or regions with higher diffuse to direct radiation solar components. The analyst should rely on accepted national and regional data to conduct her or his analysis. Also the assumptions regarding the cost and 40% higher generation of CPV are highly variable depending on the specific tracking platform and the architecture of the cell assemblies. The degradation rates for CPV are also based on relatively limited data points compared with other PV system, and may be statistically insignificant. The daily run time for the diesel and gas turbine generators has been selected to achieve annual energy generation comparable to the PV options.

Solution

The assumptions for each system option are shown in Table 8.4 along with the LCOE for each system as determined using Equation 8.10 and the assumptions.

ECONOMIC CONSIDERATIONS

Table 8.4 List of Assumptions Relating to First-Year Costs of Several 1-MW Electrical Generation Options, Including Resulting LCOE after 25 Years of Operation

System	Disc Rate	kWh Ann Inflation Rate	Initial Cost US$	Init Ann Gen (MWh)	Ann Degrad	Annual O&M	Annual O&M Inflation	LCOE (US$/kWh)
1-MW Si PV fixed array	3%	3%	2.5M	1500	0.5%	$12,000	4%	0.111
1-MW Si PV one-axis tracking	3%	3%	3.75M	1800	0.5%	$24,000	4%	0.144
1-MW Si PV two-axis tracking	3%	3%	4.8M	2000	0.5%	$32,000	4%	0.168
1-MW CPV	3%	3%	6.2M	2100	0.5%	$36,000	4%	0.202
1-MW diesel	3%	3%	0.347M	2880	0.0%	$336,700	4%	0.200
1-MW gas turb	3%	3%	0.50M	2880	0.5%	$196,500	4%	0.123

The results of the analysis of the fixed array system using Equation 8.10 are tabulated in Table 8.5 to show how the results in Table 8.4 are obtained. Note that less run time results in higher LCOE, since the initial cost now must be made up for by charging more per unit for a smaller amount of total energy production. Thus, gas turbines used as peaking generators may have only 5%–10% capacity factor on an annual basis, whereas the unit in the example has a 33% annual capacity factor.

8.3 BORROWING MONEY

8.3.1 Introduction

Sometimes the desire to own something causes the potential owner to realize that money does not grow on trees. While some money comes from paychecks, usually larger sums of money come from borrowing from banks or other lending institutions. The question at this point for the engineer-turned-economist is whether borrowed money is any different from paycheck money or money from a savings account, a mattress, or other form of liquid asset. For example, if the money to purchase one of the refrigerators of Example 8.1 had to be borrowed rather than taken from a wallet, would that affect the LCC of the refrigerator?

Once again, tables are readily available for looking up the annual payments on a loan of C_o dollars taken out at 100i% annual interest over a period of n years. For that matter, it is possible to purchase a calculator that will automatically yield the answer when it has been given the terms of the loan. An engineer, of course, will want to know how the numbers are obtained.

8.3.2 Determination of Annual Payments on Borrowed Money

To satisfy this curiosity, consider Table 8.6, representing the principal, interest, total payments, and principal balance at the end of the kth year for repayment of

Table 8.5 Example of LCOE Analysis for 1-MW Fixed Crystalline Si Array

Year	Discount Rate	Initial Cost ($\times 10^3$)	OM ($\times 10^3$)	NPV OM ($\times 10^3$)	Residual ($\times 10^3$)	Total kWh ($\times 10^6$)	NPV kWh ($\times 10^6$)
0		2500					
1	0.03		12	12	0	1.50	1.5
2	0.03		12.48	12.1165	0	1.49	1.45
3	0.03		12.98	12.2341	0	1.49	1.40
4	0.03		13.498	12.3529	0	1.48	1.35
5	0.03		14.038	12.4728	0	1.47	1.31
6	0.03		14.600	12.5939	0	1.46	1.26
7	0.03		15.184	12.7162	0	1.46	1.22
8	0.03		15.791	12.8396	0	1.45	1.18
9	0.03		16.423	12.9643	0	1.44	1.14
10	0.03		17.080	13.0902	0	1.43	1.10
11	0.03		17.763	13.2173	0	1.43	1.06
12	0.03		18.473	13.3456	0	1.42	1.03
13	0.03		19.212	13.4752	0	4.41	0.991
14	0.03		19.981	13.6060	0	1.41	0.957
15	0.03		20.780	13.7381	0	1.40	0.924
16	0.03		21.611	13.8715	0	1.39	0.893
17	0.03		22.476	14.0062	0	1.38	0.863
18	0.03		23.375	14.1421	0	1.38	0.833
19	0.03		24.310	14.2794	0	1.37	0.805
20	0.03		25.282	14.4181	0	1.36	0.778
21	0.03		26.294	14.5581	0	1.36	0.751
22	0.05		27.345	14.6994	0	1.35	0.726
23	0.05		28.439	14.8421	0	1.34	0.701
24	0.05		29.577	14.9862	0	1.34	0.677
25	0.05		30.760	15.1317	0	1.33	0.654
Total		2500	499.75	337.698	0	35.3	25.5
LCOE	0.111						

an n-year loan at 100i% annual interest, where C_o represents the amount borrowed. Note that during any of the years, it is not yet known how much will be paid on the principal in order to repay all of the principal in n years. The challenge is to arrive at an appropriate formula for equal total annual payments over the period of the loan.

Table 8.6 Breakdown of Portions of Loan Payment Allocated to Principal and Interest

Year	Pmnt on Prin	Interest Payment	Total Payment	Balance of Principal
1	A_1	iC_o	$A_1 + iC_o$	$C_o - A_1$
2	A_2	$i(C_o - A_1)$	$A_2 + i(C_o - A_1)$	$C_o - A_1 - A_2$
3	A_3	$i(C_o - A_1 - A_2)$	$A_3 + i(C_o - A_1 - A_2)$	$C_o - A_1 - A_2 - A_3$
n	A_n	$i(C_o - \cdots - A_{n-1})$	$A_n + i(C_o - \cdots - A_{n-1})$	0

ECONOMIC CONSIDERATIONS

In Table 8.6, A_k represents the amount paid on the principal after the kth year. To pay the principal fully in n years requires that the sum of the annual payments on principal must add up to the loan amount, C_o.

Setting the total payments of each year to be equal yields a solution for A_1. For example,

$$A_1 + iC_o = A_2 + iC_o - iA_1,$$

which yields

$$A_2 = A_1(1+i).$$

In general,

$$A_n = A_{n-1}(1+i) = A_1(1+i)^{n-1}.$$

Next, letting $x = (1 + i)$ and summing all the payments toward principal, yields the amount borrowed:

$$A_1 + A_1 x + A_1 x^2 + \cdots + A_1 x^{n-1} = A_1(1 + x + \cdots + x^{n-1}) = C_o.$$

But Equations 8.6 and 8.8 have shown that

$$1 + x + x^2 + \cdots + x^{n-1} = \frac{1-x^n}{1-x},$$

which yields

$$A_1 = \frac{C_o(1-x)}{(1-x^n)}. \tag{8.11}$$

Now all that remains is to add the interest payment for the first year to the principal payment, given by Equation 8.11, to get the total payment for the first year, which will be equal to the total payment for each succeeding year. Proceeding yields

$$\text{ANN PMT} = \frac{C_o(1-x)}{(1-x^n)} + C_o i = C_o \left(\frac{i}{(1+i)^n - 1} + i \right).$$

Simplifying this result yields, finally,

$$\text{ANN PMT} = C_o i \left(\frac{(1+i)^n}{(1+i)^n - 1} \right). \tag{8.12}$$

Usually, payments are made monthly rather than annually. In this case, one need only to note that rather than n payments, there will be 12n payments, and the monthly interest rate will simply be the annual rate divided by 12. Doing so converts Equation 8.12 into an equation for monthly payments. Obviously, if one wants to split hairs even more finely, Equation 8.12 could be modified for weekly, daily, or hourly payments. Equation 8.12 is also an equation that might be conveniently tabulated for various values of i and n, but again it is left to the reader to use either a computer or a programmable calculator to generate the numbers that apply to the problem at hand.

8.3.3 The Effect of Borrowing on LCC

Depending on the nature of a purchase, it is interesting to compare whether the cost of borrowing money will render the purchase undesirable. For example, if money is borrowed to purchase something that will provide a return on the investment, it may make economic sense to borrow the money. This is the standard criterion for commercial loans. The better the return, the better the reason to borrow the money. But what if it is necessary to borrow the money for the initial cost of a system that does not have an obvious return on investment? Does it make sense to borrow for something having a greater first cost if it results in higher loan repayment costs? Although there are several ways to evaluate the worthiness of a purchase, once again, people sometimes borrow money to purchase things that do not have a measurable monetary return on investment, such as automobiles, simply because they want them.

As a simple example, suppose it is possible to spend $2000 on a maintenance-free system that will last for 25 years and will reduce an electric bill by a certain amount per year. The exact system might be insulation, good windows, a solar water heater, or any number of energy efficiency measures. Suppose also that money is borrowed to purchase the system and the period of the loan is 25 years. This would be the case if the system is purchased for a dwelling at the time of construction and the item is included in the 25-year mortgage. If the mortgage rate is 4%, then Equation 8.12 shows the additional annual mortgage payments would be $147.16. If the annual savings on the electric bill exceed $147.16, then it is worth borrowing the money.

One should note, however, that the additional $147.16 mortgage payment will be constant over the life of the loan. In an inflationary environment, this means that even though the electric bill savings may not be $147.16 the first year, after a certain number of years, the annual savings may well exceed $147.16, making the investment worth considering.

What if the money is not borrowed? What if the $2000 is at hand and available for investing? What would be the return on investment if it is spent on one of the items of the previous example? The simple answer would say the return is equal to the resulting savings less than the resulting operating or maintenance costs. This works fine for the first year, but for successive years to yield a more precise estimate, the time value of money must be taken into account. Assuming the value of the quantity saved increases at the inflation rate, i, and the discount rate is d, the present worth of the savings accrued over n years will be given by Equation 8.5. As a result, the PW of the annual savings over n years is given by Equation 8.8. Hence,

ECONOMIC CONSIDERATIONS

Equation 8.8 can be applicable in either an expenditure mode or in a savings mode. By converting costs and savings to LCC, it is easy to compare savings with costs to determine whether to make the purchase.

Assuming that a system must be acquired to do something, and assuming that the system with the least LCC has been identified, it is simply a matter of deciding whether to borrow money or to use money on hand, if, indeed, the money is on hand. If the money is not on hand, then the only option is to borrow. If the money is on hand, then the criterion is whether the lending rate is less than or more than the discount rate. If money can be borrowed at a rate less than what it can earn, then it makes sense to borrow for the acquisition and to invest the money that might have been used for the purchase. This is a choice often made by the buyer of a new automobile. Should the savings account be used or should the money be borrowed? The answer depends on the relative interest on savings versus the interest on the loan.

8.4 PAYBACK ANALYSIS

As mentioned in the introduction to this chapter, some people are not interested in a renewable energy system unless they know how long it will take the system to pay for itself. Often these same people are owners of expensive automobiles who did not even think about how long it would take the expensive automobile to pay for itself. In other cases, the potential owners are business people who are interested in "greening" their business, but only if the "greening" has a reasonable payback period.

Once again, determining a payback period depends on predicting the future, since the idea is to install a system that will generate enough electricity in the first few years to offset all of the installation costs that remain after all incentive payments have been accounted for. The question is what will be the value of the electricity. The part of the future that is relatively predictable is the annual available sunlight and the system performance. The hard part is to figure out what will happen with electric rates and discount rates over the life of the system. So the process involves some guessing.

One way to take some of the uncertainty out of the guessing is to establish lower and upper limits of assumed utility cost escalation rates and then analyze the resulting payback periods. It is also possible to attempt to take into account the time value of money into the analysis, but this adds even more uncertainty to the analysis, so this analysis will be based on as many certainties as possible.

EXAMPLE 8.5

Assume that the installed cost of a 50-kW straight grid-connected PV system is $2.50/W. Also assume that this system will qualify for a 30% investment tax credit and that the cost of the system can be depreciated at the rate of 20%/year for 5 years. For system performance, assume NREL SAM PVWatts predicts an annual system production of 64,000 kWh to the grid. Assume the money is borrowed at 4% over 15 years to pay for the system and that the overall utility cost/kWh is $0.12. A more sophisticated analysis would separate out the energy and the demand portions of the electric bill, but this analysis simply divides the total bill by the total kWh. Perform a payback analysis on the system. To simplify the analysis a bit, assume the discount rate equals the inflation rate. Also, since utility rates have fallen in some areas and risen in other areas over the past 5 years, this analysis will not include any increase or decrease in utility rates.

Solution

First determine the annual loan repayment from Equation 8.12. Also note that the interest on the loan will be tax deductible for a business, but this will need to be handed over to the accountant at the end of the year. It will not be considered in this analysis. Using Equation 8.12, the annual loan repayment amount is found to be $11,243.

Next, the annual system kWh production is determined to be 64,000, which has a first-year value of $7700. At this rate, it might appear that the system will never pay for itself, but it is still useful to construct a table that looks at income versus expenses on an annual basis over the lifetime of the system, which will be assumed to be 30 years. Note that this exercise is perfectly suited to a spreadsheet analysis, where as many variables can be included as suits the fancy of the engineer and/or the accountant, including such items as Renewable Energy Credits (RECs) (see Section 8.5.1), tax refunds on interest paid, replacement system components, such as inverter, system maintenance, and so on. Table 8.7 shows the simplified results. In the table, note that the annual depreciation is considered an operating expense, so the assumption is that at the end of the year after all expenses are included, the owner will still have tax liability.

An examination of Table 8.7 shows that over the first 5 years, the combination of 30% investment tax credit plus rapid depreciation more than offset the loan repayment value, so the fact that the value of the system electricity is less than the loan repayment is offset by the tax credit and rapid depreciation. By the end of Year 5, after the system has been fully depreciated, the cumulative value of savings is about $145,000 depending upon the utility escalation rate, which has not been considered. Then, beginning in Year 6, with no further depreciation or tax credits, the system begins to lose about $3500 each year for a while. So the cumulative value of savings begins to fall, but never goes below zero. After Year 15, there are no further loan payments, so the value of the electricity is now the only contributing factor to the system cumulative savings. Even though this analysis has been considerably simplified, it should at least convince the reader that it is worth going through the exercise, even including additional assumptions (wild guesses?), if so desired.

8.5 EXTERNALITIES

8.5.1 Introduction

What happens if something owned and operated by one party causes damage to something owned by another party? A quick response would be that the entity causing the damage would be liable, and that the cost of repair of the damaged property should be the responsibility of the party doing the damage. The problem is complicated in many instances, however, when more than one party is responsible for

ECONOMIC CONSIDERATIONS

Table 8.7 Payback Analysis for Example 8.4

Year	1	2	3	4	5	6
			Costs			
Loan pmt	11,243	11,243	11,243	11,243	11,243	11,243
			Credits			
Inv Tax Cr	37,500					
Ann Depr	25,000	25,000	25,000	25,000	25,000	
Elect value	7700	7700	7700	7700	7700	7700
Net/year	58,957	21,457	21,457	21,457	21,457	−3543
Cum gain	58,957	80,414	101,871	123,328	144,785	141,242
Year	7	8	9	10	11	12
			Costs			
Loan pmt	11,243	11,243	11,243	11,243	11,243	11,243
			Credits			
Inv Tax Cr						
Ann Depr						
Elect value	7700	7700	7700	7700	7700	7700
Net/year	−3543	−3543	−3543	−3543	−3543	−3543
Cum gain	137,699	134,156	130,613	127,070	123,527	119,984
Year	13	14	15	16	17	18
			Costs			
Loan pmt	11,243	11,243	11,243	0	0	0
			Credits			
Inv Tax Cr						
Ann Depr						
Elect value	7700	7700	7700	7700	7700	7700
Net/year	−3543	−3543	−3543	7700	7700	7700
Cum gain	116,441	112,898	109,355	117,055	124,755	132,455

creating the cause of the damage, and when more than one party suffers damage as a result of the cause. And things get even more complicated when debate ensues over whether the alleged cause is really the cause. Classic examples of such situations are the link between smoking and cancer and the link between burning of fossil fuels, acid rain, and global warming.

Much debate has taken place regarding the liability of tobacco companies for alleged firsthand, smoke-induced cancer and even regarding the alleged incidence of secondhand, smoke-induced cancer. Until recent large judgments against the tobacco industry, the cost of producing cigarettes did not include a component to underwrite the cost of paying the judgments. Yet, it has been argued that tobacco is the cause of billions of dollars in medical bills, none of which have been paid out of tobacco revenues. Thus, in the past, the medical bills incurred by those exposed to tobacco smoke have been treated as externalities by the tobacco industry, while presently, these bills have become direct cost components of doing business.

There is now reasonable agreement that one cause of acid rain is the burning of fuel that contains sulfur, such as petroleum or coal. As a result, many power plants are now equipped with elaborate scrubbers that remove sulfur and other pollutants from smokestack emissions.

The Environmental Protection Agency has established limits on the amounts of various pollutants that may be contained in smokestack emissions. Limits also exist for regions, so that if a region has met its limit, then no further burning may take place unless one of the burners can be made cleaner. In some cases, when a company has not reached its emission limit, the company will sell or trade the remaining allowed emissions with another burner. As a result, monetary value is evolving for certain emissions, albeit in a bit of a roundabout way. The monetary value, however, does not relate directly to the damage done in either the form of acid rain damage or the cost of respiratory diseases. The cost of scrubbers, however, appears as a system cost during the LCC process.

Recently, the concept of REC has been introduced to the marketplace. The REC, once called "green tag," is equivalent to 1000 kWh of electricity produced by renewable sources, and RECs are now traded among energy producers. Once a renewable energy source has sold its RECs, no further claims can be made for the elimination of greenhouse gases, since selling the RECs allows another party to produce the greenhouse gases instead. Sometimes it is less expensive for a generation facility to purchase RECs from a renewable generator than to clean up the CO_2 emissions of its own generation facility.

Another factor that needs consideration in establishing cost is the effect of various forms of subsidies. Some subsidies may enter the picture as direct costs, while others may appear as externalities.

8.5.2 Subsidies

When performing an LCC, sometimes the cost of a component, a fuel or the operation of a system may be affected by a subsidy or subsidies. For example, the cost of military presence in a region to ensure the steady flow of a fuel from the region is never included in the selling price of the product. Mineral depletion allowances, however, can be factored into the selling price of a product. In other cases, governments have been known to offer price supports in order to ensure competitive prices in a world market. Tariffs, in effect, are a form of subsidy, since they ensure that domestic production will be sold at a profit, thus not competing directly with less costly products from outside a country.

Green pricing is a form of subsidy for the acquisition of clean energy sources. An example is when the customers of a utility express a willingness to pay extra every month to ensure that a part of their energy mix comes from renewable sources, such as PV.

ECONOMIC CONSIDERATIONS

Subsidies in the form of tax breaks tend to come and go on a year-by-year basis. For a few years during the 1970s, homeowners could deduct a fraction of the cost of a domestic solar hot water system from their federal income tax. Some states initiated grant programs to encourage homeowners to install solar systems and some utilities established rebate programs for part of the cost of installing various energy conservation measures, such as more-efficient air conditioning or attic insulation. Recently, several states have created buy-down programs in which rebates are offered toward the installation of PV and other renewable energy systems.

The argument is often set forth that all competing interests must compete on a *level playing field*. The meaning is simply that with so many subsidies, some of which are obvious and others of which are hidden, it is difficult for two competing interests to engage in fair competition. This is true in many industries and particularly in the energy industry.

The reader is thus reminded that no economic analysis or comparison is complete until all forms of subsidy have been considered.

8.5.3 Externalities and PV

The cost of an electrical generation source often excludes externalities. Subsequently, if a source is cleaner from an environmental viewpoint but has a higher LCC based on parameters considered, that source may not be chosen. A proper treatment of externalities includes not only the operating externalities, but also the externalities associated with the construction and salvage or decommissioning of the facility. In both categories, PV show significant advantages over nonrenewable sources, as will be shown in Chapter 9.

PROBLEMS

8.1 Obtain the data for consumer price index, Dow Jones industrial average, and effective federal funds rate from References 1–3. Plot the data and attempt to fit curves to the data to show trends. For example, use Excel graphs with trendlines, equations, and R^2 values added to show the "goodness of fit." Compare the R^2 values for linear, exponential, and polynomial fits to your curves.

8.2 Determine the present electricity cost for which the $600 refrigerator of Example 8.1 will have the same LCC as the $800 unit, assuming all other parameters to be the same.

8.3 Next time you are in an appliance store, record the first cost and annual operating costs of several refrigerators that are comparable. Then, making reasonable assumptions about discount rates, inflation rates, and appliance lifetimes, compare the LCCs of the units. You might want to ask a salesperson for information on expected lifetime and repair costs of the units, including annual maintenance contracts.

8.4 Use Equation 8.11 to compare 12 equal monthly payments to one annual payment by modifying the equation to account for monthly interest rate versus annual rate and monthly payments versus annual payments. How does the sum of 12 monthly payments compare with a single annual payment?

8.5 Rework Example 8.2 for a system that uses 4 kWh/day. This will require double the PV array and double the batteries, as well as double the fuel. However, the cost of the generator will remain the same and the cost of the maintenance will remain the same. Assume the cost of the charge controller will not increase.

8.6 Consider the LCOE analysis of the 1 MW of the nontracking PV shown in the Excel example. The analysis assumes that all the equipment are rated and guaranteed 25-year-life. However, current inverter technology is guaranteed for 10 years only with the expectation that inverters may need to be replaced every 10 years. Assuming the total cost of labor and equipment change out is fixed at $200/kW, determine how the LCOE would change for inverter replacements at Year 10 and at Year 20.

8.7 CPV is more sensitive to tracking as the modules need to be within 0.5° alignment for proper focus through lenses. The lenses should also be clean for optimum optical conditions. The cost of OM is therefore higher since modules need to clean of dust, cell cooling module fins need to be free of sand and debris, and continuous recalibration of the tracker is essential. In fact, based on national data, CPV needs 10 times more water for module washing than PV.

Considering that water availability may be an issue in the coming decades, perform a sensitivity analysis comparing the LCOE shown above for CPV with a case where water cost escalates. Assuming that the cost of water comprises 90% of the OM annual cost, draw a graph to show the impact of various escalation scenarios on the LCOE using the following:

Case 1: the base case of Example 8.4
Case 2: 7% inflation rate for the first 10 years and 10% for the last 15 years
Case 3: 10% inflation rate for the first 10 years and 15% for the last 15 years
Case 4: 15% inflation rate for the first 10 years and 25% for the last 15 years

8.8 Calculate a set of reasonable conditions on interest rate, term of loan, and cost per installed kW, for a PV system that generates an average of 10 kWh/day so that the annual loan repayment can be recovered if the value of the electricity generated is $0.20/kWh.

8.9 The value of electricity during utility peaking hours is $0.25/kWh and the money for a utility interactive system with a 30-year expected lifetime can be borrowed at an interest rate of 4%. Calculate the installed cost per kW for the system that will result in annual loan payments equal to the value of electricity produced by the system if the electricity is produced during utility peak hours.

8.10 The feed-in tariff (FIT) has been a means of providing an incentive for consumers to install PV systems that has been widely accepted, particularly in Germany and Spain. The idea is to pay the PV electric producer a rate per kWh produced that is equal to (or, sometimes, greater than) the loan repayment on the system. If the installed cost of a 5-kW grid-connected system is $3/W, if the federal investment tax credit of 30% of the system cost is applied to the system cost, if the system is installed south-facing with a 25° array tilt in Bakersfield, CA, if the NREL SAM PVWatts model can be used to estimate system performance, and if the balance of the system cost after the investment tax credit can be borrowed at 4%/year for 20 years

a. Determine the expected annual kWh generation.
b. Determine the payment in $/kWh needed to enable the system to earn enough annually to equal the value of the loan repayment.
c. Look up the present Bakersfield electric rate (or assume it to be $0.15/kWh). Then assume the rate will increase at 4%/year. Determine the year when

the electric rate in $/kWh will equal the rate of payment to the PV system owner.

8.11 Net metering is when a utility pays the same rate for electricity sold back to the utility as is charged to the consumer for purchased electricity. Assume a straight grid-connected 5-kW PV system can be installed for $2/W after all incentive payments are accounted for. Also assume the system is installed in an area where it will produce 1500 kWh/kW/year at the inverter output. Then assume that the present utility electric rate is $0.14/kWh and that the rate appears to be increasing at 4%/year. If money can be borrowed at 4%/year
   a. Calculate the annual loan repayment if the loan is for 20 years.
   b. Calculate the annual kWh production of the system and the value of this electricity at the present utility rate.
   c. Construct a bar graph that shows the difference between the annual value of the electricity generated and the annual loan repayment rate for a 30-year period. Keep in mind that the loan payments cease after 20 years.
   d. Construct the equivalent of the integral of the bar graph of Part c that shows the cumulative return on investment for the system.

8.12 Assume that the system of Problem 8.9 is installed using a single-axis tracking array to increase the annual kWh production by 25%. The additional cost of the tracking array mount increases the cost after incentives to $3/W.
   a. If all the other conditions of Problem 8.9 are the same for this case, evaluate Parts a–d for the tracking system.
   b. Which system appears to be the best choice in terms of payback?

8.13 A battery storage system is to be designed to provide a storage capacity of somewhere between 440 and 555 Ah at 48 V at a C/20 discharge rate. Four battery types are under consideration:

Battery	Volts	Capacity (Ah)	Type	Lifetime (Year)	Weight (lb)	Cost Each ($)
A	6	220	Flooded	5	56	66
B	12	255	AGM	8	168	356
C	12	180	AGM	8	135	301
D	12	555	Gel	12	564	1941

Assume $i = 2\%$, $d = 5\%$, and a 24-year system lifetime. Perform an LCC for the four battery types and discuss other considerations that may influence the choice of batteries. If the site were a homeowner, what would you recommend? Why? If the site were a remote communication system, what would you recommend? Why?

REFERENCES

1. Consumer Price Index Data from 1913 to 2016 US Inflation Calculator, http://www.usinflationcalculator.com/inflation/consumer-price-index-and-annual-percent-changes-from-1913-to-2008.

2. Dow Jones Indexes, http://www.djaverages.com.
3. Historical Changes of the Target Federal Funds and Discount Rates, http://newyorkfed.org/markets/statistics/dlyrates/fedrate.html.
4. Short, W., Packey, D., and Holt, T., *A Manual for the Economic Evaluation of Energy Efficiency and Renewable Energy Technologies*, National Renewable Energy Laboratory, Golden, CO, March 1995.
5. http://energy.gov/eere/sunshot/sunshot-vision-study (accessed July 27, 2016).

SUGGESTED READING

42 U.S.C. § 7401 et. seq. (Clean Air Act).
Markvart, T., Ed., *Solar Electricity*, John Wiley & Sons, Chichester, UK, 1994.
National Appliance Energy Conservation Act of 1987 (PL 100–12, March 17, 1987, 101 § 103).

CHAPTER 9

Externalities and Photovoltaics

9.1 INTRODUCTION

Pollution is an insidious by-product of civilization that threatens the quality and integrity of our air, soil, water, infrastructure, and most importantly our climate. The planet has a remarkable ability to cleanse itself, with significant feedback loops in place for restoration of equilibrium. One example is photosynthesis, in which the oxygen–carbon dioxide cycle is sustained by regeneration of oxygen from the carbon dioxide of respiration and oxidation. Another is digestion, in which bacteria process organic matter into a form that enables the matter to enrich soils or to nourish living bodies. Unfortunately, human seems to have figured out ways to upset the balance in each of these examples, as well as many others.

In electrical engineering terms, the feedback processes can be characterized in control terms by observing the locations of the poles of the processes. If the poles are in the left half plane, then the process is stable. If the poles are in the right half plane, then the process is unstable. Stable implies that the process will tend to correct itself. Unstable implies unbounded increase.

While many processes appear to be stable, examples of unstable processes have also been noted. In particular, increased CO_2, the most commonly known greenhouse gas, from nonpolar regions causes warming in the polar regions, resulting in thawing of permafrost regions, in which large quantities of CH_4, another greenhouse gas, are trapped. Liberation of the CH_4 causes additional warming, resulting in additional liberation of CH_4, and so forth. The question is whether another natural process exists to counteract this cycle. The process itself is unstable.

The accumulated impact of human activity is at a point where serious questions are now being asked about the ability of the earth to recover from the results of this activity. One might argue that as long as the processes remain stable, the pollution is at acceptable levels for the planet to sustain life. The challenge becomes one of defining acceptable limits on pollution and the resulting changes in natural cycles caused by human activity.

9.2 EXTERNALITIES

Externalities were mentioned briefly in Section 8.5 as factors resulting from energy or other manufacturing that are not incorporated directly into the selling price of a product. That is, once the smoke is out of the smokestack, it is no longer a fiscal responsibility of the smoke producer. In this chapter, externalities will be explored in greater detail. In particular, environmental effects, health and safety issues, and subsidies will be discussed in the context of comparing energy produced by PV means with energy produced by other means. In general, the externalities discussed are not associated with costs that appear in a life-cycle cost analysis of an energy source. The reader can expect that in the future, greater attention will be paid to those issues currently considered as externalities, as means for attaching monetary value to these issues are developed and adopted. Some readers may find the research necessary to affix monetary value to externalities of sufficient interest and challenge that they may choose to pursue this interesting and important area to a greater extent.

Perhaps one of the more challenging considerations relating to externalities is the debate as to whether they really merit consideration. Acid rain and global climate change, for example, are two phenomena that have been the subjects of a great deal of discussion over the past few decades.

Acid rain is claimed to result from the emissions of sulfur and nitrogen oxides by fossil-fuel burners. As these oxides are dissolved in raindrops, they turn into weak acids that have been blamed for the changes in pH in lakes and soils and for the etching away of building structures. The societal costs of these effects are borne by those affected rather than those who benefit from cheaper electricity.

The scientific community has reached overwhelming consensus on specific greenhouse gases that cause global climate change and has linked the increase in concentrations of these gases to human activity. Yet, there remain skeptics who either do not believe in the phenomena or otherwise claim that insufficient information is available to confirm the cause–effect relationships with 100% certainty. Unfortunately, these differences have resulted in serious polarization of political beliefs, with climate change deniers predominantly elected by one major political party in the United States and politicians who accept the scientific data predominantly elected by the other major political party. Then there are those who accept climate change, but hypothesize that it may be a good thing.

For example, Moore [1], an economist, argued in 1998 that policies to reduce the emission of greenhouse gases "may be unnecessary, would be inordinately expensive and would lead to worldwide recession, rising unemployment, civil disturbances, and increased tension between nations …." He noted that "even if significant warming were to occur, public policymakers could, at the time it became evident, launch programs to adapt to the change, such as building dikes, increasing air conditioning, and aiding farmers and ecosystems to adjust to the new weather." He goes on to observe that during colder periods, more people die from exposure to cold than the number who die from exposure to intense heat during warmer periods, suggesting that the potential consequences may not be as dire as predicted by those concerned about global warming. During the 2016 U.S. election cycle, climate change has been a contentious issue among officials and candidates for office who have proposed programs to reduce greenhouse gas emissions and their opponents who have been labeled as "climate deniers" by organizations dedicated to reduction of greenhouse gas emissions.

The history of climate concern, with extensive bibliography, is nicely documented by Paterson [2]. The greenhouse properties of the atmosphere were noted as early as 1827 by Fourier, who is perhaps better known to engineers for the Fourier series or the law of conduction. During the nineteenth century, discussions continued about the causes of climate change, and in 1872, the International Meteorological Organization was formed. In 1896, Arrhenius published an article in which he calculated that the temperature of the earth would rise about 5°C if atmospheric carbon dioxide were to double. In 1908, he suggested that industrial carbon dioxide emissions might result in a noticeable change in atmospheric carbon dioxide levels within the next few centuries. He continued to observe that perhaps this would be good, since it would warm up some of the cold regions of the planet and make them more useful for agricultural and other purposes. An alternative viewpoint was that there was no need for concern over carbon dioxide emissions, since the oceans would remove excess carbon dioxide and maintain the delicate balance between atmospheric oxygen and carbon dioxide.

The first formal significant consensus on global warming and its relation to the production of specific greenhouse gases probably came out of the Conference on the Physical Basis of Climate and Climate Modeling, held in 1974. The Intergovernmental Panel on Climate Change (IPCC), in its 1990 report, identified carbon dioxide, methane, chlorofluorocarbons (CFCs), and nitrous oxide as gases that will enhance the greenhouse effect and result, on the average, in an additional warming of the surface of the earth [2].

In this chapter, a general overview of relative environmental effects of energy sources is presented, followed with observations of specific areas of concern for a variety of PV sources and how to best minimize these potential adverse effects. The reader is encouraged to acknowledge that externalities must be viewed from economic as well as from political perspectives and that scientific, economic, and political considerations often cause the decision-making process to become merged into a complicated maze. One need only to observe the controversy over what to do with nuclear waste to appreciate the interrelationships of these three areas.

9.3 ENVIRONMENTAL EFFECTS OF ENERGY SOURCES

9.3.1 Introduction

Although the methodology of assigning precise dollar values to carbon dioxide emission, particulate emission, nitrogen oxide emission, sulfur emission, and, in general, the adverse environmental effects of the construction, operation, and decommissioning of various energy sources is complex, it is generally possible to assign relative comparative values to most of these areas. Table 9.1 shows a matrix

Table 9.1 Relative Environmental Effects of a Variety of Renewable and Nonrenewable Energy Sources

Negligible/Significant = 1 Significant = 2 Significant/Large = 3 Large = 4	SO$_x$ and NO$_x$	CO$_2$	CH$_4$	Health	Particulates	Heavy Metals	Catastrophies	Waste Disposal	Visual Intrusion	Noise	Land Requirements
Passive solar energy									1		
Photovoltaics					1	1		1	1		1
Wind									3	1	1
Biomass	1		3	1	1	1		1	1	1	3
Geothermal	1	1	1	1			1	2	1	1	
Hydro							2		3		3
Tidal							1		3		1
Water waves							1		1		
Coal	4	4	2	1	2	2	1	2	2	1	3
Oil	3	4	1	1	2	1	2	1	1		1
Natural gas	1	4	3	1			2		1		1
Nuclear	1	1		1			2	3	2		1

Source: Baumann, A. and Hill, R., *Proc. 10th EC Photovoltaics Solar Energy Conference*, Kluwer, Dordrecht, 1991, 834–837, with kind permission from Kluwer Academic Publishers.

compiled by Baumann and Hill [3] that compares 10 negative effects for 12 different energy sources with relative significances indicated. The observations should come as no particular surprise to anyone familiar with the listed sources.

9.3.2 Air Pollution

9.3.2.1 The Clean Air Act and the U.S. Environmental Protection Agency

The U.S. Environmental Protection Agency (EPA) is required by the 1990 amendment of the Clean Air Act [4] to regulate emissions of toxic air pollutants from a published list of source categories. The EPA is charged with setting standards and with establishing and promulgating rules for reducing toxic emissions to levels that meet the standards for nitrogen oxides, ozone, sulfur oxides, lead, carbon monoxide, and particulates.

Over the years, the EPA has presented data on these six pollutants in several formats, such as nonattainment maps. Currently, the EPA publishes a wealth of data, including national and regional summaries of historic and daily levels of these substances as compared to the standards [5]. The data are encouraging, since they show that for all but ground level ozone, the atmospheric concentrations in the United States have been falling steadily since 1990 and essentially all six average below

EXTERNALITIES AND PHOTOVOLTAICS 383

established national attainment levels. For example, in 2015, NO_2 concentrations have fallen 45% and are now below the national standard, ozone has fallen 23%, but is still only slightly under the national air quality standard level, SO_2 has fallen 76% and is now well below the national standard, Pb has fallen 97%, CO concentrations have fallen 77%, and PM10 particulates have fallen 36% and are currently at approximately 40% of the national standard.

Note, however, that CO_2 and other greenhouse gases are not considered to be pollutants by these standards since most are not considered to be health hazards. But according to a 2015 World Resources Institute (WRI) report, global CO_2 emissions had increased to 35.9 gigatons in 2014, a 60% increase over 1990 emissions. Of these emissions, 42% were from coal burning, 33% from oil burning, 19% from gas, and 6% from cement production. The United States led the world in per capita CO_2 emissions in 2014 at nearly 250% of the per capita emissions of either China or the European Union. Of some encouragement is the observation that increased CO_2 production seems to have leveled in 2014 and is predicted to decline in 2015 [6].

Readers are encouraged to visit the EPA website for details of current EPA programs. The WRI website provides timely information on pollutants, including valuation methodologies and economic cost discussions. According to WRI, the IPCC has calculated that to stabilize CO_2 levels at 1997 levels would require a 60% cut in emissions, maintained at that level for the next century. Readers may also wish to review their solutions to the homework problem in Chapter 1 that dealt with CO_2 emissions from burning coal and petroleum to put the CO_2 from these sources into perspective.

Air quality is not a problem limited to the United States. As developing countries continue to develop, often pollution controls take a back seat to development and the desires of the populations to have more energy at their disposal. Many of the world's larger cities have had severe air pollution problems. In fact, a red alert for smog pollution was issued on December 8, 2015 in Beijing, China, for the first time in history, when the concentration of hazardous aerosols (smog) reached 484 μg/m^3, which exceeds the safe limit threshold by a factor of 19. During the week of November 27, Alternet reported that the hazardous aerosol concentration in Beijing had reached 666 μg/m^3 [7].

For the reader who wishes to spend a year or two in the further collection of information on air pollution, on December 9, 2015, a Web search for air pollution data yielded 46,700,000 hits.

9.3.2.2 Greenhouse Gases and the Greenhouse Effect

At this point in history, most people have heard of greenhouse gases and most people have heard somewhere that CO_2 is the major culprit. The common characteristics of greenhouse gases is that they are, for the most part, nontoxic and they produce the greenhouse effect by being transparent to visible radiation, but reflective to infrared radiation. This means, simply, that as sunlight passes through the atmosphere and is absorbed by something on the surface of the earth, the object that absorbed the relatively short-wavelength sunlight will then emit relatively longer wavelength energy in the infrared region of the spectrum. As the infrared radiation

Table 9.2 Summary of Potent Greenhouse Gases Showing Potency Compared to CO_2

Gas	$\times CO_2$	Gas	$\times CO_2$	Gas	$\times CO_2$
CO_2	1.0	CFC-113	4500	CH_4	21
CFC-13	13,000	CFC-114	7000	CCl_4	1300
N_2O	310	CFC-115	7000	CCl_2H_2	15
Halon-1211	4900	Halon-1301	4900	HFC-125	3400

Source: Battisti, R. and Corrado, A., *Energy* 30, 2005, 959.

encounters a greenhouse gas molecule, the radiation will be absorbed or reflected back to the surface of the earth. Ideally, to maintain temperature equilibrium, the earth will emit the same amount of energy that it absorbs. Not doing so causes heat to accumulate or dissipate such that the temperature of the earth rises or falls enough to be able to emit the stabilizing amount of energy according to the blackbody radiation formula as was previously discussed in Chapter 2.

Less commonly known, however, is that there are greenhouse gases that exhibit a much higher ability to trap heat than CO_2. The only reason CO_2 is better known is because there is more of it, and the amount increases every year. But some of the other culprits are also increasing, so it should be of interest to know what they are. Table 9.2 summarizes the more potent greenhouse gases, indicating their relative ability to retain heat compared to that of CO_2 [8].

Recent increases in the production of oil and natural gas through enhanced recovery efforts have also resulted in accidental emissions of CH_4 into the atmosphere. The 2015–2016 leak of 90,000 metric tons over a 4-month period from a well in Aliso Canyon, CA, has resulted in the recognition that more attention needs to be paid to minimizing the escape of this gas into the atmosphere [9].

9.3.3 Water and Soil Pollution

Historically, water pollution has perhaps received the most attention in the United States in terms of the lowering of pH of lakes in the Finger Lakes region of New York, presumably caused by acid rain from SO_2 and NO_2 emissions in the Ohio River region. The lowering of the pH as a result of the inability of the lakes to buffer the effect of the acid rain has been blamed for the loss of significant fish populations. This loss has, in turn, resulted in a loss of revenue to other sectors of society such as the sport fishing and tourist industries.

The Flint, Michigan water crisis of 2015–2016 resulting from switching water supplies to a distribution system with lead pipes to over 100,000 residents has called attention to the possibility that this is only the tip of the iceberg, with many more communities facing a similar danger [10].

Other instances of water pollution with perhaps more insidious consequences is the pollution of aquifers, lakes, rivers, and streams with solid, liquid, and biological pollutants. While aquifer pollution has not been blamed on SO_2 or NO_2 or other atmospheric gases, it has been blamed on industrial solvents and other chemicals, many of which are used to support an economy with a large energy appetite. Perhaps one of the best-known historic incidents in this category is the Love Canal, with many other similar sites as identified for cleanup by the EPA. At this point, the cleanup costs of these toxic waste dumps is being quantified, so these costs can be related back as externalities resulting from past waste-dumping practices.

It has been reported that soil acidification leads to permanently reduced productivity resulting from slower decomposition of humus and diminishing humus quality. Other important effects of soil acidification include leaching of alkali and alkaline metals and release of heavy metals into soils. With changing quality of the soil surrounding the roots of trees, it is believed that the trees may become more sensitive to gaseous air pollutants [11].

More recently, the hydraulic fracturing (fracking) method of petroleum and natural gas (methane, CH_4) extraction has caused significant concern over pollution of aquifers by fracking chemicals as well as by CH_4. In addition, escaping of CH_4 to the atmosphere, as mentioned earlier, is not felt to be adequately controlled. At 21 times greater potency as a greenhouse gas than CO_2, many environmental groups have called for a ban on fracking until it can be clearly demonstrated that aquifers will not be endangered and CH_4 will be prevented from escaping to the atmosphere. And, of course, even if successfully extracted without pollution of aquifers or loss of CH_4 to the atmosphere, combustion of each CH_4 molecule will still produce a CO_2 molecule and two H_2O molecules.

9.3.4 Infrastructure Degradation

A number of studies have shown that pollution causes premature degradation of infrastructure [12]. Stonework, carbon steel, nickel and nickel-plated steel, zinc, and galvanized steel are all highly affected by sulfur dioxide in rain. It has been estimated that in unpolluted environments, galvanized coatings last up to three times longer, and in polluted areas galvanized transmission towers require repainting nearly twice as often. Hence, a cost figure can be attached to the restoration of structural finishes. However, assigning responsibility for the degradation is somewhat more problematic in that it is difficult to determine the mix of pollution sources that cause a specific amount of degradation to any particular site, since the origin of any pollutant at any particular time will depend on the wind direction.

9.3.5 Quantifying the Cost of Externalities

9.3.5.1 The Cost of CO_2

The cost of CO_2 can be considered in two ways: the cost of controlling CO_2 and the benefit of controlling CO_2. This method is applicable to other pollutants as well, but only CO_2 will be discussed in this section.

According to the WRI [6], holding CO_2 at 1990 levels will cost approximately 1%–2% of the gross domestic product (GDP) for developed countries over the long term, with reductions below 1990 levels cost increasing to 3%. An alternate analysis, however, suggests that by reducing CO_2 levels significantly by incorporation of energy conservation measures, the cost to reduce the CO_2 levels will be less than the savings in energy costs.

If action is not taken to stop the increase of CO_2 atmospheric concentration, it has been estimated that doubling of atmospheric CO_2 will lead to an average warming of 2.5°C [6]. According to current projections, this doubling will occur during the next century if no curtailment measures are taken. In this case, it is estimated that the damage due to the warming will reach approximately 1%–1.5% of GDP per year in developed countries, with substantially more in island nations. If CO_2 levels continue to rise beyond the double point, the damage may come closer to 6% of GDP. At the Paris climate conference of November–December 2015, it was pointed out that an increase of 1°C has already occurred and a strong push has been made to agree to limit future increases to less than 1.5°C [13].

Hence, it appears certain that damage will occur if nothing is done, while it is debatable whether there will be cost or cost savings if a program to reduce CO_2 levels is pursued. Considering that CO_2 production is also generally accompanied by the production of NO_2, SO_2, and particulates, if reduction of CO_2 is also accompanied by the reduction of other pollutants, then the benefits become subject to a multiplier effect. What is also being supported here is that it may be a better strategy to focus on reducing the production of CO_2 rather than sequestering CO_2 already produced. Perhaps a carefully planned combination of these two strategies may produce the most cost-effective results.

9.3.5.2 Sequestering CO_2 with Trees

In order to quantify the cost of externalities, it is necessary to arrive at a methodology for determining the value of a ton of SO_2 or NO_2 or CO_2 or other substance. Since trees, other green plants and oceans, have maintained the balance between O_2 and CO_2 over the millennia, it would seem that a tree might be valued in terms of the number of tons of CO_2 it will remove from the atmosphere in its lifetime. Since SO_2 and NO_2 generally are removed from the atmosphere by rain, which dissolves them as weak acids, presumably a price per ton can be assigned in terms of the monetary value of any damage that may result to structures or the environment from acid rain. For example, if a building requires repair more often as a result of acid rain, this is a measurable cost, provided that the increased repair costs can be documented. If the fish die in a lake as a result of acid rain and, as a result, cause a decrease in tourism to the area, this may also constitute a documented cost if the source or sources can be identified.

The cost of CO_2 and other pollutants can be measured in several ways, including the cost of repairing the damage, the cost of controlling the damage by reducing emission of pollutants, and the cost of mitigating the damage, such as by absorbing the additional CO_2 by trees. Hodas [14] reported costs of $240 per ton of carbon for removal of CO_2 from exhaust gases, with significant technical difficulty. The scrubbed CO_2 would then be liquefied and pumped to a depth of about 500 ft in the ocean, where it would then dissolve and become available for plankton and ocean vegetation. The cost of such an operation would likely be close to $1000/kW for scrubbing and disposal. There would also be a 25% energy penalty, since the energy for the scrubbing and disposal would not be available for other end users, as well as a 22% capacity penalty and additional annual maintenance costs.

Supply- and demand-side efficiency improvements are estimated to cost in the range of $20 per ton of carbon avoided. Automobile efficiency increases also save 22 lb of CO_2 per gallon not burned, with an estimated cost of $0.53 per gallon for the effort required to increase automobile efficiency to 44 miles per gallon, assuming continued use of gasoline as a fuel [14].

It is estimated that the United States would require 1.5 billion hectares of forest to absorb its annual CO_2 emissions [14]. The problem is that the total land area of the United States is only 913 million hectares, and forest in Death Valley, CA, and quite a few other areas appears as an unlikely prospect.

The idea of planting trees to mitigate the effects of CO_2 generation was first attempted in a project proposed by Applied Energy Service. This project involved planting approximately 50 million trees in Costa Rica to mitigate the CO_2 generated by a fossil-fuel electric generator in Connecticut. The project also involved a fire protection program to save 2400 hectares of forest from fires, with an overall estimated sequestering of 387,000 metric tons of carbon per year [14].

When trees are used for CO_2 mitigation, it is necessary to consider what will be done with the trees after they mature. If they are used for firewood, then the sequestered carbon is returned to its CO_2 form. If they are used for construction, then the carbon remains sequestered. However, even if they are burned, on the assumption that something will be burned, the burning of these trees displaces burning of other trees. If the trees spared from fire are genetically diverse, then an additional value can be assigned to the preservation of diversity in the biosphere.

Perhaps an even more valuable location for trees is the urban environment, particularly in tropic or near-tropic regions. Hodas [14] observes that by planting trees near buildings, the microclimate of the buildings is cooled by several degrees, thus reducing air conditioning needs and the corresponding energy required for cooling. If the energy for cooling comes from fossil-fueled generation, then the trees reduce atmospheric CO_2 directly as a result of their photosynthesis as well as indirectly by reducing the output demands and corresponding fuel requirements of the fossil-fueled generator.

9.3.5.3 Attainment Levels as Commodities

Another interesting method of assigning value to pollutants comes as a consequence of the Clean Air Act [4] and regulated attainment levels. Companies are granted a certain allowable number of pollution units for their facilities. For example, an electrical generating utility may be allowed a certain number of tons of SO_2 emissions per year. If the utility does not generate as many tons as allowed, it may transfer the remainder of its allowed emissions to another entity. In a free enterprise society, this is not normally done as a simple, friendly gesture. Rather, air quality points are traded on the commodity exchange.

It is now possible for an individual, an organization, or a company to purchase allowances for atmospheric SO_2 and NO_2 [15]. At the end of March each year, the

EPA holds an auction for SO_2 allowances. Anyone can bid on the allowances, and the highest bidder gets the allowance. It is also possible to purchase allowances through a commodity broker or to purchase them through various environmental groups that purchase the allowances and then do not use them. Not using the allowances results in cleaner air. It is also possible to purchase NO_2 allowances from brokers or from environmental groups, but they are not currently auctioned off by the EPA.

While this practice does not affix a price to emissions based on specific environmental costs, it at least shows that emissions do have some economic cost, and the reduction of emissions can result in economic benefit. What is interesting is that the law of supply and demand dictates the price of air quality points. If strict laws are enacted to reduce pollutants, companies end up investing in pollution control technologies, such as electrostatic precipitators and scrubbers. If these investments result in sufficient reduction in emission of pollutants, then a smaller number of air quality points will be needed and a larger number will be available. This means the price will go down.

Other related actions have also been offered by energy producers in trade for permission to operate, such as improving navigation channels to reduce the likelihood of accidental fuel spills. These added costs represent the attachment of a price to certain externalities and, as a result, end up incorporating the costs of the externalities into the cost of the product.

For some products, such as gasoline, CH_4, or coal, the amount of CO_2 produced when a specified quantity is burned is well defined. Bill Nye, the Science Guy, along with other knowledgeable members of the scientific community, have suggested a carbon fee be imposed on the purchase of these fuels. The fee would then be applied to a trust fund, similar to the highway trust fund, to be used specifically for research on reducing CO_2 emissions [16].

9.3.5.4 Subsidies

Subsidies were included as components of direct costs of doing business in Chapter 8, since they are usually included in the reduction of the cost of a product, such as energy, to the consumer. However, unless a level playing field exists, the costs of two energy sources cannot be properly compared. A level playing field exists when all energy sources are subsidized equally on a kWh-to-kWh basis. The difficulty is the identification of all subsidies, since some are direct and others are indirect.

A direct subsidy, such as a depletion allowance, can be quantified in terms of dollars per kWh. An indirect subsidy, such as military presence, can also sometimes be quantified, but will normally not appear in the financial records of a company. The effects of favorable tax considerations and tariffs can also be quantified in either a direct or indirect manner.

Government-sponsored research and government purchase of systems or rebates or tax credits on systems are also forms of subsidy. Generally, government involvement is justified on the basis that once the subsidized industry is able to compete without subsidy, it will generate sufficient revenues to result in sufficient taxes to reimburse the government for its initial investment. In any case, hardly an energy source exists that does not benefit to some extent from subsidies. To date, however, no one has requested a depletion allowance on sunlight, although solar access rights have been challenged in the case where one party shades another party's solar system.

9.3.6 Health and Safety as Externalities

Litigation against tobacco companies, resulting in the award of billions of dollars in damages to states to help reimburse the costs of tobacco-related illnesses, has shown that the cost of public health has become a recognized externality, with a significant price tag. Other lawsuits have also been successfully prosecuted against a variety of forms of pollution resulting from careless or indiscriminate disposal of many forms of toxic wastes. These cases confirm the linkage between public health and its economic cost.

Public safety is also at issue when energy sources are considered. The Chernobyl, Three Mile Island, and Fukushima Daiichi nuclear accidents brought to the forefront public concern over nuclear safety. The location of large fuel tanks in highly populated areas is also of concern to public advocacy groups. Transport of various fuels is also problematic, especially when large tanker trucks and railroad tanker cars are involved in accidents that result in spilling of their contents and subsequent major efforts to clean up the damage before aquifers are affected. Such an accident occurred on June 4, 2016 along the Washington–Oregon border near Mosier, OR, which resulted in derailment of 16 of 96 tank cars, burning of four, and an oil slick on the Columbia River [17]. Even in the case of renewable energy sources, concern has been expressed about the possibility of construction accidents, since installation of renewable sources tends to be more labor intensive per installed kW than conventional sources. Manufacture of the materials used in conventional generation facilities carries with it both energy costs and materials costs, along with certain levels of exposure to hazardous or toxic materials. Similar costs are associated with the manufacture of PV cells and system components. Although these costs tend to be incorporated into the production cost of the materials, when toxic waste by-products result, they are often treated as externalities.

The balance of this chapter explores the externalities associated with the production, deployment, operation, and decommissioning of PV power systems. Particular attention is paid to the environmental concerns associated with each of the phases, recognizing that in order to produce PV systems, initially other energy sources must be exploited.

9.4 EXTERNALITIES ASSOCIATED WITH PV SYSTEMS

9.4.1 Environmental Effects of PV System Implementation

The implementation phase of each PV technology can be described in terms of a production cycle, which includes potential pollutants or hazardous waste associated with each production step. Table 9.3 summarizes areas of environmental or health concern for current PV technologies and for future technologies that may

Table 9.3 Environmental Concerns Associated with PV System Production

Technology	Concern	Relative Significance[a]	Relative Control Cost[b]	Control Strategy
All	Mining	Low	Lower	MSA[c]
All	Cleaning solvents	High	Higher	Recycle
All	Steel	Medium	Lower	OSHA, EPA
All	Aluminum	Medium	Lower	OSHA, EPA
All	Concrete	Medium	Lower	OSHA, EPA
All	Glass/breakage	Low	Lower	n/a
All	Encapsulants	Low	Lower	n/a
Crystalline Si	CO_2	0.02[d]	Lower	EPA
Multi cr Si	CO_2	0.01	Lower	EPA
Thin-film Si	CO_2	0.005	Lower	EPA
Other thin films	CO_2	0.004	Lower	EPA
Future cells	CO_2	0.002	Lower	EPA
All Si cells	Silica dust	Low	Lower	OSHA, EPA
All Si cells	CH_4	Low	Lower	Confine and recycle (C and R)
All Si cells	B_2H_6	Low	Lower	C and R
All Si cells	PH_3	Low	Lower	C and R
All Si cells	AsH_3	Low	Lower	C and R
CIS	H_2Se	Low	Lower	C and R
CIS	Cd[e]	Low	Lower	C and R
CIS	Fire (Cd)	Low	Lower	None
CdTe	Cd and CdO	Low	Lower	C and R
CdTe	Te	Low	Lower	C and R
CdTe	Fire (Cd, Te)	Low	Lower	None

[a] Compared to amount produced from equivalent amount of energy from coal-fired generation. Some items are not present in coal-fired, but are compared to other by-products of coal-fired generation such as SO_2, NO_2, and particulates.
[b] Generally lower control costs since generally less to control.
[c] Mine Safety Administration.
[d] These numbers represent ratios of CO_2 to produce a kWh PV over the system lifetime to CO_2 per kWh from burning coal.
[e] It is estimated that coal-fired stack emissions produce as much Cd per kWh as is contained in CIS per lifetime kWh. CIS Cd is recycled at end of CIS life cycle.

have promising possibilities for large-scale production. The associated CO_2 for each technology represents the CO_2 resulting from the conversion of the primary energy used to produce the PV systems, so, in effect, it represents an energy cost of production. In the future, it is conceivable that energy for PV production will come from PV or other renewable sources, thus further reducing the CO_2 and other pollutants associated with the production of PV cells. This concept is feasible since current PV systems, including cells, module materials, and BOS, will produce, over their useful lifetime, between 8 and 10 times the energy used in their fabrication [18]. It is expected that new technologies and processing techniques will lead to energy returns on investment ranging between 14 and 45.

Aside from the energy cost, each technology has its own specific toxic waste areas of concern. Some concerns are common to all technologies, such as production of the materials for support structures and encapsulants for the modules. Perhaps the most important common denominator associated with all technologies is the material used for cleaning the cells. In all cases, high levels of cleanliness are required to maximize performance, and in some cases, the solvents used are highly toxic. Fortunately, these solvents have been in common use in the semiconductor industry and careful means of control are well understood within the industry. The common means of dealing with these solvents is to confine them and then recycle them.

In the case of Si PV cells, the most significant concerns are the proper handling of the dopants for n-type and p-type. Since the semiconductor industry has been handling these components effectively, again the processes are well defined. Once the impurities are in the Si, the amounts are so insignificant that the doped Si is considered to be benign.

For copper indium diselenide (CIS) cells and the cadmium telluride (CdTe) cells, the Cd content is the primary item of concern, but since the films in these devices are generally less than 2-µm thick, the total amount of Cd and Te in these cells is very small. Another concern for CIS is the use of H_2Se for the deposit of Se into the cell, but again, the use of this highly toxic substance can be kept well under control with minimal risk of having it escape into the environment.

9.4.2 Environmental Effects of PV System Deployment and Operation

The deployment of PV systems has associated environmental and health costs similar to the deployment of other energy technologies [19]. Steel, aluminum, and concrete can be expected to be part of the structures upon which the PV arrays and associated BOS components will be mounted. Hence, the production costs and associated environmental costs, such as CO production in the reduction of iron oxides, become associated with PV deployment.

Just as construction accidents occur in any construction project, it is anticipated that PV project construction will also result in construction accidents. For the most part, materials used in construction of PV facilities are nontoxic. It is thus the responsibility of the installation contractor to ensure that the workplace and work practices follow Occupational Safety and Health Administration (OSHA) rules.

Once the PV system is installed, it quietly generates pollution-free electricity any time the sun is shining. The main environmental concern associated with the operation of PV systems is if a system containing Cd or Se or Te should be exposed to fire. Analysis has shown, however, that since these materials appear in such minute quantities in thin-film PV that anyone approaching close enough to encounter significant exposure to any of the toxins would face significantly more danger from the fire itself. For example, only 400 g of Cd is present in a 1-MW CIS PV system, whereas 5 g/m^2 of Cd is present in CdTe systems, or about 25 times the amount in CIS systems.

Other dangers associated with the installed system include, of course, the possibility of electrical shock or fire. Systems installed in accordance with the *NEC* [20] and local, regional, and/or national fire safety guidelines as required by local code enforcement jurisdictions will reduce any shock or fire hazard to a minimum, both for the general public as well as maintenance personnel. The fact that PV systems are inherently current limiting tends to reduce their potential to produce high-current arcing if they are inadvertently shorted, except for very large systems as discussed in Chapter 4. Incorporation of ground fault and arc fault protection in systems adds further protection.

Since most PV components and materials are benign, transportation of modules and other system components poses minimal environmental risk compared to the transportation of oil and gas.

A final safety consideration for installed PV systems relates to their ability to withstand high winds. Inadequately mounted PV modules in high winds may tear loose from their mounting and becoming projectiles. Installations compliant with local building code wind-loading requirements, such as ASCE 7-10 [21], present minimal risk as long as the PV modules are listed to UL 1703.

9.4.3 Environmental Impact of Large-Scale Solar PV Installations

Of all solar technologies, PV installations appear to have the least serious environmental impact. As a start, conventional PV requires the least amount of water for washing and dust control. Since PV does not require a working fluid or cooling tower (in contrast to CSP technologies such as parabolic trough, tracking dishes, or power towers), the impact on local water resources is negligible. Once a PV installation is completed, it is basically an autonomous operation with no need for round the clock operational staff, except perhaps for a token security staff. This minimizes the effect of vehicular traffic on pollution, noise, and lighting interference with wildlife habitat and movements. The following mitigation strategies should be considered when designing utility-scale PV installations:

1. PV installations use relatively less water per MW than all other solar technologies. The average annual water for washing and dust control is 24,700 m^3/400 MW, or 20 acre-ft/400 MW [22]. This impact must be mitigated. There are no chemicals used for the PV panel washing except mild soap. However, if any chemicals are used for control of vegetation, a mitigation plan should be prepared.
2. PV installations have a potential impact on topography that could result in changes to the drainage and water runoff patterns. A well-designed PV installation can minimize this impact. The EPA requires that developments greater than 20 acres (8 hectares) must comply with turbidity concentration guidelines (EPA-2007c). Soil erosion resulting from the construction activity should also be mitigated through standard erosion control practices.
3. The impact on vegetation from trimming to avoid shading of panels is difficult to mitigate. This is one of the more important decisions that must be made in siting a PV plant. The Bureau of Land Management has developed a map [23] showing the regions that would be least impacted by solar installations. To reduce operation and maintenance costs and to minimize impact on local vegetation and the wildlife that depends on tall grass or vegetation, it is best to study these maps and consult with local land use experts.

4. There is a negligible site lighting impact as PV plants do not operate at night. Lighting is limited to security.
5. PV installations should be planned to avoid crucial life stages of local wildlife such breeding or nesting seasons. If the site selected is in a path of endangered species, an expert familiar with the Endangered Species Act [24] and its guidelines should be consulted.
6. Habitat fragmentation that results from installation of fences or uninterrupted linear components such as rows of solar panels or transmission lines must be considered carefully. Fencing serves the dual purpose of security and also to protect wildlife from collision with solar panels, especially at night. But it can also result in injury or confusion for some of the species that are hardwired to follow a certain path. Fences must be designed to allow reptiles, tortoises, and other small animals to get through but also be designed to minimize impact on all local wildlife and migratory routes.
7. There are some positives, however. Fthenakis and colleagues [25–28] who have written extensively on this subject have also found that in certain regions, solar PV shading may actually improve the habitat for certain species.

The design of a utility-scale PV installation should start by cataloging a complete list of all potential impacts and developing a mitigation plan to address the more critical factors.

9.4.4 Environmental Effects of PV System Decommissioning

Certain regulatory requirements apply to the decommissioning of PV systems. Assuming that the system will be disassembled at the end of its useful lifetime, it is then necessary to consider the destination of the disassembled components. Some of the components will require disposal according to toxic waste regulations unless they can be recycled, and other components will not be subject to quite as stringent requirements.

The Resource and Conservation Recovery Act (RCRA) constitutes the primary set of rules governing wastes containing Cd, Se, Pb, Cu, or Ag provided that these wastes are considered to be discarded material and are not included in any specific exclusions [29]. The EPA defines the Toxicity Characteristic Leaching Procedure (TCLP) for 39 materials to determine whether waste products containing them may be classified as hazardous. Any waste product containing any of the 39 materials that yields a soluble concentration in excess of its TCLP limits is considered hazardous.

The California Hazardous Waste Control Law introduces an additional test for toxicity characterization, with two additional indicators. The Waste Extraction Test (WET) is used to identify non-RCRA toxic wastes. The WET threshold limits are defined as the Soluble Threshold Limit Concentration (STLC) and the Total Threshold Limit Concentration (TTLC), which is the total concentration of listed materials, in any form.

In terms of STLC and TTLC, it appears that CdTe modules may be characterized as hazardous in terms of TTLC and STLC for Cd. CIS modules may be subject to TTLC for Se and STLC for Se and Cd. Polycrystalline Si modules may have a problem with TTLC and STLC for Ag or Pb. In some cases, Pb and Cu may be a problem, and a-Si:H modules do not seem to have any toxicity problems [29,30]. If CdTe and CIS modules are recycled, they are exempt from the disposal regulations. Studies have projected recycling costs of approximately $0.01–$0.04 per watt for both materials, using relatively standard separation techniques.

PROBLEMS

9.1 a. List five examples of how humans have upset the balance of various natural processes, such as photosynthesis, digestion, and water purification.
   b. List five examples of how humans either have or are currently working to restore the natural balance.
9.2 Look up the U.S. nonattainment regions, if any, on the EPA website for the six toxic air pollutants defined in the Clean Air Act.
9.3 Analyze the validity of the assumptions in the Moore claim that global warming may be beneficial and if it turns out not to be, then we can do something about it at that time.
9.4 If recycling of CdTe PV modules will cost approximately $0.04/W and the lifetime of a CdTe module is 20 years, what is the present value of the recycling cost for a 10-kW array? Make reasonable assumptions.
9.5 Determine the prices of a ton of SO_2 and a ton of NO_2 on the commodity exchange. The EPA website lists websites and phone numbers of commodity brokers and environmental organizations that purchase and sell them.
9.6 If burning a gallon of gasoline produces 22 lb of CO_2, estimate the CO_2 that your vehicle emits in a year. Compare this with the amount that would have been generated had you been driving an electric vehicle charged from a PV array or a hybrid auto or a fuel-cell-powered auto. Take into account the origin of the H_2 used in the fuel cell.
9.7 In the CO_2 scrubbing and liquefaction scheme reported by Hodas, estimate the increase in electrical cost that would be necessary to offset the cost of the process.
9.8 Compare Article 690 in the 2014 *NEC* with Article 690 in an earlier edition, preferably prior to 2008, in terms of additional safety measures required by the most recent edition. If your local library does not have a copy of an earlier edition, a friendly local electrical contractor or electrical engineer who has been business for a long time may have one.
9.9 Referring to the water use given in Section 9.4.3 for utility-scale PV installations, assuming the modules in the 400-MW system to be 400-W modules
   a. Express the water use in terms of gallons per year per module.
   b. Assuming the array will produce 2000 kWh/kW annually, express the water use in terms of gallons per kWh.

REFERENCES

1. Moore, T. G., *Climate of Fear: Why We Shouldn't Worry about Global Warming*, The Cato Institute, Washington, DC, 1998.
2. Paterson, M., *Global Warming and Global Politics*, Routledge, London, 1996.
3. Baumann, A. and Hill, R., In *Proc. 10th EC Photovoltaics Solar Energy Conference*, Kluwer, Dordrecht, 1991, 834–837.
4. 42 U.S.C. § 7401 et. seq. (Clean Air Act).

5. U.S. Environmental Protection Agency data on current and historic levels of pollutants regulated by the Clean Air Act, www.epa.gov/airtrends.
6. World Resources Institute, Insights: WRI's Blog, http://www.wri.org/blog/2015/12/4-things-you-need-know-about-current-trends.
7. http://www.alternet.org/environment/airpocalypse-beijing-issues-first-ever-red-alert-smog-reaches-deadly-levels (accessed December 10, 2015).
8. Battisti, R. and Corrado, A., Evaluation of technical improvements of photovoltaic systems through life cycle assessment methodology. *Energy* 30, 2005, 959.
9. The *Huffington Post* online newspaper, New York, has published numerous articles on Aliso Canyon. huffingtonpost.com.
10. The *Fiscal Times* has published articles on the Flint Water Crisis as have numerous other sources. thefiscaltimes.com.
11. Guderian, R., *External Environmental Costs of Electric Power*, Springer-Verlag, Berlin, 1990, 9–24.
12. Weltschev, M., *External Environmental Costs of Electric Power*, Springer-Verlag, Berlin, 1990, 25–35.
13. *21st Conference of the Parties to the U.N. Framework Convention on Climate Change*, Paris, France, December 2015.
14. Hodas, D. R., *External Environmental Costs of Electric Power*, Springer-Verlag, Berlin, 1990, 59–78.
15. For information on SO_2 and NO_2 commodity trading, see http://www.epa.gov/airmarkt/trading/buying.html.
16. Nye, B., *Unstoppable: Harnessing Science to Change the World*, St. Martin's Press, New York, 2015.
17. Protest and oil sheen on Columbia River follow Oregon train derailment, *The Guardian*, June 4, 2016, https://www.theguardian.com/us-news/2016/jun/04/protest-oil-sheen-columbia-river-oregon-train-derailment.
18. National Renewable Energy Laboratories, Information on energy return on investment for current and new PV technologies, http://www.nrel.gov/docs/fy04osti/35489.pdf.
19. Markvart, T., Ed., *Solar Electricity*, John Wiley & Sons, Chichester, UK, 1994.
20. *NFPA 70 National Electrical Code*, 2014 Ed., National Fire Protection Association, Quincy, MA, 2013.
21. *Minimum Design Loads for Buildings and Other Structures, ASCE 7-10*, American Society of Civil Engineers, Reston, VA, 2010.
22. Frisvold, G. B. and Marquez, T., Water requirements for large scale solar energy projects in the West, *J. Contemp. Water Res. Educ.*, 151(1), 2013, 106–116.
23. U.S. Bureau of Land Management, https://catalog.data.gov/dataset/blm-solar-energy-zones (accessed July 2016).
24. 16 U.S.C. § 1531 et. seq. (Endangered Species Act).
25. Fthenakis, V. M. and Kim, H. C., Greenhouse gas emissions from solar electric and nuclear power: A life cycle study, *Energy Policy*, 35, 2007, 2549–2557.
26. Fthenakis, V. M., End of life management and recycling of PV modules, *Energy Policy*, 28, 2000, 105–108.
27. Fthenakis, V. M., Fuhrmann, M. Heiser, J., Lanzirotti, A., Fitts, J., and Wang, W., Emissions and encapsulation of cadmium in CdTe PV modules during fires, *Prog. Photovolt. Res. Appl.*, 13, 2005, 713–723.
28. Fthenakis, V. M., Green, T., Blunden, J., and Krueger, L., Large photovoltaic power plants: Wildlife impacts and benefits, In *Proc 37th IEEE Photovoltaic Spec Conf*, Seattle, WA, June 2011.

29. Fthenakis, V. M. and Moskowitz, P. D., Emerging photovoltaic technologies: Environmental and health issues update, In *NREL/SNL Photovoltaics Program Review*, AIP Press, New York, 1997.
30. Fthenakis, V., Kim, H.C., and Alsema, E., Emissions from photovoltaic life cycles, *Environ. Sci. Technol.*, 42(6), 2008, 2168.

CHAPTER 10

The Physics of Photovoltaic Cells

10.1 INTRODUCTION

By this point, it is possible that the reader, now highly skilled at PV system design, might be interested in what goes on inside the PV cell during the process of converting light energy into electrical energy. This chapter is presented with the purpose of enabling the reader to become familiar with the challenges facing those engineers and physicists who spend their lives working on processes and materials aimed at reducing the cost and increasing the efficiency of PV cells. Initially, the basics of PV energy conversion are presented, followed by discussion of present limitations of cell production and some of the ideas that have emerged toward overcoming these limitations. Most readers will have already had at least an exposure to the theory of operation of the pn junction and the semiconductor diode in an electronics course, so it will be assumed that the reader will at least be familiar with the basic diode equation.

10.2 OPTICAL ABSORPTION

10.2.1 Introduction

When light shines on a material, it is reflected, transmitted, or absorbed. Absorption of light is simply the conversion of the energy contained in the incident photon to some other form of energy, typically heat. Some materials, however, happen to have just the right properties needed to convert the energy in the incident photons to electrical energy.

When a photon is absorbed, it interacts with an atom in the absorbing material by giving off its energy to an electron in the material. This energy transfer is governed by the rules of conservation of momentum and conservation of energy. Since the zero mass photon has very small momentum compared to the electrons and holes, the transfer of energy from a photon to a material occurs with inconsequential momentum transfer. Depending on the energy of the photon, an electron may be raised to a higher energy state within an atom or it may be liberated from the atom. Liberated electrons are then capable of moving through the crystal in accordance with whatever phenomena may be present that could cause the electron to move, such as temperature, diffusion, or an electric field.

10.2.2 Semiconductor Materials

Semiconductor materials are characterized as being perfect insulators at absolute zero temperature, with charge carriers being made available for conduction as the temperature of the material is increased. This phenomenon can be explained on the basis of quantum theory, by noting that semiconductor materials have an energy bandgap between the valence band and the conduction band. The valence band represents the allowable energies of valence electrons that are bound to host atoms. The conduction band represents the allowable energies of electrons that have received energy from some mechanism and are now no longer bound to specific host atoms.

At $T = 0$ K, all allowable energy states in the valence band of a semiconductor are occupied by electrons, and no allowable energy states in the conduction band are occupied. Since the conduction process requires that charge carriers move from one state to another within an energy band, no conduction can take place when all states are occupied or when all states are empty. This is illustrated in Figure 10.1a.

As temperature of a semiconductor sample is increased, sufficient energy is imparted to a small fraction of the electrons in the valence band for them to move to the conduction band. In effect, these electrons are leaving covalent bonds in the semiconductor host material. When an electron leaves the valence band, an opening is left, which may now be occupied by another electron, provided that the other electron moves to the opening. If this happens, of course, the electron that moves in the valence band to the opening leaves behind an opening in the location from which it moved. If one engages in an elegant quantum-mechanical explanation of this phenomenon, it must be concluded that the electron moving in the valence band must have either a negative effective mass and a negative charge or, alternatively, a positive effective mass and a positive charge. The latter has been the popular description,

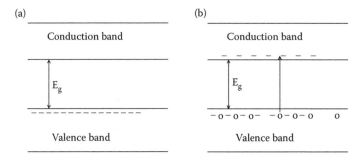

Figure 10.1 Illustration of availability of states in valence band and conduction band for semiconductor material. (a) Absolute zero. (b) Elevated temperature.

THE PHYSICS OF PHOTOVOLTAIC CELLS

and, as a result, the electron motion in the valence band is called hole motion, where "holes" is the name chosen for the positive charges, since they relate to the moving holes that the electrons have left in the valence band.

What is important to note about these conduction electrons and valence holes is that they have occurred in pairs. Hence, when an electron is moved from the valence band to the conduction band in a semiconductor by whatever means, it constitutes the creation of an electron–hole pair (EHP). Both charge carriers are then free to become a part of the conduction process in the material.

10.2.3 Generation of EHP by Photon Absorption

The energy in a photon is given by the familiar equation,

$$E = h\upsilon = \frac{hc}{\lambda} \text{(joules)}, \tag{10.1}$$

where h is Planck's constant (h = 6.63 × 10^{-34} Js), c is the speed of light (c = 2.998 × 10^8 m/second), υ is the frequency of the photon in Hz, and λ is the wavelength of the photon in meters. Since energies at the atomic level are typically expressed in electron volts (1 eV = 1.6 × 10^{-19} J) and wavelengths are typically expressed in micrometers (μm), it is possible to express hc in appropriate units so that if λ is expressed in μm, then E will be expressed in eV. The conversion yields

$$E = \frac{1.24}{\lambda} \text{(eV)}. \tag{10.2}$$

The energy in a photon must exceed the semiconductor bandgap energy, E_g, to be absorbed. Photons with energies at and just above E_g are most readily absorbed because they most closely match bandgap energy and momentum considerations. If a photon has energy greater than the bandgap, it can still produce only a single EHP. The remainder of the photon energy is lost to the cell as heat. It is thus desirable that the semiconductor used for photoabsorption have a bandgap energy such that a maximum percentage of the solar spectrum will be efficiently absorbed.

Now, referring back to the Planck formula for blackbody radiation in Chapter 2 (Equation 2.1), note that the solar spectrum peaks at $\lambda \approx 0.5$ μm. Equation 10.2 shows that a bandgap energy of approximately 2.5 eV corresponds to the peak in the solar spectrum. In fact, since the peak of the solar spectrum is relatively broad, bandgap energies down to 1.0 eV can still be relatively efficient absorbers, and in certain special cell configurations to be discussed later, even smaller bandgap materials are appropriate.

The nature of the bandgap also affects the efficiency of absorption in a material. A more complete representation of semiconductor bandgaps must show the relationship between bandgap energy and bandgap momentum. As electrons make transitions between conduction band and valence band, both energy and momentum

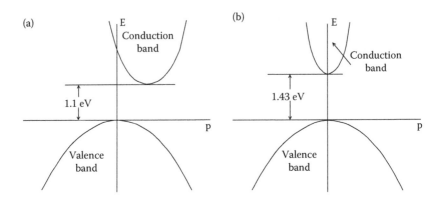

Figure 10.2 Energy and momentum diagram for valence and conduction bands (a) Si and (b) GaAs.

transfer normally take place, and both must be properly balanced in accordance with conservation of energy and conservation of momentum laws.

Some semiconducting materials are classified as direct bandgap materials, while others are classified as indirect bandgap materials. Figure 10.2 shows the bandgap diagrams for two materials considering momentum as well as energy. Note that for silicon, the bottom of the conduction band is displaced in the momentum direction from the peak of the valence band. This is an indirect bandgap, while the gallium arsenide (GaAs) diagram shows a direct bandgap, where the bottom of the conduction band is aligned with the top of the valence band.

What these diagrams show is that the allowed energies of a particle in the valence band or the conduction band depend on the particle momentum in these bands. An electron transition from a point in the valence band to a point in the conduction band must involve conservation of momentum as well as energy. For example, in Si, even though the separation of the bottom of the conduction band and the top of the valence band is 1.1 eV, it is difficult for a 1.1-eV photon to excite a valence electron to the conduction band because the transition needs to be accompanied with sufficient momentum to cause displacement along the momentum axis, and photons carry little momentum. The valence electron must thus simultaneously gain momentum from another source as it absorbs energy from the incident photon. Since such simultaneous events are unlikely, absorption of photons at the Si bandgap energy is several orders of magnitude less likely than absorption of higher energy photons.

Since photons have so little momentum, it turns out that the direct bandgap materials, such as GaAs, cadmium telluride (CdTe), copper indium diselenide (CIS), and amorphous silicon absorb photons with energy near the material bandgap energy much more readily than do the indirect materials, such as crystalline silicon. As a result, the direct bandgap absorbing material can be several orders of magnitude thinner than indirect bandgap materials and still absorb a significant part of the incident radiation.

THE PHYSICS OF PHOTOVOLTAIC CELLS

The absorption process is similar to many other physical processes, in that the change in intensity with position is proportional to the initial intensity. As an equation, this becomes

$$\frac{dI}{dx} = -\alpha I, \quad (10.3)$$

with the solution

$$I = I_o e^{-\alpha x}, \quad (10.4)$$

where I is the intensity of the light at a depth x in the material, I_o is the intensity at the surface, and α is the absorption constant. The absorption constant depends on the material and on the wavelength. Equation 10.4 shows that the thickness of material needed for significant absorption needs to be several times the reciprocal of the absorption constant. This is important information for the designer of a PV cell, since the cell must be sufficiently thick to absorb the incident light. In some cases, the path length is increased by causing the incident light to reflect from the front and back surfaces while inside the material until it ultimately generates an EHP. Figure 10.3 shows the dependence of absorption constant on wavelength for several materials. Observe that at energies below the bandgap energy, no absorption takes place. The material is transparent to these low-energy photons. At energies above the bandgap, the absorption constant increases relatively slowly for indirect bandgap semiconductors and increases relatively quickly for direct bandgap materials.

In any case, when the photon is absorbed, it generates an EHP. The question, then, is what happens to the EHP?

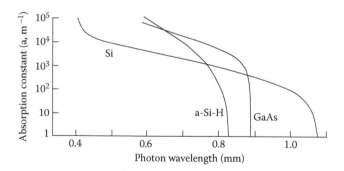

Figure 10.3 Dependence of absorption constant on wavelength for several semiconductors. (Adapted from Yang, E. S., *Microelectronic Devices*, McGraw-Hill, New York, 1988. Reproduced with permission of the McGraw-Hill Companies.)

10.2.4 Photoconductors

Once an EHP is generated, it becomes a question of how long the EHP lasts before the conduction electron returns to the valence band. Remember that the creation of an EHP does not imply that the electron will remain in the conduction band and the hole will remain in the valence band. Thermal equilibrium in a semiconductor comprises a constant generation and recombination of EHPs, so that, on the average, the population of electrons and the population of holes remain constant. In fact, the product of the concentration of holes and the concentration of electrons in thermal equilibrium is a constant, which depends on temperature and the bandgap energy of the semiconductor, along with a few other parameters unique to each semiconductor. If n_o represents the thermal equilibrium concentration of electrons per cm³ and if p_o represents the thermal equilibrium concentration of holes per cm³, then, in thermal equilibrium, with n_o and p_o at any single location in the material,

$$n_o p_o = n_i^2(T) = KT^3 e^{-(E_g/kT)}, \tag{10.5}$$

where n_i represents the intrinsic carrier concentration, K is the semiconductor-material dependent, as will be discussed later, E_g is the bandgap energy, k is the Boltzmann's constant, and T is the temperature in K. Because of the temperature dependence of n_i, it is customary to plot n_i versus 1000/T as in Figure 10.4, which shows how n_i varies with temperature for several semiconductor materials. Note that materials with smaller bandgap energies have higher intrinsic carrier concentrations. Table 10.1 shows bandgap energies and several other properties of some common semiconductors used in PV cells.

Electrons and holes generated as a result of optical absorption bring the material into a state of nonthermal equilibrium. The photon-generated excess electrons

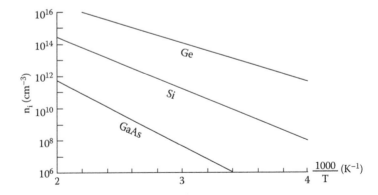

Figure 10.4 Temperature dependence of intrinsic carrier concentration for several common semiconductors. (Adapted from Streetman, B. G., *Solid State Electronic Devices*, 4th Ed., Prentice Hall, Englewood Cliffs, NJ, 1995. Reproduced with permission.)

THE PHYSICS OF PHOTOVOLTAIC CELLS

Table 10.1 Properties of Several Common Semiconductors at Room Temperature

Material	E_g (eV)	n_i (cm⁻³)	μ_n (cm²/Vs)	μ_p (cm²/Vs)	ε_r	Melting Point (°C)
Si	1.11	1.5×10^{10}	1350	480	11.8	1415
Ge	0.67	2.5×10^{13}	3900	1900	16	936
GaAs	1.43	2×10^{6}	8500	400	13.2	1238
CdS	2.42		340	50	8.9	1475
CdTe	1.48		1050	100	10.2	1098

Source: Yang, E. S., *Microelectronic Devices*, McGraw-Hill, New York, 1988; Streetman, B. G., *Solid State Electronic Devices*, 4th Ed., Prentice Hall, Englewood Cliffs, NJ, 1995.

and holes remain, on the average, for a time, τ, which is defined as the excess carrier lifetime. The recombination of EHPs is a statistical process in which some EHPs recombine in times shorter than the carrier lifetime and some take longer to recombine.

EHPs, then, are subject to a generation rate, measured in number per cm³ per second, which is proportional to the incident photon flux, and are subject to a recombination rate that is proportional to the departure of np from its equilibrium value. Under steady-state conditions, these two rates are equal.

The conductivity of a material is proportional to the density of free charge carriers and is given by

$$\sigma = q(\mu_n n + \mu_p p), \tag{10.6}$$

where μ_n and μ_p represent, respectively, the mobility of electrons and the mobility of holes in the material. Mobility is simply a measure of how easily the particles can move around in the material when an electric field is present. Mobilities are included in the material characteristics presented in Table 10.1.

Equation 10.6 clearly indicates that if shining a light on a piece of material can create excess charge carriers, then the conductivity of the material will increase, causing a corresponding decrease in the resistance of the material. The material becomes a light-sensitive resistor. A common material used for such devices is cadmium sulfide (CdS), which is used in many of the light sensors that turn lights on after dark. The most sensitive photoconductors are materials that have long lifetime of excess EHPs. However, if the problem is to detect short bursts of light occurring at a high repetition rate, it is necessary to have any excess population of electrons and holes quickly die out as soon as the light source is removed. The trade-off between speed and sensitivity is similar to the trade-off between bandwidth and gain for an amplifier as described by the familiar gain–bandwidth product.

Note that even though EHPs are generated in the host material, the material remains passive because the generated EHPs have random thermal velocities. This means that no net current flow results from their creation and since no separation of charges occurs, no voltage is produced. With no resulting voltage or current, the only effect of the creation of the additional charge carriers is the reduction in resistance of the host material. The next step, then, is to figure out a way to get work out of these photon-generated charges.

10.3 EXTRINSIC SEMICONDUCTORS AND THE PN JUNCTION

10.3.1 Extrinsic Semiconductors

Up to this point, the semiconductors discussed have been intrinsic semiconductors, meaning that the populations of holes and electrons have been equal. A somewhat more formal definition of an intrinsic semiconductor takes into account differences in electron and hole mobilities and defines intrinsic semiconductors as materials for which the Fermi level energy is at the center of the bandgap. Since Fermi levels have not been discussed and since electron and hole mobilities are generally close enough to keep the Fermi level quite close to the center of the bandgap, the equal carrier definition will suffice for the following discussion. The reader is referred to a text on semiconductor device physics for more details on Fermi levels [1,2].

At $T = 0$ K, intrinsic semiconductors have all covalent bonds completed with no leftover electrons or holes. If certain impurities are introduced into intrinsic semiconductors, there can be leftover electrons or holes at $T = 0$ K. For example, consider silicon, which is a group IV element, which covalently bonds with four nearest neighbor atoms to complete the outer electron shells of all the atoms. At $T = 0$ K, all the covalently bonded electrons are in place, whereas at room temperature, about one in 10^{12} of these covalent bonds will break, forming an EHP, resulting in minimal charge carriers for current flow.

If, however, phosphorous, a group V element, is introduced into the silicon in small quantities, such as one part in 10^6, four of the valence electrons of the phosphorous atoms will covalently bond to the neighboring silicon atoms, while the fifth valence electron will have no electrons with which to covalently bond. This fifth electron remains weakly coupled to the phosphorous atom, readily dislodged by temperature, since it requires only 0.04 eV to excite the electron from the atom to the conduction band. At room temperature, sufficient thermal energy is available to dislodge essentially all of these extra electrons from the phosphorous impurities. These electrons thus enter the conduction band under thermal equilibrium conditions, and the concentration of electrons in the conduction band becomes nearly equal to the concentration of phosphorous atoms, since the impurity concentration is normally on the order of 10^8 times larger than the intrinsic carrier concentration.

Since the phosphorous atoms donate electrons to the material, they are called *donor* atoms and are represented by the concentration, N_D. Note that the phosphorous, or other group V impurities, *do not add holes* to the material. They only add electrons. They are thus designated as n-type impurities.

THE PHYSICS OF PHOTOVOLTAIC CELLS

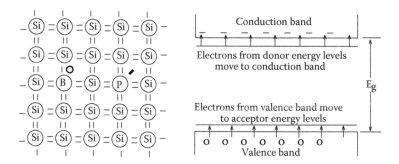

Figure 10.5 Acceptor and donor impurities in Si.

On the other hand, if group III atoms such as boron are added to the intrinsic silicon, they have only three valence electrons to covalently bond with nearest silicon neighbors. The missing covalent bond appears the same as a hole, which can be released to the material with a small amount of thermal energy. Again, at room temperature, nearly all of the available holes from the group III impurity are donated to the conduction process in the host material. Since the concentration of impurities will normally be much larger than the intrinsic carrier concentration, the concentration of holes in the material will be approximately equal to the concentration of impurities.

Historically, group III impurities in silicon have been viewed as electron acceptors, which, in effect, donate holes to the material. Rather than being termed hole donors, however, they have been called *acceptors*. Thus, acceptor impurities donate holes, but no electrons, to the material and the resulting hole density is approximately equal to the density of acceptors, which is represented as N_A. Figure 10.5 shows the effects of donor and acceptor impurities on the intrinsic material along with the positions of the energy levels of the impurities in the bandgap of the material.

Equation 10.6 shows that adding a mere one part per million of a donor or acceptor impurity can increase the conductivity of the material by a factor of 10^8 for silicon. Equation 10.5 shows that in thermal equilibrium, if either an n-type or a p-type impurity is added to the host material, then the concentration of the other charge carrier will decrease dramatically, since it is still necessary to satisfy Equation 10.5. In extrinsic semiconductors, the charge carrier with the highest concentration is called the *majority carrier* and the charge carrier with the lowest concentration is called the *minority carrier*. Hence, electrons are majority carriers in n-type material, and holes are minority carriers in n-type material. The opposite is true for p-type material.

If both n-type and p-type impurities are added to a material, then whichever impurity has the higher concentration will become the dominant impurity. However, it is then necessary to acknowledge a net impurity concentration that is given by the difference between the donor and acceptor concentrations. If, for example, $N_D > N_A$, then the net impurity concentration is defined as $N_d = N_D - N_A$. Similarly, if $N_A > N_D$, then $N_a = N_A - N_D$.

10.3.2 The PN Junction

10.3.2.1 Drift and Diffusion

When charged particles are placed in an electric field, they are exposed to an electrostatic force. This force accelerates the particles until they undergo a collision with another component of the material that slows them down. They then accelerate once again and collide again. The process continues, with the net result of the charge carrier's achieving an average velocity, either in the direction of the electric field for positive charges, or opposite the electric field for negative charges. It should be noted that this average, or *drift* velocity, is superimposed on the thermal velocity of the charge carrier. Normally, the thermal velocity is much larger than the drift velocity, but the thermal velocity is completely random so the net displacement of the charge carriers is zero. Another way to consider the thermal velocity is that at any instant, it is equally likely that a particle will be moving in any direction.

Drift current, then, is simply the component of current flow due to the presence of an electric field and is described by the familiar equation,

$$\vec{J} = \sigma \vec{E}. \tag{10.7}$$

This is simply the vector form of Ohm's law, where J is the current density in A/cm² when σ is expressed in $\Omega^{-1}\text{cm}^{-1}$ and E is measured in V/cm.

Diffusion is that familiar process by which random thermal motion of particles causes them to ultimately distribute themselves uniformly within a space. Whenever particles are in thermal motion, they will tend to move from areas of greater concentration to areas of lesser concentration, simply because at any point, the probability of motion in all directions is equal. Suppose, for example, that Regions A and B are adjoining, as shown in Figure 10.6. Suppose also that all particles in both regions are experiencing random thermal motion and that the concentration of certain particles,

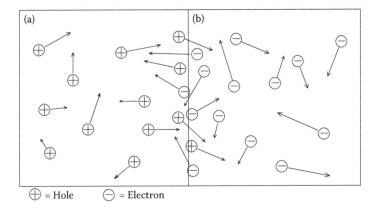

Figure 10.6 Random thermal motion and diffusion for electrons and holes. (a) More holes than electrons and (b) more electrons than holes.

z, in Region A is greater than the concentration of z-particles in Region B. At any instant, half the z-particles in Region A will be moving toward B, and half the z-particles in B will be moving toward A. Since there are more z-particles in A, the net motion of z-particles from A to B will continue until the concentrations in A and B are equal. If the particles are holes, this net movement from regions of greater concentration to regions of lesser concentration constitutes a flow of current that can be described in one dimension by the equation

$$J_p = -qD_p \frac{dp}{dx}, \qquad (10.8)$$

and if the particles are electrons,

$$J_n = qD_n \frac{dn}{dx}. \qquad (10.9)$$

In Equations 10.8 and 10.9, the change in concentration with position is known as the concentration gradient and D_p and D_n are the hole diffusion constant and the electron diffusion constant, respectively. The minus sign in Equation 10.8 accounts for the fact that if the gradient is negative, the holes flow in the positive x-direction. The lack of the minus sign in Equation 10.9 accounts for the fact that if electrons flow in the positive x-direction, the associated current is in the negative x-direction. In each case, $q = 1.6 \times 10^{-19}$ coulomb.

At this point, all that is necessary is to make two additional observations. The first observation is that donor and acceptor atoms become donor and acceptor ions when they give up their electron or hole to the host material. These ions are fixed in position in the host by covalent bonds. The second observation is that in either n-type material or in p-type material, any point in the bulk material will have charge neutrality. That is, the net charge present at any point is zero due to positive charges being neutralized by negative charges. However, when n-type and p-type materials are joined to form a pn junction, something special happens at the boundary.

10.3.2.2 Junction Formation and Built-In Potential

Although n-type and p-type materials are interesting and useful, the real fun starts when a junction is formed between n-type and p-type materials. The pn junction is treated in gory detail in most semiconductor device textbooks. Here, the need is to establish the foundation for the establishment of an electric field across a pn junction and to note the effect of this electric field on photo-generated EHPs.

Figure 10.7 shows a pn junction formed by placing p-type impurities on one side and n-type impurities on the other side. There are many ways to accomplish this structure. The most common is the diffused junction.

To form a diffused pn junction, the host material is grown with impurities, so it will be either n-type or p-type. The material is either grown or sliced into an appropriate thickness. Then the material is heated in the presence of the opposite impurity,

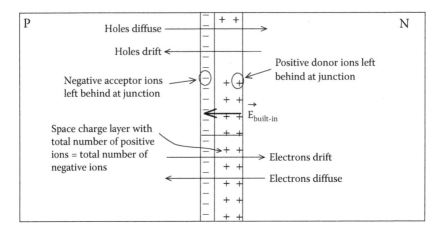

Figure 10.7 The pn junction showing electron and hole drift and diffusion.

which is usually in the form of a gas. This impurity will diffuse into the host material at a level that exceeds the host impurity level, but will only penetrate a small distance into the host material, depending on how long the host material is left at the elevated temperature. The result is a layer of material of one dominant impurity on top of the remainder of the material, which is doped with the other dominant impurity.

When the two materials are brought together, the first thing to happen is that the conduction electrons on the n-side of the junction notice the scarcity of conduction electrons on the p-side, and the valence holes on the p-side notice the scarcity of valence holes on the n-side. Since both types of charge carrier are undergoing random thermal motion, they begin to diffuse to the opposite side of the junction in search of the wide open spaces. The result is diffusion of electrons and holes across the junction, as indicated in Figure 10.7.

When an electron leaves the n-side for the p-side, however, it leaves behind a positive donor ion on the n-side, right at the junction. Similarly, when a hole leaves the p-side for the n-side, it leaves a negative acceptor ion on the p-side. If large numbers of holes and electrons travel across the junction, large numbers of fixed positive and negative ions are left at the junction boundaries. These fixed ions, as a result of Gauss' law, create an electric field that originates on the positive ions and terminates on the negative ions. Hence, the number of positive ions on the n-side of the junction must be equal to the number of negative ions on the p-side of the junction.

The electric field across the junction, of course, gives rise to a drift current in the direction of the electric field. This means that holes will travel in the direction of the electric field and electrons will travel opposite the direction of the field, as shown in Figure 10.7. Notice that for both the electrons and for the holes, the drift current component is opposite the diffusion current component. At this point, one can invoke Kirchhoff's current law to establish that the drift and diffusion components for each charge carrier must be equal and opposite, since there is no net current flow through the junction region. This phenomenon is known as the law of detailed balance.

By setting the sum of the electron diffusion current and the electron drift current equal to zero and recalling from electromagnetic field theory that

$$E = -\frac{dV}{dx}, \qquad (10.10)$$

it is possible to solve for the potential difference across the junction in terms of the impurity concentrations on either side of the junction. Proceeding with this operation yields

$$-q\mu_n n \frac{dV}{dx} + qD_n \frac{dn}{dx} = 0, \qquad (10.11)$$

which can be rewritten as

$$dV = \frac{D_n}{\mu_n} \frac{dn}{n}. \qquad (10.12)$$

Finally, recognizing the Einstein relationship, $D_n/\mu_n = kT/q$, which is discussed in solid-state physics textbooks, and integrating both sides from the n-side of the junction to the p-side of the junction yields the magnitude of the built-in voltage across the junction to be

$$V_j = \frac{kT}{q} \ln \frac{n_{no}}{n_{po}}. \qquad (10.13)$$

It is now possible to express the built-in potential in terms of the impurity concentrations on either side of the junction by recognizing that $n_{no} \cong N_D$ and $n_{po} \cong (n_i)^2/N_A$. Substituting these values into Equation 10.13 yields, finally,

$$V_j = \frac{kT}{q} \ln \frac{N_A N_D}{n_i^2}. \qquad (10.14)$$

At this point, a word about the region containing the donor ions and acceptor ions is in order. Note first that outside this region, electron and hole concentrations remain at their thermal equilibrium values. Within the region, however, the concentration of electrons must change from the high value on the n-side to the low value on the p-side. Similarly, the hole concentration must change from the high value on the p-side to the low value on the n-side. Considering that the high values are really high, that is, on the order of $10^{18}/cm^3$, while the low values are really low, that is, on the order of $10^2/cm^3$, this means that within a short distance of the beginning of the ionized region, the concentration must drop to significantly below the equilibrium value. Because the concentrations of charge carriers in the ionized region are so low, this region is often termed the *depletion region*, in recognition of the depletion of mobile charge carriers in the region. Furthermore, because of the charge due to the ions in this region, the depletion region is also often referred to as the *space charge layer*. For the balance of this text, this region will simply be referred to as the junction.

The next step in the development of the behavior of the pn junction in the presence of sunlight is to let the sun shine in and see what happens.

10.3.2.3 The Illuminated PN Junction

Equation 10.4 governs the absorption of photons at or near a pn junction. Noting that an absorbed photon releases an EHP, it is now possible to explore what happens after the generation of the EHP. Those EHPs generated within the pn junction will be considered first, followed by the EHPs generated outside, but near, the junction.

If an EHP is generated within the junction, as shown in Figure 10.8 (Points B and C), both charge carriers will be acted upon by the built-in electric field. Since the field is directed from the n-side of the junction to the p-side of the junction, the field will cause the electrons to be swept quickly toward the n-side and the holes to be swept quickly toward the p-side. Once out of the junction region, the optically generated carriers become a part of the majority carriers of the respective regions, with the result that excess concentrations of majority carriers appear at the edges of the junction. These excess majority carriers then diffuse away from the junction toward the external contacts, since the concentration of majority carriers has been enhanced only near the junction.

The addition of excess majority charge carriers to each side of the junction results in either a voltage between the external terminals of the material or a flow of current in the external circuit or both. If an external wire is connected between the n-side of the material and the p-side of the material, a current, I_ℓ, will flow in the wire from

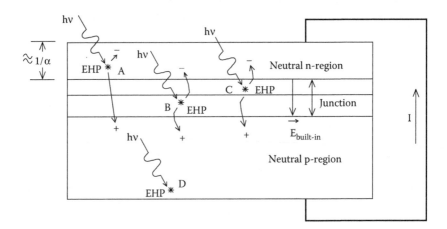

Figure 10.8 Illuminated pn junction showing desirable geometry and the creation of EHPs.

the p-side to the n-side. This current will be proportional to the number of EHPs generated in the junction region.

If an EHP is generated outside the junction region, but close to the junction (with "close" yet to be defined, but shown as Point A in Figure 10.8), it is possible that due to random thermal motion, the electron, the hole, or both will end up moving into the junction region. Suppose, for example, that an EHP is generated in the n-region close to the junction. Then suppose the hole, which is the minority carrier in the n-region, manages to reach the junction before it recombines. If it can do this, it will be swept across the junction to the p-side and the net effect will be the same as if the EHP had been generated within the junction, since the electron is already on the n-side as a majority carrier. Similarly, if an EHP is generated within the p-region, but close to the junction, and if the minority carrier electron reaches the junction before recombining, it will be swept across to the n-side where it is a majority carrier. So what is meant by close?

Clearly, the minority carriers of the optically generated EHPs outside the junction region must not recombine before they reach the junction. If they do, then, effectively, both carriers are lost from the conduction process, as in Point D in Figure 10.8. Since the majority carrier is already on the correct side of the junction, the minority carrier must therefore reach the junction in less than a minority carrier lifetime, τ_n or τ_p.

To convert these times into distances, it is necessary to note that the carriers travel by diffusion once they are created. Since only the thermal velocity has been associated with diffusion, but since the thermal velocity is random in direction, it is necessary to introduce the concept of minority carrier diffusion length, which represents the distance, on the average, which a minority carrier will travel before it recombines. The diffusion length can be shown to be related to the minority carrier lifetime and diffusion constant by the formula

$$L_m = \sqrt{D_m \tau_m}, \qquad (10.15)$$

where m has been introduced to represent n for electrons or p for holes. It can also be shown that on the average, if an EHP is generated within a minority carrier diffusion length of the junction, the associated minority carrier will reach the junction. In reality, some minority carriers generated closer than a diffusion length will recombine before reaching the junction, while some minority carriers generated farther than a diffusion length from the junction will reach the junction before recombining.

Hence, to maximize photocurrent, it is desirable to maximize the number of photons that will be absorbed either in the junction or within a minority carrier diffusion length of the junction. The minority carriers of the EHPs generated outside this region have a higher probability of recombining before they have a chance to diffuse to the junction. If a minority carrier from an optically generated EHP recombines before it crosses the junction and becomes a majority carrier, it, along with the opposite carrier with which it recombines, is no longer available for conduction. Furthermore, the combined width of the junction and the two diffusion lengths should be several multiples of the reciprocal of the absorption constant, α, and the junction should be relatively close to a diffusion length from the surface of the material upon which the photon impinges, to maximize collection of photons. Figure 10.8 shows this desirable geometry. The engineering design challenge then lies in maximizing α, as well as in maximizing the junction width and minority carrier diffusion lengths.

10.3.2.4 The Externally Biased PN Junction

In order to complete the analysis of the theoretical performance of the pn junction operating as a PV cell, it is useful to look at the junction with external bias. Figure 10.9 shows a pn junction connected to an external battery with the internally generated electric field direction included. If Equation 10.13 is recalled, taking into account that the externally applied voltage, with the exception of any voltage drop in the neutral regions of the material, will appear as opposing the junction voltage, the equation becomes

$$V_j - V = \frac{kT}{q} \ln \frac{n_n}{n_p}. \tag{10.16}$$

Note that the only difference between Equations 10.13 and 10.16 is that the electron concentrations on the n-side and on the p-side of the junction are no longer expressed as the thermal equilibrium values. This will be the case only when the externally applied voltage is zero. However, under conditions known as low injection levels, it will still be the case that the concentration of electrons on the n-side will remain close to the thermal equilibrium concentration. For this condition, Equation 10.16 becomes

$$V_j - V = \frac{kT}{q} \ln \frac{N_d}{n_p}. \tag{10.17}$$

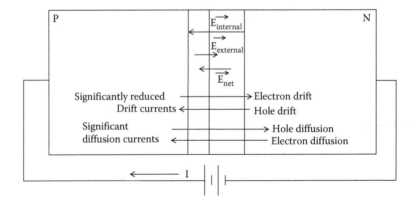

Figure 10.9 The pn junction with external bias.

THE PHYSICS OF PHOTOVOLTAIC CELLS

Since V_j can be calculated from Equation 10.14, the quantity of interest in Equation 10.17 is n_p, the concentration of minority carriers are at the edge of the junction on the p-side. Equation 10.17 can thus be solved for n_p with the result

$$n_p = N_D e^{-(q(V_j - V)/kT)} = N_D e^{(-qV_j/kT)} e^{(qV/kT)}. \qquad (10.18)$$

Next, note that the thermal equilibrium value of the minority carrier concentration occurs when $V = 0$. If the excess minority carrier concentration is now defined as $n'_p = n_p - n_o$, and if n_o is subtracted from Equation 10.17, the result is

$$n'_p(0) = n_p(0) - n_o(0) = N_D e^{(-qV_j/kT)} (e^{(qV/kT)} - 1) = n_{po}(e^{(qV/kT)} - 1). \qquad (10.19)$$

What happens to these excess minority carriers? They diffuse toward a region of lesser concentration, which happens to be away from the junction toward the contact, as shown in Figure 10.10.

If the contact is a few diffusion lengths away from the junction, it can be shown, by solving the well-known continuity equation, that the distribution of excess minority electrons between the junction and the contact will be

$$n'_p(x_p) = n'_p(0)\cosh\frac{x_p}{L_n} - n'_p(0)\ctnh\frac{W_p}{L_n}\sinh\frac{x_p}{L_n}. \qquad (10.20)$$

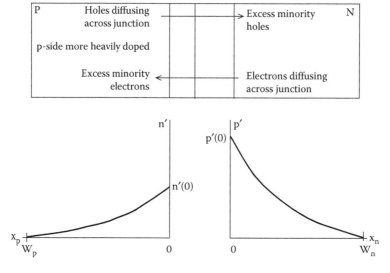

Figure 10.10 Excess minority carrier concentrations in neutral regions of pn junction device.

A similar expression can be obtained for the concentration of excess minority holes in the neutral region of the n-side of the device. To obtain an expression for the total device current, Equations 10.8 and 10.9 are used for each side of the device, noting that the gradient in the total carrier concentration is given by the gradient of the excess carrier concentration, since the spatial variation in the excess concentration will far exceed any spatial variation in the equilibrium concentration. Combining the results for electron and hole currents yields the total current in the device as it depends on the externally applied voltage along with the indicated device parameters. The result is, if A represents the cross-sectional area of the pn junction and adjoining regions,

$$I = I_n + I_p = qA\left(\frac{D_n n_{po}}{L_n}\operatorname{ctnh}\frac{W_p}{L_n} + \frac{D_p p_{no}}{L_p}\operatorname{ctnh}\frac{W_n}{L_p}\right)(e^{(qV/kT)} - 1). \quad (10.21)$$

Equation 10.21 is, of course, the familiar diode equation that relates diode current to diode voltage. Note that the current indicated in Equation 10.21 flows in the direction opposite to the optically generated current described earlier. Letting qA(nasty expression) = I_o and incorporating the photocurrent component into Equation 10.21 finally yields the complete equation for the current in the PV cell to be

$$I = I_\ell - I_o(e^{(qV/kT)} - 1). \quad (10.22)$$

Those readers who remember everything they read will recognize Equation 10.22 to be the same as Equation 3.1. Note that the current of Equation 10.22 is directed out of the positive terminal of the device, so that when the current and voltage are both positive, the device is delivering power to the external circuit.

10.4 MAXIMIZING PV CELL PERFORMANCE

10.4.1 Introduction

Equation 10.22 indicates, albeit in a somewhat subtle manner, that to maximize the power output of a PV cell, it is desirable to maximize the open-circuit voltage, short-circuit current, and fill factor of a cell. Recalling the plot of Equation 10.22 from Chapter 3, it should be evident that maximizing the open-circuit voltage and the short-circuit current will maximize the power output for an ideal cell characterized by Equation 10.22. Real cells, of course, have some series resistance, so there will be power dissipated by this resistance, similar to the power loss in a conventional battery due to its internal resistance. In any case, recalling that the open-circuit voltage increases as the ratio of photo current to reverse saturation current increases, a desirable design criteria is to maximize this ratio, provided that it does not proportionally reduce the short-circuit current of the device.

THE PHYSICS OF PHOTOVOLTAIC CELLS

Fortunately, this is not the case, since maximizing the short-circuit current requires maximizing the photocurrent. It is thus instructive to look closely at the parameters that determine both the reverse saturation current and the photocurrent. Techniques for lowering series resistance will then be discussed.

10.4.2 Minimizing the Reverse Saturation Current

Beginning with the reverse saturation current as expressed in Equation 10.21, the first observation is that the equilibrium minority carrier concentrations at the edges of the pn junction are related to the intrinsic carrier concentration through Equation 10.5. Hence,

$$p_{no} = \frac{n_i^2}{N_D} \text{ and } n_{po} = \frac{n_i^2}{N_A}. \tag{10.23}$$

So far, no analytic expression for the intrinsic carrier concentration has been developed. Such an expression can be obtained by considering Fermi levels, densities of states, and other quantities that are discussed in solid-state devices textbooks. Since the goal here is to determine how to minimize the reverse saturation current, and not to go into detail of quantum-mechanical proofs, the result is noted here with the recommendation that the interested reader consult a good solid-state devices text for the development of the result. The result is

$$n_i^2 = 4\left(\frac{2\pi kT}{h^2}\right)^3 (m_n^* m_p^*)^{3/2} e^{-(E_g/kT)}, \tag{10.24}$$

where m_n^* and m_p^* are the electron effective mass and hole effective mass, respectively, in the host material and E_g is the bandgap energy of the host material. These effective masses can be greater than or less than the rest mass of the electron, depending on the degree of curvature of the valence and conduction bands when plotted as energy versus momentum as in Figure 10.2. In fact, the effective mass can also depend on the band in which the carrier resides in a material. For more information on effective mass, the reader is encouraged to consult the references listed at the end of this chapter.

Now, using Equation 10.15 with Equation 10.23 and Equation 10.24 in Equation 10.21, the following final result for the reverse saturation current is obtained:

$$I_o = \left(4qA\left(\frac{2\pi kT}{h^2}\right)(m_n^* m_p^*)^{3/2} e^{-(E_g/kT)}\right)$$
$$\times \left(\frac{1}{N_A}\sqrt{\frac{D_n}{\tau_n}} \operatorname{ctnh} \frac{\ell_p}{\sqrt{D_n \tau_n}} + \frac{1}{N_D}\sqrt{\frac{D_p}{\tau_p}} \operatorname{ctnh} \frac{\ell_n}{\sqrt{D_p \tau_p}}\right). \tag{10.25}$$

Since the design goal is to minimize I_o while still maximizing the ratio $I_\ell:I_o$, the next step is to express the photocurrent in some detail so the values of appropriate parameters can be considered in the design choices.

10.4.3 Optimizing Photocurrent

In Section 10.3.2, the photocurrent optimization process was discussed qualitatively. In this section, the specific parameters that govern the absorption of light and the lifetime of the absorbed charge carriers will be discussed, and a formula for the photocurrent will be presented for comparison with the formula for reverse saturation current. In particular, minimizing reflection of the incident photons, maximizing the minority carrier diffusion lengths, maximizing the junction width, and minimizing surface recombination velocity will be discussed. The PV cell designer will then know exactly what to do to make the perfect cell.

10.4.3.1 Minimizing Reflection of Incident Photons

The interface between air and the semiconductor surface constitutes an impedance mismatch, since the electrical conductivities and the dielectric constants of air and a PV cell are different. As a result, part of the incident wave must be reflected in order to meet the boundary conditions imposed by the solution of the wave equation on the electric field, E, and the electric displacement, D.

Those readers who are experts at electromagnetic field theory will recognize that this problem is readily solved by the use of a quarter-wave matching coating on the PV cell. If the coating on the cell has a dielectric constant equal to the geometric mean of the dielectric constants of the cell and of air and if the coating is one-quarter wavelength thick, it will act as an impedance-matching transformer and minimize reflections. Of course, the coating must be transparent to the incident light. This means that it needs to be an insulator with a bandgap that exceeds the energy of the shortest wavelength light to be absorbed by the PV cell. Alternatively, it needs some other property that minimizes the value of the absorption coefficient for the material, such as an indirect bandgap. Several of these coatings are listed in Table 10.2.

It is also important to realize that a quarter wavelength is on the order of 0.1 μm. This is extremely thin, and may pose a problem for spreading a uniform coating of this thickness. And, of course, since it is desirable to absorb a range of wavelengths, the antireflective coating will be optimized at only a single wavelength. Despite these problems, coatings have been developed that meet the requirements quite well.

An alternative to antireflective coatings now commonly in use with Si PV cells is to manufacture the cells with a textured front surface, as shown in Figure 10.14. Textured front and back surfaces and their contribution to the capture of photons will be discussed later in this section in conjunction with Figure 10.14. The bottom line is that a textured surface acts to enhance the capture of photons and also acts to prevent the escape of captured photons before they can produce EHPs. Furthermore, the textured surface is not wavelength dependent as is the antireflective coating.

THE PHYSICS OF PHOTOVOLTAIC CELLS

Table 10.2 Some Antireflective Coatings Useful in PV Cell Production

Material	Index of Refraction
Al_2O_3	1.77
Glasses	1.5–1.7
MgO	1.74
SiO	1.5–1.6
SiO_2	1.46
Ta_2O_5	2.2
TiO_2	2.5–2.6

Source: Hu, C. and White, R. M., *Solar Cells: From Basic to Advanced Systems*, McGraw-Hill, New York, 1983. Reproduced with permission of the McGraw-Hill Companies.

10.4.3.2 Maximizing Minority Carrier Diffusion Lengths

Since the diffusion lengths are given by Equation 10.15, it is necessary to explore the factors that determine the diffusion constants and minority carrier lifetimes in different materials. It needs to be recognized that changing a diffusion constant may affect the minority carrier lifetime, so the product needs to be maximized.

Diffusion constants depend on scattering of carriers by host atoms as well as by impurity atoms. The scattering process is both material dependent and temperature dependent. In a material at a low temperature with a well-defined crystal structure, scattering of charge carriers is relatively minimal, so they tend to have high mobilities. Figure 10.11a illustrates the experimentally determined dependence of the electron mobility on temperature and on impurity concentration in Si.

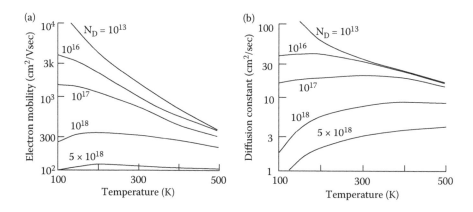

Figure 10.11 Temperature and impurity dependence of (a) electron mobility and (b) diffusion constant in silicon.

The Einstein relationship shows that the diffusion constant is proportional to the product of mobility and temperature. This relationship is shown in Figure 10.11b. So, once again, there is a trade-off. While increasing impurity concentrations in the host material *increases* the built-in junction potential, increasing impurity concentrations *decreases* the carrier diffusion constants.

The material of Figure 10.11 is single crystal material. In polycrystalline or amorphous material, the lack of crystal lattice symmetry has a significant effect on the mobility and diffusion constant, causing significant reduction in these quantities. However, if the absorption constant can be made large enough for these materials, the corresponding decrease in diffusion length may be compensated for by the increased absorption rate.

When an electron and a hole recombine, certain energy and momentum balances must be achieved. Locations in the host material that provide for optimal recombination conditions are known as recombination centers. Hence, the minority carrier lifetime is determined by the density of recombination centers in the host material.

One type of recombination center is a crystal defect, so that as the number of crystal defects increases, the number of recombination centers increases. This means that crystal defects reduce the diffusion constant as well as the minority carrier lifetime in a material.

Impurities also generally make good recombination centers, especially those impurities with energies near the center of the bandgap. These impurities are thus different from the donor and acceptor impurities that are purposefully used in the host material, since donor impurities have energies relatively close to the conduction band and acceptor impurities have energies relatively close to the valence band.

Minority carrier lifetimes also depend on the concentration of charge carriers in the material. An approximation of the dependence of electron minority carrier lifetime on carrier concentration and location of the trapping energy within the energy gap is given by

$$\tau_n = \frac{n'\left[n+p+2n_i \cosh\frac{E_t-E_i}{kT}\right]}{CN_t(np-n_i^2)}, \quad (10.26)$$

where C is the capture cross section of the impurity in cm^3/second, N_t is the density of trapping centers, and E_t and E_i are the energies of the trapping center and the intrinsic Fermi level, respectively. In most materials, the intrinsic Fermi level is very close to the center of the bandgap. Under most illumination conditions, the hyperbolic term will be negligible compared to the majority carrier concentration and the excess electron concentration as minority carriers in p-type material will be much larger than the electron thermal equilibrium concentration. Under these conditions, for minority electrons in p-type material, Equation 10.26 reduces to

$$\tau_n \cong \frac{1}{CN_t}. \quad (10.27)$$

Hence, to maximize the minority carrier lifetime, it is necessary to minimize the concentration of trapping centers and to be sure that any existing trapping centers have minimal capture cross sections.

10.4.3.3 Maximizing Junction Width

Since it has been determined that it is desirable to absorb photons within the confines of the pn junction, it is desirable to maximize the width of the junction. It is therefore necessary to explore the parameters that govern the junction width. Perhaps the reader recalls similar discussions in a previous electronics class.

An expression for the width of a pn junction can be obtained by solving Gauss' law at the junction, since the junction is a region that contains electric charge. Solution of Gauss' law, of course, is dependent upon the ability to express the spatial distribution of the space charge in mathematical or, at least, in graphical form. Depending on the process used to form the junction, the impurity profile across the junction can be approximated by different expressions. Junctions formed by epitaxial growth or by ion implantation can be controlled to have impurity profiles to meet the discretion of the operator. Junctions grown by diffusion can be reasonably approximated by a linearly graded model. The interested reader is encouraged to consult a reference on semiconductor devices for detailed information on the production of various junction impurity profiles.

The junction with uniform concentrations of impurities is convenient to use to obtain a feeling for how to maximize the width of a junction. Solution of Gauss' law for a junction with uniform concentration of donors on one side and a uniform concentration of acceptors on the other side yields solutions for the width of the space charge layer on each side of the junction. The total junction width is then simply the sum of the widths of the two sides of the space charge layer. The results for each side are

$$W_n = \left[\frac{2\varepsilon N_A}{qN_D(N_A + N_D)} \right]^{1/2} (V_j - V)^{1/2} \tag{10.28}$$

and

$$W_p = \left[\frac{2\varepsilon N_D}{qN_A(N_D + N_A)} \right]^{1/2} (V_j - V)^{1/2}. \tag{10.29}$$

Before combining these two results, it is interesting to note that the width of the junction on either the p-side or the n-side depends on the ratio of the impurity concentrations on each side. Again, since Gauss' law requires equal numbers of charges on each side of the junction, the side with the smaller impurity concentration will need to have a wider space charge layer to produce enough impurity ions to balance out the impurity ions on the other side. The overall width of the junction can now be determined by summing Equations 10.28 and 10.29 to get

$$W = \left[\frac{2\varepsilon(N_D + N_A)}{qN_A N_D}\right]^{1/2} (V_j - V)^{1/2}. \qquad (10.30)$$

At this point, it should be recognized that the voltage across the junction due to the external voltage across the cell, V, will never exceed the built-in voltage, V_j. The reason is that as the externally applied voltage becomes more positive, the cell current increases exponentially and causes voltage drops in the neutral regions of the cell, so only a fraction of the externally applied voltage actually appears across the junction. Hence, there is no need to worry about the junction width becoming zero or imaginary. In the case of PV operation, the external cell voltage will hopefully be at the maximum power point, which is generally between 0.5 and 0.6 V for silicon.

Next, observe that as the external cell voltage increases, the width of the junction decreases. As a result, the absorption of photons decreases. This suggests that it would be desirable to design the cell to have the largest possible built-in potential to minimize the effect of increasing the externally applied voltage. This involves an interesting trade-off, since the built-in junction voltage is logarithmically dependent on the product of the donor and acceptor concentrations (see 10.10b), and the junction width is inversely proportional to the square root of the product of the two quantities. Combining Equations 10.14 and 10.30 results in

$$W = \left[\frac{2\varepsilon(N_D + N_A)}{qN_D N_A}\right]^{1/2} \left(\frac{kT}{q} \ln \frac{N_A N_D}{n_i^2} - V\right)^{1/2}. \qquad (10.31)$$

Note now that maximizing W is achieved by making either $N_A \gg N_D$ or by making $N_D \gg N_A$. For example, if $N_A \gg N_D$, then Equation 10.31 simplifies to

$$W = \left[\frac{2\varepsilon}{qN_D}\right]^{1/2} \left(\frac{kT}{q} \ln \frac{N_A N_D}{n_i^2} - V\right)^{1/2}, \qquad (10.32)$$

or, if $N_D \gg N_A$, then

$$W = \left[\frac{2\varepsilon}{qN_A}\right]^{1/2} \left(\frac{kT}{q} \ln \frac{N_A N_D}{n_i^2} - V\right)^{1/2}. \qquad (10.33)$$

Another way to increase the width of the junction is to include a layer of intrinsic material between the p-side and the n-side as shown in Figure 10.12. In this *pin* junction, there are no impurities to ionize in the intrinsic material, but the ionization still takes place at the edges of the n-type and the p-type material. As a result, there is still a strong electric field across the junction and there is still a built-in potential

THE PHYSICS OF PHOTOVOLTAIC CELLS

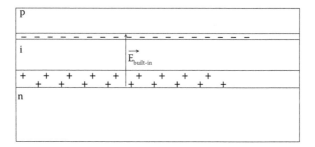

Figure 10.12 The pin junction.

across the junction. Since the intrinsic region could conceivably be of any width, it is necessary to determine the limits on the width of the intrinsic region.

The only feature of the intrinsic region that degrades performance is the fact that it has a width. If it has a width, then it takes time for a charge carrier to traverse this width. If it takes time, then there is a chance that the carrier will recombine. Thus, the width of the intrinsic layer simply needs to be kept short enough to minimize recombination. The particles travel through the intrinsic region with a relatively high drift velocity due to the built-in electric field at the junction. Since the thermal velocities of the carriers still exceed the drift velocities by several orders of magnitude, the width of the intrinsic layer needs to be kept on the order of about one diffusion length.

10.4.3.4 Minimizing Surface Recombination Velocity

If an EHP is generated near a surface, it becomes more probable that the minority carrier will diffuse to the surface. Since photocurrent depends on minority carriers diffusing to the junction and ultimately across the junction, surface recombination of minority carriers before they can travel to the junction reduces the available photocurrent. When the surface is within a minority carrier diffusion length of the junction, which is often desirable to ensure that generation of EHPs is maximized near the junction, minority carrier surface recombination can significantly reduce the efficiency of the cell.

Surface recombination depends on the density of excess minority carriers, in this case, as generated by photon absorption, and on the average recombination center density per unit area, N_{sr}, on the surface. The density of recombination centers is very high at contacts and is also high at surfaces in general, since the crystal structure is interrupted at the surface. Imperfections at the surface, whether due to impurities or to crystal defects, all act as recombination centers.

The recombination rate, U, is expressed as number/cm²/second and is given by

$$U = cN_{sr}m', \qquad (10.34)$$

where m′ is used to represent the excess minority carrier concentration, whether electrons or holes, and c is a constant that incorporates the lifetime of a minority carrier at a recombination center. Analysis of the dimensions of the parameters in Equation 10.34 shows that the units of cN_{sr} are cm/second. This product is called the *surface recombination velocity*, S. The total number of excess minority carriers recombining per unit time and subsequent loss of potential photocurrent is thus dependent on the density of recombination centers at the surface and on the area of the surface. Minimizing surface recombination thus may involve reducing the density of recombination centers or reducing the density of minority carriers at the surface.

If the surface is completely covered by a contact, then little can be done to reduce surface recombination if minority carriers reach the surface, since recombination rates at contacts are very high. However, if the surface is not completely covered by a contact, such as at the front surface, then a number of techniques have been discovered that will result in passivation of the surface. Silicon oxide and silicon nitrogen passivation are two methods that are used to passivate silicon surfaces.

Another method of reducing surface recombination is to passivate the surface and then only allow the back contact to contact the cell over a fraction of the total cell area. While this tends to increase series resistance to the contact, if the cell material near the contact is doped more heavily, the ohmic resistance of the material is decreased and the benefit of reduced surface recombination offsets the cost of somewhat higher series resistance. Furthermore, an E-field is created that attracts majority carriers to the contact and repels minority carriers. This concept will be explored in more detail in Section 10.5.2.

10.4.3.5 A Final Expression for the Photocurrent

An interesting exercise is to calculate the maximum obtainable efficiency of a given PV cell. Equation 10.4 indicates the general expression for photon absorption. Since the absorption coefficient is wavelength dependent, the general formula for overall absorption must take this dependence into account. The challenge in design of the PV cell and selection of appropriate host material is to avoid absorption before the photon is close enough to the junction, but to ensure absorption when the photon is within a minority carrier diffusion length of either side of the junction. Figure 10.13 [3] shows a typical photon absorption profile for the n-region and the p-region of a cell with the p-region at the surface.

Since direct bandgap materials tend to have larger absorption constants than indirect materials such as Si, it is relatively straightforward to capture photons in these materials. In Si, however, several clever design practices are used to increase absorption. Since some photons will travel completely across the cell to the back of the cell without being absorbed, if the back of the cell is a good reflector, the photons will be reflected back toward the junction. This is easy to do, since the back contact of the Si cell covers the entire back of the cell. However, rather than having a smooth back surface that will reflect photons perpendicular to the surface, the back surface is textured so the incident photons will be scattered at angles, thus increasing the path length.

The front of the cell, however, can be covered with an antireflective coating to maximize transmission of photons into the material. Hence, a similar scheme is

THE PHYSICS OF PHOTOVOLTAIC CELLS

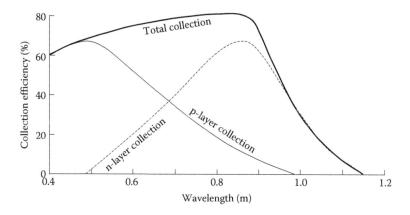

Figure 10.13 Photon collection efficiency versus wavelength and location in cell for a typical PV cell with p-layer at the surface. (Adapted from Yang, E. S., *Microelectronic Devices*, McGraw-Hill, New York, 1988. Reproduced with permission of the McGraw-Hill Companies.)

necessary to capture the photons in the material if, after bouncing off the back surface, they are still not absorbed. Once again, a textured surface will enhance the probability that a photon will undergo internal reflection, since the dielectric coefficient, and, hence, the index of refraction of the host material, is greater than that of the antireflective coating. This is analogous to when ripples on water prevent a person below the surface from seeing anything above the surface. When the surface is smooth, it is possible to see objects above the water, provided that the angle of view is sufficiently close to the perpendicular. Figure 10.14 shows a cell with textured front and back surfaces and the effect on the travel of photons that enter the host material.

The foregoing discussion can be quantified in terms of cell parameters in the development of an expression for the photocurrent. Considering a monochromatic

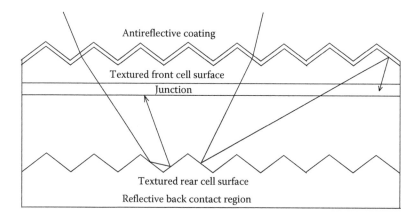

Figure 10.14 Maximizing photon capture with textured surfaces.

photon flux incident on the p-side of a p⁺n junction (the + indicates strongly doped), the following expression for the hole component of the photocurrent can be obtained. The expression is obtained from the solution of the diffusion equation in the neutral region on the n-side of the junction for the diffusion of the photon-created minority holes to the back contact of the cell.

$$\Delta I_{\ell p} = \frac{qAF_{ph}\alpha L_p}{\alpha^2 L_p^2 - 1} \left[\frac{S\cosh(w_n/L_p) + (D_p/L_p)\sinh(w_n/L_p) + (\alpha D_p - S)e^{-\alpha w_n}}{S\sinh(w_n/L_p) + (D_p/L_p)\cosh(w_n/L_p)} - \alpha L_p \right]. \tag{10.35}$$

It is assumed that the cell has a relatively thin p-side and that the n-side has a width, w_n. In Equation 10.35, F_{ph} represents the number of photons per cm² per second per unit wavelength incident on the cell. The effect of the surface recombination velocity on the reduction of photocurrent is more or less clearly demonstrated by Equation 10.35. The mathematical whiz will immediately be able to determine that small values of S maximize the photocurrent, and large values reduce the photocurrent, while one with average math skills may need to plug in some numbers.

Equation 10.35 is thus maximized when α and L_p are maximized and S is minimized. The upper limit of the expression then becomes

$$\Delta I_{\ell p} = -qAF_{ph}, \tag{10.36}$$

indicating that all photons have been absorbed and all have contributed to the photocurrent of the cell.

Since sunlight is not monochromatic, Equation 10.35 must be integrated over the incident photon spectrum, noting all wavelength-dependent quantities, to obtain the total hole current. An expression must then be developed for the electron component of the current and integrated over the spectrum to yield the total photocurrent as the sum of the hole and electron currents. This mathematical challenge is clearly a member of the nontrivial set of math exercises and is not included as a homework problem. Yet, some have persisted at a solution to the problem and have determined the maximum efficiencies that can be expected for cells of various materials. Table 10.3 [4,5] shows the theoretical optimum efficiencies for several different PV materials.

10.4.4 Minimizing Cell Resistance Losses

Any voltage drop in the regions between the junction and the contacts of a PV cell will result in ohmic power losses. In addition, surface effects at the cell edges may result in shunt resistance between the contacts. It is thus desirable to keep any such losses to a minimum by keeping the series resistance of the cell at a minimum and the shunt resistance at a maximum. With the exception of the cell front contacts, the procedure is relatively straightforward.

Most cells are designed with the front layer relatively thin and highly doped, so the conductivity of the layer is relatively high. The back layer, however, is generally

THE PHYSICS OF PHOTOVOLTAIC CELLS

Table 10.3 Theoretical Conversion Efficiency Limits for Several PV Materials at 25°C

Material	E_g	η_{max} (%)
Ge	0.6	13
CIS	1.0	24
Si	1.1	27
InP	1.2	24.5
GaAs	1.4	26.5
CdTe	1.48	27.5
AlSb	1.55	28
a-Si:H	1.65	27
CdS	2.42	18

Source: Rappaport, P., *RCA Review*, 20, September 1959, 373–379; Zweibel, K., *Harnessing Solar Power*, Plenum Press, New York, 1990.

more lightly doped in order to increase the junction width and to allow for longer minority carrier diffusion length to increase photon absorption. There must therefore be careful consideration of the thickness of this region in order to maximize the performance of these competing processes.

If the back contact material is allowed to diffuse into the cell, the impurity concentration can be increased at the back side of the cell, as illustrated in Figure 10.15. This is important for relatively thick cells, commonly fabricated by slicing single crystals into wafers. The contact material must produce either n-type or p-type material if it diffuses into the material, depending on whether the back of the cell is n-type or p-type.

In addition to reducing the ohmic resistance by increasing the impurity concentration, the region near the contact with increased impurity concentration produces

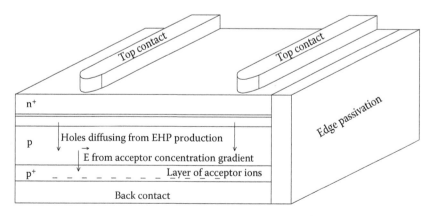

Figure 10.15 PV cell geometry for minimizing losses from cell series and shunt resistance.

an additional electric field that increases the carrier velocity, thus producing a further equivalent reduction in resistance. The electric field is produced in a manner similar to the electric field that is produced at the junction.

For example, if the back material is p-type, holes from the more heavily doped region near the contact diffuse toward the junction, leaving behind negative acceptor ions. Although there is no source of positive ions in the p-region, the holes that diffuse away from the contact create an accumulated positive charge that is distributed through the more weakly doped region. The electric field, of course, causes a hole drift current, which, in thermal equilibrium, balances the hole diffusion current. When the excess holes generated by the photoabsorption process reach the region of the electric field near the contact, however, they are swept more quickly toward the contact. This effect can be viewed as the equivalent of moving the contact closer to the junction, which, in turn, has the ultimate effect of increasing the gradient of excess carriers at the edge of the junction. This increase in gradient increases the diffusion current of holes away from the junction. Since this diffusion current strongly dominates the total current, the total current across the junction is thus increased by the heavily doped layer near the back contact.

At the front contact, another balancing act is needed. Ideally, the front contact should cover the entire front surface. The problem with this, however, is that if the front contact is not transparent to the incident photons, it will reflect them away. In most cases, the front contact is reflecting. Since the front/top layer of the cell is generally very thin, even though it may be heavily doped, the resistance in the transverse direction will be relatively high because of the thin layer. This means that if the contact is placed at the edge of the cell to enable maximum photon absorption, the resistance along the surface to the contact will be relatively large.

The compromise, then, is to create a contact that covers the front surface with many tiny fingers, as shown in Figure 10.15. This network of tiny fingers, which, in turn, are connected to larger and larger fingers, is similar to the configuration of the capillaries that feed veins in a circulatory system. The idea is to maintain more or less constant current density in the contact fingers, so that as more current is collected, the cross-sectional area of the contact must be increased. This subject is covered in more detail in the next chapter.

Finally, shunt resistance is maximized by ensuring that no leakage occurs at the perimeter of the cell. This can be done by nitrogen passivation or simply by coating the edge of the cell with insulating material to prevent contaminants from providing a current path across the junction at the edges.

10.5 EXOTIC JUNCTIONS

10.5.1 Introduction

Thus far, only relatively simple junctions have been considered. PV cell performance can be enhanced significantly by incorporating a variety of more sophisticated junctions, including, but not necessarily limited to, graded junctions, heterojunctions,

THE PHYSICS OF PHOTOVOLTAIC CELLS

Schottky junctions, multijunctions, and tunnel junctions. This section provides an introduction to these junctions so the reader will better understand the various junction types in the specific devices to be presented in the next chapter.

10.5.2 Graded Junctions

All the formulations to this point have related to the abrupt junction, in which the impurity concentration is constant to the junction and then abruptly changes to the opposite impurity concentration. While these junctions do exist, graded junctions are more common in materials for which the junction is fabricated by diffusion of impurities from the surface. The fabrication process for obtaining a linearly graded impurity profile as shown in Figure 10.16 will be discussed in some detail in the next chapter.

The significance of the graded junction is that majority carrier transport beyond the junction is improved by an additional electric field component resulting from the decreasing impurity concentration from surface to junction, as shown in Figure 10.16. The origin of this electric field component is the same as the origin of the additional field near the back contact when the back metallization is allowed to diffuse into the back of the cell. Each impurity atom donates either an electron or a hole and becomes ionized. If there is a gradient in impurity concentration, then the mobile carriers will diffuse in the direction of the gradient, leaving behind fixed impurity ions that serve either as the origin or termination of electric field lines.

An expression for the electric field in terms of the impurity concentration can be obtained by equating the diffusion current to the resulting drift current needed to balance the net current to zero when no external circuit is available for current flow. Assuming the majority carriers to be electrons in a heavily doped surface region, setting the sum of electron drift and diffusion current to zero results in

$$q\mu_n nE + qD_n \frac{dn}{dx} = 0, \tag{10.37}$$

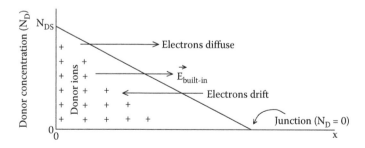

Figure 10.16 Diffusion and drift directions and electric field resulting from linearly decreasing impurity profile.

which can be solved for E with the result

$$E = -\frac{D_n}{\mu_n}\frac{1}{n}\frac{dn}{dx} = -\frac{kT}{q}\frac{1}{N_D}\frac{dN_D}{dx}. \qquad (10.38)$$

This additional E-field provides a drift component to the electron flow away from the junction toward the contact that effectively increases the velocity of the electrons toward the contact, resulting in lower resistive losses in the electron transport process toward the contact. Problem 10.7 provides an opportunity for calculation of the E-field present if the impurity concentration decreases linearly from surface to junction.

10.5.3 Heterojunctions

One of the problems identified with the optical absorption process is the absorption of higher energy photons close to the surface of the cell. Most of the EHPs generated by these higher energy photons are lost to recombination when they are created more than a diffusion length from the junction.

This phenomenon can be mitigated to some extent by the use of a heterojunction. A heterojunction is simply a composite junction composed of two materials with closely matched crystal lattices, so the bandgap near the surface of the material is greater than the bandgap near the junction. The higher bandgap region will appear transparent to photons with lower energies, so these photons can penetrate to the junction region where the bandgap is less than the incident photon energy. In the region of the junction, they can generate EHPs that will be collected before they recombine.

Heterojunctions are sometimes made between two n-regions or two p-regions as well as between n-region and p-region. The behavior of the heterojunction is dependent upon the crystal lattices, work functions, impurity doping profiles, and energy band properties of each semiconductor material, to the extent that discussion of any particular junction would probably not be applicable to a different junction. Readers interested in specific junctions, such as Ge:GaAs or AlGaAs:GaAs, will be able to obtain more information from journal and conference publications.

In some cases, materials cannot be made either n-type or p-type. Use of a heterojunction with a material that can be made to complete a pn junction is another important use of heterojunctions. For example, p-type CdS is yet to be produced. But a thin n-type CdS layer on top of a CIS structure will produce a pn junction effect, so CdS is often used as a part of a thin-film structure to produce a thin, heavily doped n-region near the front surface of the cell.

10.5.4 Schottky Junctions

Sometimes, when a metal is contacted with semiconductor material, an ohmic contact is formed and sometimes a rectifying contact is formed. It all depends on the relative positions of the work functions of the two materials. Figure 10.17 shows

THE PHYSICS OF PHOTOVOLTAIC CELLS

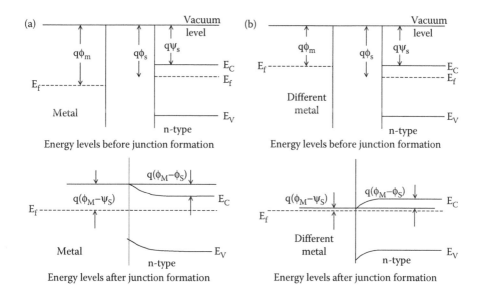

Figure 10.17 Energy levels in metal semiconductor junctions. (a) $qf_s < qf_m$ and (b) $qf_s > qf_m$.

situations where the work function of the metal, $q\phi_m$, is greater than that of the n-type semiconductor, $q\phi_s$, and where the work function of the metal is less than that of the n-type semiconductor material. The work function is simply the energy difference between the Fermi level and the vacuum level and the Fermi level is that energy where the probability of occupancy by a mobile carrier is 0.5. For the semiconductor, the energy difference between the bottom of the conduction band and the vacuum level is designated as $q\psi_s$.

When the metal and semiconductor are joined, electrons flow either from metal to semiconductor or from semiconductor to metal, depending on which has the higher Fermi level. In the case where $q\phi_s < q\phi_m$, as indicated in Figure 10.17a, electrons will flow from semiconductor to metal in a manner similar to diffusion across a pn junction from n-type to p-type. As the electrons leave, they leave behind positive donor ions, just as in the pn junction. Since the materials are in contact, at the point of contact the probability of electron occupancy must be the same for each material, which requires the Fermi levels of each material to align. Since the semiconductor surface in contact with the metal is depleted of electrons, the Fermi level must move farther away from the conduction band, because, as the Fermi level moves closer to the conduction band, more electrons will appear in the conduction band and vice versa.

The donor ions on the n-side of the contact thus become the origin of an electric field directed from the semiconductor to the metal. This field causes a drift of electrons from metal to semiconductor to balance the diffusion from semiconductor to metal. Note that holes cannot diffuse from metal to semiconductor, since the metal supports only electrons. Thus, the built-in potential is only about half what it might be if holes could also diffuse across the junction.

In Figure 10.17b, where $q\phi_s > q\phi_m$, the electrons diffuse from the metal to the semiconductor, thus increasing the concentration of electrons on the semiconductor side of the junction. This diffusion, in effect, causes the equivalent of heavier doping of the n-type semiconductor and enables current to flow easily in either direction across the junction. This is therefore an ohmic junction, whereas the junction of Figure 10.17a is a rectifying junction. External contacts to a PV cell need to be ohmic to prevent unnecessary voltage drop at the contact, whereas the rectifying contacts are useful for other purposes.

For the rectifying junction, a positive external voltage from metal to semiconductor reduces the built-in field, enabling the electrons that have diffused to the metal to continue flowing in the external circuit, as shown in Figure 10.18a. Note that as the Fermi level on the semiconductor side is drawn closer to the conduction band, more electrons can diffuse to the metal, resulting in significant current flow from metal to semiconductor. The reduced built-in field reduces drift components in the opposite direction in a manner similar to the forward bias condition of a conventional pn junction.

When the external voltage is negative, the built-in field is as shown in Figure 10.18b. Although this might be expected to result in significant electron flow from metal to semiconductor, this does not occur, since the metal electrons must overcome the barrier between the metal Fermi level and the semiconductor conduction band. Only a few of the metal electrons are sufficiently energetic to do so, so the result is a relatively small reverse saturation current, and thus a rectifying contact. The contact is rectifying even though the relatively strong E-field in the semiconductor region would be capable of enhancing drift current if replacement mobile charges were available.

For PV action, photons need to generate EHPs on the semiconductor side of the junction within a minority carrier diffusion length of the junction, so the minority carrier can be swept out of the region to the other side of the junction by the built-in field.

It has been shown that if the semiconductor material is n-type and $q\phi_m < q\phi_s$, the junction becomes ohmic, with bidirectional current conduction. When the

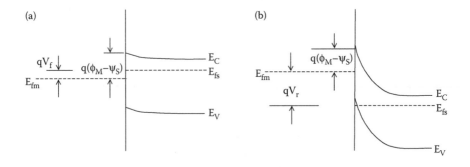

Figure 10.18 Rectifying metal-n-type semiconductor junction under forward-biased and reverse-biased conditions. (a) Forward biased. (b) Reverse biased.

THE PHYSICS OF PHOTOVOLTAIC CELLS 431

semiconductor is p-type, the opposite is true. That is, for $q\phi_m > q\phi_s$, the contact is ohmic, and for $q\phi_m < q\phi_s$, the contact is rectifying.

These conditions must thus be met when ohmic contacts are to be made to semiconductor materials. This is why certain metals such as Mo, Al, and Au are acceptable for ohmic contacts on some materials but not on others.

The Schottky barrier junction is relatively straightforward to fabricate, but is not very efficient as a PV cell since it has a relatively smaller open-circuit voltage than a conventional pn junction due to diffusion currents flowing in only one direction across the junction. On the other hand, the lower voltage drop across the junction when it is conducting results in lower power dissipation, making the Schottky diode convenient for use as a bypass diode.

10.5.5 Multijunctions

Since photon energy is most efficiently absorbed when it is near the bandgap, a clever way to absorb more photons is to stack junctions of different bandgaps. By starting near the front surface with a relatively larger bandgap material, the higher energy photons can be absorbed relatively efficiently at this junction. Then a smaller bandgap junction, perhaps followed by an even smaller bandgap junction, will enable lower energy photons to be absorbed more efficiently. Since the junctions are in series, they must produce equal currents. This is an interesting challenge, since the current at any of the multiple junctions depends more strongly on the percentage of photons with energies close to the bandgap of a specific junction. In a terrestrial environment, as time of day changes, different wavelengths are present to a greater or lesser extent, resulting in an unbalance in currents generated among multiple junctions. On the other hand, in an extraterrestrial environment, the incident spectrum is constant, allowing for optimal performance of the multijunction structure.

Figure 10.19 shows three junctions in series, each of somewhat different material in order to achieve three different bandgaps. The illustration may seem very logical, but there is an additional problem that is encountered when more than one junction is connected in series. That is simply the fact that although the p-to-n direction of

Figure 10.19 Three junctions in tandem (series), showing opposing pn junctions.

the first junction may be forward biased, this makes the n-to-p junction between the first and second pn junctions become reverse biased. The reverse bias across these junctions causes unnecessary voltage drop equivalent to the drop across blocking diodes that are sometimes inserted in strings of PV modules. Hence, a means must be devised to eliminate the effect of these reverse-biased junctions. Fortunately, the tunnel junction can eliminate this problem.

10.5.6 Tunnel Junctions

Tunnel junctions take advantage of the Heisenberg uncertainty principle, that is,

$$\Delta p \Delta x \geq \hbar, \tag{10.39}$$

where Δp and Δx are, respectively, the uncertainty in particle position and the uncertainty in particle momentum and $\hbar$ is Planck's constant divided by 2π. In a tunnel junction, the junction width is so extremely narrow that with relatively small uncertainty in particle momentum, it cannot be specified with certainty that the particle is on one side or the other side of the junction. This phenomenon is known as quantum-mechanical tunneling. The process takes place without any change in energy, since the particle essentially tunnels through the potential barrier produced by the tunnel junction. Noting Equation 10.30, it is evident that if impurity concentrations are very large on each side of the junction, the junction width will be quite small and the built-in potential will be relatively large, according to Equation 10.14. As the junction is forward biased, the junction width becomes even smaller, per Equation 10.30. Tunneling occurs when the electrons in the conduction band on the n-side rise to energy levels adjacent to empty states on the p-side, provided that the junction is sufficiently narrow, as illustrated in the energy band diagram of Figure 10.20.

Thus, the multiple pn junctions of the multijunction configuration can be accomplished by incorporating p^+n^+ tunnel junctions between each junction to eliminate the reverse-biased pn junctions. This is shown in Figure 10.21.

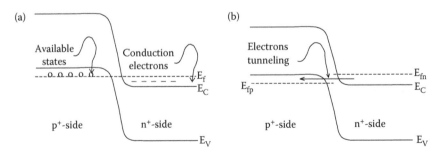

Figure 10.20 Tunnel junction showing conditions necessary for tunneling. (a) Junction in equilibrium. (b) Junction slightly forward biased.

THE PHYSICS OF PHOTOVOLTAIC CELLS

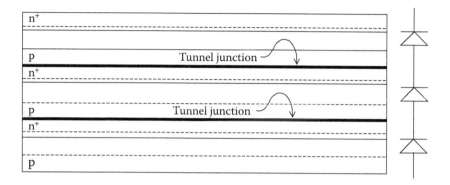

Figure 10.21 Separation of multiple pn junctions with tunnel junctions.

PROBLEMS

10.1 What can be said about the bandgap of glass? Why is silicon opaque? For what range of wavelengths will silicon appear transparent?

10.2 Given the arguments about direct versus indirect bandgap materials and the ability of these materials to absorb photons, offer an argument about the optimal bandgap structure of a material that will *emit* photons efficiently. Note that photon emission involves a transition of an electron from the conduction band to the valence band.

10.3 What would the bandgap need to be for an infrared LED? What about a red LED? What about a green LED? What about a blue LED? Would you expect the materials with these bandgaps to be direct bandgap materials or indirect bandgap materials?

10.4 Compare the average thermal velocity of an electron, v_{th}, at room temperature, if $\frac{1}{2}mv_{th}^2 = \frac{1}{2}kT$, where m is the electron mass, k is Boltzmann's constant, and T is the absolute temperature, with the drift velocity of an electron having a mobility of 10^3 cm²/Vs in an electric field of 1000 V/cm. The drift velocity equals the product of the mobility and the electric field. At what electric field strength will the drift velocity equal the thermal velocity?

10.5 Use Equation 10.12 to solve for the concentration of electrons as a function of position between the neutral n-side edge of the junction and the edge of the neutral p-side of the junction. Plot this concentration function on a logarithmic scale along with the concentration of ionized impurities in the space charge layer.

10.6 Solve the concentration of holes in the space charge layer. Show that in the space charge layer in thermal equilibrium, that $np = n_i^2$ at every position, x.

10.7 Show that Equation 10.14 can also be obtained by setting the net hole current across the junction equal to zero.

10.8 If a material has $\mu_n = 8600$ cm²/Vs, $\mu_p = 400$ cm²/Vs, and $\tau_n = \tau_p = 10$ ns, would you design a PV cell with the light incident on the n-side or the p-side of the material? Approximately what should be the distance between the edge of the space charge layer and the back contact? Explain why.

10.9 Show that Equation 10.30 results from adding Equation 10.28 to Equation 10.29.

10.10 Show that combining Equations 10.14 and 10.30 results in Equation 10.31.

10.11 Assume $N_A = 10^{20}$ cm⁻³ and $N_D = 10^{17}$ cm⁻³. Assume room temperature.

a. Calculate the built-in junction potential for Si and GaAs.
b. Calculate the junction width for Si and for GaAs under short-circuit conditions.

10.12 Use Equation 10.37 to calculate the electric field in the n-region adjacent to the junction if the donor concentration is given by $N_D(x) = N_S(1 - bx)$, where x is distance measured from the surface of the cell and b is a constant.

10.13 Calculate the widths of the following abrupt junctions in Si at room temperature with no external bias applied.
a. $N_A = 10^{16}$ and $N_D = 10^{20}$.
b. $N_A = 10^{21}$ and $N_D = 10^{21}$.

Next, calculate the widths of each junction if a forward bias of 0.3 V is applied, with all the external voltage appearing across the junction. Sketch the energy band picture for Part b.

REFERENCES

1. Yang, E. S., *Microelectronic Devices*, McGraw-Hill, New York, 1988.
2. Streetman, B. G., *Solid State Electronic Devices*, 4th Ed., Prentice Hall, Englewood Cliffs, NJ, 1995.
3. Hu, C. and White, R. M., *Solar Cells: From Basic to Advanced Systems*, McGraw-Hill, New York, 1983.
4. Rappaport, P., Harnessing the sun's energy, *RCA Review*, 20, September 1959, 373–379.
5. Zweibel, K., *Harnessing Solar Power*, Plenum Press, New York, 1990.

CHAPTER 11

Evolution of Photovoltaic Cells and Systems

11.1 INTRODUCTION

In Chapter 10, the basic theory of PV cells was presented without regard to any specific cell technology. This chapter will cover some of the fabrication processes associated with current cells along with discussions of the operation of a variety of cells, some of which are commonly in use, some of which have gone by the wayside, and others of which are still in the experimental phases. But the world of PV is experiencing more than changes in cell technology. It is also in the midst of a systems paradigm shift as large-scale system deployment is now well-underway.

Progress in PV research and development is moving so rapidly that by the time this chapter is in print, some of it will likely be outdated. As an example of the extent of PV research currently underway, the 2016 IEEE Photovoltaic Specialists Conference had a total of 992 presentations, and the IEEE Photovoltaic Specialists Conference is only one of many worldwide PV conferences. Hence, in addition to introducing the reader to a few of the technologies of 2016, it is also the goal of this chapter to provide the reader with the intellectual tools needed to read and understand current and future literature. Many of the sources of information on PV technology are listed in the reference section of this chapter.

In general, cell fabrication begins with the refining and purification of the cell base materials. After extremely pure materials are available, then the pn junction or its equivalent must be formed. In some multiple-layer cells, more than one junction is formed along with various fascinating isolation steps.

The single-crystal silicon cell has made its mark on history. Whether it will continue to make its mark on the future will depend on reducing the amount of energy consumed in the production of the cell in order to maintain its cost competitiveness over other technologies. Its fabrication and characteristics will be discussed first. During the discussion of the single-crystal silicon cell, important processes, such as crystal growth and diffusion, will be discussed. These basic processes in many cases are applicable to the fabrication of other types of PV cells.

Perhaps the most promising near-term PV cells will consist of thin films, although the authors have been observing these forecasts since the 1990s. Crystalline cells are generally considered to be the first generation of PV cells and thin films are considered to be the second generation. The third generation, yet to be commercialized, will have production costs in the range of $0.20–$0.50/W and will have efficiencies between 31% and 74%, according to Green in 2009 [1]. Certain materials have direct bandgaps with energies near the peak of the solar spectrum, along with relatively high absorption constants and the capability of being fabricated with pn junctions. These films are not single-crystal devices, so there are limitations to carrier mobilities and subsequent device performance. However, in spite of the nonsingle-crystal structures, laboratory cell conversion efficiencies of 23.3% have been achieved [2].

Thin films are advantageous because only a minimal amount of material is required to deposit a film with a thickness of 1 or 2 μm. Figure 11.1 [3] shows the photon current versus optical path length for common thin-film materials. Note that within 1–2 μm all the materials approach photon current saturation, whereas crystalline Si requires significantly greater thickness for full photon absorption. Examples of thin-film PV cells that are currently in commercial production are amorphous silicon (a-Si), copper indium gallium diselenide (CIGS), and cadmium telluride (CdTe). Gallium arsenide has been used for high-performance crystalline cells and has also been the subject of thin-film experimentation. And, with the dawn of the age of nanotechnology, PV applications have been given more than a cursory bit of attention.

Availability of cell component materials is also of interest if many gigawatts of PV power are to be obtained from any technology. Discussion of each technology includes commentary on the availability and refining of the cell component materials. Of all the thin-film materials, availability of indium may become a limiting

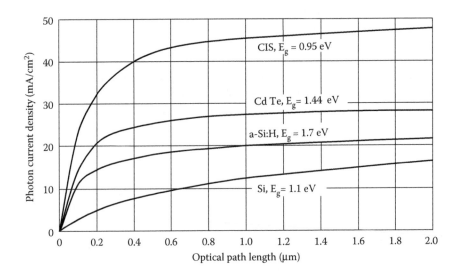

Figure 11.1 Photon current versus optical path length for thin-film materials, compared with crystalline silicon, standard test conditions. (Adapted from Tarrant, D. E. and Gay, R. R., *Research on High Efficiency, Large Area CuInSe2-Based Thin Film Modules, Final Technical Report*, NREL, Golden, CO, April 1995.)

factor in the production of CIS PV cells, but this limit will not be approached until 200 GW of cells have been produced [4]. Annual production of In would limit CIS cell production to about 4 GW per year. For comparison purposes, 2014 worldwide electrical generation capacity was just over 5000 GW [5–9], with 1850 GW from all renewable sources, including hydro. Other thin-film materials, although scarce, are sufficiently abundant for the production of many gigawatts of PV cells at reasonable cost, provided that certain production shortcomings that limit cell performance can be overcome.

11.2 SILICON PV CELLS

The first silicon PV cells were of the single-crystal variety. Single-crystal cells are the most efficient and most robust of the silicon PV cell family, but are also the most energy intensive in their production. For this reason, other varieties of silicon cells have been developed. Polycrystalline cells are somewhat less efficient, but are less energy intensive. a-Si cells are even less energy intensive, but have lower conversion efficiencies. However, since a-Si is a direct bandgap material with an energy gap larger than pure silicon, it has a much larger absorption constant than crystalline silicon, with peak absorption at a wavelength closer to the peak of the solar spectrum, and is thus a suitable material for thin-film cells. The instability mechanism present in early versions of a-Si is reasonably well-understood, and means for overcoming the instability are now in common use. Additional efficiency increases have been achieved for a-Si by the use of multijunction cell structures.

Thin Si cells, such as crystalline silicon on glass (CSG), represent a compromise between crystalline cells and amorphous cells. These relatively recent introductions to the Si cell family use novel techniques for photon capture and minimization of top surface area blocked by contacts. Good efficiencies have been achieved in lab scale devices.

11.2.1 Production of Pure Silicon [10]

The first thing that must be done in the production of a silicon cell is to find some silicon. Although nature has not seen fit to leave behind large quantities of relatively pure silicon in a form similar to the large deposits of carbon that have been mined for many years and burned in furnaces and worn on fingers, silicon has been provided in abundant quantities mixed with oxygen. Indeed, more than half the crust of the earth is composed of this silica compound, so all that is needed is to mine it and remove the oxygen and any other impurities. Generally quartz is the preferred starting point in the production of pure silicon due to fewer impurities in this material.

Of course, this is not done with zero energy. Separation of oxygen from silicon is quite energy intensive, requiring the reduction of SiO_2 with carbon in a carbon arc furnace. Regrettably, the CO_2 byproduct of this process is a greenhouse gas, so the production of pure Si for PV cells begins with the production of a greenhouse gas along with the silicon. Fortunately, over the lifetime of the cell, the amount of CO_2

from the reduction process is significantly less than the amount of CO_2 that would be produced if fossil fuels were used instead of PV cells to generate the same amount of electrical energy. Furthermore, as carbon capture technology advances, it is possible that the CO_2 from the production of Si will ultimately be sequestered and kept out of the atmosphere. The energy cost of this step, once at approximately 50 kWh/kg of metallurgical grade (99.0% pure) silicon, had dropped to 14–16 kWh/kg by 2011 [11] and continues to decrease as more-efficient processing steps are developed.

A purity of 99% means impurity levels of 1 part in 100. This means that additional refining needs to take place, since the purity of solar-grade silicon needs to be in the neighborhood of 1 part in 10^6 or so, depending on how efficient the cell is to be. Unwanted impurities lead to crystal defects and trapping levels that affect the cell output. Hence, the next step in the refining process is to remove these impurities. The traditional chemical method has been to react the metallurgical grade silicon with either hydrogen or chlorine to form either SiH_4 (silane) or $SiHCl_3$ (trichlorosilane).

The more common reaction has been to combine the Si with HCl in a fluidized bed reaction. This reaction, however, produces a combination of Si–H–Cl compounds, such as SiH_2Cl_2, $SiHCl_3$, and $SiCl_4$ along with chlorides of impurities. Each of these liquids has its own boiling point, so it is relatively straightforward to use fractional distillation to separate the $SiHCl_3$ from the other components. Although it looks simple on paper, the energy required in this step is relatively high. The next step is to remove the silicon from the trichlorosilane. This is done by reacting the trichlorosilane with hydrogen at a high temperature, producing polycrystalline silicon and HCl. Once again, improvements in process efficiency have led to reductions in the energy cost, which in 1999 was close to 200 kWh/kg of silicon. At this point, the silicon achieves the electronic grade level of purity (99.999999%).

An alternative metallurgical process is now becoming popular that yields solar-grade silicon with somewhat more impurities than electronic grade material, but still more than adequate for PV cells. This process is currently capable of producing acceptable material at an energy cost of approximately 32 kWh/kg [12].

11.2.2 Single-Crystal Silicon Cells

11.2.2.1 Fabrication of the Wafer

To produce single-crystal silicon from the polycrystalline material, the silicon must be melted and recrystallized. This is done by dipping a silicon seed crystal into the melt and slowly withdrawing it from the melt with a slight twisting motion. As the silicon is pulled from the melt, it reaches a somewhat higher purity level since the remaining impurities tend to remain behind in the melt. Under properly controlled conditions for solidification and replenishment of the melt, crystals as large as 8 inch (20.3 cm) in diameter and 6 ft (1.83 m) long can be readily grown. To produce n-type or p-type crystals, boron or arsenic can be introduced to the melt in whatever quantities are needed to produce the desired doping levels. Other Type III or Type V materials may also be introduced, but generally the diffusion constants or activation energies of these materials are less desirable from a PV performance perspective. Indium, for example, is a Group III impurity, but at room temperature, because of the higher activation energy of indium, a significant percentage of the indium atoms do not ionize, and thus are unable to donate holes to the host crystal. Phosphorous is a Group V impurity that has satisfactory ionization properties, but also has a higher diffusion constant in silicon, so that if the host material is heated, phosphorous atoms will migrate in the host material at a faster rate than arsenic atoms.

Special saws are then used to cut the ingots into wafers. Since circular wafers mounted in a module leave a large amount of empty space between the wafers, oftentimes the edges of the wafers are trimmed to make the wafers closer to square. The wafers are approximately 0.01 inch (0.254 mm) thick and are quite brittle. The sawing causes significant surface damage, so the wafers must next be chemically etched to restore the surface. In order to achieve the textured finish as described in the last chapter, it is possible to use a preferential etching process, so that after etching, the wafer surface is textured and relatively defect free. Again, the wafering process is quite energy intensive and leaves behind kerf loss that then needs to be recycled.

Over the years, several methods of preparing Si wafers without going through the wafering process have been attempted, but they have also encountered cost roadblocks and are seldom used in 2016. Work continues on the ideal way to produce Si wafers without having to first produce single-crystal ingots or multicrystalline blocks and then saw them into wafers.

11.2.2.2 Fabrication of the Junction [13]

The next step in the production of the single-crystal silicon cell is to create the pn junction. In electronic semiconductors, pn junctions can be created by diffusion, epitaxial growth, and ion implantation. These methods work fine for junctions having areas in the 10^{-12} m^2 range, but for PV cells with junction areas in the 10^{-2} m^2 range, the time it takes to produce a given area of junction becomes very important. Hence, due to cost constraints, pn junctions in crystalline silicon PV cells are made primarily by diffusion.

Impurity atoms diffuse into silicon in a manner similar to the diffusion of holes and electrons across regions of nonuniform impurity density. The concentration as a function of distance from the surface, x, and time is determined by the solution of the familiar diffusion equation

$$\frac{\partial N}{\partial t} = D \frac{\partial^2 N}{\partial x^2}, \qquad (11.1)$$

where N represents the density of the impurity and D is the diffusion constant associated with the specific impurity. It should be noted that D is highly temperature dependent, especially close to the melting temperature of the host material. At room temperature, diffusion is negligible, but in the neighborhood of 1000°C, diffusion becomes appreciable.

The solution to Equation 11.1 depends upon the boundary conditions imposed on the impurity. Most diffusion processes begin by holding the surface concentration of impurities constant for a fixed time. Then the impurity source is removed so the total number of impurities now remains constant, equal to the number of impurity atoms that diffused into the host during the constant surface concentration step of the process. The first step, known as the *predeposition*, leads to the well-known complimentary error function solution, that is,

$$N(x,t) = N_o \operatorname{erfc} \frac{x}{2\sqrt{Dt}}, \qquad (11.2)$$

where N_o is the concentration of impurities at the surface of the wafer. Suppose the substrate is uniformly doped p-type material and the impurities being diffused are n-type. The net impurity concentration at any depth, x, then, is the difference between the n-type and the p-type concentrations, or,

$$N(x,t) = N_o \operatorname{erfc} \frac{x}{2\sqrt{Dt}} - N_A. \qquad (11.3)$$

When the net impurity concentration is positive, the material is n-type, and when the net impurity concentration is negative, the material is p-type. The junction is located where the net impurity concentration is zero, which represents the transition between n-type and p-type. The depth of the junction can thus be determined by setting $N(x,t) = 0$ and solving for x, yielding the result

$$x = 2\sqrt{Dt}\operatorname{erfc}^{-1}\frac{N_A}{N_o}. \qquad (11.4)$$

This will not normally be the final junction depth, since after the predeposition step, the impurity concentration at the wafer surface is close to the solid solubility limit for the host material, which shortens the minority carrier diffusion length in this region. This preliminary junction depth estimate is useful, however, because it leads to a determination of the amount of time needed in the predeposition step to produce a junction at a depth less than the final desired junction depth, which will ultimately be 5–10 times the initial junction depth.

To reduce the surface concentration, a *drive-in* diffusion is carried out next. The drive-in diffusion is simply the continued application of heat to the material after the supply of the surface impurity is removed. The result is that the impurity continues to diffuse from the region of greater concentration to the region of lesser concentration. The impurity concentration is given, to a good approximation, by the solution of the diffusion equation under conditions of constant impurities per unit area. The assumption is that all of the impurities diffused into the material during the predeposition step remain at the surface of the material. The number of impurities per unit area can be determined by integrating Equation 11.3 from the surface into the

EVOLUTION OF PHOTOVOLTAIC CELLS AND SYSTEMS

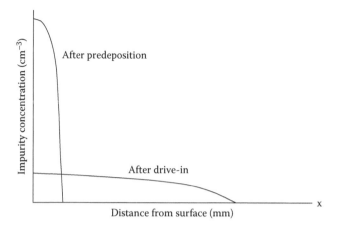

Figure 11.2 Concentrations of impurities after two diffusion steps.

material. While the actual distance of travel of the impurities is less than 1 μm, the integration is performed from 0 to infinity to yield

$$Q(t) = \int_0^\infty N(x,t)dx = 2N_o\sqrt{\frac{Dt}{\pi}}. \qquad (11.5)$$

After the drive-in step, the net impurity concentration is represented fairly accurately by the Gaussian distribution function,

$$N(x,t) = \frac{Q}{\sqrt{\pi Dt}} e^{-(x^2/4Dt)} - N_A. \qquad (11.6)$$

This expression can now be solved for the junction depth by again setting $N(x,t) = 0$. The solution is left as an exercise for the reader (Problem 11.2). It is thus a straightforward matter to control the diffusion of impurities into the wafer to produce a junction at any desired depth from the surface. Since it is desirable to have the junction within a minority carrier diffusion length of the surface, the depth can be set to achieve this goal. Note that Equations 11.3 through 11.6 also enable the cell designer to control the impurity concentration profile on the n-side of the junction. Figure 11.2 shows the impurity concentrations after the two diffusion steps have been completed.

11.2.2.3 Contacts

After creating the pn junction, the next step is to affix contacts to the cell. If the intent is to fabricate contacts that will last as long as the rest of the cell, then alligator clips are somewhat inadequate. The contacts must be low resistance and ohmic and must maintain their physical and electrical integrity through extreme temperature cycling. The contact material must be compatible with bonding of connecting wires/ribbons so the cell can be integrated into a module.

For many cells, the back contact covers the entire cell and is commonly made of evaporated aluminum that is annealed by heating the material after evaporation step is complete. The annealing process causes the aluminum to diffuse slightly into the silicon, creating a strong bond that will not break under thermal cycling. Since aluminum is a Group III element, it also acts as a p-type donor and produces a heavily doped p-region adjoining the contact. This heavily doped p-region creates an impurity gradient that produces a resulting electric field that accelerates holes toward the back contact.

The front contact is considerably more challenging for several reasons. First of all, if the front contact is opaque to light, then it will block photons from entering the cell and being absorbed. This means the area of the front contact must be minimized. On the other hand, if the area is too small, then the resistance of the contact increases. Since the junction is only a small distance below the front surface, if the contact material is annealed, it is possible that it will diffuse into the silicon and short out the junction. For that matter, even if the contact does not diffuse across the junction initially, this can occur over time under operation at higher temperatures and shorten the lifetime of the cell. Hence, the design constraints on the front contact are that it must be small and ohmic and present low series resistance while not threatening the junction.

The series resistance at the front contact of a cell depends on the distance the charge carrier must travel through the host material to the contact in addition to the resistance of the contact itself. The conductivity of a material is given by

$$\sigma = q(\mu_n n + \mu_p p). \tag{11.7}$$

It is convenient to consider the resistance between two points on an ohms-per-square basis. Figure 11.3 shows a piece of material with dimensions $\ell \times \ell \times t$. The resistance between the two ends of the material is given by

$$R = \frac{1}{\sigma} \frac{\ell}{\ell t} = \frac{1}{q(\mu_n n + \mu_p p)t} \text{ ohms/square.} \tag{11.8}$$

For a typical thin n-type silicon region with $n \approx 10^{20} \gg p$, $\mu_n \approx 100$, and a junction depth of approximately 1 µm (=t), the resistance per square becomes $R = 6.25\ \Omega$.

Now suppose two contacts are each 2 cm long and spaced 0.5 cm apart, as shown in Figure 11.4. The problem is to approximate the resistance between bulk and contact. Since the worst-case resistance is from the midpoint between the two contacts, the longest distance to a contact is 0.25 cm. Over the 2-cm distance, a total of eight squares, each 0.25 cm on a side, can be inserted between the midpoint and one of the contacts and another eight can be inserted between the midpoint and the other contact. Hence, there is the equivalent of 16 squares in parallel, each of which has

EVOLUTION OF PHOTOVOLTAIC CELLS AND SYSTEMS

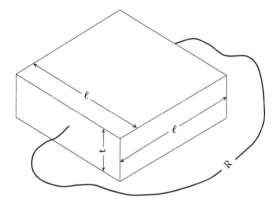

Figure 11.3 Determination of resistance per square.

a resistance of 6.25 Ω. The resulting resistance of the 16 parallel squares is thus 6.25/16 = 0.39 Ω.

Suppose further that the cell is approximately 10 cm in diameter and generates a total of 2.5 A under standard illumination. Assuming uniform generation of current, this means that each cm² generates approximately 0.032 A. Since the area between the two contacts is 1 cm², approximately 32 mA is generated in this region that needs to be carried by the contacts after it reaches the contacts. In general, a contact will be collecting current from both sides, so that each cm of contact length will collect approximately 16 additional mA of current, if the spacing is maintained at 0.5 cm.

An estimate of the power loss to the series cell resistance can be obtained from the worst-case resistance and the total current flowing through this resistance, with the result that $P = I^2R = 0.4$ mW. The power generated in this region, assuming a maximum power voltage of approximately 0.55 V, is $P = IV = 0.032 \times 0.55 = 16.5$ mW. Hence, approximately 2.4% of the power generated is lost to the ohmic resistance of the cell between the generation point and the contact in this worst-case example. Since most of the charge carriers have a smaller distance of travel to the

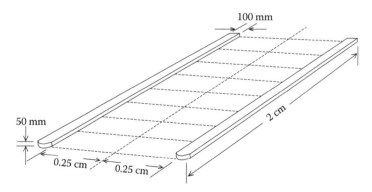

Figure 11.4 Determination of resistance between bulk and contacts.

contact, the resistance experienced will be less, and the overall power loss will be less than 2% between bulk and contact.

To determine power losses in the contacts, it is necessary to determine the resistance of the contact. Aluminum has a bulk resistivity, ρ, of 2.7×10^{-6} Ω-cm. Assuming a contact with typical thickness of 50 µm and width of 100 µm, the resistance per centimeter length is $R = \rho/A = 0.054$ Ω/cm. Hence, the contact with 2 cm length will have a resistance of 0.108 Ω.

Since the current in the contact increases linearly from zero at the beginning to 32 mA at the end, the average current in the contact is 16 mA. However, to more accurately express the ohmic losses in the contact, the losses should be integrated over the length of the contact. Choosing any point, x, along the contact, the current at point x will be $I(x) = 0.016x$, and the differential resistance of this portion of the contact will be 0.054dx Ω, since the resistance of the contact is 0.054 Ω/cm of length. The result of this calculation is

$$P = \int_0^2 d(I^2 R) = \int_0^2 (0.016x)^2 0.054 dx = 3.7 \times 10^{-5} \text{ W}. \quad (11.9)$$

This is only about 10% of the power loss over the cell surface. Hence, most of the series resistance is due to the surface resistivity of the bulk cell material. However, if the contact is only 5 µm thick, the resistance increases by a factor of 10 and the power loss is comparable to the bulk losses. These calculations illustrate a means that may be used to determine optimal spacing between top contacts as well as the dimensions of the top contacts to keep power losses to a minimum while still enabling maximum photon absorption.

As the current in the contacts continues to increase, the cross-sectional area of the contact must increase to enable the contact to carry the total current within acceptable voltage drop limitations. This is analogous to the circulatory system of plants, in which leaves contain capillaries that extend over the leaf to provide transport for nutrients.

The front contacts can be fabricated by several means, including evaporation in a vacuum chamber or silk screening with a paste. For fine lines, the lines may be defined with photoresist as is done in the production of small geometry electronic semiconductors.

If aluminum is the contact material, in order to make an ohmic contact to the n-type material, it is necessary to anneal the contact for approximately 5 minutes at a temperature of approximately 450°C. To prevent transport of the aluminum through the junction, it is advisable to use a small percentage of silicon in the aluminum or to first evaporate a very thin (≈ 0.01 µm) layer of titanium or chromium to act as a barrier to the aluminum diffusing into the junction.

11.2.2.4 Antireflective Coating (ARC)

After affixing the contacts to the material, an antireflective coating may be applied to the cell, normally by evaporation, since the coating must be so thin.

A quarter-wavelength coating has a thickness of approximately 0.15 µm. These coatings are commonly used for photographic lenses to increase their speed by reducing reflection and simultaneously increasing transmission. Typical coatings are listed in Table 10.2. As long as the photon wavelength is relatively close to the quarter-wavelength constraint, transmission in excess of 90% can be achieved for the cell surface.

Because antireflective coatings optimize transmission at only one wavelength, textured cell surfaces are becoming more common for enhancing light trapping over the full spectrum.

11.2.2.5 Modules

The cells must then be mounted in modules, with interconnecting wires or metal foil "ribbons." The interconnects are generally either ultrasonically bonded or soldered to the cell contacts. The cells are then mounted to the module base and encapsulated with a glass or composite. The encapsulant must be chosen for long life in the presence of ultraviolet radiation as well as possible degradation from other environmental factors. Depending on the specific module environment, such as blowing sand, salt spray, acid rain, or other not-so-friendly environmental component, the encapsulant must be chosen to minimize scratching, discoloration, cracking, or any other damage that might be anticipated. Earlier encapsulants, such as ethylene vinyl acetate (EVA), tended to discolor as a result of exposure to high levels of ultraviolet radiation and higher temperatures. Other encapsulants had a tendency to delaminate under thermal stresses. Modern materials now seem to have overcome these problems [14].

After the cells are encapsulated in modules, they are ready to produce electricity very reliably for a long time, normally in excess of 20 years.

11.2.3 A High-Efficiency Si Cell with All Contacts on the Back [15]

Figure 11.5 shows the cross section of the SunPower A-300 monocrystalline Si cell. In 2009, cell efficiencies of 22% were achieved and 20.3% efficiency was achieved for modules [1]. Basically, the cell consists of a substrate of n-type Si that has alternating positive and negative contacts to alternating n^+- and p-type contact areas that are diffused into the substrate. As EHPs are generated in the substrate, with a few being generated in the pn junction regions, the electrons are attracted to the n-type contact and the holes are attracted to the p-type contact as a result of the E-fields across the pn junction and between the n^+ contact and the substrate.

Note that the region between the n^+ contact and the weaker n-type substrate essentially constitutes a graded junction, where electrons flow from the more heavily doped contact region to the less heavily doped substrate. As a result, net positive charges are left behind in the n^+ region such that an electric field is created that is directed toward the substrate from the contact region. This field enhances the drift of electrons from the photon-generated EHPs from the substrate to the contact region.

The p-regions create pn junctions with the substrate with resulting E-fields directed from the substrate toward the p-region, as is the case for conventional

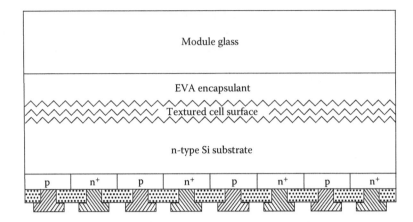

Figure 11.5 SunPower all back contact A-300 cell showing encapsulant and module glass. (Adapted from Kim, D. and Bunea, G., *The Journey of Sunpower to 20% Module Efficiency*, SunPower Corporation, 2011, http://www.avsusergroups.org/joint_pdfs/2012_2Kim.pdf.)

pn junctions. The substrate doping is weaker than the p-region doping, so that the junction region is predominantly in the substrate. This field enhances the drift of minority holes from photon-generated EHPs in the substrate, across the junction into the p-regions. The cell surface is textured to optimize photon capture. The lack of front contacts on the cell allows a greater level of photon penetration into the substrate layer, since the surface is not reflecting.

In 2005, the surface polarization effect was announced for these cells. When modules that used these cells were installed in the field with negative grounds, it was noticed that the cell module performance degraded over time much faster than anyone would expect for a crystalline Si technology. Further analysis revealed that even though the glass and EVA encapsulant are essentially insulators, it was still possible for minute currents to flow *from* the more positive EVA *to* the grounded module frames via the EVA and glass pathway, equivalent to charging a capacitor. Current flowing *from* the EVA means electrons flowing *to* the EVA from the module frame.

If the EVA layer is considered to be analogous to the gate region of a MOSFET, then the accumulation of negative charges in the EVA layer is equivalent to a negative charge on the gate of the MOSFET. This negative charge, in turn, attracts positive charge from the substrate, namely, minority carrier holes, to the cell surface rather than to the p-type contact. These holes are then lost to recombination with substrate electrons, which is the equivalent of an EHP being generated more than a minority carrier diffusion length from the pn junction.

Fortunately, this phenomenon is easily avoided, and reversible, if the module is operated with a positive ground rather than a negative ground. This results in positive charge build-up in the EVA layer rather than negative charge build-up. The resulting E-field from EVA to substrate repels holes rather than attracts them, so the hole collection efficiency is actually enhanced somewhat by operating the module with a positive ground. So the system designer simply needed to be sure to use system power conditioning components that can operate with a positive grounded array.

By 2011, however, SunPower had introduced the E20® modules with >20% efficiency that would function equally well with negative grounding or in an ungrounded array for use with transformerless inverters, without any evidence of polarization effects [16]. By 2015, a randomly selected X-Series module was tested at NREL with the result of module efficiency of 22.8% [17].

11.2.4 Multicrystalline Silicon Cells

At this point, it should be clear that the production of single-crystal Si cells is highly energy intensive. The large amount of energy used in the wafering process includes the single-crystal Si lost when the crystals are sawed into wafers. Further loss occurs if the round wafers are trimmed to approximate a square to fill a greater percentage of a module with cells. The question arises whether a means can be conceived for growing single-crystal wafers or perhaps a compromise can be achieved by manufacturing wafers that approximate single-crystal material.

One compromise involves pouring molten Si into a crucible and controlling the cooling rate. The result is not single crystalline material, since no seed crystal is used, but the multicrystalline Si obtained by this process has a square cross section and is sufficiently close to the single-crystal ideal that small cell efficiencies in the range of 20% have been obtained and module efficiencies of 15.5% have been achieved [1]. It is still necessary to saw the ingots into wafers, but the wafers are square, so no additional sawing is needed, as was the case with the round Si ingots. This process increases the production rate per kilogram of material by reducing the kerf loss.

Essentially, the same processes are used with multicrystalline wafers that are used with single-crystal wafers. Multicrystalline modules are characterized by the cells completely filling the modules, with the cells having a sort of speckled appearance resulting from the departure from single-crystal structure. Since the multicrystalline cells still maintain the basic crystalline properties, the 1.1-eV indirect bandgap results in the need for thicker cells with surface texturing to provide for maximum photon capture, as in the case of single-crystal cells.

Since multicrystalline cells are currently in the production phase, it is reasonable to project efficiency increases and cost per watt reductions as research and development progresses.

11.2.5 Other Thin Silicon Cells

As this text goes to press in 2016, significant advances have been made in the area of thin Si cells. The same technologies that were reported as having promise at the IEEE 29th Photovoltaic Specialists Conference in 2002 were also included on the 2009 IEEE 34th Photovoltaic Specialists Conference list of promising technologies. Thin crystalline cell technology has been pursued with the hope that cell efficiency can be improved through the use of crystalline Si and that cell manufacturing cost can be lowered by using less material in the construction of the cell. A number of processes, including thin Si on ceramics [18], thin-film CSG [19], and epitaxial growth of Si on existing crystalline Si with subsequent removal of the epitaxially grown cell from the existing Si substrate (the PSI process) [20], are described in the 2009 Conference Proceedings.

The CSG process has produced modules of areas in the range of 480–900 cm^2 since 1998. Efficiencies rose from 2% in 1999 to 8% in 2002 and approached 20% in 2014 [2]. The structure consists of textured glass, antireflective coating, n$^+$pp$^+$ structure, resin insulator, and metallization that dips into the structure from the back surface to form back and front cell contacts. The front contact does not require a transparent conducting oxide (TCO), so losses from the TCO are eliminated. TCOs will be discussed in more detail later in this chapter.

The modules consist of cells that are monolithically connected, and due to isolated metal interconnects that produce a fault-tolerant structure, performance of the module is not degraded as the size is scaled up. The production cost of this module in May of 2002 was $1.95/W.

In 2009, other emerging thin Si technologies included microcrystalline and nanocrystalline materials, both of which were receiving continued attention at the 2016 IEEE Photovoltaic Specialists Conference.

11.2.6 Amorphous Silicon Cells

11.2.6.1 Introduction

a-Si has no predictable crystal structure. Its atoms are located at more or less random angles and distances from each other. As a result, many of the potential covalent bonds in the silicon are not completed, resulting in silicon valence electrons looking for another electron to pair up with. These lonely electrons cause a large number of equivalent impurity states in the bandgap. In addition, the noncrystalline nature of the material results in very low values for electron and hole mobilities. The impurity states typically resulted in trapping of mobile carriers, so the combination of impurity states and diminished transport properties at first rendered a-Si as a rather poor semiconductor material.

But persistent solid-state physicists confirmed that the incomplete bonds could be passivated with hydrogen, with its single electrons looking for friendly, single electrons, with which to covalently bond. This significantly reduced the number of impurity states in the bandgap. With n-type and p-type impurities, a pn junction could be formed. Furthermore, by incorporating an intrinsic layer between p$^+$ and n-type material, a reasonable EHP generation region could be created with reasonable transport properties resulting from minimized impurity scattering in the passivated intrinsic material. Another positive feature of the a-Si:H system is that it has a direct bandgap close to 1.75 eV, resulting in a high absorption coefficient and qualifying a-Si:H as a good potential candidate for a thin-film PV material.

11.2.6.2 Fabrication

Figure 11.6 shows the structure of a basic a-Si:H cell. Fabrication of the cell begins with the deposition of a TCO layer on a glass substrate. The TCO, typically n+ SnO, constitutes the front contact of the cell. Next, a very thin layer of p+ a-Si:H is deposited, usually by plasma decomposition of SiH_4. The degenerate n-type TCO and the degenerate p-layer form a tunnel heterojunction. After the p+ layer, a slightly n-type intrinsic layer is deposited, followed by a stronger n-layer and finally a back contact, usually of Al, is deposited.

Initial operation of the basic cell with a relatively wide intrinsic layer was found to result in cell degradation when the cell was operated under sunlight conditions. The degradation was accounted for by the Staebler–Wronski effect, which explains the degradation in terms of increased density of scattering and trapping states in the intrinsic layer in proportion to photon exposure.

To mitigate the effects of the wider i-layer, cells were designed with narrower i-layers, but stacked, as shown in Figure 11.7a. These stacked cells require inclusion of

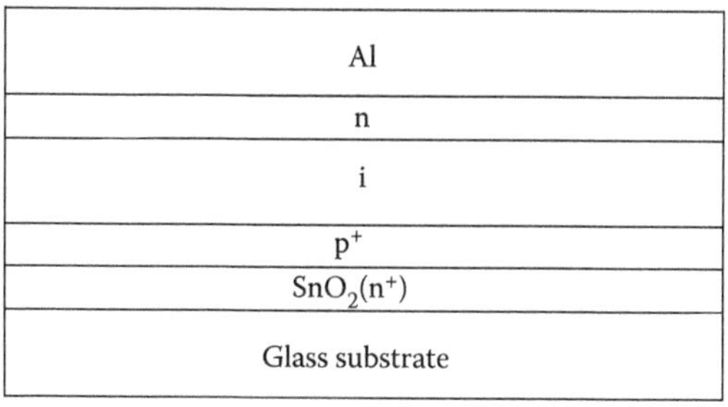

Figure 11.6 Basic a-Si:H cell structure.

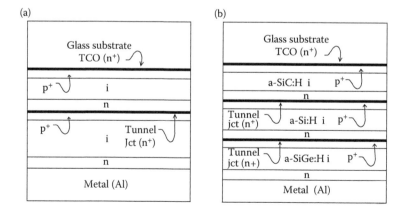

Figure 11.7 Stacked a-Si:H (a) with Si only and (b) with SiGe and SiC layers.

tunnel junctions to prevent the blocking action of the pn junctions of adjacent cells, as discussed in Section 10.5.6. These cells have shown improved long-term performance.

The concept of stacked cells has been extended one step further by recognizing that C and Ge also bond reasonably well with a-Si to produce either a-SiC:H or a-SiGe:H. The C alloy has a higher bandgap, thus enhancing absorption in the blue range, while the Ge alloy has a smaller bandgap, thus enhancing absorption in the infrared range. The net result is an overall performance improvement over the a-Si:H cell. The structure is shown in Figure 11.7b, noting again the need for tunnel junctions between cell layers.

Another interesting variation in cell structure is shown in Figure 11.8 [21], where a stainless steel substrate has been used to produce a flexible cell. Rolls of a-Si cells for use in roofing and other building-integrated applications are currently on the market [21]. Figure 11.8 also shows a-Si on a very lightweight polymer substrate, about 2 mils thick, that has been proposed for extraterrestrial applications where minimal weight is important and where stresses on the structure are minimal [22]. Both types of cell have been reported to have efficiencies in the 10% range.

The cell was produced on stainless steel by a proprietary roll-to-roll process. Note that the back surface incorporates an Al/ZnO textured film between the n-layer and the stainless steel to enhance photon trapping. The bandgap of an a-SiGe:H cell is dependent upon the fraction of Ge in the mix. In this structure, the top a-Si:H intrinsic layer has a bandgap of approximately 1.8 eV. The middle a-SiGe:H cell has a Ge fraction of approximately 15% in the intrinsic layer, resulting in a bandgap of approximately 1.6 eV, and the bottom a-SiGe:H cell has an intrinsic layer Ge fraction of approximately 45%, resulting in a bandgap for this layer of approximately 1.4 eV [23].

The challenge in creating the PV-on-polymer structure is keeping the polymer from deforming during heating portions of the processing. To do so, the process

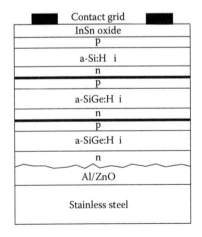

Figure 11.8 a-Si:H cell on stainless steel and polymer substrates. (Adapted from Guha, S. et al., *Proc. 26th IEEE PV Spec. Conf.*, 1997, 607–610; Huang, J. et al., *Proc. 26th IEEE PV Spec. Conf.*, 1997, 699–702. © 1997, IEEE.)

involves using a silicone gel between the bottom of the polymer and a heat sinking material, so the polyimide can be protected from heat stresses.

11.2.6.3 Cell Performance

Table 10.3 shows a theoretical maximum efficiency for a-Si of 27%, whereas in the late 1990s, large-scale efficiencies for a-Si devices were reported in the 10% range, with lab cell stable efficiencies for triple-junction devices of approximately 14% [24]. As of 2009, these records remained in place [1]. In 2015, the cell efficiency of single junction lab devices had increased to 13.6% [2]. Considering that initial efficiencies of only a few percentage were achieved, these efficiencies represent significant progress.

With the worldwide efforts in place to continually improve cell processing, along with the incorporation of novel and practical applications that can justify the use of cells with efficiencies <15%, continued progress can be expected with amorphous and thin-film silicon cells, despite the bankruptcy of one of the major a-Si:H manufacturers.

When cells are stacked, they must be designed so that absorbing layers produce equal photocurrent, since the layers are approximately ideal current sources connected in series. The increased efficiency of stacked (series) cells is thus due to more complete photon capture, with EHPs being effectively separated at the junctions, resulting in higher overall cell open-circuit voltages resulting from the combined series junctions.

The efficiency achieved by increasing the number of series junctions is also affected by the air mass to which the cell is exposed, since certain wavelengths are subjected to greater atmospheric absorption as the air mass changes. Thus, a terrestrial cell may have high efficiency at noon, but earlier or later, when the sun travels through more atmosphere, thus increasing the air mass, the performance of one or more sections of a stacked cell may be degraded, thus reducing overall cell current. Optimal performance of stacked cells is thus achieved when they are designed for extraterrestrial use.

Overall cell efficiency is also limited by the wavelength range over which the ARC will effectively minimize reflections, since the ARC optimal thickness is one-fourth wavelength. This limits the number of effective series junctions.

11.3 GALLIUM ARSENIDE CELLS

11.3.1 Introduction

The 1.43-eV direct bandgap, along with a relatively high absorption constant, makes GaAs an attractive PV material. Historically, high production costs once limited the use of GaAs PV cells to extraterrestrial and other special purpose uses, such as in concentrating collectors. Recent advances in concentrating technology, however, enable the use of significantly less active material in a module, such that cost-effective, terrestrial devices are now commercially available. In this section, the basic GaAs cell and its components will be discussed. Concentrating PV will be discussed in more detail in Section 11.6.4.7.

Production of pure gallium arsenide requires first the production of pure gallium and pure arsenic. The two materials are then combined to form GaAs. Most modern GaAs cells consist of thin films of GaAs grown on substrates such as Ge by an assortment of film growth processes.

11.3.2 Production of Pure Cell Components

11.3.2.1 Gallium [25]

Gallium was predicted by Mendeleev and first discovered by spectroscopic analysis by Lecoq de Boisbaudran in 1875. De Boisbaudran then separated out Ga from its hydroxide by electrolysis in the same year. Ga is a metal, which liquefies just above room temperature, but has a very high boiling point. It is found as a trace element in diaspore, sphalerite, germanite, bauxite, and coal, all of which contain Zn, Ge, or Al, in addition to the Ga. In fact, coal flue dusts can contain up to 1.5% Ga, although most contain less than 0.1%. The relative abundance of Ga is comparable to the abundance of Pb and As, but the percentage composition of Ga in any naturally occurring mineral rarely exceeds 1%. The most important source of Ga is bauxite, even though this ore contains only 0.003%–0.01% Ga.

Gallium can be extracted by many different methods, depending on the host material. For example, in bauxite, the weight ratio of Al to Ga is approximately 8000:1. In the Bayer process for Al extraction, bauxite is first mixed with a NaOH solution. The solution is autoclaved, diluted, and decanted, at which point a red mud is removed from the material. Decomposition follows, which results in the removal of hydrated alumina along with extracts having Al:Ga ratios of approximately 200:1. This Al:Ga mixture is then subjected to evaporation and is again mixed with NaOH and fed back into the mixing, autoclave, dilution, decantation, and decomposition steps again.

Semiconductor-grade Ga, with a purity of 99.999+%, is obtained by a variety of physical and chemical processes. These methods include chemical treatments with acids or gases at high temperatures and physicochemical methods, such as filtration of fused metal, heating in vacuum, dissolving again, and subjecting to further electrolysis or crystallization as monocrystals. It is also possible to react Ga with Cl, fractionally distill the solution until the desired purity is reached, and then recover the Ga and reconvert it to metal.

11.3.2.2 Arsenic [26]

Arsenic has been known since 1250 AD. In 1649, Schroeder published two methods of preparing the element. Arsenic is found in many forms in nature, including sulfides, arsenides, sulfarsenides, oxides, and arsenates. The most common source of As is FeSAs. When FeSAs is heated, the As sublimes, leaving FeS behind. Arsenic oxidizes rapidly if heated and, along with its compounds, is very poisonous. Arsenic is typically marketed in its arsenic trioxide form, which can be obtained at various purity levels by resublimation. This form is most commonly used in the manufacture of insecticides.

Semiconductor-grade As can be obtained by reducing a chemically purified compound with a highly pure solid or gas. One such highly pure form is arsine (AsH_3), which is also highly poisonous and requires very special precautions. If elemental As is desired, the AsH_3 can be decomposed by heating, resulting in highly pure elemental As. A large percentage of use of As in semiconductors involves the decomposition of AsH_3 at high temperatures.

11.3.2.3 Germanium [25]

The existence of Ge was predicted by Mendeleev in 1870, and it was isolated by Winkler in 1886. Only a small number of minerals contain Ge in appreciable quantities, including (AgGe)S, (AgSnGe)S, (CuZnAsGe)S, and (CuFeAsGe)S. Most Ge, however, is recovered from Pb–Zn–Cu ores in Africa.

The recovery process involves heating the ore under reducing conditions. This vaporizes Zn and Ge so they can be oxidized and collected. The fumes are then leached with H_2SO_4 and the pH is then gradually increased. As the pH increases to 3, Ge is 90% precipitated, whereas Zn begins to precipitate at pH of 4. If Zn and Ge are both precipitated at a pH of 5, a 50:1 Zn:Ge ratio solution will produce a Zn–Ge precipitate containing close to 10% Ge. If MgO is used as the base, then a Mg–Ge precipitate containing about 10% Ge is obtained.

The next step is to react the precipitate with strong HCl, which causes $GeCl_4$ to form. The $GeCl_4$ can then be fractionally distilled and reacted with H_2O to form GeO_2. The GeO_2 can then be reduced with H_2 to obtain pure Ge. The resulting Ge can then be zone refined to obtain semiconductor-grade Ge.

11.3.3 Fabrication of the Gallium Arsenide Cell

Crystalline gallium arsenide is somewhat more difficult to form than silicon, since gallium and arsenic react exothermically when combined. The most common means of growing GaAs crystals is the liquid encapsulated Czochralski (LEC) method. In this method, the GaAs crystal is pulled from the melt. The melted GaAs must be confined by a layer of liquid boric oxide. The biggest challenge is to create the GaAs melt in the first place. Several means have been developed, such as first melting the Ga, then adding the boric oxide, and then injecting the As through a quartz tube [27].

Most modern GaAs cells, however, are prepared by epitaxial growth of a GaAs film on a suitable substrate. Figure 11.9 shows one basic GaAs cell structure. The cell begins with the growth of an n-type GaAs layer on a substrate, typically Ge, but more recently, single-crystal GaAs, using the metal organic chemical vapor deposit (MOCVD) process followed by an epitaxial lift-off (ELO) step [28]. Then a p-GaAs layer is grown to form the junction and collection region. The top layer of p-type GaAlAs has a bandgap of approximately 1.8 eV. This structure reduces minority

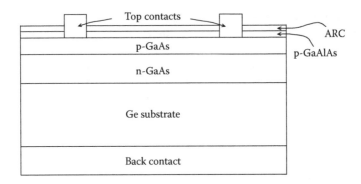

Figure 11.9 Structure of a basic GaAs cell with GaAlAs window and Ge substrate.

carrier surface recombination and transmits photons below the 1.8 eV level to the junction for more-efficient absorption.

A number of other GaAs structures have been reported, including cells of other III–V compounds. Figure 11.10 shows a cascaded AlInP/GaInP/GaAs structure grown by molecular beam epitaxy (MBE) [29] and an InP cell fabricated with the organo-metallic vapor phase epitaxy (OMVPE) process [30] (Table 11.1).

The epitaxial growth process involves passing appropriate gases containing the desired cell constituents over the surface of the heated substrate. As the gases contact

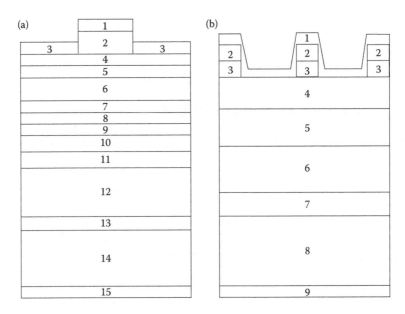

Figure 11.10 (a) Cascaded AlInP/GaInP/GaAs cell and (b) InP cell. Properties of the regions of the cells are summarized in Table 11.1. (Adapted from Lammasniemi, J. et al., *Proc. 26th IEEE PV Spec. Conf.*, 1997, 823–826; Hoffman, R. et al., *Proc. 26th IEEE PV Spec. Conf.*, 1997, 815–818 © 1997, IEEE.)

EVOLUTION OF PHOTOVOLTAIC CELLS AND SYSTEMS

Table 11.1 Summary of Composition of Regions of Cells of Figure 11.10

	AlInP/GaInP/GaAs Cell			InP Cell		
Region	Material	Thickness	Doping	Material	Thickness	Doping
1	Au/Ni/Ge	Contact		MgF$_2$/ZnS	110–55 nm	ARC
2	SiO$_2$/SiN$_x$	ARC		Au-Ge	2–3 µm	Front grid
3	GaAs	600 nm	n = 8E18	InGaAs	0.1–0.5 µm	
4	AlInP	25 nm	n = 2E18	InP	50–100 nm	p = 1E18
5	GaInP	75 nm	n = 1–4E18	InP	100–200 nm	p = 1E17
6	GaInP	400 nm	p = 5–500E16	InP	1.5–4.0 µm	n = 1E17
7	AlInP	25 nm	p = 5E18	InP	250–500 nm	n = 1E18
8	GaAs	10 nm[a]	p = 1E20	InP	400 µm	n > 1E18
9	GaAs	10 nm[a]	n = 8E18	Au-Ge	Contact	
10	GaInP	50 nm	n = 2E18			
11	GaAs	100 nm	n = 1E18			
12	GaAs	3500 nm	p = 1E17			
13	GaInP	100 nm	p = 1E19			
14	GaAs	Substrate				
15	Au/Pt/Ti	Contact				

Source: Lammasniemi, J. et al., *Proc. 26th IEEE PV Spec. Conf.*, 1997, 823–826; Hoffman, R. et al., In *Proc. 26th IEEE PV Spec. Conf.*, 1997, 815–818. © 1997, IEEE.
[a] Tunnel junction.

the substrate, the H or CH$_3$ attached to the In or P or Ga is liberated and the In, P, or Ga attaches to the substrate. Hence, to grow a layer of p-type InP, a combination of trimethyl indium, phosphine (for P), and diethyl zinc (for acceptor impurities) are mixed in the desired proportions and passed over the heated substrate for a predetermined time until the desired layer thickness is obtained. The process is repeated with different mixes of gases to form the other layers at the desired thickness.

GaAs cells remain expensive to fabricate and are thus used primarily for extraterrestrial applications and in concentrating systems.

11.3.4 Cell Performance

Cells fabricated with III–V elements are generally extraterrestrial quality. In other words, they are expensive, but they are high-performance units. Efficiencies in excess of 20% are common for single junction cells and efficiencies of cells fabricated on more-expensive GaAs substrates had exceeded 34% in 2009 [31]. An important feature of extraterrestrial-quality cells is the need for them to be radiation resistant. Cells are generally tested for their degradation resulting from exposure to healthy doses of 1 MeV or higher energy protons and electrons. Degradation is generally less than 20% for high exposure rates.

Extraterrestrial cells are sometimes exposed to temperature extremes, so the cells are also cycled between −170°C and +96°C for as many as 1600 cycles. The cells also need to pass a bending test, a contact integrity test, a humidity test, and a high temperature vacuum test, in which the cells are tested at a temperature above 140°C in vacuum for 168 hours [32].

Fill factors in excess of 80% have been achieved for GaAs cells. Single-cell open-circuit voltages are generally between 0.8 and 0.9 V.

The design of stacked cells depends upon the air mass under which the device is intended to operate. In order to ensure equal photon-generated current in each absorption layer, layer thickness needs to be adjusted for the air mass under which operation is anticipated, because air masses do not attenuate the entire spectrum proportionately, as shown in Figure 2.2.

In particular, if a cell absorbs efficiently at $\lambda = 0.9$ μm, significantly more photons are available at this wavelength at AM 0 than at AM 1. Thus, for operation at AM 1, the absorber width would need to be increased to generate a photocurrent comparable to that which a narrower layer would generate at AM 0. This consideration is particularly important when optimizing cell performance at AM 0, but can also be relevant for cells designed for use in regions where exposures to AM 1.5 or AM 2.0 may occur for long periods of cell operation.

Recently, interest has been shown in growing epitaxial III–V compounds on a crystalline Si cell as substrate material, resulting in a multijunction device. Several experiments were reported on at the 2016 43rd Photovoltaic Specialists Conference. More will be discussed in Section 11.6.4.7.

The bottom line is that III–V cells are excellent extraterrestrial performers, and it appears that continued research will improve performance, reduce mass, and reduce cost of the cells. The extent to which III–V cells will be able to compete with other technologies for terrestrial applications will depend on the degree of future improvements in all technologies, including reliable tracking devices, and heat sinks that are required for concentrating systems. This mystery will likely unfold before the eyes of any reader born after the fabrication of the first commercial PV module. The first commercial terrestrial module was a concentrating module, since concentrations up to 1000 suns can increase cell efficiency beyond 40% and such concentrations also minimize the use of expensive cell material and use less-expensive concentrating and cooling materials instead.

11.4 CIGS CELLS

11.4.1 Introduction

The first CIS PV cell, without the Ga, was reported in 1974 by a group at Bell Laboratories [33]. Copper indium diselenide was chosen as a potential PV material because of its attractive direct bandgap (1.0 eV), its very high optical absorption coefficient, and its potentially inexpensive preparation. Furthermore, the cell components are available in adequate quantities and the manufacturing, deployment, and decommissioning of the technology fall within acceptable environmental constraints.

While current technology involves the use of Cd as a cell component, the total quantity used in a cell is very small, and efforts are underway to identify alternate, less-toxic materials to replace the Cd. The challenge with CIS, as with other thin-film technologies, is to prepare a cell with appropriate electric field for collection of photon-generated EHPs. Then ohmic contacts need to be affixed, with the front contact being highly conducting, but transparent to incident photons. The final device structure must be stable and must be encapsulated to ensure long module lifetime.

Unlike Si, which has been studied intensely for decades and is well-understood by the scientific community, the fundamentals of CIS are less well-understood. For example, energy bands in Si have been studied extensively and detailed explanations are available in nearly any solid-state devices text. CIS, on the other hand, has only recently received significant attention. As basic research on the properties of CIS has led to increased understanding of the fundamental properties of the material, significant device improvements have been made, including increasing the efficiency of a Cu(In,Ga)(Se,S) lab cell to 22.3% [34].

After more than 10 years of field testing and continued work on product development, commercial CIGS modules are now available. One creative development has been the encapsulation of the CIGS cells in cylindrical geometry, in sizes similar to standard 4-ft fluorescent lamps [35]. To add to the utility of these novel modules, the module frames were designed to impose minimal wind load so that they can be installed without penetration on flat roof surfaces. Furthermore, on reflective roof surfaces, the cylindrical cells can also collect sunlight reflected off the roof surface. Regrettably, even though this technology had proven itself in field installations, it fell victim to market forces when the price of silicon-based modules dropped to a small fraction of the production cost of this technology, resulting in the manufacturer filing for bankruptcy.

11.4.2 Production of Pure Cell Components

11.4.2.1 Copper [36]

Of all the elements used in the production of semiconductors, except for silicon, copper is probably the most abundant. It is known from prehistoric times. It occasionally occurs in pure form in nature, but also occurs in many minerals, such as cuprite, malachite, azurite, chalcopyrite, and bornite. The sulfides, oxides, and carbonates are the most important sources of Cu. The refining process involves smelting, leaching, and electrolysis. Since Cu is used in so many applications and is found in so many locations in so many different compounds, many different methods of Cu refining have been developed. The reader is referred to Reference 36 for examples.

Copper can be refined to semiconductor grade by electrolysis of a solution of copper sulfate and sulfuric acid. The anode is made of approximately 99% pure Cu and the solution is maintained so that almost everything that plates out on the cathode is Cu. The chemical reactions at anode and cathode are, respectively,

$$\text{Anode} \quad Cu \rightarrow Cu^{2+} + 2e^- \qquad (11.10)$$

and

$$\text{Cathode} \quad Cu^{2+} + 2e^- \rightarrow Cu. \qquad (11.11)$$

11.4.2.2 Indium [37]

Indium is most commonly found with zinc, but is also found with iron, lead, and copper ores. Little use was made of In until it was found to be useful in certain semiconductors. Most commercial In is now obtained from the flue dusts and residues from smelting lead and zinc. As late as 1924, less than an ounce (28.3 g) of pure In existed, even though it is rather widely scattered throughout the world, albeit in very small concentrations in host minerals.

Several means of extracting In from its natural states are available. In the presence of lead and zinc, the material can be melted and treated with chlorine gas or other chlorine source. This removes the Zn and In as chlorides, provided that the temperature is low enough to prevent evaporation of the In. The chloride slag is then leached with dilute sulfuric acid, which causes the In to precipitate along with Zn dust. The next step is to melt the In–Zn mixture and remove the Zn with Cl.

Indium is refined to semiconductor grade by further physical or chemical separation techniques, such as zone refining and fractional distillation of liquid In compounds.

11.4.2.3 Selenium [38]

Selenium is a Group VI element and, as a result, is very similar to sulfur in many of its chemical properties, which accounts for its common occurrence with CuS. In its natural form, it combines naturally with 16 other elements and is a major component of 39 mineral species and a minor component of another 37 mineral species. Since Se is always combined with other elements, such as S, there are no identified "reserves" of Se. Certain native plant species preferentially absorb Se and are sometimes indicators of the presence of the element. Elemental Se is relatively nontoxic, but many Se compounds are very poisonous. Selenium is recovered primarily from the anode muds from the refining of copper. It is also recovered from flue dusts from processing copper sulfide ores.

A number of methods of recovery of Se are available, depending on the starting material and the desired end products.

11.4.2.4 Cadmium [39]

Cadmium was discovered in 1817 [25] by Friedrich Stromeyer in zinc carbonate and by K. S. L. Hermann in a specimen of zinc oxide. The common denominator is that cadmium is commonly found embedded in zinc compounds, although cadmium sulfide is found in Scotland, Bohemia, and Pennsylvania.

Since zinc is routinely refined, the leftover cadmium from the process is generally recovered, mixed with carbon, and redistilled to yield an enriched dust. This process is repeated several times. Then hydrochloric acid is mixed with the cadmium dust, and zinc is added, resulting in the precipitation of the cadmium. After several repetitions of this process, electrolysis is finally used to deposit the final material. Cadmium is a Group II metallic element, and along with solutions of its compounds, it is highly toxic.

11.4.2.5 Sulfur [40]

Sulfur is abundant in the crust of the earth, both in elemental and in combined forms. Elemental sulfur tends to occur in layers above salt domes. Abundant supplies of elemental sulfur have been discovered in many locations around the world. It has been used for thousands of years in many applications.

Elemental sulfur is generally recovered by pumping superheated water down a well to melt the sulfur, after which the molten sulfur is extracted from the well. The elemental sulfur is then generally reacted with a suitable element to enable the sulfur to be carried in a gaseous state for vapor deposition onto a substrate.

11.4.2.6 Molybdenum [41]

Molybdenum, a Group VI metal, is an important metal for making ohmic contacts, particularly as the back contact in a CIS cell. The metal was first prepared by P. J. Hjelm in 1782. It has high strength and high corrosion resistance and is often used in alloys, particularly in the production of stainless steel. It retains its strength at high temperatures better than most other metals.

Most Mo is mined in the United States, Canada, and Chile. Molybdenum occurs principally in MoS_2, but the concentration of MoS_2 in ores is rather small. About a ton of ore must be mined, crushed, and milled to recover about 4 lb of Mo. After the ore is crushed, the MoS_2 is recovered in relatively high concentration by floatation.

Roasting then drives off the sulfur and oxidizes the Mo to MoO_3. Reduction with H_2 at high temperature yields Mo in powder form with purity in the 99% range. The powder can be reacted with halogens to form compounds such as MoF_6 and $MoCl_5$, which are suitable for vapor phase deposition of Mo.

11.4.3 Fabrication of the CIS Cell

While it is possible to produce both n-type and p-type CIS, homojunctions in the material are neither stable nor efficient. A good junction can be made, however, by creating a heterojunction with n-type CdS and p-type CIS.

The ideal structure uses near-intrinsic material near the junction to create the widest possible depletion region for collection of generated EHPs. The carrier diffusion length can be as much as 2 μm, which is comparable with the overall film thickness. Figure 11.11 [42] shows a basic ZnO/CdS/CIGS/Mo cell structure, which was in popular use in 2004. Since then, improvements in the TCO and processing steps have led to more-efficient cells, but the basic cell structure is still the subject of incremental improvements. Again, CIGS technology is advancing rapidly as a

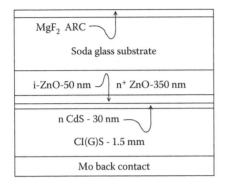 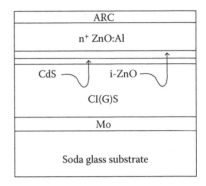

Figure 11.11 Typical CIGS thin-film structures. (Adapted from Ullal, H. S., Zweibel, K., and von Roedern, B., In *Proc. 26th IEEE PV Spec. Conf.*, 1997, 301–305.)

result of the Thin-Film Photovoltaics Partnership Program and the U.S. Department of Energy SunShot Program [43], so by the time this paragraph is read, the structure of Figure 11.11 may be suitable only for history books and general discussion of the challenges encountered in thin-film cell development.

Nearly, a dozen processes have been used to achieve the basic cell structure of Figure 11.11. The processes include rf sputtering, reactive sputtering, chemical vapor deposition, vacuum evaporation, spray deposition, and electrodeposition (ED). Sometimes these processes are implemented sequentially and sometimes they are implemented concurrently.

In the physical vapor deposition (PVD) process, which was used to achieve a record laboratory cell efficiency, the constituent elements are deposited under a relatively high vacuum of 10^{-6} Torr. In the PVD process, the four elements can be simultaneously evaporated, they can be sequentially evaporated followed by exposure to Se, or they can be sequentially evaporated in the presence of Se. The soda lime glass substrate is maintained between 300°C and 600°C during the evaporation process.

The front ohmic contact is straightforward, and ZnO generally works well. The trick is to achieve sufficiently high conductivity without absorbing any of the incident photons. Often the ZnO is applied in two layers. The layer on the glass is high-concentration n-type, with a very thin intrinsic layer in contact with the CdS. The more heavily doped layer has high conductivity, in the neighborhood of 4 Ω/square, and the narrow intrinsic layer acts as a passivation layer between the thin CdS layer and the TCO, but is narrow enough to allow efficient transport of electrons to the TCO.

11.4.4 Cell Performance

When new-technology PV cells are developed, normally small area cells are fabricated first to determine whether it appears practical to extend the technology to larger area cells and modules. In 2008, the highest performance achieved for a CIGS laboratory cell was 19.4%. Scaling up to the production of modules with areas up to

3459 cm^2 resulted in decrease in efficiency to 13.5% [1]. In 2015, a 22.3% efficiency was achieved for a lab cell and the 2014 record of 23.3% was achieved for a concentrator lab cell [2]. The increased efficiency resulted from improvements in the CIS absorber layer and in the junction formation process [44].

Clearly, the challenge in module development is to overcome the factors that result in degradation of cell performance, and clearly, these challenges are being undertaken ambitiously. To do so requires understanding of the factors that cause degradation. Some of these considerations include general device design, contact grid design, antireflection coatings, and the sheet resistance of window layers.

An example of the trade-offs involved in scaling up a technology is the TCO. For a laboratory-scale device with an area of 1 cm^2 or so, a TCO layer can be relatively thin, with a sheet resistivity of 15 Ω/square, since the relatively small current from the cell will experience minimal voltage drop through this contact. The thin window absorbs a minimal amount of incident radiation so the CIS absorber layer can achieve maximum conversion efficiency. However, as the cell is made larger, the sheet resistance of the TCO must be reduced to prevent voltage drop at the contact and corresponding degradation of fill factor and cell efficiency. The price paid for lower sheet resistance is a greater amount of absorption of incident photons by the transparent contact.

Series connection of cells can also cause cell performance degradation. Figure 11.12 [45] shows how individual cells have been replaced by minimodules that are monolithically connected. This monolithic connection process eliminates the need for separate fabrication processes for interconnecting cells.

After deposition of the Mo, it is scribed to separate adjacent cells, creating cells with a length of about 4 ft and a width of a fraction of an inch. Next, after deposition of the CIS cell components, the CIS is scribed. Finally, after deposition of the TCO, the TCO and CIS are scribed. While this process may appear to be straightforward, it can also be appreciated that the first cut needs to remove all of the Mo, but none of the glass. The second and third cuts must leave the Mo intact. Considering thickness

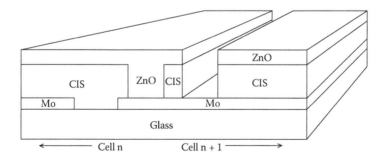

Figure 11.12 Siemens monolithic method of series cell connection. (Adapted from Tarrant, D. E. and Gay, R. R., *Thin-Film Photovoltaic Partnership Program—CISBased Thin Film PV Technology*, Phase 2 Technical Report, October 1996–October 1997, NREL Report NREL/SR-520-24751, Golden, CO, May 1998.)

of a few micrometers or less for these layers, the process falls somewhat short of being classified as trivial.

By incorporating Ga into the CIS mixture, the bandgap of the material can be increased beyond 1.1 eV. This movement of the bandgap energy closer to the peak of the solar spectrum increases conversion efficiency in this wavelength region and leaves lower-energy photons for capture by the free carrier absorption process in the TCO layer while the higher energy photons are converted to EHPs. The result is increasing the cell open-circuit voltage from approximately 0.4 V to as high as 0.68 V, with fill factors approaching 80% for laboratory cells. Experiments with adding sulfur to the selenium have also resulted in cell performance improvements.

Exposure to more intense light has actually caused efficiencies to increase slightly (concentrator cell). Exposure to elevated temperatures has resulted in loss of efficiency, but light soaking restored the modules to original efficiency levels.

11.5 CDTE CELLS

11.5.1 Introduction

In theory, CdTe cells have a maximum efficiency limit close to 25%. The material has a favorable direct bandgap and a large absorption constant, allowing cells of a few micrometer thickness. By 2001, efficiencies approaching 17% were being achieved for laboratory cells, and module efficiencies had reached 11% for the best large-area (8390 cm^2) module [46]. Efforts were then focused on scaling up the fabrication process to mass produce the modules, with the result of achieving a production cost of less than $1.00/W in 2008. This cost included an escrow account to be used for recycling the materials at the end of module life [47].

Although tellurium is not as abundant as other cell components, cadmium and tellurium are both available in sufficient quantities for the production of many gigawatts of arrays. The amount of cadmium poses a possible fire hazard and some concern at the time of decommissioning of the modules, as indicated in Chapter 9. Analysis of the concerns, however, has shown that the Cd of the cells would be recycled at decommissioning time and that the danger of burns from any fire far exceeds the danger of contact with any Cd released from heating of the modules. Means for recycling CdTe modules exist that result in recovery of glass, $CdCO_3$, electrolytically refined Te, and clean EVA at a cost of less than $0.04/W [48].

Purification of all of the components of the CdTe cell has been discussed so far except for the Te. After a brief summary of the extraction and purification of Te, cell fabrication and cell performance will be discussed.

11.5.2 Production of Pure Tellurium [40]

Tellurium is a Group VI metallic element, which was discovered by Muller in Transylvanian gold ore in 1782 and first extracted and identified by M. H. Klaproth in 1798. It is found as tellurides of copper, lead, silver, gold, iron, and bismuth and is widely distributed over the surface of the earth, although its percentage in the earth's crust is very small. The primary sources of tellurium for production are leftovers from copper and lead refining, where tellurium and selenium both appear in very small quantities.

When copper is produced by electrolysis, the tellurium is precipitated along with other impurities in the copper to become a sludge at the anode. Several alternative chemical means are then used to separate the tellurium from the other impurities in the sludge. The first step is to produce tellurium dioxide, which precipitates out of solution while the other impurities remain dissolved. Alternatively, other impurities may first be precipitated, leaving behind higher concentrations of tellurium in the form of tellurous acid. If the oxide is produced, then the oxide is reduced to form elemental tellurium.

Elemental tellurium remains contaminated with iron, copper, tin, silver, lead, antimony, and bismuth and can be further purified by low-pressure distillation, where the heavier metals remain in the residue. Selenium, however, is volatile and remains a contaminant in the distilled tellurium. Further purification can be achieved by dissolving the tellurium in strong nitric acid. Diluting and boiling hydrolyzes the tellurium to a precipitate form. The precipitate is separated, washed, dissolved in hydrochloric acid, and reduced with sulfur dioxide.

This relatively pure tellurium can be brought to the ultrahigh purity state by zone refining in an inert gas atmosphere, and single crystals can be grown by either the Czochralski or Bridgman methods.

Tellurium is classified as "probably" toxic, and reasonable care is recommended in its handling.

11.5.3 Production of the CdTe Cell

Figure 11.13 [49] shows a typical CdTe cell structure. The cell begins with a glass superstrate with a transparent, conducting oxide layer about 1 μm thick, a thin CdS

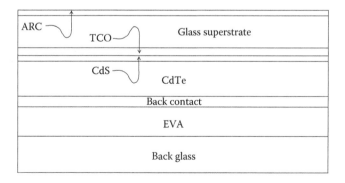

Figure 11.13 Basic structure of a CdTe PV cell. (Adapted from Ullal, H. S., Zweibel, K., and von Roedern, B., In *Proc. 26th IEEE PV Spec. Conf.*, 1997, 301–305.)

buffer layer about 0.1 µm thick, a CdTe layer a few µm thick, and a rear contact of Au, Cu/Au, Ni, Ni/Al, ZnTe:Cu, or (Cu, HgTe).

The TCO layer has been fabricated with SnO, InSnO, and Cd_2SnO_4. The Cd_2SnO_4 has been shown to exhibit better conductivity and better transparency [50] than the other TCOs and may end up as the preferred TCO. The Cd_2SnO_4 layer is deposited by combining CdO and SnO_2 in a 2:1 proportion in a target material, which is then deposited by means of radio frequency magnetron sputtering onto the glass superstrate.

The thin n-type CdS layer has been deposited by the MOCVD process as well as by other thin-film deposition techniques. The layer needs to be annealed prior to deposition of the CdTe in order to reduce CdS surface roughness, thus reducing defects on the CdS/CdTe boundary. This is normally accomplished in air at approximately 400°C for about 20 minutes.

An interesting challenge existed with regard to the CdS layer, since the layer is so thin. During deposition of the CdTe or subsequent heat treatment of the cell, intermixing of the CdS and CdTe can occur at the boundary. This can result in junctions between the CdTe and the TCO layer, which causes significant reductions in cell open-circuit voltage. Several methods of minimizing this mixing were proposed [51] and this problem has now been overcome. Recently, it has been discovered that replacing the CdS layer with CdSO as the n-window layer improves the bandgap and results in a higher cell efficiency [52].

Numerous methods have been used to deposit the CdTe layer, including atmospheric pressure chemical vapor deposition (APCVD), atomic layer epitaxy (ALE), close-spaced sublimation (CSS), ED, laser ablation, PVD, screen printing (SP), spray, sputtering, and MOCVD [31]. The CdTe layer is subjected to a heat treatment in the presence of $CdCl_2$ for about 20 minutes at about 420°C (788°F). This treatment enhances grain growth in the CdTe layer to reduce grain boundary trapping effects on minority carriers. Nonheat-treated CdTe cells tend to have open-circuit voltages less than 0.5 V, while after heat treating, V_{OC} can exceed 0.8 V.

Another important step in optimizing cell performance is to ensure stability of the back contact. Before the back contact is applied, the CdTe is etched with nitric-phosphoric (NP), resulting in a layer of elemental Te at the back CdTe surface. The elemental Te produces a more stable contact between the p-type CdTe and the back metal [53].

Final processing of modules involves encapsulation of the back of the cell with a layer of EVA between the metallization and another layer of glass.

11.5.4 Cell Performance

No fewer than nine companies have shown an interest in commercial applications of CdTe. As of 2001, depending on the fabrication methodology, efficiencies close to 17% had been achieved for small area cells (≈ 1 cm^2), and 11% on a module with an area of 8390 cm^2 [31]. Furthermore, sufficient experience had been logged with large-scale production to identify areas in which improvement was needed [54]. These areas included improving the design, operation, and control of a CdTe reactor;

increasing the understanding of the fundamental properties of CdTe films; maintaining uniformity of materials and device properties over large areas, including the interdiffusion of CdTe and CdS and the back contact; obtaining stability of the back contact; and addressing any environmental concerns over Cd. Since Te availability may limit cell production, increasing performance with thinner layers of CdTe is also desirable. Reducing the thickness of the CdTe layer to 0.5 μm will allow for four to five times the cell area, provided that efficiency can be maintained or increased. Just as in the case of other thin-film arrays, area-related costs limit the minimal cost per watt of CeTe arrays, so increase in efficiency remains important to minimize the overall cost per watt.

Fill factors for lab cells have been obtained in the 65%–75% range [55], with the slope of the cell J–V curve at V_{OC} in the range of 5 Ω-cm^2. Short-circuit current densities upward from 25 mA/cm^2 were obtained by using a thin buffer layer of CdS along with a thin insulating TCO layer between the heavily doped TCO and the CdS layer [31].

Experimental CdTe arrays in sizes up to 25 kW were deployed in California, Ohio, Tunisia, Colorado, and Florida for testing purposes [31]. Soon after it was shown that no degradation was observable after 2 years, production-scale manufacturing began. CdTe modules are now being manufactured and marketed at the rate of more than 3 GW annually [56] for utility-scale projects and the magic $1.00/W production cost barrier has now been broken for these modules, with 2015 production costs for 16% efficient modules now at approximately $0.51/W [43].

11.6 EMERGING TECHNOLOGIES

11.6.1 New Developments in Silicon Technology

While progress continues on conventional Si technology, new ideas are also being pursued for crystalline and amorphous Si cells. The goal of Si technology has been to maintain good transport properties, while improving photon absorption and reducing the material processing cost of the cells. When the thickness of Si is reduced, some sort of substrate material is required to maintain sufficient physical strength for the cell. Two substrates under investigation in 2002 were ceramic and graphite, both having led to cell efficiencies in the 10% range [57]. Diffusion of contaminants such as Mg, Mn, and Fe from ceramic substrates has been shown to limit performance of Si on ceramics, and methods of introducing barrier layers at the Si-ceramic interface have been shown to limit the introduction of contaminants into the Si [58].

Cost reductions have been achieved by reducing the thickness of the cells, reducing the cost of polysilicon, improvements in metallization, process engineering, and economies of scale [43].

Another interesting opportunity for cost reduction in Si cell production is to double up on processing steps. For example, a technique has been developed for simultaneously diffusing boron and phosphorous in a single step, along with growing a passivating oxide layer [57].

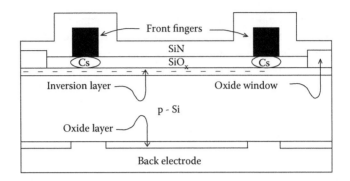

Figure 11.14 Structure of an MIS-IL Si cell. (Adapted from Metz, A. et al., In *Proc. 26th IEEE PV Spec. Conf.*, 1997, 31–34.)

As an alternative to the pn junction approach to Si cells, MIS-IL (metal insulator semiconductor inversion layer) cells have been fabricated with 18.5% efficiency [59]. The cell structure is shown in Figure 11.14. The cell incorporates a point-contacted back electrode to minimize the rear surface recombination, along with Cs beneath the MIS front grid and oxide window passivation of the front surface to define the cell boundaries. Further improvement in cell performance can be obtained by texturing the cell surfaces.

The MIS-IL cell uses the top SiO_x layer as a tunnel junction. The presence of positive charges in the SiO_x layer creates the electric field from oxide to p-Si, thus creating the inversion layer at the top of the p-type material. This field then separates the EHPs just as the E-field at a pn junction separates the carriers.

Other groups have worked on cells with both contacts on the back, in order to eliminate the shadowing of the front surface by the front contacts [55]. The ACE Designs project, funded by the European Community, resulted in the development of three types of rear contact Si cells: metallization wraparound (MWA), metallization wrap-through (MWT), and emitter wrap-through (EWT). A laser-grooved, buried grid (LGBG) process applied to the MWA technology was estimated to have the best potential for lowest-cost rear-contacted cells [60]. Some of these methods are now in commercial production with module efficiencies >20%.

New developments in surface texturing may also simplify the process and result in additional improvement in Si device performance. Discovery of new substrates and methods of growing good quality Si on them is also an interesting possibility for performance improvement and cost reduction for Si cells.

Another idea that has been investigated is to combine crystalline and amorphous Si into a tandem cell arrangement to take advantage of the different bandgaps of the two materials in increasing absorption efficiency [61]. Work in Japan produced a 1-cm^2 tandem cell with an open-circuit voltage of 1.4 V, a fill factor of 71.9%, and an efficiency of 14.5% [62]. The cell is composed of a textured TCO on a glass substrate, followed by an a-Si:H cell, an interlayer, a microcrystalline Si cell, and a back contact/reflector. The interlayer is incorporated to produce some reflection of the incident photons back into the a-Si to better match the current densities of the two cells. This heterojunction with intrinsic thin layer (HIT) technology is now commercially available in two types of modules, one of which is a standard, single-sided module and the other is a bifacial module that absorbs photons from the front and the back sides. In 2008, the cell efficiency approached 19% and the module efficiency was rated at 15.7% [63]. By 2015, efficiencies of HIT laboratory cells had reached 25.6% [2].

Still another interesting process for producing crystalline, thin-film Si cells involves the epitaxial growth of very thin crystalline cells on existing crystalline cells [64]. The growth takes place at a temperature that does not melt the existing cell, and thus the epitaxially grown cell can be "peeled" off the existing cell and mounted on its own substrate, usually glass. By fabricating the new cell on a cell with a textured surface, the new cell also will have a textured surface. The overall thickness of the epitaxially grown cell is less than 20 μm, and the epitaxial growth process is convenient for adding n-type impurities and then switching to p-type impurities to produce the pn junction of the cell. The thin layers justify epitaxial growth as the mechanism for cell production.

Since new ideas will continue to emerge as interest in Si PV technology continues to grow, the interested reader is encouraged to attend PV conferences and to read the conference publications to stay up-to-date in the field.

11.6.2 CIS-Family-Based Absorbers

Figure 11.15 [65] shows the theoretical maximum efficiency of a solar absorber as a function of bandgap. Table 11.2 [66] shows the bandgaps of a family of CIS-type materials. Much is yet to be learned about inhomogeneous absorbers and composite absorbers composed of combinations of these various materials. The possibility of multijunction devices is also being explored.

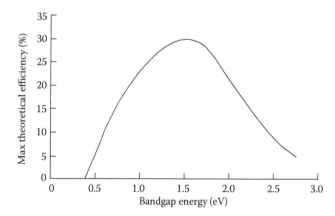

Figure 11.15 Theoretical maximum PV cell efficiency versus bandgap. (Adapted from Granata, J., Sites, J. R., and Tuttle, J. R., In *Proc. 14th NREL PV Program Review, AIP Conf. Proceedings 394*, Lakewood, CO, 1996, 621–630.)

Table 11.2 Bandgaps of CIS-Related Materials

Material	Bandgap (eV)
$CuInSe_2$	1.05
$AgInSe_2$	1.24
$CuInS_2$	1.56
$CuGaSe_2$	1.67
$AgGaSe_2$	1.69
$AgInS_2$	1.95
$CuGaS_2$	2.33
$AgGaS_2$	2.56

Source: Tarrant, D. E. and Gay, R. R., *Research on High Efficiency, Large Area CuInSe2 Modules, Final Technical Report*, NREL, Golden, CO, April 1995.

Efficiency of large-area devices is critically dependent on spatial uniformity of the absorber layers and electrodes as well as on the performance of cell interconnects. Hence, with any of the listed absorber possibilities, the key to performance will be in the engineering of reliable and reproducible module processing techniques.

Even an issue as mundane as the quality and cleanliness of the glass substrate can have a significant effect on module performance. Polishing the glass with CeO_2 has been shown to increase module performance [66]. Furthermore, it appears that if a certain amount of Na diffuses from the glass to the absorber, the absorber performance is increased [65].

Meanwhile, work is underway to reduce the material usage in the production of CIS modules in order to further reduce production costs. Examples of reduction of material use include halving the width of the Mo contact layer, reduction in the use of H_2S and H_2Se, and a reduction in ZnO, provided that a minimum thickness can be maintained [67].

11.6.3 Other III–V and II–VI Emerging Technologies

It appears that compound tandem cells will receive appreciable emphasis in the III–V family of cells over the next few years. For example, $Ga_{0.84}In_{0.16}As_{0.68}P_{0.32}$, lattice matched to GaAs, has a bandgap of 1.55 eV and may prove to be an ideal material for use under AM 0 conditions, since it also has good radiation resistance [68]. Cells have been fabricated with $Al_{0.51}In_{0.49}P$ and $Ga_{0.51}In_{0.49}P$ window layers, with the best 1 cm^2 cell having an efficiency of just over 16%, but having a fill factor of 85.4%. The best performing window was the AlInP.

Mechanical stacking of materials having different lattice constants has also been proposed [69]. Since multijunction devices are used primarily in concentrating systems, an update on this technology will be presented in Section 11.6.4.7.

Cell efficiencies can be increased by concentrating sunlight on the cells. Although the homojunction cell efficiency limit under concentration is just under 40%, quantum well (QW) cells have been proposed to increase the concentrated efficiency beyond the 40% level [70]. In QW cells, intermediate energy levels are introduced between the host semiconductor's valence and conduction bands to permit absorption of lower-energy photons. These levels must be chosen carefully so that they will not act as recombination centers, however, or the gains of EHPs from lower-energy incident photons will be lost to the recombination processes. Laboratory cells have shown higher V_{OC} resulting from a decrease in dark current for these cells. Intense study is underway with the goal of understanding the electronic processes that take place within these cells [71].

11.6.4 Other Technologies

11.6.4.1 Thermophotovoltaic Cells

To this point, discussion has been limited to the conversion of visible and near-infrared spectrum to EHPs. The reason is simply that the solar spectrum peaks out in the visible range. However, heat sources and incandescent light sources produce radiation in the longer infrared regions, and in some instances, it is convenient to harness radiated heat from these processes by converting it to electricity. This means using semiconductors with smaller bandgaps, such as Ge. More exotic structures, such as InAsSbP, with a bandgap of 0.45–0.48 eV, have also been fabricated. The InAsSbP can be fabricated as p-type and n-type. The pn junction is grown on a substrate of InAs [72].

Theoretical work continues in the TPV area in 2016. Conventional TPV systems are based upon heat from engines and furnaces. But it is also possible to create a solar TPV system, where the heat comes from the sun and heats a material that subsequently delivers heat to TPV cells [73]. Reuse of photons is also being explored by another group. Thin-film PV cells can be fabricated with high reflectivity of photons having energies below the bandgap of the PV material. These photons are then reflected back to the emitter, resulting in reheating of the source [74].

11.6.4.2 Intermediate Band Solar Cells

In all cells described to this point, absorption of a photon has resulted in the generation of a single EHP. If an intermediate band material is sandwiched between two ordinary semiconductors, it appears that it may be possible for the material to absorb two photons of relatively low energy to produce a single EHP at the combined energies of the two lower-energy photons. The first photon raises an electron from the valence band to the intermediate level, creating a hole in the valence band, and the second photon raises the electron from the intermediate level to the conduction band. The trick is to find such an intermediate band material that will "hold" the electron until another photon of the appropriate energy impinges upon the material. Such a material should have half its states filled with electrons and half empty in order to optimally accommodate this electron-transfer process. It appears that III–V compounds may be the best candidates for implementation of this technology. Theoretical maximum efficiency of such a cell is 63.2% [75].

11.6.4.3 Supertandem Cells

If a large number of cells are stacked with the largest bandgap on top and the bandgap of subsequent cells decreasing, the theoretical maximum efficiency is 86.8% [76]. By 2008, a 1-cm^2 four-junction cell had been fabricated with an efficiency of 35.4%. The maximum theoretical efficiency of this cell structure is 41.6% [77]. The highest efficiency laboratory cells in 2015 were four-junction concentrator cells with efficiencies that had reached 46% [2]. Perhaps one day one of the readers of this paragraph (or one of their great-great grandchildren) will fabricate a cell with the maximum theoretical efficiency and incorporate it into a module with 80% efficiency that will sell for less than $0.15/W. Presently, supertandem cells have not been commercialized due to the high cost of fabrication that leads to unfavorable LCOE for terrestrial uses. Extraterrestrial use, however, away from atmospheric absorption lines, is a more promising possible use for these cells, provided they meet the harsh environmental adaptation requirements of extraterrestrial space.

11.6.4.4 Hot Carrier Cells

The primary loss mechanism in PV cells is the energy lost in the form of heat when an electron is excited to a state above the bottom of the conduction band of a PV cell by a photon with energy greater than the bandgap. The electron will normally drop to the lowest energy available state in the conduction band, with the energy lost in the process being converted to heat. Hence, if this loss mechanism can be overcome, the efficiency of a cell with a single junction should be capable of approaching that of a supertandem cell. One method of preventing the release of this heat energy by the electron is to heat the cell, so the electron will remain at the higher energy state. The process is called *thermoelectronics* and is currently being investigated [76] as mentioned in Section 11.6.4.1.

11.6.4.5 Optical Up- and Down-Conversion

An alternative to varying the electrical bandgap of a material is to reshape the energies of the incident photon flux. Certain materials have been shown to be capable of absorbing two photons of two different energies and subsequently emitting a photon of the combined energy. Other materials have been shown to be capable of absorbing a single high-energy photon and emitting two lower-energy photons. These phenomena are similar to up and down-conversion in communications circuits at radio frequencies.

By the use of both types of materials, the spectrum incident on a PV cell can be effectively narrowed to a range that will result in more-efficient absorption in the PV cell. An advantage of this process is that the optical up- and down-converters need not be a part of the PV cell. They simply need to be placed between the photon source and the PV cell. In tandem cells, the down-converter would be placed ahead of the top cell and the up-converter would be integrated into the cell structure just ahead of the bottom cell.

11.6.4.6 Organic PV Cells [77]

Even more exotic than any of the previously mentioned cells is the organic cell. In the organic cell, electrons and holes are not immediately formed as the photon is absorbed. Instead, the incident photon creates an *exciton*, which is a bound EHP. In order to free the charges, the exciton binding energy must be overcome. This dissociation occurs at the interface between materials of high electron affinity and low ionization potential [77]. Figure 11.16 [78] shows a basic organic PV (OPV) cell structure. The cell consists simply of an organic material, usually a polymer/plastic, sandwiched between a high work function metal and a lower work function metal, in order to create an electric field across the organic absorber.

One of the challenges in the development of OPV cells is getting the charge carriers out of the absorber and to the contacts, which is more-or-less equivalent to getting the charge carriers out of a pn junction and into the bulk regions of a conventional PV cell. The absorbers that have been tried have generally absorbed most of the incident photons in a submicron thickness, but have been hampered by very short carrier lifetimes and correspondingly short diffusion lengths. As a result, a large majority of the charges recombine within the absorber before they can travel to a contact [78]. The result is a solar cell with efficiency of about 1%.

Researchers have found that by using two different organic materials to form the active layer, transport properties and thus cell efficiency can be improved. The first layer consists of a heterojunction where the second junction material, usually fullerene, will act as an n-type semiconductor and the polymer layer will act as a p-type semiconductor. Thus, when a photon creates an EHP or an exciton, the electron is accepted by the fullerene and the hole is accepted by the polymer, thus separating the exciton and moving the electron and hole toward the contacts. This structure is still limited by short diffusion lengths, so to overcome this problem, the heterojunction is distributed throughout the absorber to enable separation of the charge carriers throughout the absorber. As of 2015, the best reported OPV lab cell efficiency was 11.5% [2].

OPV systems are currently at the advanced demonstration phase, with some products for sale. Most applications have been directed toward building-integrated

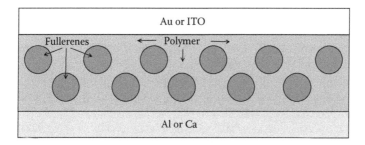

Figure 11.16 Basic structure of an OPV cell. (Adapted from University of Houston, Institute for Nanotechnology, Organic Solar Cells, http://ine.uh.edu/research/organic-solar-cells/index.php, 2016.)

PV (BIPV), in which, for example, the material has been integrated into cement and metal structural components, as well as glass and irregularly shaped fabrics used in backpacks and tents [79].

11.6.4.7 Concentrating PV Cells [80]

Concentrating PV (CPV) cells and systems are designed to focus large amounts of sunlight onto a small area cell. Work has been done up to 1000 suns, which is 1000 times the intensity of the sun, or, 1,000,000 W/m^2. This range of concentration is generally broken down into a 300–1000 suns range (high concentration, or HCPV) and <100 suns (low concentration, or LCPV). HCPV cells are typically III–V compounds with multijunctions, and LCPV cells are typically Si-based cells, which may be single junction or sometimes heterojunction devices.

As mentioned earlier, since III–V cells are more expensive, it is advantageous to use smaller cells with less-expensive concentrating structure to focus the incident sunlight on the cells. For high concentration, however, this requires using the direct beam of the sunlight via double-axis trackers, which means the technology is most cost-effective in areas such as deserts that experience a large fraction of direct sunlight. In fact, standard test conditions for concentrating cells have been adopted using 800 W/m^2 as standard incident radiation, since the diffuse component of sunlight is eliminated. For LCPV, either single-axis or double-axis trackers are used.

CPV has advantages and disadvantages. Perhaps one of the most attractive features of HCPV is the fact that the dual-axis trackers must be spaced apart far enough to avoid shading among the trackers. This allows for the land between the tracking stands to be used for agriculture, since the racking is generally implemented with a single mounting pole for each tracker. Although the tracking must be quite precise, recent significant progress on tracker design has made this possible.

Another important feature of tracking is that it results in maximum possible energy production throughout the day, especially during late-afternoon utility peaking hours. Furthermore, III–V cells tend to be less degraded by ambient temperatures, since their output power has a smaller temperature dependence than Si devices. As the efficiency of concentrator cells continues to increase, as discussed in earlier sections, the cost-effectiveness of HCPV is also enhanced. As of 2015, an HCPV module efficiency of 38.9% was achieved.

One disadvantage of CPV is that they only perform well in regions having direct sunlight of at least 2000 kWh/m^2/year, with 2500 kWh/m^2/year a preferred level, which pretty much limits their use to desert environments.

Presently, most CPV systems use Fresnel lenses to focus the direct solar beam onto the multijunction III–V cell. The most commonly used cell in recent large commercial (MW) systems is the triple-junction, lattice-matched Ga$_{0.50}$In$_{0.50}$ P/Ga$_{0.99}$In$_{0.01}$As/ Ge cell. The limiting factor on this cell is that the Ge layer absorbs nearly twice as much of the incident radiation as the other layers, creating a mismatch of current sources in series.

Another disadvantage of the cell assembly is that dust and sand interfere with passive cooling of the assembly, leading to temperature rise and efficiency losses. It has also been observed that when CPV modules are mounted on large trackers, the trackers bend, resulting in angular distortion between cells and thus overall less than optimal photon capture.

In 2012, over 120-MW of CPV, mostly HCPV, was installed. These systems were generally contracted while the price of Si modules was in the $3–$4/W range. As the price of Si modules plummeted after 2012, installations of CPV also dropped to under 20 MW in 2015, while installation of Si technology skyrocketed. This is a typical example of the risk involved in investing in new technology in competition with well-established older technology.

11.6.4.8 Perovskites

Perovskites, hardly on the radar screen in 2008, are an exciting new technology. The efficiency of laboratory cells has increased from about 14% in 2013 to 22.1% in 2016 [2].

The basic perovskite structure is shown in Figure 11.17 [81]. A close look shows that this structure might be classified as a body-centered, face-centered cubic structure, since the A-component appears in the body center, the B-components appear at the cell corners, and the X-components appear in the centers of the cube faces. Typically, the A-component is a large positively charged ion, which might be organic. The B-component is a somewhat smaller, positively charged ion, and the C-component is a small, negatively charged ion. The basic structure is recognized as ABX_3, where A has a single positive charge, B has a double positive charge, and X has a single negative charge [81].

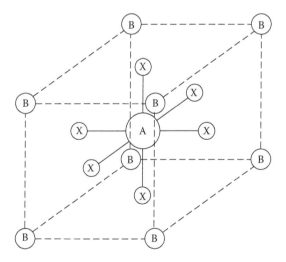

Figure 11.17 Basic structure of a perovskite cell. (Adapted from Perovskites and perovskite solar cells: An introduction, www.ossila.com/pages/perovskites-and-perovskite-solar-cells-an-introduction (accessed June 2016).)

From the figure, one can deduce that all of the A-component belongs to the cell, thus adding a charge of +2. The B-components, however, are each shared with seven other cells that join at the corners, so that each contributes only one-eighth of its charge to the cell and seven-eighths to the other seven cells. Thus, the eight corner (B) components contribute a total of +1 charge to the cell. The six X-components are each shared with one adjacent cell, such that half their total charge is shared with adjoining cells and the remaining half can be claimed by the original cell, thus resulting in a contribution of −3 charge to the cell, resulting in a zero net charge for the cell.

For this structure to act as a photon absorber, it has been found that good absorption efficiency can be achieved by choosing an organic cation, such as methylammonium $(CH_3NH_3)^+$ to fill the A-position, lead Pb^{2+} for the B-positions, and a halide or halide combination, such as I^-, Cl^-, Br^-, or a combination of them, such as $xI^-(1-x)Cl^-$, for the X-positions [81]. The exact absorption process is still not well-understood [81,82]. For example, it has not yet been confirmed that the process involves creation of excitons, such as in organic devices, or EHPs, as in traditional devices. What is understood is that the sorts of charge selective interface layers as used in organic devices also act as efficient collectors of electrons and holes from the perovskite layer.

Figure 11.18 [81] shows one implementation of a perovskite cell. Depending upon specific cell components, the absorber layer can be a fraction of a μm thick and still harvest a large fraction of the incident photons. One group of researchers has found that electron–hole diffusion lengths can exceed 1 μm in an organometal trihalide perovskite absorber if the absorber is of the mixed halide triiodide variety [82]. With a perovskite layer thickness of ≈400 nm, this ensures that a very large fraction of carriers generated in the absorber layer will find their way to the charge selective interface layers, as opposed to traditional OPV cells that have very short diffusion lengths for charge carriers.

With lab cell efficiencies now exceeding 20% along with low temperature processing in air, this technology appears to be very promising for commercialization

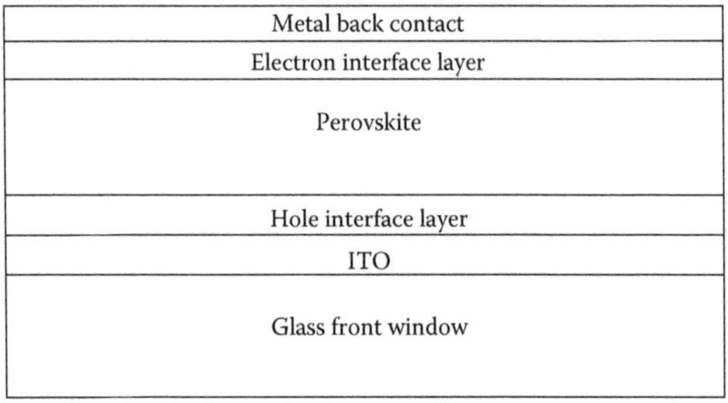

Metal back contact
Electron interface layer
Perovskite
Hole interface layer
ITO
Glass front window

Figure 11.18 Perovskite layers. (Adapted from Perovskites and perovskite solar cells: An introduction, www.ossila.com/pages/perovskites-and-perovskite-solar-cells-an-introduction (accessed June 2016).)

at some point in the near future, as soon as the production costs can be reduced to compare with existing commercial technologies and as soon as instabilities in the performance of the devices can be overcome. Significant work in underway to identify the source(s) of instability and some progress has been made. One group has reported that the device interfaces and mobile ions play a major role in light-induced instability, regardless of whether the cell structure is p–i–n or n–i–p [83]. The next step, of course, is to figure out how to stabilize these processes and to determine whether any other instabilities may remain.

11.6.4.9 Quantum Dot and Dye-Sensitive Solar Cells [79]

Worth mentioning are two additional thin-film solar cell technologies: quantum dot and dye-sensitized cells.

In a basic quantum dot structure, light impinges on a top TCO electrode, then a blocking layer, which is selected to attract electrons, then a TiO_2 film, then a quantum dot light absorbing layer of II–VI material, such as PbS, PbSe, CdS, or CdSe, then a hole transport layer, and a back electrode. EHPs are created in the quantum dot layer. As of 2016, the maximum certified cell efficiency was 11.3%. This technology is still in its infancy, with some enthusiasts noting that it is possible that a combination of quantum dot and perovskites may produce the PV source of the future. For the moment, the technology is at the lab cell level, with no modules reported to date. Long-term testing will also be needed to prove acceptable stability of the devices.

Dye-sensitized solar cells have a front TCO contact on top of TiO_2 particles that are coated with a dye. The light is absorbed by the dye, which, in turn, reacts with the TiO_2 to create EHPs. Differences in work functions then attract the electrons to the TCO and the holes to the counter (back) electrode. Maximum certified cell efficiency as of 2016 was 11.9%. Dye-sensitive PV cell work has been going on since the early 1990s.

The vision for quantum dot and dye-sensitive technologies is used in building-integrated applications (BIPV), because of the flexibility of the materials. However, it appears that both will remain under experimentation for the next few years.

11.7 NEW DEVELOPMENTS IN SYSTEM DESIGN

11.7.1 Micro Grids

Currently, utility-interactive inverters must incorporate anti-islanding algorithms so that they will shut down if the grid shuts down, regardless of whether other PV inverters are connected to the grid that might tend to fool the inverter into thinking they are the utility grid. But what if a situation might exist where it is important to maintain a PV island in the midst of a utility outage? One might imagine a remote town or village with a single feeder connection to the utility grid. If the grid goes down, should the entire town go down, or, alternatively, be transferred to individual or collective nonutility generation? What about setting up PV systems such that islanding is allowed for the systems in the town, but still disconnects the town or neighborhood from the utility feed until utility power is restored? The question, yet to be answered, includes determining the extent to which battery backup will be required, the means of connecting alternate, nonutility generation sources, whether inverters should act as current sources or, perhaps, voltage sources, and how to minimize any sort of system losses. Revisions to IEEE 1547 are now allowing ride through for certain situations.

Clearly, for islanding to be effective against a utility outage, adequate storage needs to be available. Perhaps just as important as maintaining power availability during a utility outage is having available stored energy to return to the grid to help smooth the utility demand curve while the grid is active, resulting in dual function of the storage capacity. This leads to the next topic, which deals with how to optimize grid performance.

11.7.2 Smart Grids

The best sources of renewable energy tend to be in places that are not densely populated, such as sunny and dry regions in Southwestern United States to windy regions along the coasts and in the region between Texas and the Dakotas. As a result, the ability to move electricity generated in these regions to highly populated areas is lacking. In order to achieve goals of 20% or 25% of all electricity used coming from renewable sources, it will be necessary to develop a grid structure that will enable efficient transmission of power from regions where renewable energy is most efficiently generated to regions where it is needed for consumption, which, in effect, means having more West–East grid connections. It has been reported that because of transmission congestion, consumers in Eastern United States pay $16.5 billion annually in higher electricity prices [84]. Furthermore, it is estimated that by updating our electricity infrastructure, Americans can save $80 billion in unnecessary power plant construction over the next 20 years [85].

Smart grids are a work in progress. Many of the characteristics that smart grids will need have been identified, such as updating regulations to discourage utilities from purchasing obsolete equipment, developing open standards, providing consumers with energy management tools and time-of-use pricing, and providing utilities with incentives for selling less electricity. Advanced load management to better conform load profiles to electric generation capacity as well as using consumer generation and storage and utility-scale storage to supplement grid needs will also likely be components of the smart grid. The grid will likely be able to communicate with plug-in electric vehicles such that the vehicles will be able to deliver power to the grid when the grid needs power. The electric vehicle batteries will charge when the grid has excess power, or, perhaps, when the vehicle need is given a higher priority.

Communications will be the heart of the smart grid, as generation and consumption capacity and needs will be able to communicate, with the result of more-efficient management of the transfer of power from generation to load. It is likely that many of the lessons learned from the Internet and cell phone networks will be applicable to the smart grid communication network, with the difference being that rather than multiplexing packets of information, large packets of power will be efficiently dispatched over extra-high-voltage transmission lines as well as over neighborhood distribution lines.

With advanced communications, the "smart garage" concept also becomes possible [86], in which storage methods in the garage, such as direct battery storage or the batteries of a battery-backup PV system, can dispatch power to the grid if excess energy is available and is requested by the grid.

Thus, as new ideas are brought to market in PV, wind, and other renewable sources, it is highly likely, if not absolutely certain, that communication with a smart grid will be an integral part of the design of the systems.

As more and more of the capacity of the proposed smart grid system is provided by PV and wind sources, each of which has its own elements of predictability and unpredictability, utilities are expressing concern over their responsibilities to the grid when renewable source availability is curtailed, such as late-afternoon utility peaking hours when PV output is beginning to decrease. It might seem obvious that energy storage must be the key to overcoming the unpredictability of PV and wind, but it may not be obvious where to incorporate the storage in the most effective manner.

A recent study has concluded that at least one well-suited location for storage is at customer locations where the customer is paying utility demand charges, particularly when the customer demand coincides with when the grid demand peaks [87]. The reason is that if a customer is able to significantly reduce their demand during utility daily peaks, their investment in storage will be at least partially offset by their savings in demand charges, in addition to their savings in utility energy charges, if time of use (TOU) rates apply. The higher the avoided demand charges, the quicker the payback on the storage investment. The study demonstrated that typical internal rates of return (IRR) in California, Michigan, and Massachusetts on storage investments by customers on demand rate schedules is approximately 10%. In New York, the IRR was closer to 20%.

In the future, even more attractive IRRs should be possible as the cost of storage continues to decrease. For the popular LFP cells, the cost per kWh of storage is expected to drop from $1000 in 2015 to $650 in 2020 [87]. Even though lead-acid storage costs are less, it appears that life-cycle costs for the LFP units may be more attractive. And, as discussed in Chapter 3, developments of other larger-scale storage technologies, such as flow cells, are advancing rapidly. At some point, it will likely make good economic sense for utilities to use retired fossil-fuel power plant sites as sites for massive energy storage, for which they will be able to include in their rate bases.

Another interesting alternative to physical or chemical storage is based upon the fact that in Eastern United States, when late-afternoon electric demand increases, say, at 5 p.m., PV systems in Western United States are still producing near-maximum power. In fact, if some of these PV arrays were to be west-facing, they would peak out even later in the day, meaning 1–3 hours later in Central and Eastern Time Zones. This suggests an argument for additional West–East utility transmission capacity to augment other storage methods. While the technical argument is relatively simple, the economics of such a proposal needs additional study.

11.7.3 Inverter Performance Enhancement

As utility-scale PV systems grow larger and larger, interesting questions arise over these systems. For example, how will the grid react if all of a sudden, clouds cause the output of a 50-MW system to drop within seconds to 10 MW or less. Or, for that matter, what if the opposite should happen? Will it be necessary to incorporate some sort of storage at the array output so the inverter output will be more stable when cloud cover is anything but stable? Will the grid have sufficient damping capability to prevent rapid PV power changes from affecting grid stability? In effect, this amounts to applying either an impulse or a step function excitation to the grid and then observing the response. Will the grid be smart enough to deal with fluctuations in supply and load?

Then the question arises as to whether an inverter should always be required to operate at unity power factor. Generally, large-scale utility rotating generation must supply reactive power as well as real power. As discussed in Chapter 4, IEEE 1547 is currently being revised to allow for inverters to operate at power factors different from unity, under control of the utility to which the inverter output is being delivered. Fortunately, it should be a straightforward exercise for an inverter to be programmed to determine the optimal phase angle for current injection into the grid. This would include having the inverter output supply leading power factor current to cancel out lagging power factor loads when such a power factor adjustment would lower utility line losses. Not only is this an interesting electrical issue, but it will probably also lead to some interesting billing questions, such as whether the PV generator should be credited for any role it may play in reducing utility line losses.

On a smaller scale, present dc-coupled battery-backup utility-interactive systems encounter greater losses than straight grid-connected systems. Improving the overall electrical throughput efficiency of these systems will also present a challenge to the designers of inverters and charge controllers. The competition, of course, lies in ac-coupled systems that reduce losses to battery charging and discharging, not unlike the race between crystalline silicon and thin-film technologies for the most cost-effective life-cycle energy production.

11.7.4 Module Performance Enhancement

Over the years, a number of creative ideas have been incorporated into increasing module performance. In particular, the problem of partial shading of modules has been addressed in a number of ways, and there is no doubt that additional creative ideas will be applied to this challenge. The partial shading problem is exacerbated by the fact that typically, cells and modules have been connected in series. If PV cells acted as voltage sources, this would not necessarily be a major item of concern, because of the series connection. However, the maximum power point of cells is generally at the transition in the performance curve between close to ideal current source and close to ideal voltage source.

While the traditional solution to this problem has been the incorporation of bypass diodes into the module internal interconnections, other ideas have also been tried, such as hooking up the cells in parallel and then using a boost dc–dc converter to increase the voltage at the module connection point. This is an interesting concept, since the voltage could be boosted to a fixed figure that would allow for parallel connection of modules acting as voltage sources or, if the module output current were to be controlled, then the modules could be connected in series, with each module contributing exactly the same current to the array and each module having its own internal MPPT function.

A method recently reported [88] involved replacing bypass diodes with differential power processors (DPPs), such that the voltage across a set of series-connected cells would not be reduced to the forward-biased voltage of the bypass diode. Instead, the DPPs would adjust the voltage of the shaded cell string such that the current of this string would match the current of the unshaded sections of the module. Simulations showed definite improvement over bypass diodes, provided that the cells associated with each internal DPP were carefully defined.

11.8 SUMMARY

Regardless of the technology or technologies that may result in low-cost, high-performance PV cells, it must be recognized that the life-cycle cost of a cell depends on the cells having the longest possible, maintenance-free lifetime. Thus, along with the developments of new technologies for absorbers, development of reliable encapsulants and packaging for the modules will also merit continued research and development activity.

Every year engineers make improvements on products that have been in existence for many years. Automobiles, airplanes, electronic equipment, building materials, and many more common items see improvements every year. Even the yo-yo, a popular children's toy during the 1940s and 1950s, came back with better-performing models. Hence, it should come as no surprise to the engineer to see significant improvements and scientific breakthroughs in the PV industry well into this millennium. The years ahead promise exciting times for the engineers and scientists working on the development of new PV cell and system technologies, provided that the massive planning and execution phases can be successfully undertaken.

PROBLEMS

11.1 Assume a PV module is to have dimensions of 37.5 inch × 63.2 inch. Also assume that 8-inch round cells are available.
   a. Calculate the percentage of the module area that can be covered with circular cells.
   b. If the cells operate with 18% efficiency and if there is no mismatch loss among cells, calculate the overall module efficiency if the round cells are used.
   c. Assume the cells are "squared up" by sawing off the edges to obtain cells that measure 6 inch between the straight sides. Calculate the percentage fill of these "squared up" cells and repeat the module efficiency calculation.
   d. What is the percentage loss of material in the "squaring up" process?

11.2 Determine the expression for the depth of the junction after the drive-in diffusion step.

11.3 Sketch the impurity distribution profile between the junction and the back contact of a single-crystal silicon cell to show how the annealed aluminum of the back contact creates an accelerating E-field.

11.4 What volume and weight of tellurium is needed to produce a square meter of CdTe thin film with a thickness of 1.0 μm? Assume the Cd and Te occupy equal volumes within the film.

11.5 What volume and weight of indium is required to fabricate a 1-MW PV CIS array if the CIS layer thickness is 1.0 μm, assuming that 18% of the layer volume is due to the In. Assume an array efficiency of 17% and standard test conditions.

11.6 The energy payback time (EPBT) for a PV technology is defined as the time needed for the completed cell to generate an amount of electrical energy equal to the energy required for producing the cell. Refer to current literature and tabulate the EPBT for the most efficient crystalline Si cells, multicrystalline Si cells, thin Si (microcrystalline or nanocrystalline) cells, and amorphous Si cells.

11.7 Calculate or look up the EPBT for CIGS cells, CdTe cells, and GaAs or other III–V cell. If you do the calculation, list your assumptions.

REFERENCES

1. Green, M., Third generation photovoltaics: Assessment of progress over the last decade, In *Proc. 34th IEEE PV Spec. Conf.*, 2009.
2. National Renewable Energy Laboratories, Best Research-Cell Efficiencies, lab cell efficiency record history for all technologies up to 2016, http://www.nrel.gov/ncpv/images/efficiency_chart.jpg.
3. Tarrant, D. E. and Gay, R. R., *Research on High Efficiency, Large Area CuInSe2-Based Thin Film Modules, Final Technical Report*, NREL, Golden, CO, April 1995.
4. Fthenakis, V., Recycling workshop 2, In *Proc. 34th IEEE PV Spec. Conf.*, 2009.
5. *World Statistics, Nuclear Energy around the World*, NEI Knowledge Center, Nuclear Energy Institute, Washington, DC, July 2015, http://www.nei.org/Knowledge-Center/Nuclear-Statistics/World-Statistics.
6. Statista, The statistics portal, http://www.statista.com/statistics/217256/global-installed-coal-power-generation-capacity/. The intro page shows projected 2015 coal-generated electricity capacity.
7. Statista, The statistics portal, http://www.statista.com/statistics/217252/global-installed-power-generation-capacity-of-natural-gas/. The intro page shows projected 2015 natural-gas-generated electricity capacity.
8. Beiter, P., US DOE, NREL, EERE, Global renewable energy, all forms, http://www.nrel.gov/docs/fy16osti/64720.pdf.
9. The Shift Project Data Portal, http://www.tsp-data-portal.org/Historical-Electricity-Generation-Statistics#tspQvChart.
10. Markvart, T. Ed., *Solar Electricity*, John Wiley & Sons, Chichester, UK, 1994.
11. http://www.pveducation.org/pvcdrom/manufacturing/refining-silicon (accessed July 2016).
12. Xakalashe, B. S. and Tangstad, M., Silicon processing: From quartz to crystalline silicon solar cells, In *Southern African Pyrometallurgy*, Vol. 83, March 2011, Ed. R. T. Jones and P. den Hoed, http://pyrometallurgy.co.za/Pyro2011/Papers/083-Xakalashe.pdf.

13. Streetman, B. G., *Solid State Electronic Devices*, 4th Ed., Prentice Hall, Englewood Cliffs, NJ, 1995.
14. Bohland, J., Accelerated aging of PV encapsulants by high intensity UV exposure, In *Proc. 1998 Photovoltaic Performance and Reliability Workshop*, Cocoa Beach, FL, November 3–5, 1998.
15. Sunpower discovers the "Surface Polarization Effect" in high efficiency solar cells, Sunpower Corp Internal Document, personal communication with SunPower representative, May 2009.
16. Kim, D. and Bunea, G., *The Journey of Sunpower to 20% Module Efficiency*, SunPower Corporation, 2011, http://www.avsusergroups.org/joint_pdfs/2012_2Kim.pdf.
17. DeGraaff, D., *Update: SunPower Solar Panel Sets New World Record for Efficiency*, SunPower Blog, Oct 8, 2015, http://us.sunpower.com/blog/2015/10/08/sunpower-offers-customers-most-efficient-solar-panels-commercially-available-today/.
18. DelleDonne, E. et al., In *Proc. 29th IEEE PV Spec. Conf.*, 2002, 82–85.
19. Basore, P.A., In *Proc. 29th IEEE PV Spec. Conf.*, 2002, 4952.
20. Brendel, R. et al., In *Proc. 29th IEEE PV Spec. Conf.*, 2002, 86–89.
21. Guha, S. et al., In *Proc. 26th IEEE PV Spec. Conf.*, 1997, 607–610.
22. Huang, J. et al., In *Proc. 26th IEEE PV Spec. Conf.*, 1997, 699–702.
23. Guha, S. and Yang, J., In *Proc. 29th IEEE PV Spec. Conf.*, 2002, 1070–1075.
24. Yang, J. et al., In *Proc. 26th IEEE PV Spec. Conf.*, 1997, 563–568.
25. Standen, A., Executive Ed., *Kirk-Othmer Encyclopedia of Chemical Technology*, 2nd Ed., Vol. 10, John Wiley & Sons, New York, 1968.
26. Standen, A., Executive Ed., *Kirk-Othmer Encyclopedia of Chemical Technology*, 2nd Ed., Vol. 2, John Wiley & Sons, New York, 1968.
27. Williams, R., *Modern GaAs Processing Methods*, Artech House, Norwood, MA, 1990.
28. Alta Devices, Sunnyvale, CA, http://www.altadevices.com/technology-overview/.
29. Lammasniemi, J. et al., In *Proc. 26th IEEE PV Spec. Conf.*, 1997, 823–826.
30. Hoffman, R. et al., In *Proc. 26th IEEE PV Spec. Conf.*, 1997, 815–818.
31. Kazmerski, L. L., In *Proc. 29th IEEE PV Spec. Conf.*, 2002, 21–27.
32. Brown, M. R. et al., In *Proc. 26th IEEE PV Spec. Conf.*, 1997, 805–810.
33. Shay, J. L., Wagner, S., and Kasper, H. M., *Appl. Phys. Lett.*, 27(2), 1975, 89.
34. Kamada, R. et al., In *Proc 43rd IEEE Photovoltaic Spec Conf.*, Solar Frontier K. K., Atsugi, Kanagawa, Japan, 2016.
35. Optimized Photovoltaic Systems for Commercial Rooftops, http://www.solyndra.com/Products/Optimized-PV.
36. Standen, A., Executive Ed., *Kirk-Othmer Encyclopedia of Chemical Technology*, 2nd Ed., Vol. 6, John Wiley & Sons, New York, 1968.
37. Standen, A., Executive Ed., *Kirk-Othmer Encyclopedia of Chemical Technology*, 2nd Ed., Vol. 11, John Wiley & Sons, New York, 1968.
38. Standen, A., Executive Ed., *Kirk-Othmer Encyclopedia of Chemical Technology*, 2nd Ed., Vol. 17, John Wiley & Sons, New York, 1968.
39. *The Encyclopedia Brittanica*, 15th Ed., Encyclopedia Brittanica, Inc., Chicago, IL, 1997.
40. Standen, A., Executive Ed., *Kirk-Othmer Encyclopedia of Chemical Technology*, 2nd Ed., Vol. 19, John Wiley & Sons, New York, 1968.
41. Standen, A., Executive Ed., *Kirk-Othmer Encyclopedia of Chemical Technology*, 2nd Ed., Vol. 13, John Wiley & Sons, New York, 1968.
42. Konagai, M., In *Proc. 29th IEEE PV Spec. Conf.*, 2002, 38–43.

43. Woodhouse, M. et al., *On the Path to Sunshot, The Role of Advancements in Solar Photovoltaic Efficiency, Reliability, and Costs*, National Renewable Energy Laboratory, Golden, CO, NREL/TP-6A20-65872, http://www.nrel.gov/docs/fy16osti/65872.pdf.
44. Kamada, R. et al., In *Proc. 43rd IEEE PV Spec Conf.*, 2016.
45. Tarrant, D. E. and Gay, R. R., *Thin-Film Photovoltaic Partnership Program—CIS-Based Thin Film PV Technology*, Phase 2 Technical Report, October 1996–October 1997, NREL Report NREL/SR-520-24751, Golden, CO, May 1998.
46. Kazmerski, L. L., In *Proc 29th IEEE PV Spec. Conf.*, 2002, 21–27.
47. Fthenakis session on recycling First Solar person at *Proc 34th IEEE PV Spec. Conf*, 2009, Fthenakis presentation.
48. Bohland, J. et al., In *Proc. 26th IEEE PV Spec. Conf.*, 1997, 355–358.
49. Ullal, H. S., Zweibel, K., and von Roedern, B., In *Proc. 26th IEEE PV Spec. Conf.*, 1997, 301–305.
50. Wu, X. et al., In *Proc. 26th IEEE PV Spec. Conf.*, 1997, 347–350.
51. McCandless, B. E. and Birkmire, R. W., In *Proc 26th IEEE PV Spec. Conf.*, 1997, 307–312.
52. Chen, Y. et al., In *Proc. 43rd IEEE PV Spec. Conf.*, 2016.
53. Levi, D. H. et al., In *Proc. 26th IEEE PV Spec. Conf.*, 1997, 351–354.
54. Birkmire, R. W., In *Proc. 26th IEEE PV Spec. Conf.*, 1997, 295–300.
55. Zhou, C. Z. et al., In *Proc. 26th IEEE PV Spec. Conf.*, 1997, 287–290.
56. *Fraunhofer Institute for Solar Energy Systems, ISE, with support of PSE AG*, Photovoltaics Report, June 6, 2016, https://www.ise.fraunhofer.de/de/downloads/pdf-files/aktuelles/photovoltaics-report-in-englischer-sprache.pdf.
57. Krygowski, T., Rohatgi, A., and Ruby, D., In *Proc 26th IEEE PV Spec. Conf.*, 1997, 19–24.
58. Slaoui, A. et al., In *Proc. 29th IEEE PV Spec. Conf.*, 2002, 90–93.
59. Metz, A. et al., In *Proc. 26th IEEE PV Spec. Conf.*, 1997, 31–34.
60. Schönecker, A. et al., In *Proc. 29th IEEE PV Spec. Conf.*, 2002, 106–109.
61. Kuznicki, Z. T., In *Proc. 26th IEEE PV Spec. Conf.*, 1997, 291–294.
62. Yamamoto, K. et al., In *Proc. 29th IEEE PV Spec. Conf.*, 2002, 1110–1113.
63. HIT Double™ Bifacial Photovoltaic Modules data sheet download from http://sanyo.com/solar/.
64. Brendel, R. et al., In *Proc. 29th IEEE PV Spec. Conf.*, 2002, 86–89.
65. Granata, J., Sites, J. R., and Tuttle, J. R., In *Proc. 14th NREL PV Program Review, AIP Conf. Proceedings 394*, Lakewood, CO, 1996, 621–630.
66. Tarrant, D. E. and Gay, R. R., *Research on High Efficiency, Large Area CuInSe2 Modules, Final Technical Report*, NREL, Golden, CO, April 1995.
67. Weiting, R., In *Proc. 29th IEEE PV Spec. Conf.*, 2002, 478–483.
68. Jaakkola, R. et al., In *Proc. 26th IEEE PV Spec. Conf.*, 1997, 891–894.
69. Sharps, P. R. et al., In *Proc. 26th IEEE PV Spec. Conf.*, 1997, 895–898.
70. Barnham, K. W. J. and Duggan, G., A new approach to high-efficiency multi-band-gap solar cells, *J. Appl. Phys.*, 67, 1990, 3490–3493.
71. Corkish, R. and Honsberg, C., In *Proc. 26th IEEE PV Spec. Conf.*, 1997, 923–926.
72. Khvostikov, V. P. et al., In *Proc. 29th IEEE PV Spec. Conf.*, 2002, 943–946.
73. Dupre, O., Vaillon, R., and Green, M., In *Proc. 43rd Photovoltaic Spec. Conf.*, 2016.
74. Scranton, G. et al., In *Proc. 43rd IEEE Photovolataic Spec Conf.*, 2016.
75. Luque, A. et al., In *Proc. 29th IEEE PV Spec. Conf.*, 2002, 1190–1193.
76. Green, M. A., In *Proc. 29th IEEE PV Spec. Conf.*, 2002, 1330–1334.
77. Zahler, J. M. et al., In *Proc. 29th IEEE PV Spec. Conf.*, 2002, 1029–1032.

78. University of Houston, Institute for Nanotechnology, Organic Solar Cells, http://ine.uh.edu/research/organic-solar-cells/index.php, 2016.
79. Jacoby, M., The future of low cost solar cells, *Chem. Eng. News*, 94(18), 2016, 30–35.
80. Philipps, S.[1], Bett, A.[1], Horowitz, K.[2], and Kurtz, S.[2], Current Status of Concentrator Photovoltaic (CPV) Technology, Version 1.2, February, 2016. [1]Fraunhofer Institute for Solar Energy Systems ISE in Freiburg, Germany; [2]National Renewable Energy Laboratory (NREL) in Golden, CO.
81. Perovskites and perovskite solar cells: An introduction, www.ossila.com/pages/perovskites-and-perovskite-solar-cells-an-introduction (accessed June 2016).
82. Stranks, S. D. et al., Electron-hole diffusion lengths exceeding 1 micrometer in an organometal trihalide perovskite absorber, *Science*, 342(6156), 2013, 341–344.
83. Dalal, V. et al., In *Proc 43rd Photovoltaic Spec Conf.*, 2016.
84. Hendricks, B., Wired for progress, *Solar Today*, 23(4), 2009, 26, solartoday.org.
85. Berst, J., For a smart grid, look to smart states, *Solar Today*, 23(4), 2009, 30, solartoday.org.
86. Burns, C. M., The smart garage (V2G): Guiding the next big energy solution, *RMI Solut. J.*, 1(1), 2008, 22.
87. Halbe, A., Yeaman, D. and Ellington, D., In *Proc. 43rd IEEE Photovoltaic Spec. Conf.*, 2016.
88. Galtieri, J. and Krein, P., In *Proc. 43rd IEEE Photovoltaic Spec. Conf.*, 2016.

Appendix: Design Review Checklist

The purpose of this textbook has been to prepare the engineer for the design of PV systems. The purpose of this Appendix is to provide a checklist against which the system designer can verify that sufficient information has been provided in the design to meet the needs of whoever is given the tasks of design review, installation, inspection, operation, and maintenance of the designed system. The following format, based on recommendations by the Florida Solar Energy Center and local building inspection officials, is used by VB Engineering, Inc., for preparation of permit application packages for design review and permitting. If design projects are assigned for a course that uses this book, it is recommended that the designs contain the information listed in this appendix.

RECOMMENDED INFORMATION TO SUBMIT FOR PV INSTALLATION PERMIT

1. Permit application forms for the code jurisdiction issuing the installation permit.
2. Notice of Commencement and Release of Lien (Florida requirement; check local requirements).
3. System description (about 1 letter-size page long), including
    a. Whether the system is utility interactive
    b. Whether the system has batteries
    c. DC array rating
    d. Number of source circuits
    e. Number of modules per source circuit
    f. Number of inverters, inverter ratings, and inverter compliance with UL 1741 if applicable
    g. Whether fossil-fuel generator is included in system
    h. Nominal battery storage in kWh
    i. System block diagram showing power flows
    j. Brief description of how system operates
    k. Statement of expected system performance, for example, average kWh/day from NREL SAM
4. Site plan showing locations of all new equipment, existing electrical service equipment, and point of utility connection.
5. Electrical riser diagram showing how the new PV system connects into the existing service.
6. Standby load summary if battery-backup system.
7. Electrical schematic diagram(s) showing all wire sizes and types, junction boxes, conduit sizes, disconnects, overcurrent protection, and major equipment wiring connections.
8. Certification of component listings (UL, CE, etc.), including manufacturers' data sheets for all major equipment, such as modules, array mounts, charge controllers, inverters, and batteries.
9. Site information for wind loading, including wind load in psf.

10. Structural drawings showing means of attachment of PV array to roof, if roof mounted, and means of attachment of array to other structures if not roof mounted.
11. Electrical calculations showing ampacity, temperature ratings, conduit fill deratings, voltage drop calculations for all dc and ac conductors, and ambient temperature effects on PV array output voltage. Sizes of disconnects and overcurrent protection should also be included, as well as input and output voltages and currents and voltage and power specifications for all major pieces of equipment.
12. Structural calculations to show necessary screw thread penetrations into trusses for roof-mounted arrays, along with calculated maximum pull on attachment points and safety factors included in calculations. If other than roof mount, show calculations to indicate that the structure can withstand all forces that are significant in the area, such as wind, snow, seismic, etc.
13. Code references for electrical and structural calculations (*NEC*, ASCE 7, etc.) and code references relating to the point of utility connection and instructions for labeling.
14. Certification that array layout meets local fire code requirements. In fact, it is a good idea to have information on local fire code requirements prior to commencement of array design, so the design will not need to be altered for fire code compliance after completion of the design.

Index

A

AA, *see* Aluminum Association
Absorbed glass mat batteries (AGM batteries), 63, 238
Absorption stage, 81
AC,
    disconnects and overcurrent protection, 142–143, 151–152
    inverter power, 133
    voltage sources, 173–174
Acceptor, 405
AC-coupled battery-backup systems, 280
    120-V battery-backup inverter with 240-V straight grid-connected inverter, 283–285
    120/208-V three-phase AC-coupled system, 285–286
    120/240-V battery-backup inverter, 281–283
    PV systems, 235–236
Accumulation, 8–9
Acid rain, 380, 384, 386
Active trackers, 220
AC wiring,
    voltage drop calculations, 162–163
    wire and overcurrent protection sizing, 161–162
    wire sizing, disconnects, and overcurrent protection, 171–172
Aerodynamic forces, 204
Aerodynamic wind loading, 203
Aesthetics, 125–126, 220–221
AFC, *see* Alkaline fuel cell
AFCI, *see* Arc fault circuit interrupter
AGM batteries, *see* Absorbed glass mat batteries
AGS, *see* Automatic generator start
Ah, *see* Ampere-hours
Air mass = 1 (AM 1), 26
Air pollution,
    Clean Air Act and EPA, 382–383
    greenhouse gases and effect, 383–384
Air quality, 383, 387–388
AISC, *see* American Institute of Steel Construction
AISI, *see* American Iron and Steel Institute
Albedo radiation, 25
ALCC, *see* Annualized LCC
ALE, *see* Atomic layer epitaxy
Alkaline fuel cell (AFC), 71
Alloy steel, 195

Altitude effects, 101
Aluminum, 196–198, 442, 444
    oxide, 116, 197
    wire, 115
Aluminum Association (AA), 201
AM 1, *see* Air mass = 1
American Institute of Steel Construction (AISC), 201
American Iron and Steel Institute (AISI), 201
American Society of Civil Engineers (ASCE), 127, 201, 202, 207–208
American Society of Mechanical Engineers (ASME), 202
American Society of Metals (ASM), 201
American Society of Testing and Materials (ASTM), 202
Amorphous material, 418
Amorphous silicon cells (a-Si cells), 436, 448; *see also* Single-crystal silicon cells
    a-Si:H system, 448–449
    cell performance, 451
    fabrication, 449–451
Ampere-hours (Ah), 60, 244
Amplitude stability, 101–102
Analemma, 30–31
Annealing process, 442
Annualized LCC (ALCC), 364–365
Annual payments on borrowed money, 367
    breakdown of portions of loan payment, 368
    interest payment, 369
    programmable calculator, 370
Anode, 311
    anodic metal, 193
    resistance to ground in standard 1000 Ω-cm soil, 313
Anti-islanding algorithms, 475–476
Antireflective coating (ARC), 416, 444–445
APCVD, *see* Atmospheric pressure chemical vapor deposition
Applied Energy Service, 387
Aquifer pollution, 384–385
ARC, *see* Antireflective coating
Arc discharge, 106
Arc fault circuit interrupter (AFCI), 97, 105, 134
Arc fault protection, 105–106, 142, 151
Arc flash calculations,
    AC voltage sources, 173–174
    DC current sources, 174–176
    source of arc, 172

487

Array
  cooling process, 214
  design, 339–341
  mounting equipment, 246–247
  power rating, 303
  selection, 145–149
  tilt, 328–330, 351–352
Array layout, 166–168, 270, 272–273
  array performance, 273–275
  flat-roofed buildings, 270–271
Array mounting system design, 210
  aesthetics, 220–221
  array cooling process, 214
  building integration considerations, 212–213
  costs and durability of array-roof configurations, 213
  designing, 210
  enhancing array performance, 213
  ground-mounted arrays, 218–220
  high wind speed case, 227–229
  irradiance enhancement, 213–214
  lower wind speed case, 229–231
  minimizing installation costs, 210–212
  protection from vandalism, 214–215
  roof-mounted arrays, 215–218
  shading, 214
Array sizing, 132–133, 164–165, 239–240, 244–245, 320–321, 328–330
  determination, 308–309
  procedure, 351–352
Arsenic (As), 452–453
ASCE, *see* American Society of Civil Engineers
ASCE 7-10, 224
ASCE 7 wind load analysis tabular method, 223
  analytical procedures, 224–225
  ASCE 7-10, 223–224
  roof zones, 225
  table look-up method, 225–227
  wind pressures for exposure category, 226
a-Si cells, *see* Amorphous silicon cells
ASM, *see* American Society of Metals
ASME, *see* American Society of Mechanical Engineers
Asphalt enamels, 196
Associated system electronic components, 78; *see also* Photovoltaic systems (PV systems)
  charge controllers, 78–82
  inverters, 86–97
  LCBs, 82–86
  MPPTs, 82–86
ASTM, *see* American Society of Testing and Materials
Atmosphere effect on sunlight, 25–27
Atmospheric concentration, 386
Atmospheric pressure chemical vapor deposition (APCVD), 464
Atomic layer epitaxy (ALE), 464
Automatic generator start (AGS), 270
Average daily PV system load, 350
Average power, determination of available, 316–317
Azimuth angle, 31, 32, 39–40

B

Back shields, 214–215
Balance of system (BOS), 104, 136, 149, 303–304; *see also* BOS components
  DC and AC disconnects and overcurrent protection, 142–143, 151–152
  estimating system annual performance, 152
  final system electrical schematic diagram, 144–145, 153
  ground fault and arc fault protection, 142
  point of utility connection, 143–144, 152
  rapid shutdown, 141–142
  rapid shutdown, ground fault, and arc fault protection, 151
  wire and conduit from rooftop junction box to inverter, 138–141
  wiring from array to rooftop junction box, 137–138, 149–151
Ballpark accuracy, 205
Battery
  capacity of, 60
  connections, 286–291
  C–Zn batteries, 57
  requirements, 317
  sizing, 238–239, 318–320
  technologies, 68–70
  voltage, 302
Battery-backup grid-connected PV systems, 235, 295
  AC-coupled battery-backup PV systems, 235–236
  AC-coupled battery-backup systems, 280–286
  array sizing, 239–240
  battery-backup design basics, 237
  18-kW batter y-backup system, 267–280
  120/240-V battery-backup system, 257–267
  inverter sizing, 238
  load determination, 237–238
  single inverter 120-V battery-backup system, 241–257
Battery-to-inverter current, 332, 348
Battery charging

INDEX

489

efficiency factor, 313
process, 254
Battery connections, 286
  lead-acid connections, 286–291
  other battery systems, 291
Battery management system (BMS), 65, 338
Battery selection, 242–244, 260–261, 275, 302, 337–338
  procedure, 351
  PV-powered mountain cabin, 327–328
Battery storage
  mechanism, 49
  requirements, 308
  system, 304
Bauxite, 452
Bayer process, 452
Beam radiation, 25
Bending stresses, 189
BIPV, *see* Building integrated photovoltaics
BIPVs, *see* Integrated mounting arrays
Black-and-white pyranometer, 35
Blackbody radiators, 306
BMS, *see* Battery management system
Boltzmann's constant, 402
Borrowing money, 367
  effect of borrowing on LCC, 370–371
  determination of annual payments on borrowed money, 367–370
BOS, *see* Balance of system
BOS components, 103, 107, 113; *see also* Balance of system (BOS)
  ground fault, arc fault, surge, and lightning protection, 105–106
  grounding devices, 107
  inverter bypass switches, 106
  items, 333–334
  of PV system, 103–104
  rapid shutdown, 107–108
  selection, 321–322, 332–334, 348
  source circuit combiner boxes, 106–107
  switches, circuit breakers, fuses, and receptacles, 104–105
  wire, circuit breaker, and switch selection, 332–333
BOS selection and completion of design, 246, 262, 275; *see also* Balance of system (BOS)
  array mounting equipment, 246–247
  equipment grounding conductor, 253, 266
  grounding electrode conductor, 253, 266
  rapid shutdown system, 266–267
  rooftop junction box, 247–248, 262–263
  source-circuit combiner box and surge arrestor, 248, 263

wire and circuit breaker sizing—AC side, 251, 265
wire and circuit breaker sizing—DC side, 248–251, 263–264
wiring of standby loads, 251–253, 265–266
Bridgman method, 463
BTU economics, 12–13
Buck–boost converter, 85
Buckling, 222–223
Building codes, 202
  requirements, 202–203
Building integrated photovoltaics (BIPV), 126, 217–218, 471–472
  applications, 475
  products, 212
Bulk voltage, 81
Bypass diodes, 54
to protecting shaded cells, 55

C

Cadmium (Cd), 458–459
Cadmium sulfide (CdS), 403
Cadmium telluride (CdTe), 400, 436
Cadmium telluride cells (CdTe cells), 391, 462
  of pure tellurium production, 462–463
  performance, 464–465
  production, 463–464
California Energy Commission (CEC), 134
California Hazardous Waste Control Law, 393
Carbon,
  capture technology, 438
  steel, 195
Carbon dioxide (CO_2), 25, 379, 381, 384, 390
  cost, 385–386
  emissions, 381, 383
  sequestering with trees, 386–387
Cascaded AlInP/GaInP/GaAs cell, 454, 455
Cast alloys, 197
Cathode, 311
Cathodic protection system, 296, 311; *see also* Refrigeration system
  flow of current and charge carriers in electrolyte system, 311
  system design, 312–315
CdTe cells, *see* Cadmium telluride cells
CEC, *see* California Energy Commission
Cell
  cell resistance losses, minimizing, 424–426
  fill factor, 52
  maximum power, 52
  power, 52
  variation of cell electrolyte, 63

CFCs, *see* Chlorofluorocarbons
Charge controllers, 78, 245–246, 296
    charging considerations, 79–82
    discharging considerations, 82
    and inverter selection, 309
    programming, 253–256
    PV-powered mountain cabin, 330
    refrigeration system, 321
    selection, 303, 309, 344–346
    source-circuit combiner box and selection, 258–260
    Thevenin equivalent circuit for battery, 79
Chemical corrosion, 193–195
Chloride slag, 458
Chlorofluorocarbons (CFCs), 381
CIGS cells, *see* Copper indium gallium diselenide cells
Circuit breakers, 104–105
    AC side, 251, 265
    DC side, 248–251, 263–264
    selection, 332–333, 346–348
CIS, *see* Copper indium diselenide
Clean Air Act, 382–383
"Climate deniers," 380–381
Close-spaced sublimation (CSS), 464
Coal-tar, 196
Code compliance, 131
Codes and standards, 114, 201–202
    aesthetics, 125–126
    electromagnetic interference, 126
    IEEE standard 1547-2003, 119–125
    NEC, 114
    other issues, 125
    structural considerations, 127–128
    surge protection, 126–127
Coefficient of utilization (CU), 306
Color temperature of light, 305
Column buckling, 190–191
Combiner boxes, 165–166
Commodities, attainment levels as, 387–388
Comparator output, 93
Compound interest, 6–7
Computational methods of determining shading, 40–41
Computer programs, 296
Concentrating PV system (CPV system), 366
    CPV cells, 472–473
Conductivity of material, 442
Conductors
    equipment grounding, 253, 266
    grounded coductors, 105, 140
    grounding, 105, 140, 322
    grounding electrode, 253, 266
    properties of copper, 116

Connected load, 318
Constraints, 200
Consumer price index, 357, 358
Contact(s), 441
    annealing process, 442
    conductivity of material, 442
    high-efficiency Si cell, 445–447
    material, 425
    power losses in, 444
    resistance between bulk and contacts, 443
    resistance per square, 443
    worst-case resistance, 443–444
Copper (Cu), 457–458, 463
    copper conductors, properties of, 116
Copper indium diselenide (CIS), 400, 456
    cells, 391, 394
    CIS-family-based absorbers, 467–468
    fabrication of CIS cells, 459–460
Copper indium gallium diselenide cells (CIGS cells), 436, 456
    cell performance, 460–462
    CIGS technology, 459–460
    commercial CIGS modules, 457
    fabrication of CIS cell, 459–460
    production of pure cell components, 457–459
    Siemens monolithic method of series cell connection, 461
Corona discharge, 106
Corrected load, 318, 319
Corrected load energy, 340
Corrosion
    crevice, 194
    dry, 195
    erosion, 194
    Galvanic, 193–194
    intergranular, 194
    stress corrosion cracking, 195
Cost constraints, 200
Cost quantification of externalities,
    attainment levels as commodities, 387–388
    cost of CO_2, 385–386
    sequestering CO_2 with trees, 386–387
    subsidies, 388–389
CPV system, *see* Concentrating PV system
Crevice corrosion, 194
Critical loads, 73
Crystalline cells, 435–436
Crystalline gallium arsenide, 453
Crystalline silicon on glass (CSG), 437, 448
CSG, *see* Crystalline silicon on glass
CSS, *see* Close-spaced sublimation
CU, *see* Coefficient of utilization

Cumulative present worth factor, 359
Curtain-wall glazing techniques, 217
C–Zn batteries, 57
Czochralski method, 463

D

Data monitoring, 131
Days of autonomy, 75–76
DC
    current sources, 174–176
    disconnects and overcurrent protection, 142–143, 151–152
    splicing mechanism, 247–248
    wiring, 159–161
DC/AC ratio, 240
DC-coupled battery-backup, systems, 280–281
    utility-interactive system, 478
DC-coupled systems, 235–236
Dead loads, 204
Declination, 29
Decomposition, 452
Deep-discharge lead-acid batteries, 62
Depletion region, 410
Design review checklist, 485–486
Dezincification, 194
DH, *see* Hours of daylight
Diesel fuel, 359
Differential power processors (DPPs), 479
Diffusion, 406–407, 440
    of contaminants, 465
    diffused junction, 407
    diffuse radiation, 25
    equation, 439
Direct beam normal incidence (DNI), 366
Direct conversion of sunlight to electricity, 13–15
Direct mounting arrays, 218
Direct radiation, 25
Direct subsidy, 388
Discharge rate, 60
Disconnect selection, 346–348
Discount rate (DR), 356, 365
DNI, *see* Direct beam normal incidence
Donor atoms, 404
Donor impurities in Si, 405
Doubling time, 7–8
DPPs, *see* Differential power processors
DR, *see* Discount rate
Drift, 406–407
Drive-in diffusion, 440–441
Dry corrosion, 195
Dye-sensitive solar cells, 475

E

Earth
    orbit and rotation of, 29–31
    solar constant for, 28
Economic considerations, 355
    borrowing money, 367–371
    externalities, 372–375
    LCC, 356–367, 368
    payback analysis, 371–372, 373
ED, *see* Electrodeposition
Eddy current losses, 318
Effective wind area (EWA), 227, 228, 230
EHPs, *see* Electron–hole pairs
18-kW battery-backup system; *see also* 120/240-V battery-backup system; Single inverter 120-V battery-backup system
    battery and BOS selection, 275
    charge controllers, 270
    final design, 277–280
    inverter selection, 268–270
    using inverters in parallel, 267
    module selection and array layout, 270–275
    roof, 267–268
    standby loads, 268
    wire sizing, 275–277
Einstein relationship, 418
Einstein's formula, 23–24
Elastic limit, 187
Electrical
    engineering, 379
    generating system, 365
    load, 99–100
    production, 331–332
Electrochemical stack, 68
Electrodeposition (ED), 460
Electrolysis of water, 70
Electromagnetic field theory, 416
Electromagnetic interference (EMI), 119, 126
Electromotive force (emf), 124
Electron–hole pairs (EHPs), 50–51, 399, 403
    generation by photon absorption, 399–401
Electronic semiconductors, 439
Electroplating system, 311
ELO step, *see* Epitaxial lift-off step
emf, *see* Electromotive force
EMI, *see* Electromagnetic interference
Emitter wrap-through (EWT), 466
Endangered Species Act, 393
Energy
    demand, 2
    density of hydrogen, 70
    units, 15–17

Energy sources, environmental effects of, 381
  air pollution, 382–384
  health and safety as externalities, 389
  infrastructure degradation, 385
  quantifying cost of externalities, 385–389
  renewable and nonrenewable energy sources, 382
  soil pollution, 384–385
  water pollution, 384–385
Energy storage, 57
  battery technologies, 68–70
  fuel cell, 71
  hydrogen storage, 70–71
  lead-acid storage battery, 58–64
  lithium-ion battery technologies, 64–66
  nickel-based battery systems, 66–68
  storage options, 71–72
  types of rechargeable batteries, 57
Energy use patterns
  2013 worldwide per capita GD P *vs.* per capita kW, 4
  current world, 2
  distribution of energy users (2013), 4
  energy equity, 4–5
  free-trade agreements, 5
  global energy production mix (2013), 3
  growth of worldwide energy production, 3
Environmental Protection Agency, 374
EPA, *see* U.S. Environmental Protection Agency
Epitaxial growth, 454–455, 467
Epitaxial lift-off step (ELO step), 453
Equipment grounding conductor, 253, 266
Erosion corrosion, 194
Error function, 11
Ethylene vinyl acetate (EVA), 445, 446
Euler's formula, 191
EVA, *see* Ethylene vinyl acetate
Evolution of PV cells and systems; *see also* Photovoltaic systems (PV systems)
  CdTe cells, 462–465
  CIGS cells, 456–462
  CIS-family-based absorbers, 467–468
  CPV cells, 472–473
  crystalline cells, 435–436
  developments in silicon technology, 465–467
  fabrication processes, 435
  gallium arsenide cells, 451–456
  hot carrier cells, 470
  IEEE Photovoltaic Specialists Conference, 435
  III–V and II–VI emerging technologies, 468–469
  intermediate band solar cells, 469
  new developments in system design, 475–479
  optical up-and down-conversion, 470
  OPV cells, 471–472
  perovskites, 473–475
  quantum dot and dye-sensitive solar cells, 475
  supertandem cells, 470
  thermophotovoltaic cells, 469
  thin-film materials, 436–437
EWA, *see* Effective wind area
EWT, *see* Emitter wrap-through
Exciton, 471
Exotic junctions, 426
  graded junctions, 427–428
  heterojunctions, 428
  multijunctions, 431–432
  Schottky junctions, 428–431
  Tunnel junctions, 432–433
Exponential environment, resource lifetime in, 9–10
Exponential function, 8–9
Exponential growth, 6
  accumulation, 8–9
  compound interest, 6–7
  doubling time, 7–8
  resource lifetime in exponential environment, 9–10
Exposure, 224
Externalities, 355–356, 372, 380–381
  air pollution, 382–384
  environmental effects of energy sources, 381
  health and safety as, 389
  infrastructure degradation, 385
  liability of tobacco companies, 373
  and PV, 375
  with PV systems, 389–394
  quantifying cost of, 385–389
  renewable and nonrenewable energy sources, 382
  soil pollution, 384–385
  subsidies, 374–375
  water pollution, 384–385
Externally biased PN junction, 412–414
Extraterrestrial cells, 455–456
Extraterrestrial solar spectrum, 24–25
Extrinsic semiconductors, 404–405

F

Fabrication
  a-Si cells, 449–451
  of CIS cell, 459–460
  Fabrication of gallium arsenide cells, 453–455

INDEX

of junction, 439–441
process, 427, 435
of wafer, 438–439
Fan, 297–298
Federal Communications Commission (FCC), 126
Federal funds rate, 357
Fermi level, 429, 430
    energy, 404
    intrinsic, 418
Field measurement of shading objects, 38–40
Fill factor (FF), 52, 465
Flexural stresses, *see* Bending stresses
Float voltage, 81
Floor plan of mountain cabin, 324
Florida Solar Energy Center, 485
Flow batteries, 68
Foot-candle, 304
Foot-pound (ft-lb), 15, 16
Force, 186–187
    acting on PV arrays, 203
    dead loads, 204
    live loads, 204
    snow loads, 209–210
    structural loading considerations, 203–204
    wind loads, 204–209
48-V system voltage, 327
Fossil-fuel(ed)
    generation, 387
    generator connection options, 256–257
    generators, 235
Four-pole generators, 99
Fracking method, *see* Hydraulic fracturing method
Frequency stability, 101
Fresnel lenses, 472
ft-lb, *see* Foot-pound
Fuels, 389
    cell, 71
    types, 100
Functional requirements for PV system, 198–199
Fuses, 104–105

G

Gallium (Ga), 452
Gallium arsenide (GaAs), 400
Gallium arsenide cells, 451; *see also* Silicon PV cells
    cascaded AlInP/GaInP/GaAs structure, 454
    cell performance, 455–456
    direct bandgap, 451–452
    fabrication of, 453–455
    pure cell components production, 452–453
Galvanic corrosion, 193–194
Gas discharge lamps, 306
Gasoline, 100
Gassing phenomenon, 59
Gaussian function, 9, 11
    distribution function, 441
Gauss' law, 419
GDP, *see* Gross domestic product
Gel battery, 238, 243
Gel cells, 63
Generators, 336; *see also* Inverters
    altitude effects, 101
    amplitude stability, 101–102
    backup source for standalone system, 97
    comparison of PV inverters to mechanically rotating, 123
    connection port, 344–345
    cost and maintenance information, 103
    efficiency *vs.* electrical load, 99–100
    frequency stability, 101
    fuel types, 100
    maintenance, 102–103
    noise level, 102
    operating characteristics, 98
    operating hours and operating cost, 342–344
    overload characteristics, 102
    power factor considerations, 102
    rotation speed, 98–99
    selection, 103, 342
    type of starting, 102
    types and sizes, 97–98
    waveform harmonic content, 101
Germanium (Ge), 453
GFDI, *see* Ground fault detection and interruption
Global climate change, 380
Global radiation, 25
Global warming, 27, 380
Glow discharge, *see* Corona discharge
Google Earth©, 40
Google Sketch-Up©, 40
Graded junctions, 427–428
"Green" source, 237
Greenhouse,
    effect, 381, 383–384
    gases, 27, 379, 380–381, 383–384
Green pricing, 374
Green tag, 374
Grid, 257
Grid-connected inverters, 97
Grid-connected system, 295, 302

Grid-connected utility-interactive PV systems,
    113–114
  array installation, 130
  array sizing, 132–133
  balance of system, 136–145
  codes and standards, 114–128
  design considerations, 128
  design of microinverter-based system,
    154–157
  design of system, 132, 145–154
  determining system energy output, 128–130
  installation considerations, 132
  inverter selection, 133–134
  inverter selection and mounting, 130–131
  module selection, 134–136
  nominal 20 kW system design, 157–163
  nominal 500-kW system design, 163–176
  system commissioning, 176–178
  system performance monitoring, 178–179
Gross domestic product (GDP), 385–386
Grounded fault protection, 105
Ground fault, 151
Ground fault detection and interruption (GFDI),
    97, 116
  device, 323
  protection, 171
Ground fault protection, 105, 142
Grounding,
  conductors, 105, 140, 322
  electrode conductor, 253, 266
  fault protection, 105
Ground-mounted arrays, 218; *see also* Roof-
    mounted arrays
  pole mounting arrays, 218–219
  rack mounting arrays, 218, 219
  tracking-stand mounting arrays, 219–220
Group III impurity, 439
Group V impurity, 439

H

H-bridge, 87, 91
  multilevel H-bridge modified sine wave
    inverter, 90, 95
HCPV cells, *see* High concentration PV cells
Health and safety as externalities, 389
Heating water, 325
Heisenberg uncertainty principle, 432
Heterojunctions, 428
Heterojunction with intrinsic thin layer
    technology (HIT technology),
    466–467
High-efficiency Si cell, 445–447
High concentration PV cells (HCPV cells), 472

High head pump, 300
HIT technology, *see* Heterojunction with intrinsic
    thin layer technology
Hole motion, 399
Hooke's law, 188
Horizontal mounting, 41
Hot carrier cells, 470
Hot-dip galvanizing, 196
Hour angle, 32
Hours of daylight (DH), 32
Hubbert, M. King, 10–11
Hubbert's Gaussian model, 10
  error function, 11
  Hubbert's predictions for U. S. and world
    petroleum production, 11
  Hubbert's theory, 12
120-V battery-backup inverter with 240-V straight
    grid-connected inverter, 283–285
120/208-V three-phase AC-coupled system,
    285–286
120/240-V battery-backup system; *see also*
    18-kW battery-backup system
  on available roof space, 257
  battery selection, 260–261
  BOS selection and completion of design,
    262–267
  grid, 257
  inverter selection process, 260
  module selection and source circuit design,
    257–258
  source-circuit combiner box and charge
    controller selection, 258–260
  standby loads determination, 260–261
  with 240-V straight grid-connected inverter,
    281–283
Hybrid-powered residence, 334–349
  array design, 339–341
  battery selection, 337–338
  BOS component selection, 348
  charge controller and inverter selection,
    344–346
  generator operating hours and operating cost,
    342–344
  generator selection, 342
  loads, 336–337
  propane generator energy production, 335
  PV array, 334
  total system design, 348–349
  wire, circuit breaker, and disconnect
    selection, 346–348
Hydraulic fracturing method, 385
Hydrogen storage, 70–71
Hysteresis, 318
  effect, 298

INDEX

I

IE, *see* Incident energy
IEEE 1547-2003 standard, 114
IEEE 1547 standard, 78, 119, 120, 131
    IEEE Standard 1547.4, 121
    IEEE Standard 1547.7, 121
    IEEE Standard 1547.8, 121
IEEE 29th Photovoltaic Specialists Conference, 447
IEEE 34th Photovoltaic Specialists Conference, 447
IEEE Photovoltaic Specialists Conference, 435
IEEE Standard 1547-2003, 119
    comparison of PV inverters to mechanically rotating generators, 123
    development, 119
    islanding analysis, 123–125
    specific requirements, 119–122
IEEE Standard 1584-2002, 173
Illuminated PN junction, 410–412
Incident energy (IE), 173, 175
Inclinometer, 39–40
Indirect subsidy, 388
Indium (In), 458
Inflation rate, 356, 357
Infrared radiation, 383–384
Infrastructure degradation, 385
Integrated mounting, *see* Building integrated photovoltaics (BIPV)
Intensity of sunlight, 28
Intergovernmental Panel on Climate Change (IPCC), 381, 383
Intergranular corrosion, 194
Intermediate band solar cells, 469
Internal rates of return (IRR), 477
International Meteorological Organization, 381
Intrinsic Fermi level, 418
Intrinsic semiconductors, 404
Inverters, 86, 157–158, 296; *see also* Generators
    bypass switches, 106
    features, 95–97
    input power, 322
    modified sine wave inverters, 89–91
    performance enhancement, 478
    programming, 253–256
    PWM inverters, 91–95
    selection process, 241–242, 260
    sizing, 238
    square wave inverters, 87–89
    transformerless inverters, 95
Inverter selection, 133–134, 149, 309, 344–346
    and mounting, 130–131
    PV-powered mountain cabin, 330–331
    refrigeration system, 321

IPCC, *see* Intergovernmental Panel on Climate Change
Iron, 195, 197
IRR, *see* Internal rates of return
Irradiance, 28
    enhancement, 213–214
Irradiation, 28
    maximizing on collector, 35–38
Islanding analysis, 123–125
I–V characteristics,
    I–V relationship, 299
    of PV array, 72
    PV cell I–V curve, 53
    of real and ideal PV cells, 51
    rectangular cell, 52

J

Junction fabrication, 439–441
Junction width maximization, 419–421

K

Kirchhoff's current law, 105, 408

L

Lacquers, 196
Lamp wattage and daily load, 307–308
Laser-grooved, buried grid process (LGBG process), 466
Lasers, 306
Law of detailed balance, 408
LBCI voltage, *see* Low battery cut in voltage
LBCO voltage, *see* Low battery cut out voltage
LCBs, *see* Linear current boosters
LCC, *see* Life-cycle cost
LCOE method, *see* Levelized cost of electricity method
LCPV cells, *see* Low concentration PV cells
Lead, 196
Lead-acid,
    batteries, 62, 319, 328
    system, 66
    units, 64
Lead-acid connections, 286
    battery string, 289–291
    connection of batteries, 286–288
    open delta three-phase system, 287
    SOC, 288–289
    three possible battery hookup configurations, 289
Lead-acid storage battery,
    chemistry of lead-acid cell, 58–59
    properties, 60–64

496 INDEX

Lead–antimony electrodes, 62
Lead oxide (PbO$_2$), 58
LEC method, *see* Liquid encapsulated Czochralski method
LEDs, *see* Light-emitting diodes
Less precise measurements, 35
Levelized cost of electricity method (LCOE method), 355
    analysis, 365–367
    formula, 365
Level playing field, 375
LFP, *see* Lithium iron phosphate
LGBG process, *see* Laser-grooved, buried grid process
Life-cycle cost (LCC), 355, 360
    analysis, 97, 380
    effect of borrowing, 370–371
    examples, 360–364
    LCOE analysis, 365–367, 368
    present worth factors and present worth, 357–360
    time value of money, 356–357
    unit electrical cost, 365
Light-emitting diodes (LEDs), 25
Lighting load determination, 304–306
Lightning, 106
    protection, 127
Linear current boosters (LCBs), 78, 296
Linear interpolation, 301
Liquid encapsulated Czochralski method (LEC method), 453
Lithium-ion battery technologies, 64–66
Lithium iron phosphate (LFP), 238, 296
    batteries, 335
    cell, 65
    systems, 291
Live loads, 204
Load(s), 87, 203–204, 336–337
    buckling, 222–223
    computing mechanical load and stresses, 221
    dead, 204
    determination, 237–238, 318
    live, 204
    other loads, 210
    PV-powered mountain cabin, 324–327
    snow, 209–210
    tensile stresses, 222
    wind, 204–209
    withdrawal loads, 221–222
Low-carbon steel, 195
Low battery cut in voltage (LBCI voltage), 255–256
Low battery cut out voltage (LBCO voltage), 255–256
Low concentration PV cells (LCPV cells), 472

Lumens formula, 306
Luminous efficacy, 305

M

Maintenance factor (MF), 306
Majority carrier concentration, 405
Malthusian theory, 6
MATLAB® function, 8
Maximum power point tracking (MPPT), 54, 72, 73, 130, 131, 240, 298
    charge controller, 321
    and LCBs, 82–86
MBE, *see* Molecular beam epitaxy
MBOE, *see* Million barrels of oil equivalent
MCFC, *see* Molten carbonate fuel cell
Mechanical considerations for PV system, 185
    aluminum, 196–198
    array mounting system design, 210–221
    building code requirements, 202–203
    chemical corrosion, 193–195
    column buckling, 190–191
    computing mechanical loads and stresses, 221–223
    constraints, 200
    design and installation guidelines, 201
    establishing mechanical system requirements, 198
    forces acting on PV arrays, 203–210
    functional requirements, 198–199
    mechanical system design process, 198
    non-PV materials, 185–186
    operational requirements, 199–200
    properties of materials, 185
    standards and codes, 201–202
    standoff, roof mount examples, 223–231
    steel, 195–196
    strength of materials, 190
    stress and strain, 186–189
    thermal expansion and contraction, 191–193
    trade-offs, 200–201
    ultraviolet degradation, 193–195
Mechanical system design process, 198
Medium head pump, 300
Metal insulator semiconductor inversion layer (MIS-IL), 466
Metallization wrap-through (MWT), 466
Metallization wraparound (MWA), 466
Metal organic chemical vapor deposit process (MOCVD process), 453, 464
Metal oxide varistor (MOV), 106
Methane (CH$_4$), 381, 385
Methyl-ammonium ((CH$_3$NH$_3$)$^+$), 474
MF, *see* Maintenance factor

INDEX 497

Miami NREL SAM PVWatts simulation for parking lot lighting system, 308
Micro grids, 475–476
Microinverter-based system,
    bells and whistles, 156–157
    feature of, 154–155
    system design, 155–156
Million barrels of oil equivalent (MBOE), 16
Million tons of oil equivalent (MTOE), 16
Minimum Design Loads for Buildings and Other Structures, 202
Minority carrier,
    concentration, 405
    diffusion lengths, maximizing, 417–419
    lifetimes, 418
MIS-IL, see Metal insulator semiconductor inversion layer
MOCVD process, see Metal organic chemical vapor deposit process
Modified sine wave inverters, 89–91
Module(s), 158–159, 297–298, 445
    efficiency, 55
    performance enhancement, 478–479
    selection, 134–136, 245–246, 257–258, 270, 271, 302–303
Modulus of elasticity, 188, 189
Modulus of rigidity, 188, 189
Molecular beam epitaxy (MBE), 454
Molten carbonate fuel cell (MCFC), 71
Molybdenum (Mo), 459
MOSFET, 83–85, 446
MOV, see Metal oxide varistor
MPPT, see Maximum power point tracking
MSRI, see U.S. Million Solar Roofs Initiative
MTOE, see Million tons of oil equivalent
Multicrystalline silicon cells, 447
Multijunctions, 431–432
Multiple anode adjusting factors, 314
MWA, see Metallization wraparound
MWT, see Metallization wrap-through

N

National Electrical Code® (NEC), 63, 104, 105, 113, 114, 202, 392–393
    by National Fire Protection Association, 114–115
    NEC 690. 11, 105
    NEC 690. 12, 108, 142
    NEC 690. 5, 105
    voltage drop and wire sizing, 116–119
National Energy Conservation and Policy Act, 2–3
Natural gas, 384, 385

NEC, see National Electrical Code®
Net energy, 12–13
Ni–Cd batteries, see Nickel–Cadmium Batteries
Nickel (Ni), 66; see also Lead-acid storage battery
    Ni–Cd batteries, 66–67
    nickel-based battery systems, 66
    NIMH batteries, 67–68
    Ni–Zn batteries, 67
Nickel–Cadmium Batteries (Ni–Cd batteries), 57, 66–67
Nickel–metal hydride battery (NIMH battery), 57, 67–68
Nickel–Zinc batteries (Ni–Zn batteries), 67
Nighttime losses, 131
NIMH battery, see Nickel–metal hydride battery
Nitric-phosphoric (NP), 464
Ni–Zn batteries, see Nickel–Zinc batteries
No-load power consumption, 131
NOCT, see Nominal operating cell temperature
Noise level, 102
Nominal 20 kW system design, 157–163
    annual system performance estimate, 163
    inverter, 157–158
    modules, 158–159
    PV system, 157
    system AC wiring, 161–163
    system DC wiring, 159–161
Nominal 500-kW system design,
    AC wire sizing, disconnects, and overcurrent protection, 171–172
    advantage of using multiple smaller inverters, 163
    arc flash calculations, 172–176
    array layout, 166–168
    array sizing, 164–165
    combiner boxes, 165–166
    configuring array, 164
    inverter, 163–164
    modules and array, 164
    wire sizing and voltage drop calculations, 168–171
Nominal operating cell temperature (NOCT), 54–55, 214
Non-south-facing mounting, 41–43
Nonbattery-backup system, 237
Noncritical applications, 76
Noncritical loads, 73
Noncritical operation, 75
Noncritical storage times, 76
Nonfilament light sources, 25
Nonheat-treated CdTe cells, 464
Normal curve, 11
Normal incidence pyrheliometer, 35

Northern Hemisphere, 27, 29, 37–38
NP, *see* Nitric-phosphoric
NREL, *see* U.S. National Renewable Energy Laboratory
NREL SAM PVWatts, 273–274
  model, 328, 339
  option, 239
  simulation, 316
N-type concentrations, 440
N-type impurities, 405

O

Occupational Safety and Health Administration rules (OSHA rules), 391
Off-grid residence, 334–349
  array design, 339–341
  battery selection, 337–338
  BOS component selection, 348
  charge controller and inverter selection, 344–346
  generator operating hours and operating cost, 342–344
  generator selection, 342
  loads, 336–337
  propane generator energy production, 335
  PV array, 334
  total system design, 348–349
  wire, circuit breaker, and disconnect selection, 346–348
Ohm's law, 117, 406
OM cost, *see* Operations and maintenance cost
OMVPE process, *see* Organo-metallic vapor phase epitaxy process
Open delta three-phase system, 287
Operational requirements for PV system, 199–200
Operations and maintenance cost (OM cost), 365
Optical absorption, 397
  generation of EHP by photon absorption, 399–401
  photoconductors, 402–404
  photon, 397–398
  semiconductor materials, 398–399
Optical up-and down-conversion, 470
OPV cells, *see* Organic PV cells
Orbit of Earth, 29–31
Organic PV cells (OPV cells), 471–472
Organo-metallic vapor phase epitaxy process (OMVPE process), 454
Orientation considerations, 41–43
  horizontal mounting, 41
  non-south-facing mounting, 41–43

OSHA rules, *see* Occupational Safety and Health Administration rules
Overcurrent protection, 142–143, 151–152
Overload characteristics, 102
Oxygen, 379
  ion, 58
  oxygen–carbon dioxide cycle, 379
Ozone, 25

P

PAFC, *see* Phosphoric acid fuel cell
Paints, 196
Parking lot lighting design, 306
  battery storage requirements, 308
  charge controller and inverter selection, 309
  daily load by fixture, 307–308
  determination of array size, 308–309
  final system schematic, 309–310
  lamp wattage determination, 307–308
  locations of fixtures and approximate coverage patterns, 307
  structural comment, 310–311
Passive trackers, 220
Payback analysis, 371–372, 373
Paycheck money, 367
Peak sun hours (psh), 28, 75, 316
PEMFC, *see* Proton exchange membrane fuel cell
Peoples Republic of China, 2
Performance-based systems, 179
Perovskites, 473–475
Phosphine (P), 455
Phosphoric acid fuel cell (PAFC), 71
Phosphorous, 404, 439
Photoconductors, 402–404
Photocurrent, 53
  final expression for, 422–424
  maximizing junction width, 419–421
  maximizing minority carrier diffusion lengths, 417–419
  minimizing reflection of incident photons, 416–417
  minimizing surface recombination velocity, 421–422
  optimization, 416
Photon(s)
  absorption, 399–401, 421, 422, 436
  energy, 431
  minimizing reflection of, 416–417
Photosynthesis, 27, 379, 387
Photovoltaic(s) (PV), 375
  installation permit, 485–486
  inverters, 123

INDEX

module, 54–56, 128
output circuits, 170
Photovoltaic arrays (PV arrays), 49, 56–57
   dead loads, 204
   forces acting on, 203
   installation, 130
   live loads, 204
   snow loads, 209–210
   wind loads, 204–209
Photovoltaic cells (PV cells), 14, 49, 50–54
   exotic junctions, 426–433
   extrinsic semiconductors, 404–405
   I–V characteristics of real and ideal, 51
   maximizing PV cell performance, 414
   minimizing cell resistance losses, 424–426
   minimizing reverse saturation current, 415–416
   optical absorption, 397–404
   performance of, 6
   photocurrent optimization, 416–424
   physics of, 397
   PN junction, 406–414
   Si PV cells, 391
Photovoltaic systems (PV systems), 1, 49, 185, 202, 235, 295, 355, 380; *see also* Evolution of PV cells and systems
   associated system electronic components, 78–97
   availability, 73
   BOS components, 103–108
   cells, modules, and arrays, 50
   connecting to utility grid, 113
   cost of, 113
   decommissioning, environmental effects of, 393–394
   deployment and operation, environmental effects of, 391–392
   direct conversion of sunlight to electricity with, 13–15
   energy storage, 57–72
   environmental effects of implementation, 389–391
   examples, 50
   externalities with, 389
   generators, 97–103
   large-scale solar PV installations, environmental impact of, 392–393
   loads, 72–73
   new concerns about availability, 77–78
   traditional concerns to stand-alone systems, 73–77
Physical vapor deposition process (PVD process), 460

Pin junction, 420–421
Piping
   friction loss, 299
   system, 299
Pitting, 194
Planck formula, 399
Planck's blackbody radiation formula, 23–24
Planck's constant, 24, 399
Planck's radiation law, 172
Planting trees, 387
PN junction(s), 406, 439; *see also* Exotic junctions
   built-in potential, 407–410
   diffusion, 406–407
   drift, 406–407
   with external bias, 412
   externally biased, 412–414
   illuminated, 410–412
   junction formation, 407–410
Point of common coupling (PCC), *see* Point of utility connection (PUC)
Point of utility connection (PUC), 119, 143–144, 152
Pole mounting arrays, 218–219
Pollutants, 383, 384–385, 386–388
Pollution, 379, 385, 389
   controls, 383, 388
Polycrystalline, 418, 437
Polymer electrolyte membrane fuel cell, *see* Proton exchange membrane fuel cell (PEMFC)
Population and energy demand, 2
Portable highway advisory sign, 315
   determination of average power, 316–317
   determination of battery requirements, 317
   microcontroller in system, 317
   portable fossil-fueled generator, 315
Power factor, 102
Power rating, 131
Precision measurements, 34–35
Predeposition, 440
Present worth (PW), 357–360
   factors, 357–360
Pressure forces, 205
Proportional limit, 187
Proton exchange membrane fuel cell (PEMFC), 71
psf, *see* Per square foot; Pounds per square foot
p-region doping, 446
p-type concentrations, 440
p-type impurities, 405

PUC, *see* Point of utility connection
Pulse width modulation (PWM), 72, 91
    charge controller, 313–314
    control of waveform, frequency, and amplitude, 92
    full-bridge PWM converter, 94
    inverter, 91
    PWM waveform, 92–93
    switching transition, 94–95
    three-level PWM inverter configuration, 94
Pump and PV I–V characteristics, 83
Pumping rate, 299
Pumping system
    battery selection, 302
    BOS and completion of system, 303–304
    charge controller selection, 303
    design approach for, 301
    module selection, 302–303
    pump selection, 301–302
Pump selection, 301–302
Pure silicon production, 437–438
Pure tellurium production, 462–463
PV, *see* Photovoltaic(s)
PVD process, *see* Physical vapor deposition process
PV-powered mountain cabin, 323
    array sizing and tilt, 328–330
    battery selection, 327–328
    BOS component selection, 332–334
    charge controller selection, 330
    deep-well-water pump, 323
    excess electrical production, 331–332
    floor plan of mountain cabin, 324
    inverter selection, 330–331
    load determination, 324–327
PV-powered parking lot lighting system, 304
    determination of lighting load, 304–306
    parking lot lighting design, 306–311
PV-powered water pumping system, 298
    design approach for simple pumping system, 301–304
    selection of system components, 299–300
PVWatts model, 75, 129, 302, 339
PVWatts version of NREL SAM, 163, 309, 313
PW, *see* Present worth
PWM, *see* Pulse width modulation
Pyranometer, 34–35

Q

Quantum-mechanical tunneling, 432
Quantum dot, 475

Quantum theory, 398
Quantum well cells (QW cells), 468–469

R

R^2 values, 8
Rack mounting arrays, 216–217, 218, 219
Radiated emission, 126
Rapid shutdown, 108, 131, 141–142, 151, 155, 161, 237
    system, 266–267
RCR, *see* Room cavity ratio
RCRA, *see* Resource and Conservation Recovery Act
Reactive metals, 197
Receptacles, 104–105, 327
Recombination centers, 418, 421, 422
RECs, *see* Renewable Energy Credits
Redox flow batteries (RFBs), 68
Refining process, 12, 435, 438, 457, 463
Refrigeration system, 317; *see also* Cathodic protection system
    array sizing, 320–321
    battery sizing, 318–320
    BOS component selection, 321–322
    charge controller and inverter selection, 321
    hysteresis and eddy current losses, 318
    load determination, 318
    system design, 322–323
Refrigerator compressor, 321
Renewable Energy Credits (RECs), 372
Renewable portfolio standards (RPSs), 1
Residual value (RES value), 365
Resource and Conservation Recovery Act (RCRA), 393
Resource lifetime in exponential environment, 9–10
RES value, *see* Residual value
Reverse saturation current, minimizing, 415–416
RFBs, *see* Redox flow batteries
Roll-to-roll process, 450
Romex, *see* Type NM
Roof-mounted arrays, 212, 215; *see also* Ground-mounted arrays
    array mount design, high wind speed case, 227–229
    array mount design, lower wind speed case, 229–231
    ASCE 7 wind load analysis tabular method, 223–227
    direct mounting arrays, 218
    examples, 223

INDEX

exposure, and correction factors, 231
    integrated mounting arrays, 217–218
    rack-mounted arrays, 216–217
    standoff mounting arrays, 215–216
Roof space, design of system based upon,
    array selection, 145–149
    balance of system, 149–153
    extension of design to lower wind speed region, 153–154
    inverter selection, 149
Rooftop junction box, 247–248, 262–263
    to inverter, wire and conduit from, 138–141
    wiring from array to, 137–138
    wiring from array to, 149–151
Roof zones, 224, 225
Room cavity ratio (RCR), 306
Rotation of Earth, 29–31
Rotation speed, 98–99
Rounding down, 135, 258
Rounding up, 135
RPSs, *see* Renewable portfolio standards
Rupture strengths, 188

S

SAE, *see* Society of Automotive Engineers
SAM, *see* System Advisor Model
Sandia Frequency Shift (SFS), 122
Sandia National Laboratories, 122
Sandia Voltage Shift (SVS), 122
SC, *see* Source-circuit
Scattered sunlight, 25
Scattering process, 417–418
Schottky junctions, 428–431
SCP, *see* System control panel
Screen printing (SP), 464
Selective leaching, 194
Selenium (Se), 458, 463
Semiconductor
    industry, 391
    materials, 14, 398–399, 400
    semiconductor-grade As, 453
    semiconductor-grade Ga, 452
SFS, *see* Sandia Frequency Shift
Shading, 38, 214
    computational methods of determining, 40–41
    field measurement of objects, 38–40
    Solar Pathfinder, 39
Shadow band stand, 35
Shear forces, 188
Shell Oil Company, 10–12

Siemens monolithic method of series cell connection, 461
Silane (SiH_4), 438
Silicon (Si), 436, 438, 465
    acceptor and donor impurities in, 405
    MIS-IL Si cell, 466
    new developments in Si technology, 465–467
Silicon PV cells, 437; *see also* Gallium arsenide cells
    amorphous silicon cells, 448–451
    high-efficiency Si cell, 445–447
    multicrystalline silicon cells, 447
    production of pure silicon, 437–438
    single-crystal silicon cells, 438–445
    SunPower, 446
    thin silicon cells, 447–448
Single-crystal silicon cells, 435, 438; *see also* Amorphous silicon cells (a-Si cells)
    ARC, 444–445
    contacts, 441–444
    fabrication of junction, 439–441
    fabrication of wafer, 438–439
    modules, 445
Single inverter 120-V battery-backup system; *see also* 18-kW battery-backup system
    array sizing, 244–245
    battery selection process, 242–244
    BOS selection and completion of design, 246–253
    charge controller and module selection process, 245–246
    fossil-fuel generator connection options, 256–257
    inverter selection process, 241–242
    programming inverter and charge controller, 253–256
    standby loads, 241
SiO varistor (SOV), 106
6061 aluminum alloy, 198
6063 aluminum alloy, 198
6xxx alloy series, 198
Skin friction forces, 205
"Slippery" glass surfaces, 209
Smart grids, 476–477
Smog pollution, 383
Snow loads, 209–210
Snow-shedding process, 209
SOC, *see* State of charge
Society of Automotive Engineers (SAE), 202
SOFC, *see* Solid oxide fuel cell
Soil
    acidification, 385
    pollution, 384–385
    resistivity, 312–313

Solar
    altitude, 31, 32
    constant, 24
    energy, 203
    large-scale solar PV installations, 392–393
    noon, 30, 31, 33
    Pathfinder, 39
    spectrum, 23–25
    systems, 1
Solid oxide fuel cell (SOFC), 71
Soluble Threshold Limit Concentration (STLC), 393–394
Source-circuit (SC), 169
    calculations, 168–170
    charge controller selection, 258–260
    combiner box, 106–107, 248, 258–260, 263
    design, 257–258
South-facing PV system, 41–42
Southern California Gas Company, 71
SOV, *see* SiO varistor
SP, *see* Screen printing
Space charge layer, 410
Spangle, 196
Square wave
    harmonics, 89
    inverters, 87–89
Staebler–Wronski effect, 449
Stand-alone
Stand-alone PV systems, 295
    array sizing and tilt procedure, 351–352
    battery selection procedure, 351
    cathodic protection system, 311–315
    configuration, 297–298
    design procedures, 349–352
    determination of average daily PV system load, 350
    hybrid-powered, off-grid residence, 334–349
    inverters, 96
    LCB, 296
    outdoor lighting systems, 308
    portable highway advisory sign, 315–317
    PV-powered mountain cabin, 323–334
    PV-powered parking lot lighting system, 304–311
    PV-powered water pumping system, 298–304
    refrigeration system, 317–323
    traditional concerns to, 73–77
Standard(s), 201–202
Standard candle, 305–306
Standard Test Conditions (STC), 135

Standby loads
    array sizing, 244–245
    battery selection process, 242–244
    BOS selection and completion of design, 246–253
    charge controller and module selection process, 245–246
    determination, 241, 260–261
    fossil-fuel generator connection options, 256–257
    inverter selection process, 241–242
    programming inverter and charge controller, 253–256
    single inverter 120-V battery-backup system on, 241–257
Standoff mounting arrays, 215–216
    array mount design, high wind speed case, 227–229
    array mount design, lower wind speed case, 229–231
    ASCE 7 wind load analysis tabular method, 223–227
    examples, 223–231
    Exposure C, Exposure D, and other correction factors, 231
State of charge (SOC), 321
STC, *see* Standard Test Conditions
Steel, 195–196
STLC, *see* Soluble Threshold Limit Concentration
Storage days, *see* Days of autonomy
Storage options, 71–72
Strain, 186
    force, 186–187
    pull test of bar, 187
    shear stresses, 188–189
    stress–strain curve for steel, 187–188
Strength of materials, 190
Stress corrosion cracking, 195
Stress(es), 186
    force, 186–187
    mechanical loads and, 221–223
    pull test of bar, 187
    shear stresses, 188–189
    stress–strain curve for steel, 187–188
Structural comment, 310–311
Structural loading considerations, 203–204
Submersible pump, 325
Subsidies, 374–375, 388–389
Substrate doping, 446
Sulfur (S), 459
Sun, 23
    solar spectrum, 23–25
    tracking, 31–34

INDEX 503

Sunlight
  atmosphere effect on, 25–27
  effect of atmosphere on, 25–27
  capturing, 35
  direct conversion to electricity, 13–15
  intensity of, 28
  irradiance, 28
  maximizing irradiation on collector, 35–38
  measuring, 34–35
  orbit and rotation of Earth, 29–31
  orientation considerations, 41–43
  shading, 38–41
  specifics, 27
  tracking Sun, 31–34
SunPower A-300 monocrystalline Si cell, 445
Supertandem cells, 470
Surface polarization effect, 446
Surface recombination velocity, 421–422
Surge arrestor, 248, 263
Surge power rating, 131
Surge protection, 106, 126–127
SVS, see Sandia Voltage Shift
Switches, 104–105
Switch selection, 332–333
System Advisor Model (SAM), 129
System components selection, 299–300
System control panel (SCP), 270
System design,
  cathodic protection system, 312–315
  refrigeration system, 322–323

T

Table look-up method, 225–227
TCO, see Transparent conducting oxide
Tellurium (Te), 462
Tellurium dioxide, 463
Tensile strength, 188
Tensile stresses, 222
Test for sustainability, 12–13
THD, see Total harmonic distortion
Thermal equilibrium, 402
Thermal expansion and contraction, 191–193
Thermal stresses, 191–193
Thermoelectronics, 470
Thermophotovoltaic cells, 469
Thevenin equivalent circuit of battery, 61, 79
Thin-Film Photovoltaics Partnership Program, 459–460
Thin films, 436
Thin silicon cells, 447–448
Three-phase distribution panel, 157
  annual system performance estimate, 163
  inverter, 157–158
  modules, 158–159
  PV system, 157
  system AC wiring, 161–163
  system DC wiring, 159–161
THWN-2, 347
Time of use (TOU), 477
Time value of money, 356–357
TMY, see Typical meteorological year
Torsional shear stresses, 189
Total harmonic distortion (THD), 91
Total Threshold Limit Concentration (TTLC), 393–394
TOU, see Time of use
Toughness of material, 188
Toxic air pollutants emissions, 382
Toxic waste, 312
Tracking-stand mounting arrays, 219–220
Tracking arrays, 214, 220
Trade-offs, 200–201
Transformer coupling, disadvantage of, 88
Transformerless inverters, 95
Transparent conducting oxide (TCO), 448
Trichlorosilane (SiHCl$_3$), 438
Tropic of Cancer, 29
Tropic of Capricorn, 29
TTLC, see Total Threshold Limit Concentration
Tungsten incandescent light filament, 305–306
Tunnel junctions, 432–433
Two hundred and forty-V straight grid-connected inverter
  120-V battery-backup inverter with, 283–285
  120/240-V battery-backup inverter with, 281–283
Two-pole generators, 99
Type NM (Romex), 156, 251–252
Type THHN insulation, 116
Typical meteorological year (TMY), 34, 133

U

UL 1741 standard, 119, 122, 131, 132, 134, 154–155, 164, 178, 235, 267, 281, 295, 485
Ultimate tensile strength, 188
Ultraviolet (UV), 25
  degradation, 127–128, 193–195
  radiation, 186, 195, 445
  UV-A, 195
  UV-B, 195
  UV-C, 195
Uniform attack, 193
Unit electrical cost, 365
Uplifting pressure, 208

U.S. Department of Energy SunShot Program, 459–460
U.S. Environmental Protection Agency (EPA), 382–383, 388, 393
U.S. Federal Communications Commission, 126
U.S. Million Solar Roofs Initiative (MSRI), 1
U.S. National Renewable Energy Laboratory (NREL), 129
Utility-interactive inverters, 96, 123
UV, see Ultraviolet

V

Vandalism, protection from, 214–215
Varnishes, 196
Voltage drop,
  calculations, 162–163, 168
  combiner to recombiner wiring, 170
  disconnects, GFDI, and overcurrent protection, 171
  source-circuit calculations, 168–170
  and wire sizing, 116–119
Voltage regulation (VR), 49, 101

W

Wafer, fabrication of, 438–439
Wafering process, 438–439, 447
Warranty, 131
Waste Extraction Test (WET), 393
Water loss, 63
Water pollution, 384–385
Water pumping system,
  battery selection, 302
  BOS and completion of system, 303–304
  charge controller selection, 303
  design approach for pumping system, 301
  module selection, 302–303
  pump selection, 301–302
  PV-powered, 298
  selection of system components, 299–300
Water vapor, 25
Waveform harmonic content, 101
Waxes, 196

Web-based monitoring systems, 179
West-facing PV system, 42
WET, see Waste Extraction Test
Wind loads, 203, 204
  aerodynamic forces, 204
  ASCE, 207–208
  ASCE 7 wind load analysis tabular method, 223–227
  Hughes modular array field, 206
  lower wind speed region, extension of design to, 153–154
  net pressure, 206–207
  standoff-mounted array, 208
  types, 205
  wind forces, 205–206
  worst-case scenario, 209
Wire selection, 332–333, 346–348
Wire sizing, 168, 275–277, 278
  AC side, 251, 265
  combiner to recombiner wiring, 170
  DC side, 248–251, 263–264
  disconnects, GFDI, and overcurrent protection, 171
  source-circuit calculations, 168–170
Wire to-water pumping efficiency, 325
Withdrawal loads, 221–222
World Resources Institute (WRI), 383, 385–386
Wrought aluminum alloys, 197

X

Xenon, 306

Y

Yield point, 187
Yield strength, 18
Young's modulus, see Modulus of elasticity

Z

Zinc, 193, 196
ZnO/CdS/CIGS/Mo cell structure, 459–460